POLYMER SCIENCE

POLYMER SCIENCE

FIFTH EDITION

(*In two colour*)

V R GOWARIKER
Ex-Director
Vikram Sarabhai Space Centre
Trivandrum, India

N V VISWANATHAN
Formerly
Advisor-Safety, Vikram Sarabhai Space Centre
Trivandrum, India

JAYADEV SREEDHAR
Formerly Scientist/Engineer-SF
Vikram Sarabhai Space Centre
Trivandrum, India

NEW AGE
TECHNO
PRESS

An Imprint of

NEW AGE INTERNATIONAL (P) LIMITED, PUBLISHERS
LONDON • NEW DELHI
Bangalore • Chennai • Cochin • Guwahati • Hyderabad • Kolkata • Mumbai
Visit us at www.newagepublishers.com

Published by New Age International (P) Ltd., Publishers
First Edition: 1986
Fifth Edition *(in two colour)* **2024**

GLOBAL OFFICES

• New Delhi **NEW AGE INTERNATIONAL (P) LIMITED, PUBLISHERS**
7/30 A, Daryaganj, New Delhi-110002, (INDIA), **Tel.:** (011) 23253472, 23253771, **Mob.:** 9315905300
E-mail: contactus@newagepublishers.com • Visit us at www.newagepublishers.com

• London **NEW AGE INTERNATIONAL (UK) LTD.**
27 Old Gloucester Street, London, WC1N 3AX, UK
E-mail: info@newacademicscience.co.uk • Visit us at www.newacademicscience.co.uk

BRANCHES

• Bangalore 37/10, 8th Cross (Near Hanuman Temple), Azad Nagar, Chamarajpet, Bangalore- 560 018
Tel.: (080) 26756823, **Mob.:** 9315905288, **E-mail: bangalore@newagepublishers.com**

• Chennai 26, Damodaran Street, T. Nagar, Chennai-600 017, **Tel.:** (044) 24353401, **Mob.:** 9315905309
E-mail: chennai@newagepublishers.com

• Cochin CC-39/1016, Carrier Station Road, Ernakulam South, Cochin-682 016
Tel.: (0484) 4051304, **Mob.:** 9315905289, **E-mail: cochin@newagepublishers.com**

• Guwahati Hemsen Complex, Mohd. Shah Road, Paltan Bazar, Near Starline Hotel
Guwahati-781 008, **Tel.:** (0361) 2513881, **Mob.:** 9315905296, **E-mail: guwahati@newagepublishers.com**

• Hyderabad 105, 1st Floor, Madhiray Kaveri Tower, 3-2-19, Azam Jahi Road, Near Kumar Theater
Nimboliadda Kachiguda, Hyderabad-500 027, **Tel.:** (040) 24652456, **Mob.:** 9315905326
E-mail: hyderabad@newagepublishers.com

• Kolkata RDB Chambers (Formerly Lotus Cinema) 106A, 1st Floor, S N Banerjee Road
Kolkata-700 014, **Tel.:** (033) 22273773, **Mob.:** 9315905319, **E-mail: kolkata@newagepublishers.com**

• Mumbai 142C, Victor House, Ground Floor, N.M. Joshi Marg, Lower Parel, Mumbai-400 013
Tel.: (022) 24927869, 24915415, **Mob.:** 9315905282, **E-mail: mumbai@newagepublishers.com**

• New Delhi 22, Golden House, Daryaganj, New Delhi-110 002, **Tel.:** (011) 23262368, 23262370
Mob.: 9315905300, **E-mail: sales@newagepublishers.com**

C-23-08-13747

Printed in India at Ajit Printing Press, Delhi.
Typeset at In-house, Delhi.

NEW AGE INTERNATIONAL (P) LIMITED, PUBLISHERS
7/30 A, Daryaganj, New Delhi - 110002
Visit us at **www.newagepublishers.com**
(CIN: U74899DL1966PTC004618)

PREFACE TO THE FIFTH EDITION

This book has been thoroughly read by our subject experts and efforts have been made to present an error free edition. Though the overall format is not disturbed but the presentation and the rubric has been amplified. More clarity is incorporated wherever necessary and the figures have been redrawn and coloured for better understanding. A new **Appendix D: Structures of Some Common Polymers** has been added in this Fifth Edition. Suggestions for further improvement shall be received with appreciation.

Publisher

PREFACE TO THE FIRST EDITION

We have had the privilege of looking at the polymer science and technology programmes from over 20 countries, representing a cross-section of both the developing and the developed world. Indeed, the inputs received from institutions located as far and wide—and diverse—as Argentina, Australia, China, Germany, Belgium, Hong-Kong, Finland, India, Austria, Japan, Norway, Turkey, France, Israel, Britain, USA and USSR and our numerous discussions with the academic community have charted, to our pride, the premise of this book.

One thing was apparent during these interactions: Polymer science as a discipline has long expanded its traditional influence from the "post-graduate" level to the levels of graduation and diverse industrial courses in polytechnics and university colleges. Largely because of this, a vast student community now emerging often finds many of the existing treatises on polymers (excellent as they are) rather too advanced for their comprehension.

Keeping especially this new kind of students in mind, this book is written, on the one hand, for those who would appreciate being *gently* introduced to polymers and, on the other hand, for industrial personnel who would like to develop a better insight of their normal work by being exposed to wider aspects of polymer science. Accordingly, simple analogies and illustrations have been used to explain some of the most abstract concepts. After elaborating the principles involved in the determination of, for instance, molecular weight or glass transition temperature, actual worked out examples are also added. Easy-to-follow laboratory procedures for the synthesis of polymers using different polymerisation techniques are made part of the write-up. Several Tables have been introduced in order that the book may serve to a considerable extent as a Hand Book on important polymer related parameters. General and chapterwise bibliographies have been included for further studies. The book is structured in such a way that a person with some background of organic and physical chemistry should be able to comprehand the text with clarity. The idea is to take the reader to a level from where he could easily assimilate exhaustive treatises on linear flexible polymers, listed in the bibliography.

Among the professionals and academics who not only took keen personal interest but also painstakingly reviewed many of the chapters of this book are: Dr. Robert A. Frosch, former NASA Administrator and now Vice President, General Motors Corporation, Warren, Michigan; Prof. E.M. Pearce, Director, Polymer Research Institute, Brooklyn, N.Y.; Dr.R.F. Landel and Dr. W.G. Knauss of Jet Propulsion Lab. Pasadena, California; Prof. C.G. Overberger, Director, Macromolecular Research Centre, Ann Arbor, Michigan; Prof. Bruno Vollmert, Director, Polymer Institute Der Universitat Karlsruhe, Karlsruhe; Prof. A. W. Birley, Director, Institute of Polymer Technology, Loughborough, Leicestershire; Prof. Makato Okawara, Tokyo Institute of Technology, Tokyo; Prof. D. Lal, FRS, Physical Research Laboratory, Ahmedabad; Prof. E. Fitzer, Institute fur Chemical Technik Der Universitat Karlsruhe, Karlsruhe; Prof. M. Santappa, Ex-Vice Chancellor and Prof. H. Kothandaraman of University of Madras, Madras; Dr. H.A. Monteiro, CIBA-Geigy India Ltd., Bombay; Dr. Sukh Dev, Director, Malti-Chem Research Centre, Baroda; Prof. S. Ramaseshan, Raman Research Institute, Bangalore; Dr. S. Gundiah, NCL, Poona; Dr. G.C. Joshi, IIP, Dehradun; Prof. S.R. Patel, Sardar Patel University, Vallabh Vidyanagar; Dr. V. Siddhartha, DRDO, Delhi; Dr. K.V.C. Rao, Director, Chemicals Group, VSSC, Trivandrum; Dr. V.N. Krishnamurthy, Director, Propellants Group, VSSC, Trivandrum; and Mr. K.S. Sastri, General Manager, Chemical-Engineering Complexes, VSSC, Trivandrum. The one person whose deep personal interest in this book meant so much to us is Prof. S. Dhawan, the founding Chief of the Indian Space Department and a very respected figure in the Indian aerospace programmes. We are indebted to all of them.

It is estimated that in industrially advanced countries more than 50% of the chemists employed in industries and research institutions work on polymeric systems in one form or another. Obviously, a great deal of additional human resources is required to be added year after year to polymer based industries and research institutions. To meet this need, hundreds of polytechnics, colleges and universities all over the world are, in increasingly large numbers, training people in polymer science and technology. We were excited by the material received on the programme content and the sheer number of the institutes that conduct courses in polymer science. A list of over 50 such institutions from across the world together with a sample of some typical courses is given at the end of this book. We are grateful to the authorities of these Universities/Institutions for responding to our requests and providing us with a canvass against which polymers as a discipline is viewed at various academic institutions.

M/s K.K. Gopalakrishnan, V. Vijayan and G. Yohannan did the entire secretarial work. M/s Samuel Abraham and Appukkuttan Nair did the art work. We are extremely thankful to them.

As authors, we owe deeply to the creative environments of the Vikram Sarabhai Space Centre (VSSC), Trivandrum and to our wives for their encouragement and patience during the writing of this book. One of us (VRG) also thanks Prof. Holt Ashley, Department of Aeronautics and Astronautics, Stanford University for facilitating wonderful environments for very fruitful academic encounters at Stanford.

VSSC, Trivandrum, India
Stanford University, Stanford, USA

V.R. Gowariker
N.V. Viswanathan
Jayadev Sreedhar

Acknowledgements

I. Permission granted by the following copyright holders to reproduce their copyright materials is gratefully acknowledged

Copyright holder	Journal/Book reference	Copyright year	Figure/Table in this book (Material reproduced)
1	2	3	4
Akademische Verlagsgesellschaft, Leipzig	Schulz G.V., Zeitschrift fur Physikalische Chemie. B. 46, 137 (1940) and B. 47, 155 (1940)	1940	Fig. 13.3
American Chemical Society, Washington	Mayo, F.R. and Lewis, F.M., J.Am. Chem. Soc. 66, 1594 (1944)	1944	Fig. 8.1
Dr. Dietrich Steinkopff Verlag, Darmstadt	Kanig, G., Kolloid, Z., 190, 1 (1963)	1963	Fig. 6.9
General Electric Company, Fairfield	Price, F.P. in "Growth and Perfection of Crystals" (Doremus, R.H., Roberts, B.W. and Turnbull D., Eds), John Wiley, New York, (1958) pp. 466	1958	Fig. 7.12
	Keller A., in ibid, (1958) pp. 518	1958	Fig. 7.15(a)
Huthig und Wepf Verlag, Basel	Henglein, A., Makromol Chem. 14, 128 (1954)	1954	Fig. 10.4
Hewlett Packard Company, Palo Alto	Collins, E.A., Bares J. and Billmeyer F.W., "Experiments in Polymer Science", John Wiley, New York (1973) pp. 126	—	Fig. 13.16
John Wiley & Sons, New York	Mark, H.F, in "Introduction to Polymer Science and Technology: An SPE Text Book" Eds. Kaufman, H.S., and Falcetta, J.J. (1977) pp. 12	1977	Fig. 3.8
	Billmeyer F.W., in ibid, (1977) pp. 185	1977	Fig. 13.21
	Geil, P. H., in ibid, (1977) pp. 218	1977	Fig. 7.16
	Billmeyer, F.W., "Text Book of Polymer Science" (1962) pp. 9	1962	Fig. 7.19

	Billmeyer, F.W., ibid, (1962) pp. 62	1962	Fig. 13.12 &. 13.13
	Collins, E.A., Bares, J., and Billmeyer F.W. "Experiments in Polymer Science" (1973) pp. 126	1973	Fig. 13.16
	Brandrup, J. and Immergut, E.H., "Polymer Handbook" (1966)	1966	Table 8.1
	Brandrup, J. and Immergut, E.H., Ibid, 2nd Edn. (1975)	1975	Tables 2.4-2.6, 4.1, 6.1, 8.2, 12.1 & 12.2
	Geil, P.H. "Polymer Single Crystals" (1963) pp. 179	1963	Fig.7.15(b)
	Gaylord, N.G. and Mark, H., "Linear and Stereo Regular Addition Polymers" (1959) pp. 58	1959	Fig. 7.10
	Fineman, M. and Ross, S.D., J. Polym. Sci. 5, 259 (1950)	1950	Fig. 8.2
	Eppe, R., Fischer, B.W. and Staurt, H. A., ibid, 34, 721 (1959)	1959	Fig. 7.13
	Keller, A., ibid, 36, 361 (1959)	1959	Fig. 7.14
	Natta, G. and Corradini, P., ibid., 39, 29 (1959)	1959	Fig. 7.10
	Maley, L.E., ibid, C8, 253, (1965)	1965	Fig. 13.8(b)
McGraw-Hill Inc., New York	Winding, C.C. and Hiatt, G. D., "Polymeric Materials", (1961) pp. 66	1961	Fig. 14.3
	Seymour, R.B. "Introduction to Polymer Chemistry", (1971) pp. 53	1971	Fig. 3.6 and 3.7
	1984-85 Modern Plastics Encyclopedia (1984)	1984	Table 7.1
	Modern Plastics, 47 (Number 10), 76 (1970)	1970	Fig. 1.5
Mir Publishers Moscow	Tager, A., "Physical Chemistry of Polymers", (Trans. Sobolev, D. and Bobrov., N.), (1978) pp. 424	1978	Fig. 12.5
	Tager, A., ibid (1978) pp. 496	1978	Fig. 13.25
	Strepikheyev, A., Derevitskaya, V. and Slonimsky, G., "A First Course in Polymer Chemistry", (Trans. Sobolev, D.) (1971) pp. 84	1971	Table 4.3
Plenum Press, New York	Elias, H.G. "Macromolecules", (1977) pp. 160	1977	Fig. 7.3
	Elias, H.G., ibid, (1977) pp. 735	1977	Fig. 2.4

Reinhold Publishing Corp., New York	Miller M.L., "The Structure of Polymers" (1966) pp. 292	1966	Fig. 6.7
Royal Society of Chemistry, London	Gregg, R.A. and Mayo, F.R., Discuss. Faraday Soc., 2, 328 (1947)	1947	Fig. 4.2
	Gee, G., Trans. Faraday Soc., 40, 261 (1944)	1944	Fig. 13.13
	Leitner M., ibid, 51, 1015 (1955)	1955	Fig. 7.18
	Pepper, D.C., Quart. Rev., 8, 88 (1954)	1954	Fig. 8.4
Springer-Verlag, Heidelberg	Vollmert, B., "Polymer Chemistry", Springer-Verlag, New York (1973), pp. 102	1973	Fig. 8.2
	Vollmert, B., ibid, pp.261	1973	Fig. 7.11
	Vollmert, B., ibid, pp. 355	1973	Fig. 13.27
	Vollmert, B., ibid, pp. 363	1973	Fig. 13.24
	Vollmert, B., ibid, pp. 493	1973	Fig. 12.3
	Vollmert, B., ibid, pp. 517	1973	Fig. 12.4

II. The following materials in the Public Domain are also reproduced. The source as indicated below is acknowledged:

Flory, P.J., J. Am. Chem. Soc., **58**, 1877 (1936), for Fig. 4.7 and 4.8 in this book.

Flory, P.J., ibid, 61, 3334 (1939), for Fig. 4.5 and 4.6 in this book.

Badgley, W.J. and Mark, H., in "High Molecular Weight Organic Compounds (Frontiers in Chemistry, Vol. VI), Eds. Burk, R.E. and Grummit, O., Interscience Publishers, New York (1949) for Fig. 13.12 in this book.

III. The following Scientists/Institutions/Companies have provided necessary data/photographs for inclusion in this book or have granted us permission to reproduce material from their technical literature. Their courtesy is gratefully acknowledged:

Blow Plast Ltd., Bombay for Fig. 1.3, 1.4 and 15.7.

E.I. Du Pont Nemours & Co., Willmington for Fig. 13.33.

Gundiah, S. and Co-workers, National Chemical Laboratory, Pune for Fig. 13.26; Table 13.4. 13.6 and 13.7.

Knauer & Co., Berlin for Fig. 13.17.

Perkin Elmer. Norwalk for Fig. 13.35.

Toyo Rayon Co., Tokyo for Fig. 15. 18

Vikram Sarabhai Space Centre, Trivandrum:

—Chemicals Group for Fig. 13.6, 13.9, 13.34 and 13.36; Table 4.2 and 13.5.

—Materials Group for Fig. 7.1.

Waters Associates Ltd., Singapore for Fig. 13.8(a).

CONTENTS

Preface to the Fifth Edition v
Preface to the First Edition v
Acknowledgements vii

1. The Genesis of Polymers **1–10**

1.1 A Brief History 1
1.2 What are Polymers? 5
1.3 How are Polymers Made? 7
1.4 Classification of Polymers 9

2. Chemistry of Polymerisation **11–62**

2.1 Introduction 11
2.2 Chain Polymerisation 15
2.3 Step Polymerisation 36
2.4 Miscellaneous Polymerisation Reactions 46
2.5 Polymerisation Techniques 52

3. Molecular Weight and Size **63–77**

3.1 'Average' Molecular Weight 63
3.2 Number-average and Weight-average Molecular Weights 64
3.3 Sedimentation and Viscosity-average Molecular Weights 67
3.4 Molecular Weight and Degree of Polymerisation 68
3.5 Polydispersity and Molecular Weight Distribution in Polymers 68
3.6 The Practical Significance of Polymer Molecular Weight 71
3.7 Size of Polymer Molecules 73

4. Kinetics of Polymerisation 78–101

4.1 Introduction 78
4.2 Free-radical Chain Polymerisation 78
4.3 Cationic Polymerisation 86
4.4 Anionic Polymerisation 89
4.5 Polycondensation 91

5. Chemical and Geometrical Structure of Polymer Molecules 102–112

5.1 General Remarks on Polymer Microstructure 102
5.2 Microstructures Based on the Chemical Structure 103
5.3 Microstructures Based on the Geometrical Structure 105

6. Glass Transition Temperature 113–130

6.1 What is Glass Transition Temperature? 113
6.2 Glassy Solids and Glass Transition 113
6.3 Transitions and Associated Properties 119
6.4 Factors Influencing the Glass Transition Temperature 123
6.5 Glass Transition Temperature and Molecular Weight 126
6.6 Glass Transition Temperature and Plasticisers 127
6.7 Glass Transition Temperature of Copolymers 128
6.8 Glass Transition Temperature and Melting Point 129
6.9 Importance of Glass Transition Temperature 129
6.10 Heat Distortion Temperature 130

7. Crystallinity in Polymers 131–145

7.1 Crystalline Solids and Their Behaviour Towards X-rays 131
7.2 Polymers and X-ray Diffraction 132
7.3 Degree of Crystallinity 133
7.4 Crystallisability 134
7.5 Polymer Crystallisation 134
7.6 Crystallites 136
7.7 Structural Regularity and Crystallisability 137
7.8 Other Factors Affecting Crystallisability 138
7.9 Helix Structures 139
7.10 Spherulites 140
7.11 Polymer Single Crystals 142

7.12 Folding of Chains During Crystal Formation 142
7.13 Effect of Crystallinity on the Properties of Polymers 144

8. Copolymerisation 146–160

8.1 Introduction 146
8.2 Free-radical Copolymerisation 147
8.3 Ionic Copolymerisation 157
8.4 Copolycondensation 159

9. Individual Polymers 161–200

9.1 Polyethylene 161
9.2 Polypropylene 163
9.3 Polystyrene 163
9.4 Polyacrylonitrile 164
9.5 Polymethyl Methacrylate 165
9.6 Polyesters 166
9.7 Polyformaldehyde 167
9.8 Polyparaphenylene 168
9.9 Polycarbonates 168
9.10 Polysulphone 169
9.11 Polyimides 170
9.12 Polyamides 171
9.13 Polyurethanes 173
9.14 Polyureas 173
9.15 Polyethylene Glycols and Polypropylene Glycols 174
9.16 Polyacrylic Acid 174
9.17 Polymethacrylic Acid 175
9.18 Polyvinyl Acetate 175
9.19 Polyvinyl Alcohol 176
9.20 Polyvinyl Acetals 177
9.21 Polyvinyl Carbonate 178
9.22 Polyvinyl Chloride 178
9.23 Polyvinylidene Chloride 179
9.24 Polyvinyl Fluoride and Polyvinylidene Fluoride 180
9.25 Polytetrafluoroethylene 181
9.26 Polyvinyl Carbazole, Polyvinyl Pyrrolidone and Polyvinyl Oxazolidinone 181
9.27 Polybutylene and Polyisobutylene 182

9.28	Polyisoprenes	183
9.29	Polybutadienes	184
9.30	Polychloroprene	186
9.31	Polysulphides	186
9.32	Phenol-formaldehyde Resin	188
9.33	Urea-formaldehyde and Melamine-formaldehyde Resins	190
9.34	Epoxy Polymers	193
9.35	Silicone Polymers	194
9.36	Organotin Polymers	195
9.37	Organotitanium Polymers	196
9.38	Cellulose and its Derivatives	196

10. Polymer Degradation **201–225**

10.1	What is Polymer Degradation?	201
10.2	Types of Degradation	201
10.3	Thermal Degradation	204
10.4	Mechanical Degradation	208
10.5	Degradation by Ultrasonic Waves	209
10.6	Photodegradation	210
10.7	Degradation by High-energy Radiation	213
10.8	Oxidative Degradation	215
10.9	Hydrolytic Degradation	225

11. Polymer Reactions **226–260**

11.1	Introduction	226
11.2	Hydrolysis	227
11.3	Acidolysis	228
11.4	Aminolysis	228
11.5	Hydrogenation	228
11.6	Addition and Substitution Reactions	229
11.7	Reaction of Various Specific Groups	231
11.8	Cyclisation Reactions	236
11.9	Cross-linking Reactions	238
11.10	Reactions Leading to Graft and Block Copolymers	252
11.11	Miscellaneous Reactions	256

12. Polymer Solutions 261–284

12.1 The Process of Polymer Dissolution 261
12.2 Thermodynamics of Polymer Dissolution 264
12.3 The Flory-Huggins Theory of Polymer Solutions 269
12.4 Nature of Polymer Molecules in Solution 273
12.5 Size and Shape of Macromolecules in Solution 275
12.6 Viscosity of Dilute Polymer Solutions 277
12.7 Viscosity of Concentrated Polymer Solutions 283

13. Experimental Methods 285–333

13.1 Polymer Synthesis 285
13.2 Isolation and Purification of Polymers 289
13.3 Polymer Fractionation 292
13.4 Molecular Weight Determination 300
13.5 Molecular Weight Distribution Curve 325
13.6 Determination of Glass Transition Temperature 329

14. Elastomeric, Fibre-Forming and Plastic Materials 334–347

14.1 Introduction 334
14.2 Elastomeric Materials 334
14.3 Fibre-forming Materials 336
14.4 Plastic Materials 337
14.5 Rheology of Polymeric Materials 338

15. Polymer Processing 348–370

15.1 Introduction 348
15.2 Plastics, Elastomerics and Fibres 348
15.3 Compounding 350
15.4 Processing Techniques 351

Bibliography 371
Appendix A: Table of Physical Constants 372
Appendix B: Polymer in Daily Life 373
Appendix C: A Representative List of Universities/Institutions Offering Courses in Polymer Science and Technology 376
Appendix D: Structures of Some Common Polymers 387
Index 397

1

THE GENESIS OF POLYMERS

1.1 A BRIEF HISTORY

Since the formation of the earth over 4 billion years ago, in its giant 'laboratory', elements like carbon, hydrogen, oxygen and nitrogen have been combining to form complex molecules. Such a combination some day must have triggered off the most intriguing and fascinating process called life, the material basis for whose origin was a polymer. This polymer, called protein, got synthesised in Nature from simple chemical compounds like methane, ammonia, and carbon-dioxide. The life so started, then evolved through the ages and one form became 'human'. And as it did so, almost the whole human body came to be built around the same polymer. Nature's 'intimacy' to polymers is thus a few billion years old—give or take a few million years, if you want to be meticulous!

Meanwhile, polymers had also appeared in their other natural forms like wood, cotton, cellulose, starch, etc. which man began to use. Why? Well, what is after all the most powerful motivation for all of us? The instinct to survive. Even millions of years ago, man's basic needs happened to be the same as now—he had to drape himself, had to have a shelter, and had to defend himself from the mightier. So he started using animal skin to cover his body, a structure of leaves and trees as shelter, wood clubs as weapons, and in doing so, he had furthered his association with natural polymers.

Over the many civilisations, however, human beings have been finding different ways to meet their basic requirements. For instance, the need to cover the body was met by animal skins at one time, but perhaps it is being met by polyester today. But then as now, it is only a polymer—whether natural, in the form of animal skin or synthetic, in the form of nylons and polyesters—that came forward to meet not only the genuine need, but also the wanton wants of men and women!

In the fifteenth century, a famous discoverer, Christopher Columbus, noticed the natives of South America playing a bouncing game with a solid mass collected from the exudate of a tree, that they called 'weeping wood'. While enjoying the game with the solid ball, little did they realise that what they were handling was to grow into a multi-billion dollar (or ruble or yen) rubber industry! Years later, a renowned scientist, Joseph Priestley, showed that this solid mass could rub off pencil marks from paper and, hence, the material was called rubber.

In the multifarious growth of polymers, there have been some useful 'accidents'. For instance, Christian Schonbein, a Swiss scientist, broke a beaker which contained a mixture of nitric acid and sulphuric acid. He must have obviously got panicky, for he picked up his wife's apron to mop up the mess. What a thing to do! But then, the poor man did promptly wash the apron and left it for drying near a fire-place. The cotton apron, however, soon caught fire and vanished into thin air. His wife must have admonished him, but in the midst of all that, all of a sudden, Schonbein realised what had happened: cotton, the textile material of which the apron was made, got converted into combustible 'gun cotton', a nitro derivative of a naturally occurring polymer cellulose.

The above incident took place around the mid-nineteenth century, when the elephant tusks used, among others, for making billiard balls, had somehow become scarce. There was thus a need to find a suitable substitute for ivory. Necessity being the proverbial mother of invention, John Wesley Hyatt of New York took up the challenge and developed a material which resembled ivory and was capable of being moulded in different shapes by applying heat. Hyatt named this material 'celluloid' which, as we know is a major invention contributing to our present 'plastic' age.

Meanwhile, some of the nineteenth-century researchers had encountered a 'messy' kind of problem while dealing with certain chemicals. Some reactions with these chemicals had yielded highly tacky and viscous materials which would stick thickly to the beakers and choke the valves of the glassware. But it took the genius of a scientist like Leo Baekeland to realise that through these sticky beakers a new branch of chemistry was 'sticking' out and these peculiar materials were soon to open up the doors of a great technological treasure house. In 1909, from two very common chemicals (phenol and formaldehyde), he developed a resin (which was later named after him as Bakelite), that could be moulded into hard infusible articles (Fig. 1.1). Bakelite was a forerunner of many other modern synthetic polymers. For instance, in 1912, Jacques Brandenburger introduced a famous transparent material, cellophane. Within a decade or so, several polymers started appearing in newer and newer forms with increasingly better properties from the laboratories of scientists the world over.

Fig. 1.1 Telephone instrument and other electrical components made of phenol formaldehyde (Bakelite) plastic.

Most of the synthetic polymers are of a relatively recent origin. In fact, they appeared later than the radio and the airplane did, but look at the impact they have made on our life. Whether we are looking for fancy decoration articles, textiles, building materials or packing materials, the polymers have brought to the fore a scintillating range of irresistibly attractive choices. Look at the spectrum of plastic articles (Fig. 1.2). They have become so much a part of our society that it appears almost unthinkable that we could ever do without them.

Fig. 1.2 A spectrum of plastic articles used in our daily life.

You name a desirable property—high strength, light weight, good flexibility, special electrical properties, resistance to chemicals, amenability for quick and mass production and for fabrication into complex shapes in a wide variety of colours—some polymer will almost always meet your requirement. Polymers can be converted into strong solid articles, flexible rubber-like masses, soft and resilient foams, smooth and fine fibres, clean and clear glass-like sheets, swollen, jelly-like food materials and so on. They can be used to bond objects, seal joints, fill cavities, bear loads—in fact, anything from clothing the naked to powering a space vehicle to even replacing a human organ! This should explain why this age could pass as the 'polymer' age.

Polymers have obviously not been discovered overnight. They came out of long and persevering studies by a host of motivated scientists whose work has enriched human life. Today, the overall insight into polymer science and technology is so deep that a material scientist can create an almost limitless range of new materials.

Just as an architect chooses bricks, stones, logs of woods, etc. in varying shapes, sizes and patterns to create various designs, a chemist can produce innumerable plastics, rubbers, foams, fibres, etc. by judiciously combining various chemicals to react under desired conditions. A bullet-proof material like polycarbonate which combines the transparency of glass and the strength of steel is one such example (Fig. 1.3). In future, we can expect even more formidable utility combinations to emerge as the polymer chemist becomes an even better 'architect of molecules'.

Fig. 1.3 Armour shield made of polycarbonate combines the transparency of glass and the strength of steel.

1.2 WHAT ARE POLYMERS?

What are polymers? For one thing, they are complex and giant molecules and are different from low molecular weight compounds like, say, common salt. To contrast the difference, the molecular weight of common salt is only 58.5, while that of a polymer can be as high as several hundred thousands. These big molecules or 'macromolecules' are made-up of much smaller molecules. The small molecules, which combine to form a big molecule, can be of one or more chemical compounds. To illustrate, imagine a set of rings of the same size and made of the same material. When these rings are interlinked, the chain formed can be considered as representing a polymer from molecules of the same compound. Alternatively, individual rings could be of different sizes and materials, and interlinked to represent a polymer from molecules of different compounds. The two situations are shown in Fig. 1.4.

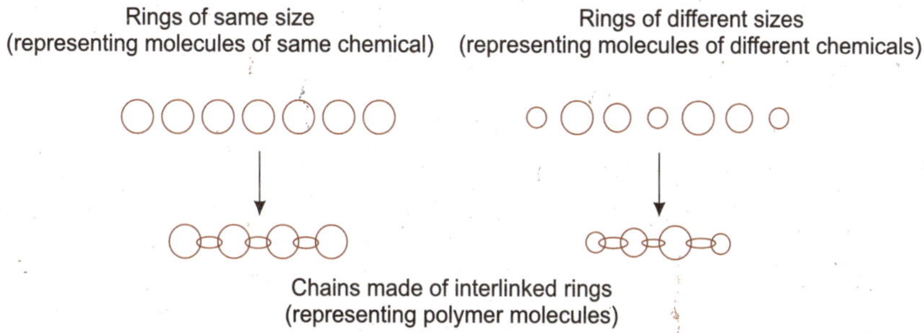

Fig. 1.4 Conceptual representation of the formation of polymer molecule by linking together of molecules of simple chemicals.

This interlinking of many units has given the polymer its name, *poly* meaning 'many' and *mer* meaning 'part' (in Greek). As an example, a gaseous compound called butadiene, with a molecular weight of 54, combines nearly 4,000 times and gives a polymer known as polybutadiene (a synthetic rubber) with about 200,000 molecular weight. The picture is simply as follows:

$$\text{butadiene} + \text{butadiene} + \cdots + \text{butadiene} \longrightarrow \text{polybutadiene}$$
$$(4{,}000 \text{ times})$$

One can thus see how a substance with as small a molecular weight as 54 grows to become a giant molecule of $(54 \times 4{,}000) \approx 200{,}000$ molecular weight. It is essentially the 'giantness' of the size of the polymer molecule that makes its behaviour different from that of a commonly known chemical compound such as benzene. Solid benzene, for instance, melts to become liquid benzene at 5.5°C and, on further heating, boils into gaseous benzene. As against this well-defined behaviour of a simple chemical compound, a polymer like polyethylene does not melt sharply at one particular temperature into a clean liquid. Instead, it becomes increasingly softer and, ultimately, turns into a very viscous, tacky molten mass. Further heating of this hot, viscous, molten polymer does convert it into various gases but they are no longer polyethylene (Fig. 1.5).

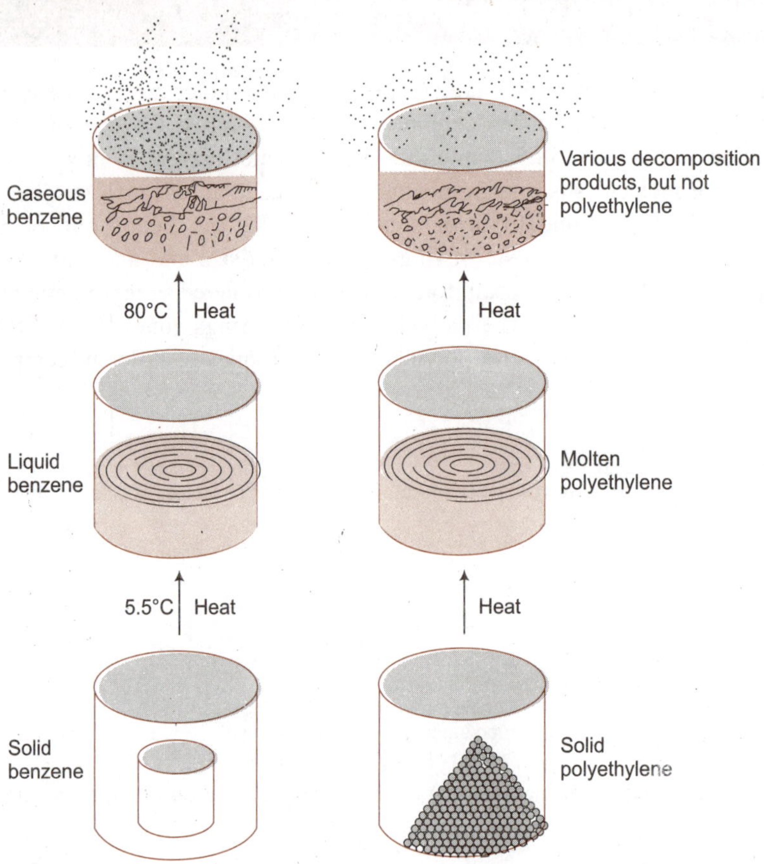

Fig. 1.5 Difference in behaviour on heating of a low molecular weight compound (benzene) and a polymer (polyethylene).

Another striking difference between the behaviour of a polymer and that of a low molecular weight compound concerns their solubility pattern. Let us take, for example, sodium chloride and add it slowly to a fixed quantity of water. The salt, which represents a low molecular weight compound, dissolves in water up to a point (called saturation point) but, thereafter, any further quantity added does not go into solution but settles at the bottom and just remains there as solid. The viscosity of the saturated salt solution is not very much different from that of water. But if we take a polymer instead, say, polyvinyl alcohol, and add it to a fixed quantity of water, the polymer does not go into solution immediately. The globules of polyvinyl alcohol first absorb water, swell and get distorted in shape and after a long time go into solution. Also, we can add a very large quantity of the polymer to the same quantity of water without the saturation point ever being reached. As more and more quantity of the polymer is added to water, the time taken for the dissolution of the polymer obviously increases and the mix ultimately assumes a soft, dough-like consistency. Another peculiarity is that, in water, polyvinyl alcohol never retains its original powdery nature as the excess sodium chloride does in a saturated salt solution. In conclusion, we can say that (*a*) the long time taken by polyvinyl alcohol for dissolution, (*b*) the absence of a saturation point, and (*c*) the increase in the viscosity are all characteristics of a typical polymer being dissolved in a solvent and these characteristics are attributed mainly to the large molecular size of the polymer. The behaviour of a low molecular weight compound and that of a polymer on dissolution are illustrated in Fig. 1.6.

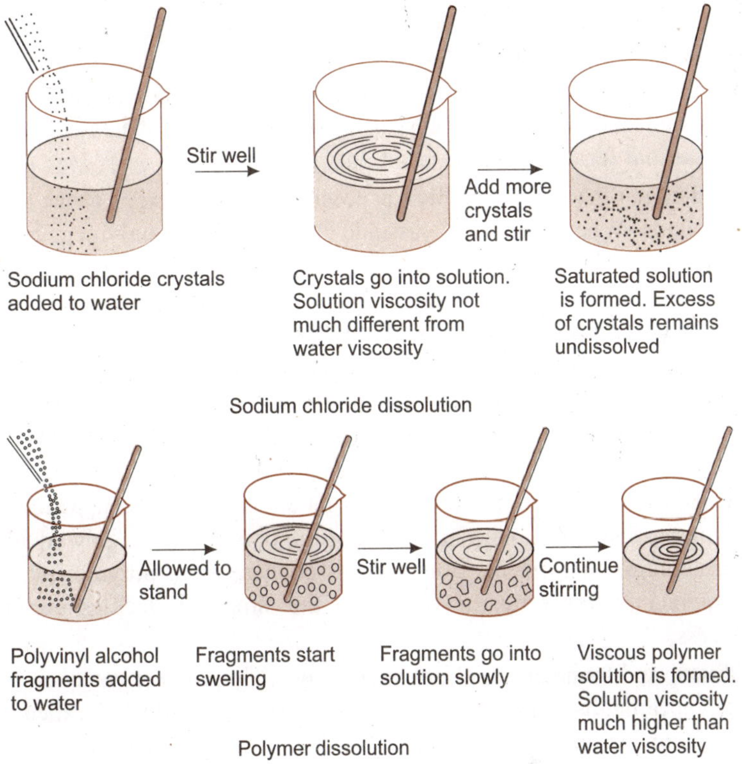

Fig. 1.6 Difference in the solubility behaviour of a low molecular weight compound (sodium chloride) and a polymer (polyvinyl alcohol).

1.3 HOW ARE POLYMERS MADE?

We have seen that a polymer is made-up of many small molecules which have combined to form a single long or large molecule. The individual small molecules from which the polymer is formed are known as *monomers* (meaning, single part) and the process by which the monomer molecules are linked to form a big polymer molecule is called 'polymerisation'.

We now give a single analogy to illustrate the polymerisation phenomenon. Some children are relaxing on a playground—some in groups and some alone. Then appears a drill master and orders them to form a line, which the children very grudgingly do. With another order, he directs each child to hold the hand of the child on either side of him or her so as to form a long row. Now there is a problem. The very mischievous ones, who are interspersed in the row, decide not to extend their left hands. What would then happen to the hand-to-hand row of the lot? Naturally, the line will extend only up to a particularly naughty boy; he having decided not to extend his one hand to the neighbouring child, the chain breaks. Well, the two hands of the child, essential for the row to grow on either side, represent the two reactive sites of the monomer, which are essential if the monomer molecule is to combine with another to grow in size and form a giant molecule of the polymer. It should be noted that the two reactive sites—or what is termed bifunctionality—of the monomer are prerequisites, if the molecular build-up process is to continue, just as the extension of both the hands of every child is essential for their hand-to-hand line to grow.

We have stated earlier that polymerisation is possible with molecules of the same or of different monomeric compounds. When molecules just add on to form the polymer, the process is called 'addition' polymerisation. The monomer species in this case retains its structural identity when it gets converted into a polymer. For example, the molecules of ethylene monomer can add on to form polyethylene, in which, the structural identity of ethylene is retained.

When, however, molecules do not just add on but also undergo some reaction in forming the polymer, the process is called 'condensation' polymerisation. Here, the two molecules (of the same or different monomers), condense to form a polymer. The condensation takes place between two reactive functional groups, like the carboxyl group (of an acid) and the hydroxyl group (of an alcohol). While forming the polymer, however, water molecules also get eliminated. We can, therefore, see that while, in 'addition' polymerisation, the molecular weight of the polymer is roughly equal to that of all the molecules which combine to form the polymer, in 'condensation' polymerisation, the molecular weight of the polymer is lesser by the weight of the simple molecules eliminated during the condensation process.

In both 'addition' and 'condensation' polymerisations, the polymer molecule formed contains a structural identity, repeating itself several times. These repeating entities are called the repeat units or the monomeric units of the polymer molecule. The size of the polymer molecule is decided by the number of repeat units present in it. This number denotes the degree of polymerisation.

For example, 5 molecules of ethylene monomer can add on to each other to form a single molecule of polyethylene, as represented in Fig. 1.7 (top portion). Here, the repeat unit is —CH_2—CH_2— and the polymer molecule contains 5 such repeat units. Hence, the degree of polymerisation is 5. Similarly, when 4 molecules of a hydroxy acid (HO—R—COOH) condense, a polyester molecule is formed. The repeat unit in this case will be —R—COO— [see Fig. 1.7 (bottom portion)]. This polyester molecule has 4 such repeat units and, hence, its degree of polymerisation is 4.

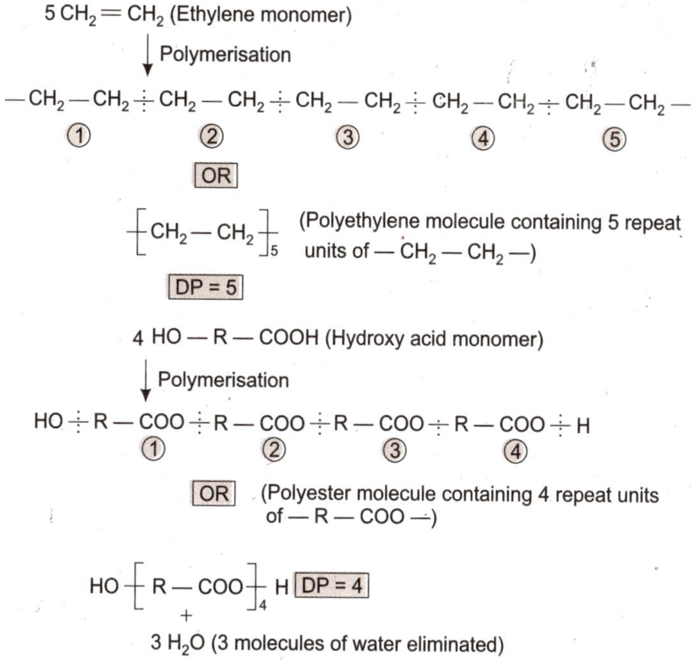

Fig. 1.7 Illustrations showing the degree of polymerisation (DP) of a polymer molecule.

1.4 CLASSIFICATION OF POLYMERS

Polymer is a generic name given to a vast number of materials of high molecular weight. These materials exist in countless forms and numbers because of a very large number and types of atoms present in their molecules. Polymers can have different chemical structures, physical properties, mechanical behaviour, thermal characteristics, etc. and can be classified in different ways, as now discussed.

1.4.1 Natural and Synthetic Polymers

Depending on their origin, polymers can be grouped as natural or synthetic. Those isolated from natural materials are called natural polymers. Typical examples are: cotton, silk, wool and rubber. Cellophane, cellulose rayon, leather and so on are, in fact, chemical modifications of natural polymers.

Polymers synthesised from low molecular weight compounds are called synthetic polymers. Typical examples are: polyethylene, PVC, nylon and terylene.

1.4.2 Organic and Inorganic Polymers

A polymer whose backbone chain is essentially made of carbon atoms is termed an organic polymer. The atoms attached to the side valencies of the backbone carbon atoms are, however, usually those of hydrogen, oxygen, nitrogen, etc. The majority of synthetic polymers are organic and they are very extensively studied. In fact, the number and variety of organic polymers are so large that when we refer to 'polymers', we normally mean organic polymers. The molecules of inorganic polymers, on the other hand, generally contain no carbon atom in their chain backbone. Glass and silicone rubber are examples of inorganic polymers.

1.4.3 Thermoplastic and Thermosetting Polymers

Some polymers soften on heating and can be converted into any shape that they can retain on cooling. The process of heating, reshaping and retaining the same on cooling can be repeated several times. Such polymers, that soften on heating and stiffen on cooling, are termed 'thermoplastics'. Polyethylene, PVC, nylon and sealing wax are examples of thermoplastic polymers. Some polymers, on the other hand, undergo some chemical change on heating and convert themselves into an infusible mass. They are like the yolk of the egg, which on heating sets into a mass, and, once set, cannot be reshaped. Such polymers, that become an infusible and insoluble mass on heating, are called 'thermosetting' polymers.

1.4.4 Plastics, Elastomers, Fibres and Liquid Resins

Depending on its ultimate form and use, a polymer can be classified as plastic, elastomer, fibre or liquid resin. When, for instance, a polymer is shaped into hard and tough utility articles by the application of heat and pressure, it is used as a 'plastic'. Typical examples are polystyrene, PVC and polymethyl methacrylate. When vulcanised into rubbery products exhibiting good strength and elongation, polymers are used as 'elastomers'. Typical examples are natural rubber, synthetic rubber, silicone rubber. If drawn into long filament-like materials, whose length is at least 100 times its diameter, polymers are said to have been converted into 'fibres'. Typical examples are nylon and terylene. Polymers used as adhesives, potting compounds, sealants, etc. in a liquid form are described as liquid resins. Commercially available epoxy adhesives and polysulphide sealants are typical examples.

Books for Further Reading

Mark, H.F., *Giant Molecules,* Time-Life Books, New York, 1966.

Rosato, D.V., Fallon, W.K. and Rosato, D.V., *Markets in Plastics,* Reinhold, New York, 1969.

Dubois, J.H., *Plastics History—USA,* Cahners, Boston, 1972.

2

CHEMISTRY OF POLYMERISATION

2.1 INTRODUCTION

The synthesising of polymers like PVC, polyethylene, polypropylene or polystyrene through the process called polymerisation is not a difficult job for the present-day polymer chemists. However, many of the polymers that we see today were totally unknown a few decades ago and, indeed, synthetic polymer chemistry itself was a speculative, rather than a well-founded experimental science. For instance, a versatile polymer called Teflon, which does not dissolve even in boiling nitric acid, came into the limelight through an unplanned experiment. Roy J. Plunkett, a scientist in Dupont, was experimenting with Freon gases, the well-known fluorocarbon refrigerants. At one stage of the experiment, he filled some cylinders with tetrafluoroethylene gas and left them to stand overnight. The next day, he saw that the pressure in one of the cylinders had dropped to a very low value. Mistaking it for a leakage of gas, he weighed the cylinder to determine the weight loss. However, there was no decrease in the weight, thereby suggesting that the gas kept under pressure had not leaked but had got transformed into a solid or liquid. On cutting open the cylinder, Plunkett discovered a white waxy material. He thus obtained the precious Teflon, much in the same way as a miner strikes diamonds while searching for coal.

Polymer synthesis is not difficult today, because of the abundant availability of literature on the subject. To synthesise a few tons of polymers like polystyrene, all we need is the appropriate quantity of the monomer and catalyst, and a suitable polymerisation reactor. Moreover, many polymer-producing techniques have been so extensively studied and experimented upon, that we can almost always get a polymer of our choice in terms of the required molecular weight, structure, crystallinity, etc.

We have to repeat at this stage, for the sake of continuity, that polymerisation is a process which allows simple low molecular weight compounds to combine and form a complex high molecular weight compound. For this, each molecule of the compound should have the capability to react at least with two other molecules of the same or some other compound. In other words, they should have a functionality of at least two. The functionality of a compound depends on the number of reactive sites it has. A compound assumes functionality because of the presence of reactive functional groups like —OH, —COOH, —NH_2, —SH, —NCO, etc. The number of such functional groups (as listed in Table 2.1) per molecule of the compound defines its functionality.

Table 2.1 Examples showing functionality of compounds

Compound	*Chemical formula*	*Number of functional groups present per molecule*				*Functionality*
		—OH	*—COOH*	*—NH_2*	*—NCO*	
I. Carboxylic acids						
Acetic acid	CH_3COOH	—	1	—	—	Monofunctional
Malonic acid	$HOOCCH_2COOH$	—	2	—	—	Bifunctional
Tricarballylic acid	$HOOCCH_2CH(COOH)$-CH_2COOH	—	3	—	—	Trifunctional
Benzoic acid	C_6H_5COOH	—	1	—	—	Monofunctional
Phthalic acid, isophthalic acid or terephthalic acid	$HOOCC_6H_4COOH$	—	2	—	—	Bifunctional
II. Alcohols						
Ethyl alcohol	CH_3CH_2OH	1	—	—	—	Monofunctional
Ethylene glycol	$HOCH_2CH_2OH$	2	—	—	—	Bifunctional
Trimethylol propane	$CH_3CH_2C(CH_2OH)_3$	3	—	—	—	Trifunctional
Pentaerythritol	$C(CH_2OH)_4$	4	—	—	—	Tetrafunctional
Benzyl alcohol	$C_6H_5CH_2OH$	1	—	—	—	Monofunctional
Salicyl alcohol	$HOC_6H_4CH_2OH$	2	—	—	—	Bifunctional
III. Amines						
Hexyl amine	$CH_3(CH_2)_4CH_2NH_2$	—	—	1	—	Monofunctional
Hexamethylene diamine	$H_2NCH_2(CH_2)_4CH_2NH_2$	—	—	2	—	Bifunctional
Aniline	$C_6H_5NH_2$	—	—	1	—	Monofunctional
Phenylene diamines (ortho, meta and para)	$H_2NC_6H_4NH_2$	—	—	2	—	Bifunctional
IV. Isocyanates						
Methyl isocyanate	CH_3NCO	—	—	—	1	Monofunctional
Hexamethylene diisocyanate	$OCN(CH_2)_6NCO$	—	—	—	2	Bifunctional
Toluene diisocyanate	$CH_3C_6H_3(NCO)_2$	—	—	—	2	Bifunctional
V. Amino acids						
Glycine	H_2NCH_2COOH	—	1	1	—	Bifunctional
Glutamic acid	$HOOCCH_2CH_2CH(NH_2)COOH$	—	2	1	—	Trifunctional
Lysine	$H_2N(CH_2)_4CH(NH_2)COOH$		—	1	2	—
Trifunctional						
VI. Hydroxy acids						
Lactic acid	$CH_3CH(OH)COOH$	1	1	—	—	Bifunctional
Malic acid	$HOOCCH_2CH(OH)COOH$	1	2	—	—	Trifunctional
Tartaric acid	$HOOC(CHOH)_2COOH$	2	2	—	—	Tetrafunctional
Gallic acid	$HOOCC_6H_2(OH)_3$	3	1	—	—	Tetrafunctional
Hydroxy glutamic acid	$HOOCCH_2CH(OH)CH(NH_2)$-$COOH$	1	2	1	—	Tetrafunctional

Some compounds, however, do not contain any reactive functional group but the presence of double or triple bonds in the molecule bestows polyfunctionality on them. For example, ethylene which can take on two atoms of hydrogen, or of halogen, has a functionality of two (bifunctional) because of the presence of the double bond:

$$CH_2{=}CH_2 + H_2 \rightarrow CH_3{-}CH_3$$

$$CH_2{=}CH_2 + Cl_2 \rightarrow CH_2Cl{-}CH_2Cl$$

Similar is the case with other homologous of ethylene (such as propylene, butylene, etc.) and vinyl compounds such as vinyl chloride ($CH_2{=}CHCl$), styrene ($CH_2{=}CHC_6H_5$), etc. Acetylene, on the other hand, has a functionality of four (tetrafunctional), since it can react with four atoms of hydrogen or halogen:

$$CH{\equiv}CH + 2H_2 \rightarrow CH_3{-}CH_3$$

$$CH{\equiv}CH + 2Cl_2 \rightarrow CHCl_2{-}CHCl_2$$

There are some other compounds in which the presence of easily replaceable hydrogen atoms imparts functionality. For example, phenol, which can undergo esterification reaction through the hydroxyl functional group, also undergoes substitution reaction at three sites, where replaceable hydrogen atoms are available for substitution. In the following two examples, phenol exhibits monofunctionality in the first case and trifunctionality in the second:

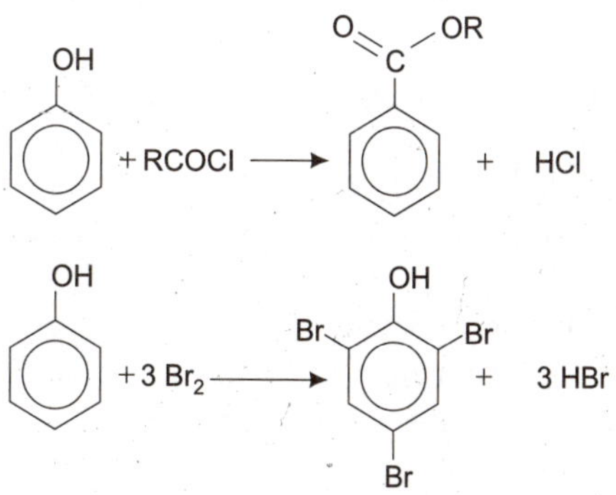

Low molecular weight compounds having a functionality of two or more are called monomers. For them to polymerise, we have only to induce suitable chemical reactions between them. Then, these monomer molecules combine (at the reactive sites or through the functional groups) to form fewer but higher molecular weight molecules. Depending on the functionality of the monomers used, we get linear, branched or three-dimensional cross-linked polymers. For example, if we take bifunctional monomers, each molecule can only add two more molecules and give a linear polymer as shown:

$$n\,{}^{\cdot}A^{\cdot} \longrightarrow {}^{\cdot}A{-}A{-}A{-}A{-}A{-}A{-}A{-}A{-}A^{\cdot}$$

(Bifunctional monomer) (Linear polymer)

• represents functionality and
— represents bond formed
through the functionality.

If, on the other hand, we take trifunctional monomers, we get a cross-linked polymer:

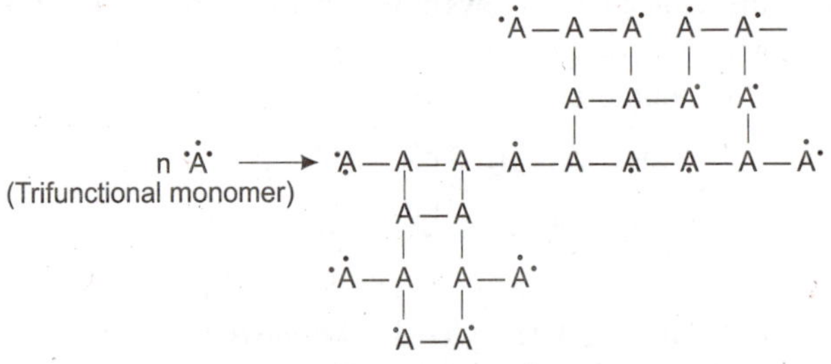

(Cross-linked polymer)

Next, if we take a mixture of bifunctional and trifunctional monomers, their ratio decides whether the polymer (resulting from the mixture of the two monomers) shall be branched or cross-linked.

Another noteworthy point is that, unlike a low molecular weight compound whose molecular weight is invariant with regard to the conditions of the reaction that forms it, the polymer molecular weight depends on the conditions of the polymerisation. For example, consider the formation of the low molecular weight compound ethyl acetate from ethyl alcohol and acetic acid:

$$C_2H_5OH + CH_3COOH \rightarrow CH_3COOC_2H_5 + H_2O$$

The conditions (such as temperature, time, and reactant concentration) of the reaction in this case will affect only the yield, but the product formed will always be ethyl acetate, a liquid with a molecular weight of 88 and a boiling point of 77°C. This is so because one molecule of ethyl acetate results from one molecule of ethyl alcohol reacting with one molecule of acetic acid; and once this ethyl acetate molecule is formed, it remains completely indifferent to (or we may call unreactive with) the other molecules of the reactants. Any further reactions between the reactant molecules, on account of any favourable reaction conditions or whatever, only produce more molecules of the same product and, hence, only the yield increases. As against this, in the polymerisation process, the product of the reaction itself can further react with additional reactant molecules or with other intermediary products. We can thus get products with a range of molecular weights, which, in turn, can further react with the reactant molecules or intermediate products to increase their molecular weight (*i.e.,* of the products).

Theoretically, this process of the molecule getting increasingly bigger would appear to be endless, resulting in the formation of a single molecule of infinite molecular weight. In practice, however, polymer molecules of definite but different molecular weights are simultaneously formed because of such factors as thermodynamic feasibility and kinetic probability. To start with, depending on the polymerisation conditions, a number of dimer products are formed and each tries to react further with fresh reactant (monomer) molecules or with other dimer products. In this way, a number of polymer molecules start growing simultaneously. The number of repeat units added to each of the growing polymer molecule is, however, not the same as the energy level of the various growing molecules and the probability of kinetic collisions are different. The net result is that all the molecules in a polymer sample produced from the same monomer and at the same reaction conditions are not of the same molecular weight. Hence, the molecular weight of the sample can only be expressed as an average of the molecular weights of all the molecules present in the sample. The average molecular weight* of a polymer that is synthesised can, however, be controlled within a certain range by adjusting the conditions of polymerisation and, hence, it is possible to get the same polymer in varying ranges of average molecular weights, say, from a liquid that can be poured, to a tough solid.

If we want to make a necklace from, say, spherical beads, we can resort to a method depending on the situation. For instance, some beads may have hooks attached to them, some may have a hole right through, and some may be just plain beads with no hook or hole. The beads having hooks can be just latched; those with a hole can be threaded, and the plain ones just glued to each other. Now, if our 'beads' are the monomers, depending on their type, functionality and other peculiarities, we can choose the pathway for the formation of the 'polymer necklace'.

The terms 'addition' and 'condensation' polymerisation, used in Chapter 1, will be recalled in this regard. These terms, incidentally, were based on the conventional classification by Carothers (1929) and have since been modified by H.F. Mark (1950) as chain polymerisation and step polymerisation. We will hereafter use the modified terms for classifying the polymerisation reactions.

2.2 CHAIN POLYMERISATION

Chain polymerisation is characterised by a self-addition of the monomer molecules, to each other, very rapidly through a chain reaction. No by-product is formed; the product has the same elemental composition as that of the monomer. The bifunctionality is provided by the double bonds present in the monomer. Compounds containing reactive double bonds can, therefore, undergo a chain polymerisation reaction. Typical examples are vinyl compounds ($CH_2{=}CHX$), allyl compounds ($CH_2{=}CH.CH_2X$), olefines ($CH_2{=}CHR$) and dienes ($CH_2{=}CR{-}CH{=}CH_2$). Since a majority of these monomers fall under the 'vinyl' category, chain polymerisation is also customarily termed 'vinyl' 'polymerisation'. Table 2.2 shows some typical monomers capable of undergoing chain polymerisation.

Chain polymerisation consists of 3 major steps, namely, initiation, propagation and termination and the process can be brought about by a free radical, ionic, or coordination mechanism. Depending on the mechanism, we have three types of chain polymerisation, as described in the forthcoming sections.

2.2.1 Free-Radical Polymerisation

The initiation of the polymer chain growth is brought about by free-radicals produced by the decomposition of compounds called initiators. The term 'chain growth' represents a process involving a continuous and very rapid addition of the monomer units to form polymer molecules or polymer chains. As more and more monomer units are added, the length of the polymer chains increases continuously and the chains grow rapidly.

Initiators

Initiators are thermally unstable compounds and decompose into products called free-radicals. If R—R is an initiator, and the pair of electrons forming the bond between the two R's, can be

*For 'average molecular weight', see Section 3.1.

Table 2.2 Monomers capable of undergoing chain polymerisation

Monomers	
Nomenclature	*Formula*
I. *Vinyl compounds*	
Acrylamide	$CH_2=CHCONH_2$
Acrylic acid	$CH_2=CHCOOH$
Acrylonitrile	$CH_2=CHCN$
Bromotrifluoroethylene	$BrFC=CF_2$
Butyl acrylate	$CH_2=CHCOOC_4H_9$
α-cyanomethyl acrylate	$CH_2=C(CN)COOCH_3$
Methacrylic acid	$CH_2=C(CH_3)COOH$
Methacrylonitrile	$CH_2=C(CH_3)CN$
Methylmethacrylate	$CH_2=C(CH_3)COOCH_3$
α-methyl styrene*	$CH_2=C(CH_3)C_6H_5$
Styrene	$CH_2=CHC_6H_5$
Nitro ethylene	$CH_2=CHNO_2$
p-nitro styrene	$CH_2=CHC_6H_4NO_2$
Tetrafluoro ethylene	$F_2C=CF_2$
Vinyl acetate	$CH_2=CHOCOCH_3$
Vinyl carbazole	$CH_2=CH(C_{12}H_8N)$
Vinyl chloride	$CH_2=CHCl$
Vinyl ethyl ether	$CH_2=CHOC_2H_5$
Vinyl ethyl ketone	$CH_2=CHCOC_2H_5$
2-Vinyl pyridine	$CH_2=CH(C_5H_4N)$
Vinylidene chloride	$CH_2=CCl_2$
II. *Allyl compounds*	
Allyl acetate*	$CH_2=CHCH_2OCOCH_3$
Allyl alcohol*	$CH_2=CHCH_2OH$
Allyl chloride*	$CH_2=CHCH_2Cl$
III. *Olefines*	
Ethylene	$CH_2=CH_2$
Propylene	$CH_2=CHCH_3$
Isobutylene	$CH_2=C(CH_2)_2$
IV. *Dienes*	
1, 3 Butadiene	$CH_2=CH—CH=CH_2$
Chloroprene	$CH_2=CH—CCl=CH_2$
Isoprene	$CH_2=CCH_3—CH=CH_2$

*These monomers are difficult to polymerise.

represented by dots, the initiator can be written as

$$R : R$$

When energy is supplied to this compound, say, in the form of heat, the molecule is split into two symmetrical components. Each component carries with it one of the electrons from the electron pair. This type of decomposition, where the molecule is split into two identical fragments, is called 'homolytic decomposition'. The two fragments, each carrying one unpaired (lone) electron with it, are called free-radicals, *i.e.*

$$R\cdot\!:\!\cdot R \longrightarrow \dot{R}$$

The decomposition of the initiator to form free-radicals can be induced by heat energy, light energy or catalysts.

A host of low molecular weight compounds comprising mainly azo compounds, peroxides, hydroperoxides, peracids and peresters are useful as initiators. Some of them are given in Table 2.3.

Table 2.3 List of free-radical initiators

Commercial name	*Chemical structure*	*Free-radicals formed*
Azo-bis-cyano valeric acid	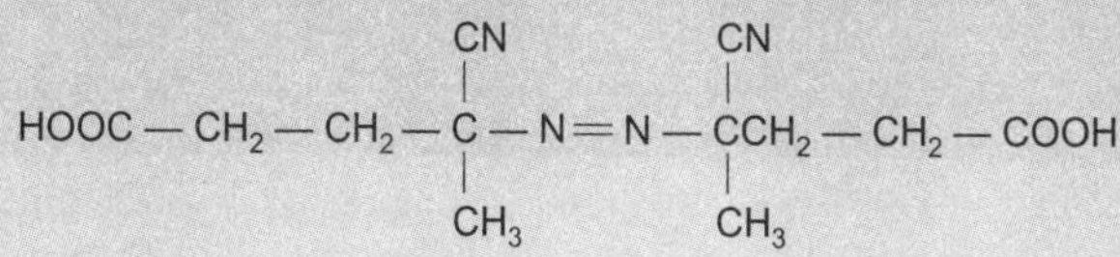	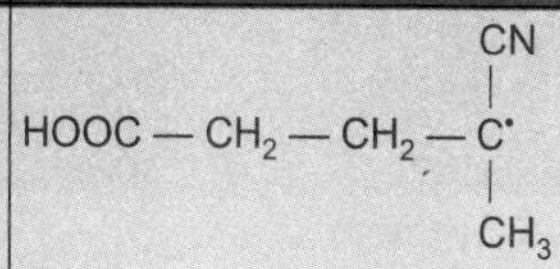
Azo-bis-diphenyl methane	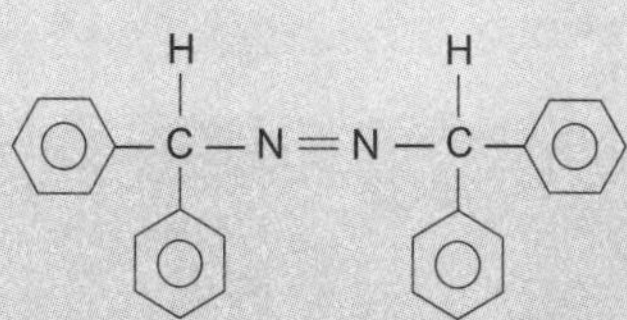	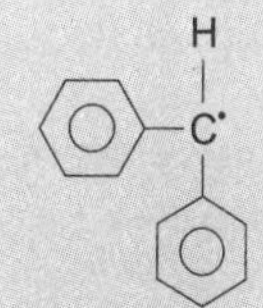
Azo-bis-methyl isobutyrate	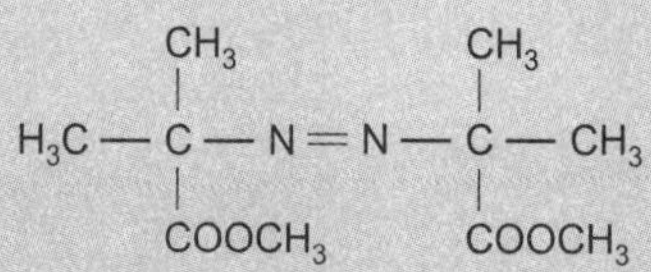	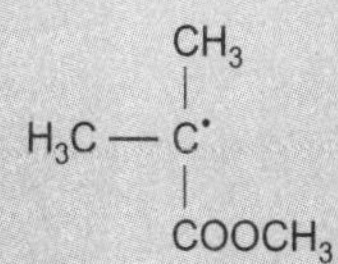
Azo-bis-isobutyronitrile	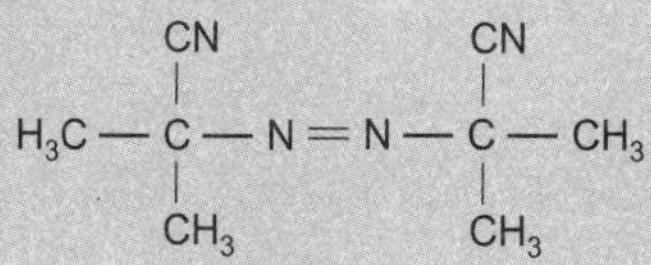	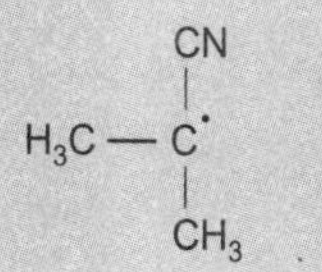
Benzoyl peroxide	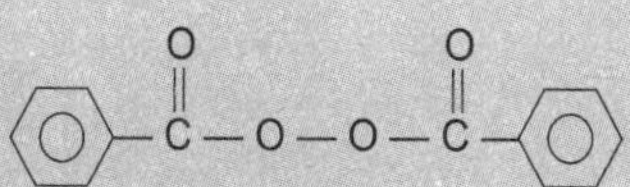	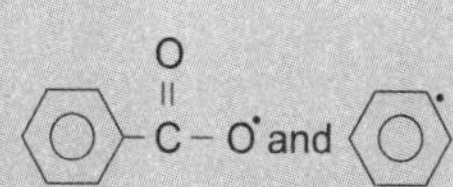
Tert, butyl peroxide	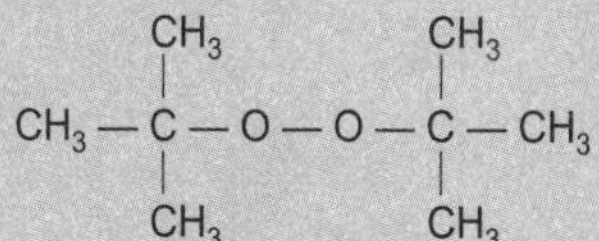	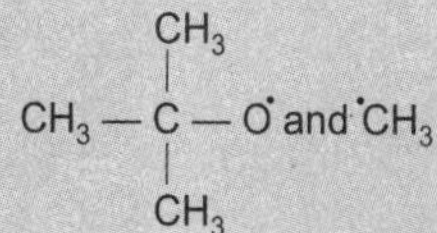
Cumyl peroxide	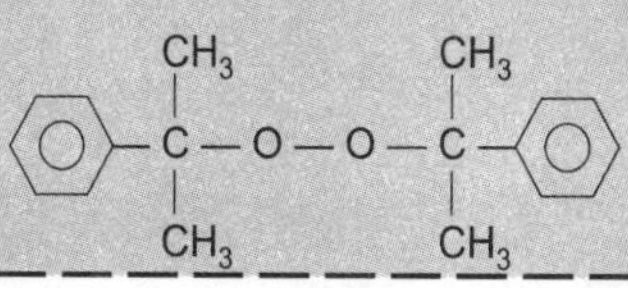	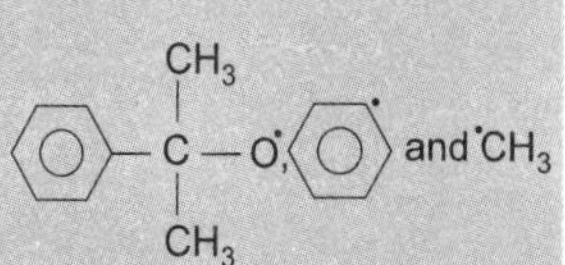

(*Contd.*)

Commercial name	Chemical structure	Free-radicals formed
Cyclo hexanone peroxide	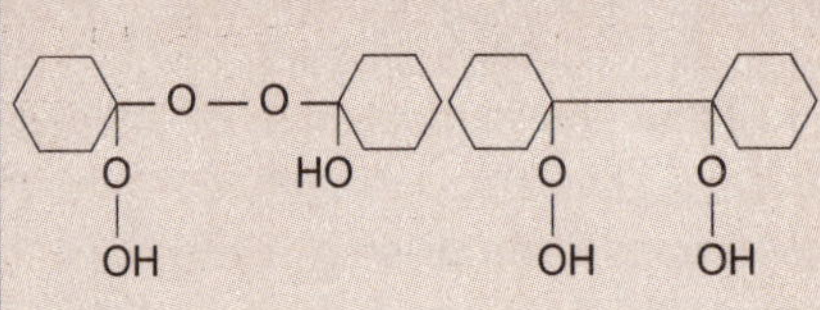	$^{\bullet}OH$ and $^{\bullet}CH_2-(CH_2)_4-COOH$
Glutaric acid peroxide	$HOOC-(CH_2)_3-C(=O)-O-O-C(=O)-(CH_2)_3-COOH$	$HOOC-(CH_2)_3-C(=O)-O^{\bullet}$ and $HOOC-(CH_2)_2-{}^{\bullet}CH_2$
Hydrogen peroxide	$HO-OH$	$^{\bullet}OH$
Lauroyl peroxide	$CH_3-(CH_2)_{10}-C(=O)-O-O-C(=O)-(CH_2)_{10}CH_3$	$CH_3-(CH_2)_{10}-C(=O)-O^{\bullet}$ and $CH_3-(CH_2)_9-{}^{\bullet}CH_2$
Methyl ethyl ketone peroxide	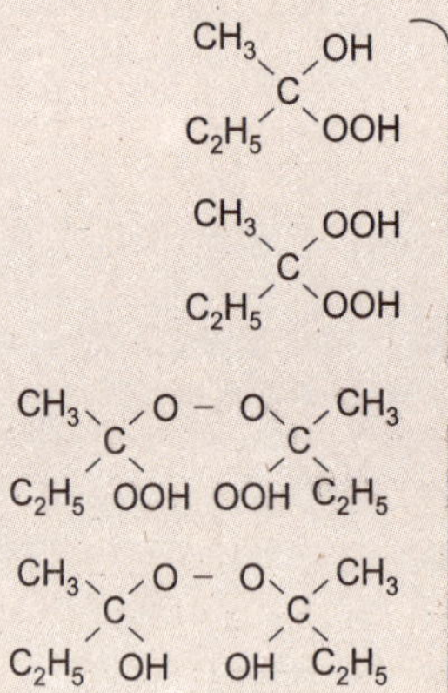 (Mixed structure)	$^{\bullet}OH$
Tert, butyl hydroperoxide	$CH_3-C(CH_3)_2-O-OH$	$CH_3-C(CH_3)_2-O^{\bullet}$, $^{\bullet}OH$ and $^{\bullet}CH_3$
Cumene hydroperoxide	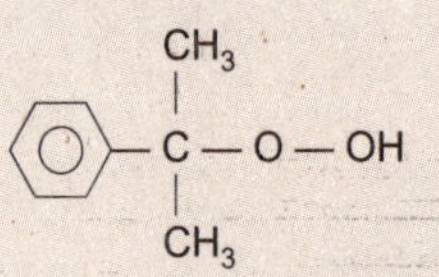	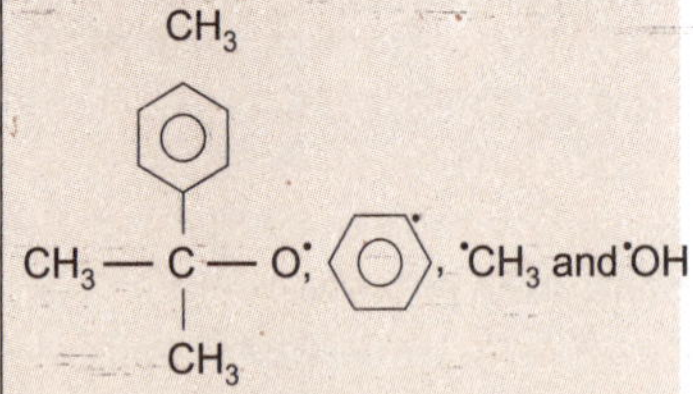
Peracetic acid	$CH_3-C(=O)-O-OH$	$CH_3-COO^{\bullet}$, $^{\bullet}CH_3$ and $^{\bullet}OH$
Perbenzoic acid	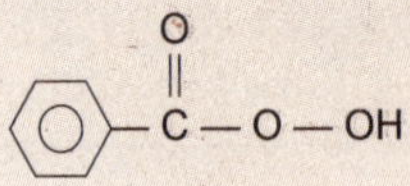	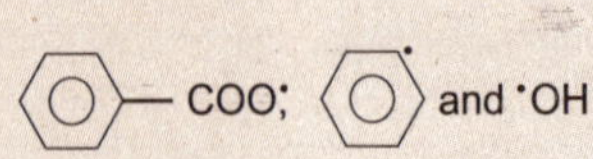

(*Contd.*)

Commercial name	*Chemical structure*	*Free-radicals formed*
Tert. butyl peracetate	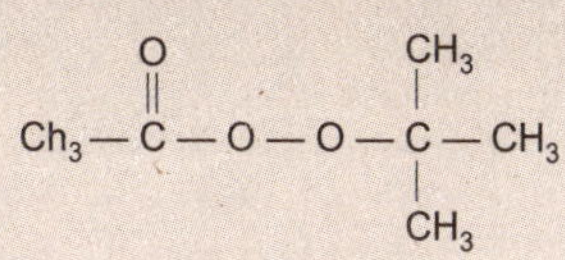	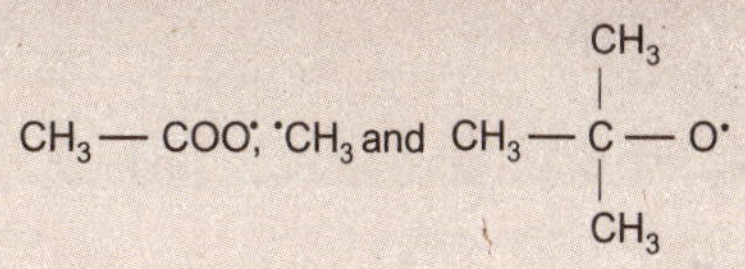
Tert. butyl perbenzoate	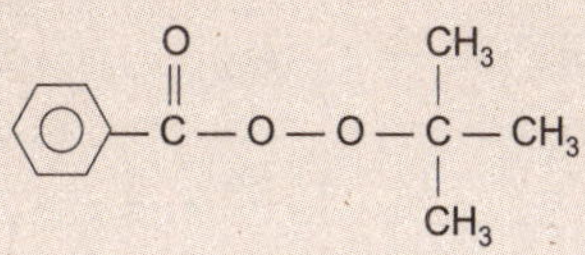	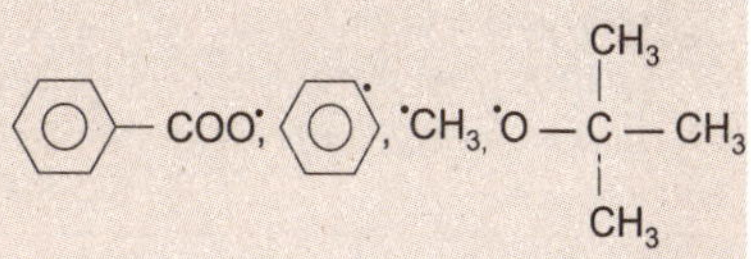
Diisopropyl percarbonate	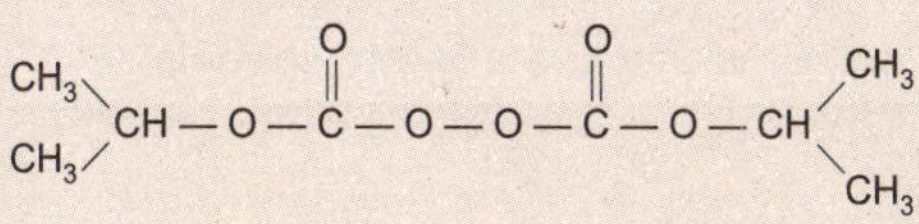	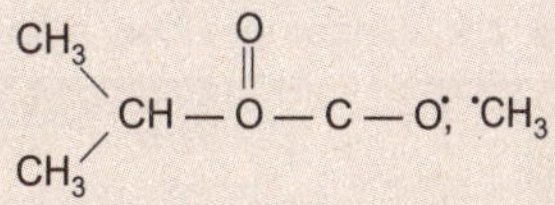
Potassium persulphate	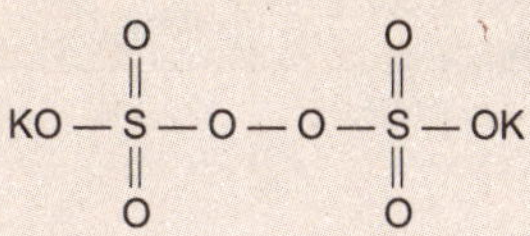	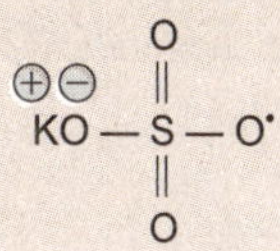

The way thermal decomposition takes place, can be illustrated by the following examples, of benzoyl peroxide and azo-bis-isobutyronitrile:

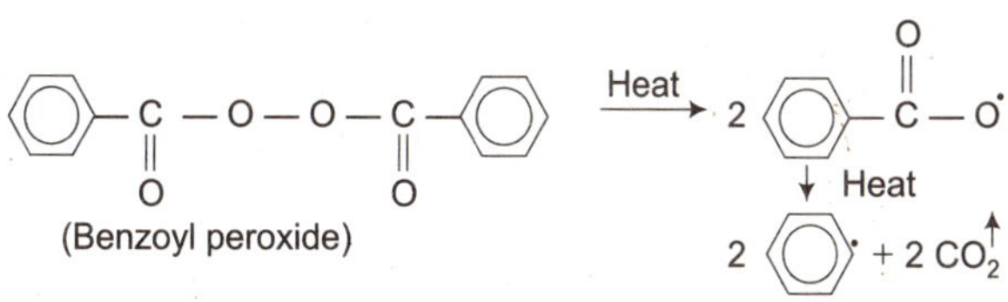

(Benzoyl peroxide)

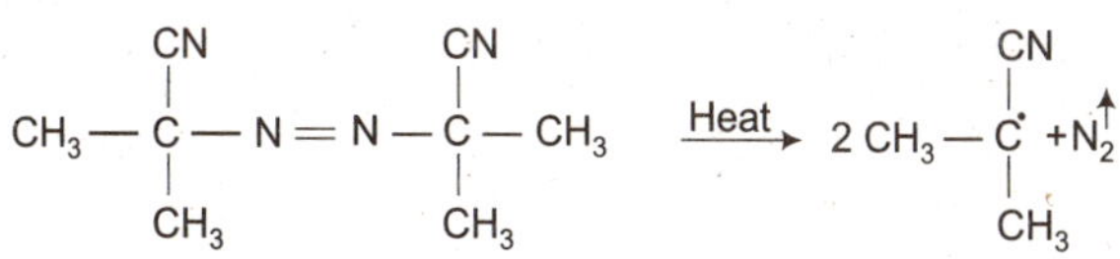

(Azo-bis-isobutyronitrile)

The rate of decomposition of these initiators depends, apart from their chemical nature, on the reaction temperature and the solvents used. This means that the same initiator in a given solvent decomposes at different rates at different temperatures; and the same initiator at a given temperature will have different decomposition rates in different solvents (see Tables 2.4 and 2.5).

Table 2.4 Decomposition behaviour of certain initiators at different temperatures*

Initiator	*Solvent*	*Time taken for 50% decomposition (half-life), $t_{1/2}$ (hours)*			
		40°C	*50°C*	*60°C*	*70°C*
Azo-bis-isobutyronitrile	Benzene	354	73	17 (60.5°C)	5 (69.5°C)
Benzoyl peroxide	Acetone	—	86	—	7
Lauroyl peroxide	Carbon tetrachloride	662	167	41	10
Diethyl peroxydicarbonate	2, 2′-oxydiethylene bis (allyl carbonate)	28	7	1.5	—

*Data from *Polymer Handbook*, Eds. Brandrup, J. and Immergut, E.H., John Wiley & Sons, New York, 1975.

Table 2.5 Decomposition behaviour of certain initiators in different solvents*

Initiator	*Temp.(°C)*	*Time taken for 50% decomposition (half-life) $t_{1/2}$ (hours)*						
		Benzene	*Toluene*	*Carbon tetrachloride*	*Ethyl acetate*	*Acetone*	*Acetonitrile*	*Dioxane*
Azo-bis-isobutyronitrile	40	354	—	895	410	—	—	—
Benzoyl peroxide	70	13	—	—	—	7	11	—
Tert. butyl peroxide	120	14	14	21	—	—	—	—
Propionyl peroxide	65	10	10	—	—	—	—	4
Lauroyl peroxide	40	392	—	662	319	—	—	—

*Data from *Polymer Handbook*, Eds. Brandrup, J. and Immergut, E.H., John Wiley & Sons, New York, 1975.

Initiators can also be decomposed by using ultraviolet light. The free-radicals formed by the photodecomposition of an initiator are the same as those formed by its thermal decomposition:

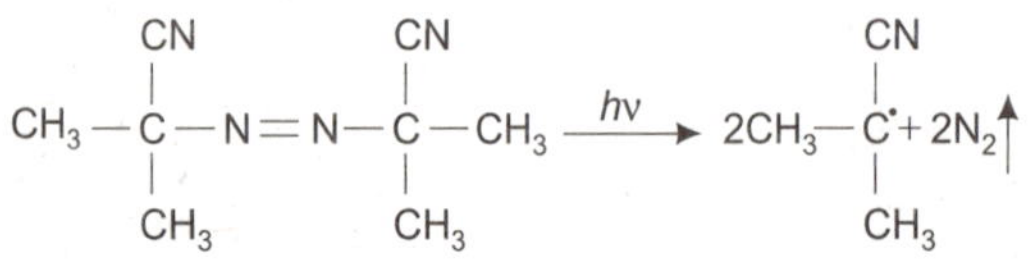

where $h\nu$ represents light energy.

The rate of decomposition, however, depends mainly on the intensity and wavelength of radiation and not so much on the temperature. The solvent controls the intensity of the radiation incident on the initiator molecule and, hence, has an influence on the rate of decomposition.

In addition to the decomposition of the initiator, ultraviolet light can also produce free-radicals by the direct excitation of the monomer molecules:

$$\underset{\text{(Monomer molecule)}}{CH_2{=}CHX} \xrightarrow{h\nu} \underset{\text{(Excited monomer molecule)}}{[CH_2{=}CHX]^*}$$

$$2[CH_2{=}CHX]^* \longrightarrow \underset{\text{(Monomer free-radicals)}}{CH_3{-}\dot{C}HX + CH_2{=}\dot{C}X}$$

It should be noted that polymerisation reactions initiated by ultraviolet light fall under the class of 'photoinitiated polymerisation'.

Initiators can also be induced to decompose into free-radicals by using suitable catalysts. An electron transfer mechanism resulting in a reduction-oxidation (redox) reaction is involved in all these decompositions. Peroxides and hydroperoxides are more susceptible to this kind of decomposition. Benzoyl peroxide, for example, can be decomposed at room temperature with an aromatic tertiary amine:

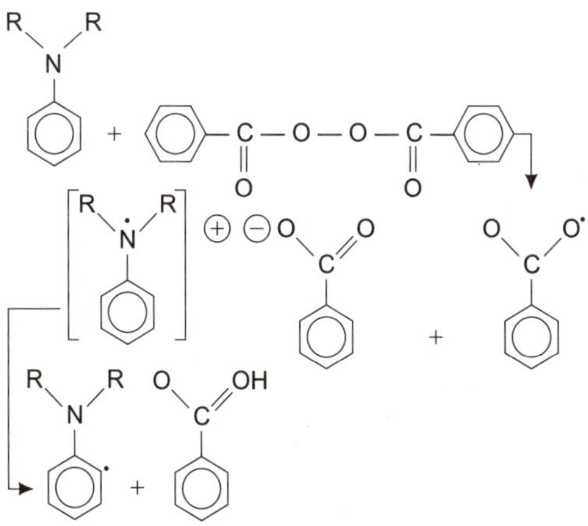

The decomposition of hydrogen peroxide by a ferrous ion and that of a hydroperoxide by a cobaltous ion are other examples:

$$H_2O_2 + Fe^{2+} \rightarrow Fe^{3+} + HO^- + H\dot{O}$$
$$R—O—OH + Co^{2+} \rightarrow R\dot{O} + Co^{3+} + HO^-$$
$$R—O—OH + Co^{3+} \rightarrow RO\dot{O} + Co^{2+} + H^+$$

Polymerisation reactions utilising such redox initiators are termed 'redox polymerisation'.

Initiation

We know that a free-radical contains a lone (unpaired) electron. A lone electron is like a prospective bachelor, always looking for another lone electron to couple with and get stabilised. A free-radical is, therefore, highly reactive and can attack any molecule which either has a lone electron or is prepared to part with one of its electrons. This is what happens in the process of initiation. The free-radical $\dot{R}$ attacks the double bond in the monomer molecule, resulting in the following chemical change:

$$\dot{R} + CH_2 = \underset{\displaystyle X}{\underset{|}{CH}} \longrightarrow R—CH_2 = \underset{\displaystyle X}{\underset{|}{\dot{C}H}}$$

This process is explained as follows. In a double bond formed between two carbon atoms, one pair of electrons exists as sigma (σ) electrons, while the other pair, occupying certain orbitals in the molecule not so close to the nucleus, exists as (π) electrons. If we denote the sigma electron pair by a dash (–), and the π electron pair by two dots, the double bond between two carbon atoms can be represented as C∸C instead of C═C. The representation of the π electron pair by two dots, by itself, is not really correct since the electrons exist as a diffuse cloud and are also constantly under spin. But since it is difficult to show an electron diffuse cloud in a formula, the notation using a pair of dots has become an accepted practice.

The π electrons are not very close to the nucleus and protrude a little away from the axis of the molecule. They are, therefore, susceptible to attack by other reactive species. When free-radicals are produced by the homolytic decomposition of the initiator, what happens is that the free-radical interferes with one of the π electrons and forms a normal pair of electrons at the sigma level; and the other electron of the erstwhile π pair is transferred to the other end of the molecule as shown:

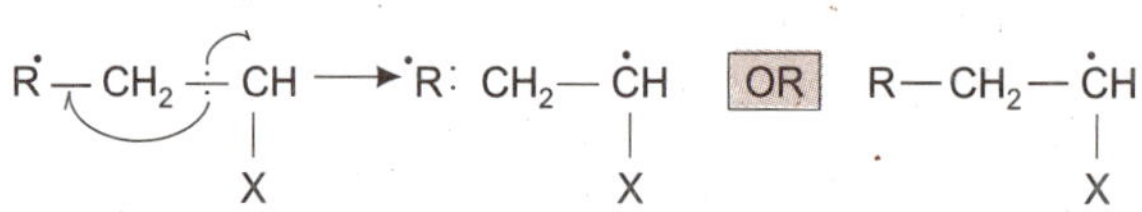

The jump of the dot in the direction of the arrows should be noted. It is as if the two π electrons of the monomer molecule 'divorce' each other and one of the 'separated' partners couples with the lone electron present in the free-radical to form a sigma bond. Now, the monomer unit is linked to the free-radical unit through a sigma bond forming a single molecule. The other electron of the original π electron pair, now deprived of the partner, becomes 'unpaired'. That is to say, the free-radical site is now shifted from the initiator fragment to the monomer unit. This process of the electron pair coming down from its π energy level to the sigma level is associated with an energy release of some 20 kilocalories, as π electrons are at a higher energy level than the sigma electrons. Hence, the free-radical attack on the monomer initiating polymerisation is an exothermic process, whereas free-radical formation by initiator decomposition is an endothermic process. This whole sequence, in which, one free-radical attacks a monomer molecule, adds the monomer molecule to itself, and while so adding, simultaneously transfers the free-radical site from itself to the monomer unit, is termed the initiation step.

Propagation

After 'initiation' comes 'propagation'. In the propagation step, the radical site at the first monomer unit attacks the double bond of a fresh monomer molecule. This results in the linking up of the second monomer unit to the first and the transfer of the radical site from the first monomer unit to the second, by the unpaired electron transfer process (as explained earlier):

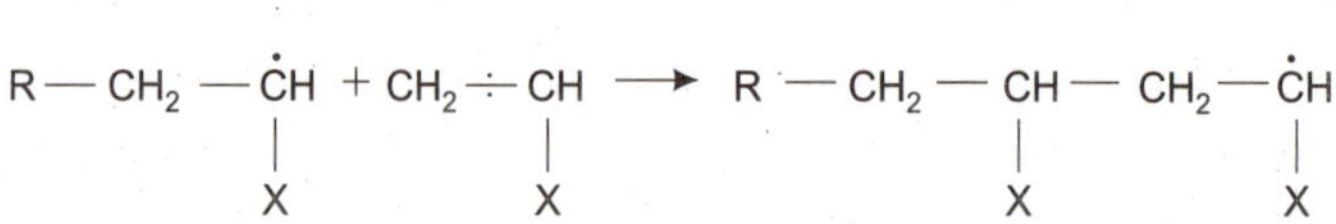

It should be noted that this chain still contains a radical site (indicated by a dot) at its end carbon atom and can, therefore, attack yet another monomer molecule (and thus take it on) with a simultaneous transfer of the radical site to the new monomer unit added:

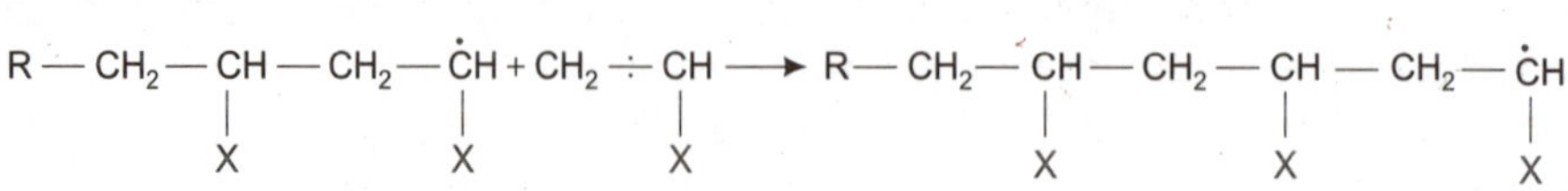

This process involving a continuing attack on fresh monomer molecules which, in turn, keep successively adding to the growing chain one after another is termed 'propagation'. The propagation lasts till the chain growth is stopped by the free-radical site being 'killed' by some impurities or by a sheer termination process, or till there is no further monomer left for attack. The structure of the growing chain can be represented by

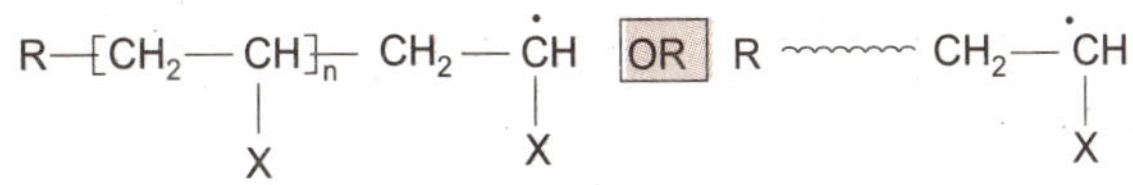

where *n* denotes the number of monomer units added up in the chain growth, and the wavy line indicates the polymer chain made of *n* number of monomeric units. The mode of addition of the incoming monomer to the growing chain can be of the head-to-tail, tail-to-tail, head-to-head or tail-to-head type. If we call the —CH_2— and the —CHX— parts of a monomeric unit its head and tail respectively, the four modes of addition can be represented as follows:

(*i*)

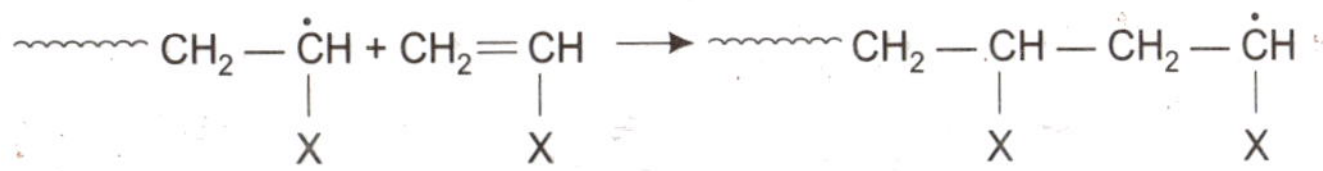

Head-to-tail

(*ii*)

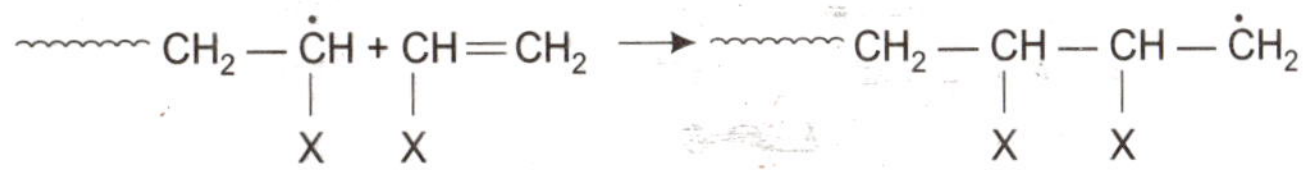

Tail-to-tail

(*iii*)

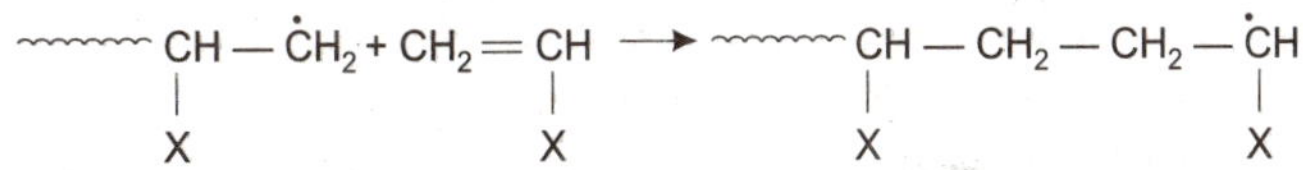

Head-to-head

(*iv*)

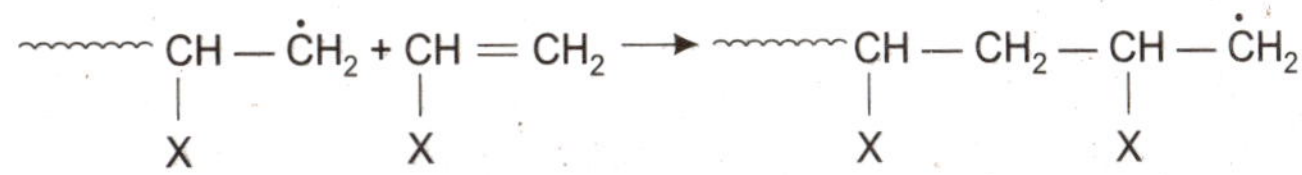

Tail-to-head

As explained earlier, each time a new monomer unit is added to the growing chain, the π electron pair cascades down to the sigma level. Whenever this happens, there is an energy release of about 20 kcal and, hence, the process of chain propagation is brought about without adding any external energy to the system. In other words, a very small amount of energy is supplied to decompose the initiator so as to form its free-radical fragments, and the polymer chain starts growing with large amounts of energy release.

Termination

After 'propagation' comes 'termination'. As you would guess, any further addition of the monomer units to the growing chain is hereafter stopped, and the growth of the polymer chain is arrested.

Since the decomposition of the initiator produces many free-radicals at the same time, each one of them can initiate and propagate the chain growth simultaneously and, hence, at any given time, there may be quite a few growing chains present in the system. Depending on factors such as temperature, time and monomer and initiator concentrations, there exists a statistical probability of the two growing chains coming close to and colliding with each other. When such a collision takes place, the following two reactions occur, resulting in the arrest of the chain growth:

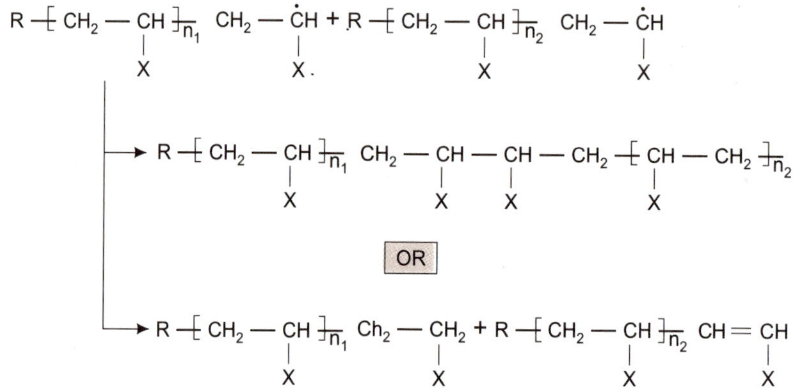

In the first case, the two growing chains unite by the coupling of the lone electron present in each chain to form an electron pair and, thus, nullify their reactiveness. Since this process involves the coupling of the two lone electrons, this kind of termination is known as 'termination by coupling'. In the second case, one H from one growing chain is abstracted by the other growing chain and utilised by the lone electron for getting stabilised, while the chain, which had donated the H, gets stabilised by the formation of a double bond. In this case, the termination process results in the formation of two polymer molecules of shorter chain length as against a single molecule of a longer chain length obtainable by the first method. This type of termination is called 'termination by dis-proportionation'.

From the foregoing termination reactions, it may be noted that the product molecules formed do not contain any free-radical site and, hence, cannot grow any further. The process of termination thus results in the deactivation of the growing chain. The deactivated chain, in fact, forms the polymer. In order to contrast it with an active growing chain, the polymer molecule formed can be referred to as a 'dead' polymer chain. We know that a polymer molecule is made-up of several repeat units. From the termination reactions given above, it may also be noted that, in addition to the several repeat units, the polymer molecule contains certain other groups as well (—R and —CH_2=CHX in this case) at the two chain ends. These groups are called the 'end groups'.

Chain Transfer

There is yet another method of chain termination, which takes place by the 'transfer reaction'. In the termination reaction brought about by coupling or disproportionation, the products formed are 'dead' or 'non-reactive', and, hence, no further chain propagation is possible. In the case of 'transfer reaction', however, while the growth of one polymer chain is stopped or arrested, forming a dead polymer, there is a simultaneous generation of a new free-radical capable of initiating a fresh polymer chain growth. This reaction takes place by the abstraction of a hydrogen atom or some other atom from the initiator, monomer or polymer or from any other species present in the system, including the solvent or any inadvertent impurity. This is represented by

$$R\!-\!\left[\,CH_2-\underset{\displaystyle X}{\underset{|}{CH}}\,\right]_n CH_2-\underset{\displaystyle X}{\underset{|}{\dot{C}H}} + R'H \longrightarrow R\!-\!\left[\,CH_2-\underset{\displaystyle X}{\underset{|}{CH}}\,\right]_n CH_2-\underset{\displaystyle X}{\underset{|}{CH_2}} + R'^{\bullet}$$

The growing chain is now terminated, but a new free-radical **R'·** is formed. **R'·** will now initiate the polymer chain growth afresh, which will be followed by the chain propagation. Here, the termination of one chain growth and the initiation of a new one takes place simultaneously. It is as if the chain growth is transferred from one site to another and hence, this phenomenon is called 'chain transfer'. We shall come across further aspects of chain transfer in Chapter 4.

Inhibitors

Inhibitors are chemical substances capable of inhibiting or killing the chain growth by combining with the active free-radicals and forming either stable products or inactive free-radicals. Hydroquinone, nitrobenzene, dinitrobenzene and benzothiazine are some of the inhibitors customarily used in the polymer industry. The inhibiting action of these compounds can be depicted by the following example:

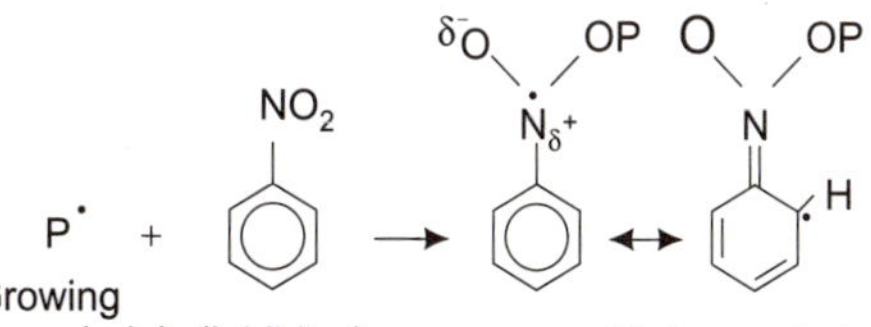

(Growing polymer chain) (Inhibitor) (Polymer chain with resonance-stabilised end group)

Here, the inhibitor, nitrobenzene, adds on the growing chain P·, forming a polymer chain with a nitrobenzene end group carrying a radical site. The nitrocompound end of the chain is, however, resonance stabilised. This resonance-stabilised free-radical end is not active enough to attack a fresh monomer molecule and add it on to the chain. No further propagation can, therefore, take place. The free-radical nature of the end group is, however, powerful enough to recombine with the radical of another growing chain and terminate the growth of the latter:

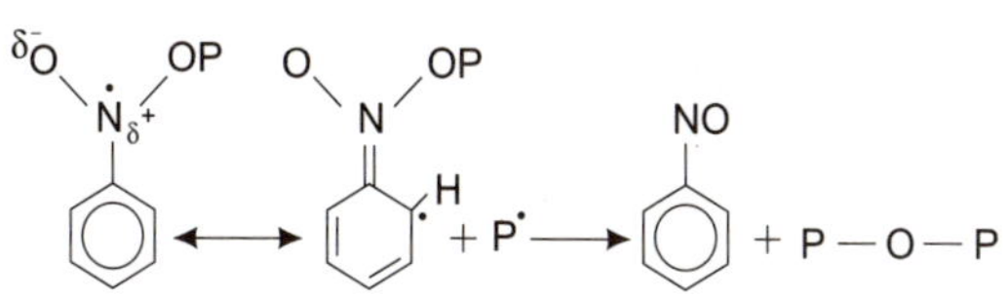

(Dead polymer molecule)

In this case, we can see that a single molecule of the inhibitor has killed two growing chains.

Some of the inhibitors such as diphenyl picryl hydrazide (DPPH) exist in the form of stable free-radicals which can easily stop the chain growth by direct coupling:

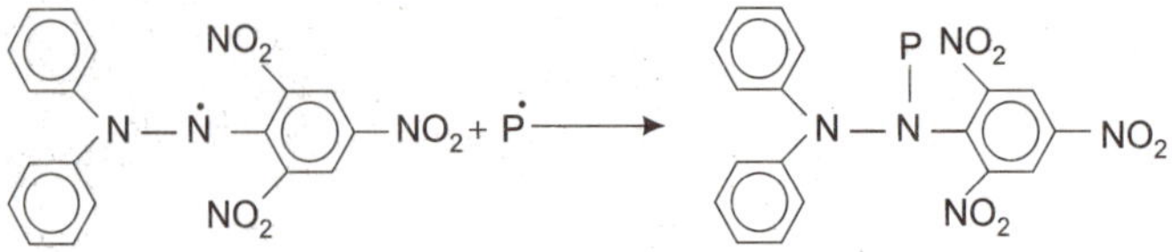

(DPPH) (Dead polymer Molecule)

Atmospheric oxygen is a good inhibitor. The inhibiting action of oxygen is due to its biradical nature, as depicted by

$$P^{\bullet} + \dot{O} - \dot{O} \rightarrow P - O - \dot{O}$$

$$P - O - \dot{O} + P^{\bullet} \rightarrow P - O - O - P$$

This is why radical polymerisation is generally carried out under a nitrogen atmosphere, *i.e.*, to avoid contact with atmospheric oxygen. A major use of inhibitors, is in the preservation of monomers, during production and storage. Monomers are produced and stored in the presence of small quantities of these inhibitors, as, otherwise, due to their high reactivity, they undergo polymerisation during production or storage. Without inhibitors, we cannot think of so many monomers being transported from one place to another and stored before actual use. Before use, naturally, monomers will have to be freed from the inhibitors. This is done either by distilling the monomer, when we can get the inhibitor-free monomer or by washing the monomer with an aqueous solution of sodium or potassium hydroxide, when the inhibitor is destroyed and removed by the aqueous phase.

In industrial practice, however, inhibitors are not removed prior to polymerisation. They are, nevertheless, killed by adding further quantities of initiators to the monomer. Inhibitors are also used in the polymer industry for the purpose of arresting the polymerisation beyond a certain conversion so as to achieve a uniform product and avoid cross-linking. When used for this purpose, inhibitors are called 'short stops'. They are normally added towards the end of polymerisation (*i.e.*, after the required conversion has been achieved), when they kill all the active radicals and arrest polymerisation.

2.2.2 Ionic Polymerisation

The ionic mechanism of chain polymerisation also involves an attack on the π electron pair of the monomer. This time, however, not by the unpaired electron of the free-radical, but by either a positive or negative ion. How? Well, imagine a system where a proton is introduced into a monomer. The proton now pulls the π electron pair towards it and the positive charge of the proton is transferred to the farther end of the monomer molecule, forming a carbonium ion. It will be noticed that, in this process, a sigma bond is formed between the proton and the monomer unit and the polymer chain growth 'initiated'. This is represented as follows:

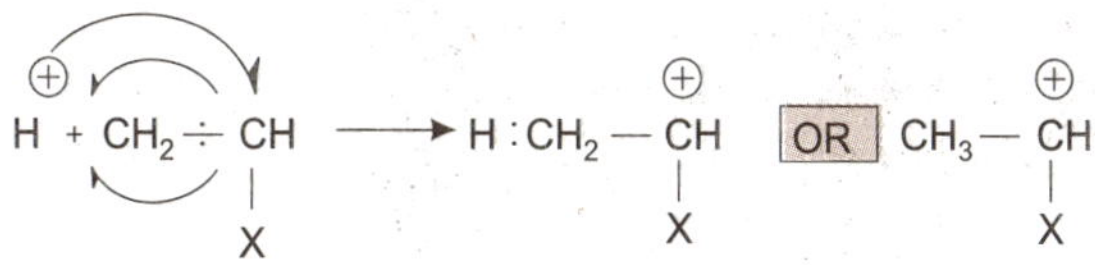

The carbonium ion (C^+) attacks the π electron pair of the second monomer molecule and pulls it over; the positive charge is transferred to the farther end of the second monomer unit:

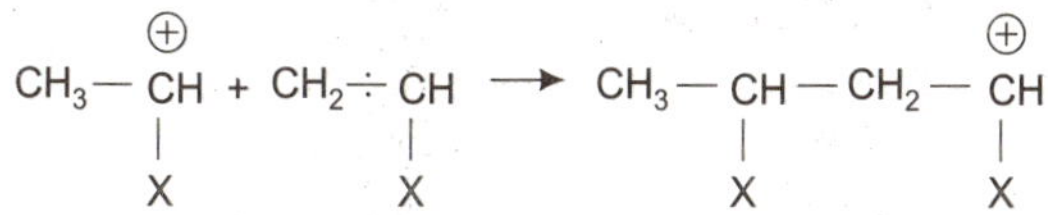

And thus a chain reaction sets in. In this chain reaction, however, no 'divorce' of the electron pair takes place, but only a displacement of the pair as a whole and the formation of a carbonium ion. Strong Lewis acids such as BF_3 (generally called the 'catalysts' in ionic polymerisation), in the presence of small amounts of water or methanol (known as the 'co-catalysts'), form hydrates which exist as ion pairs:

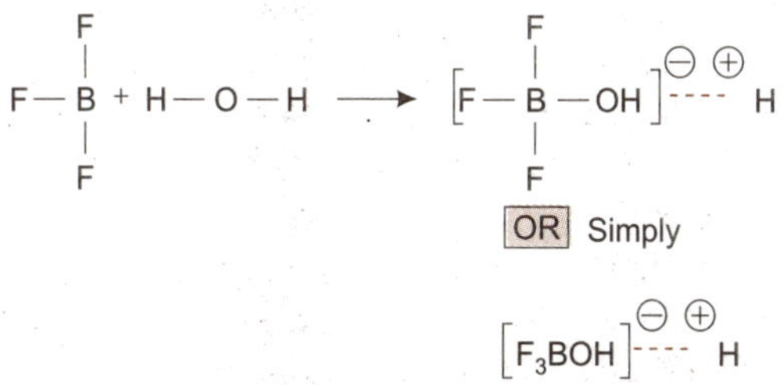

where $\overset{\oplus}{H}$ is the initiating cationic species and $[F_3BO\overset{\ominus}{H}]$ is the counter-ion.

The H^+ proton now attacks the π electron of the monomer, starts the chain growth through the carbonium ion and the $[F_3BO\overset{\ominus}{H}]$ anion faces the growing chain end:

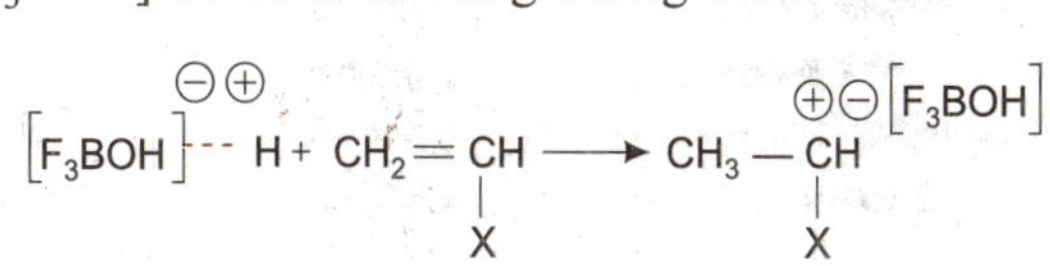

This step constitutes the initiation step of the polymerisation. The propagation reaction involves further adding up of the monomer units and simultaneous transfer of the charge to the newly added monomer unit:

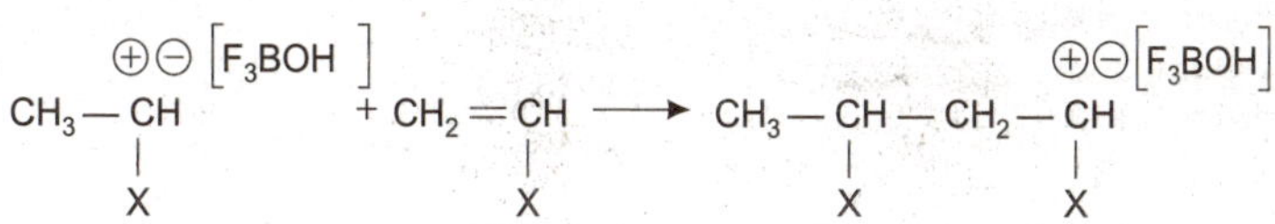

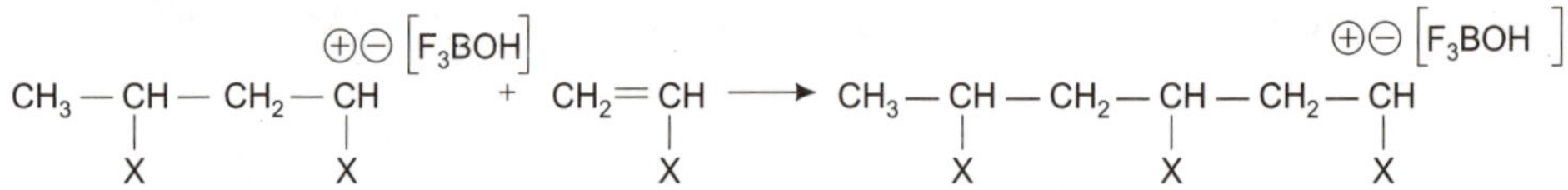

As more and more monomer units are added up, the chain keeps on growing and, in doing so, the π electron pairs of the monomer units are pulled in a direction opposite to the growth of the chain. Also, the C^+ ion keeps on moving in the direction of the chain growth and the counter-ion, all the time, moves along with the C^+ ion like a 'watch dog'.

Termination occurs when a collision between the growing C^+ ion and an anion (either from the counter-ion of another growing chain or from some deliberately added anion) takes place. The possible reactions that take place during termination involve the following two factors:

(*i*) Donation of a proton to the counter-ion (which is, in fact, the reversal of the initiation step), resulting in the formation of a double bond at the end of the growing polymer molecule and the resultant arrest of the chain growth:

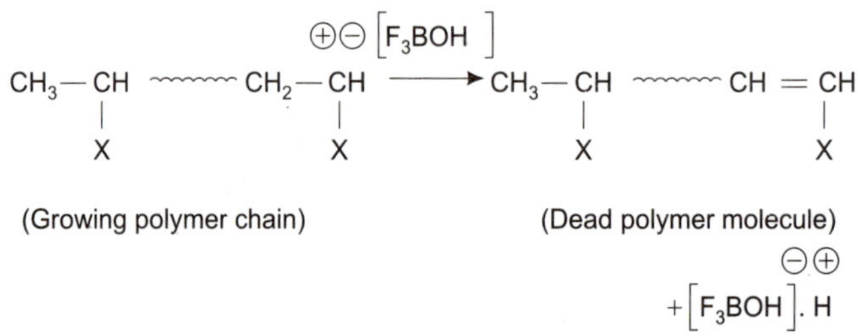

This process of donation of a proton and the re-formation of the BF_3 the hydrate is called 'ion pair precipitation'.

(*ii*) Formation of a covalent bond between the carbonium ion and the counter-ion when the termination occurs by simple 'coupling':

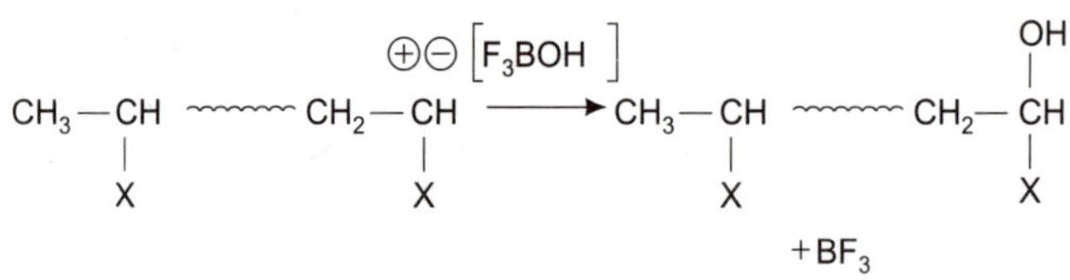

In both the cases, however, regeneration of the initiator takes place. This type of polymerisation, where the initiation is done by a proton and the propagation carried out by a carbonium ion, is called 'cationic polymerisation'. The initiators used for this purpose, in addition to BF_3, include $AlCl_3$, $SnCl_4$ and $TiCl_4$. Examples of monomers that can undergo cationic polymerisation are isobutylene, styrene, methyl styrene and many vinyl ethers.

There is another type of ionic polymerisation, called 'anionic polymerisation', where the attack on the π electron pair of the monomer is done by a negatively charged ion, an anion. Such a system has extra electrons and the resultant negative charge attacks the π electron pair, pushing it, as far away as possible, that is, to the end of the molecule. At the same time, it forms a sigma bond with the monomer molecule. The electron pair pushed to the end of the molecule results in the formation of a carbanion:

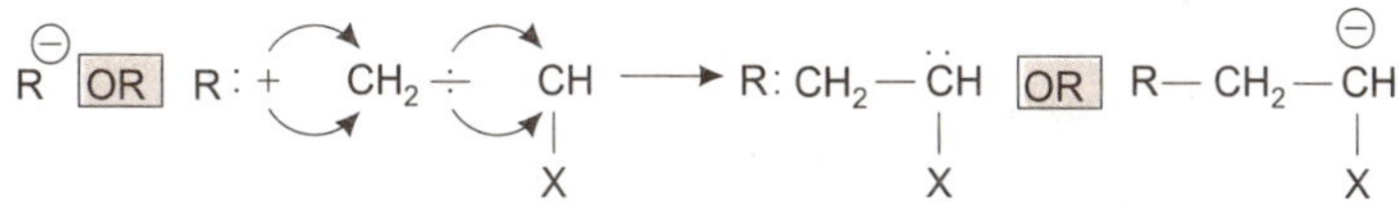

The carbanion thus formed now propagates the chain growth by attacking the second monomer unit, pushing its π electron pair further away to the end and forming a sigma bond with the new monomer unit:

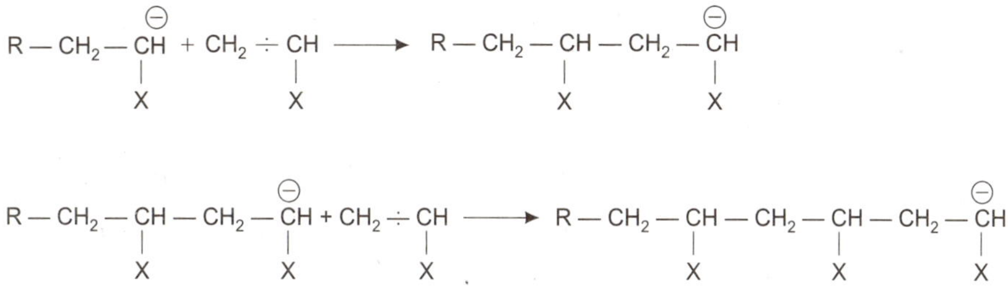

OR Simply

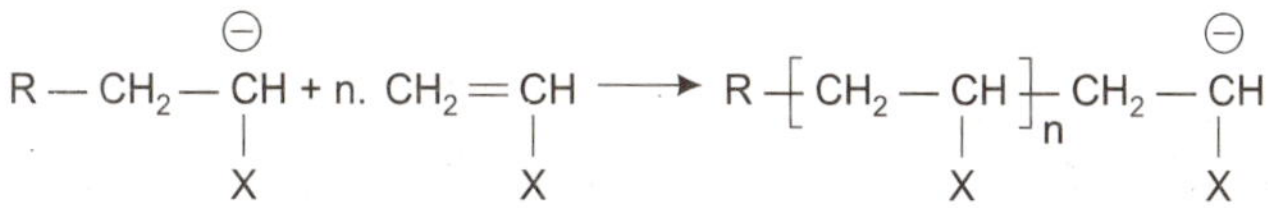

So, we have again, as in cationic polymerisation, an electron pair displacement process, where the strong negatively charged anion pushes the π electron pair of the monomer double bond down to sigma electron level. The difference between the cationic and the anionic processes is that, in the former case, the movement of the electron pair is in a direction opposite to that of the chain growth, whereas, in the latter case, it is in the same direction as that of the chain growth. Of course, in the anionic process also, an ion pair exists, with a positively charged counter-ion always running along the direction of the chain growth close to the carbanion. The general type of initiators used in this case are the organo-alkali compounds (*i.e.,* the alkyl and aryl derivatives of alkali metals) such as butyl lithium, triphenyl methyl potassium and ethyl sodium. Alkali metal amides, alcoxides and hydroxides can also be used for the anionic polymerisation process. Monomers capable of undergoing anionic polymerisation include butadiene, isoprene, styrene and acrylonitrile.

The termination in anionic polymerisation is, however, not usually a spontaneous process, and unless some impurities are present or some strongly ionic substances are added deliberately, invariably, no termination occurs. Thus, if polymerisation is carried out under controlled conditions and impurities are avoided, the reaction proceeds till all the monomer is consumed. In this case, while there is no more monomer left to be polymerised, the carbanions at the chain ends remain potentially active. What this means is that if a fresh quantity of monomer is added the polymerisation again goes on until all the freshly added monomer is consumed. Again, the chain ends are like a gasping tiger, ready to drink away any 'monomer blood' that may be supplied. Studies have shown that polymerisation can be restarted in this manner, even after weeks, by adding fresh monomer. It is as if the polymer is a 'living' entity, and in fact, the polymers produced by this technique are called 'living polymers'. They are 'alive' and keep on growing, as long as we supply a fresh quantity of the monomer just as many of the living organisms grow as long as food is available. The 'living polymerisation technique', is useful for many applications. Block, copolymers, for example, are prepared by employing this technique (see Section 11.10).

2.2.3 Coordination Polymerisation

Polymerisation reactions, especially of olefines and dienes, catalysed by organo-metallic compounds, fall under the category of coordination polymerisation. The first step in the polymerisation process is the formation of a monomer-catalyst complex between the organo-metallic compound and the monomer:

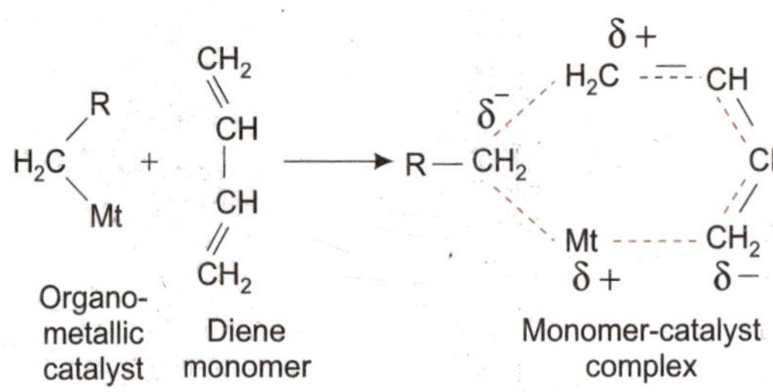

where Mt represents transition metals such as Ti, Mo, Cr, V, Ni or Rh.

In the formation of this monomer-catalyst complex, a coordination bond is involved between a carbon atom of the monomer and the metal of the catalyst. Hence, the polymerisation effected by such catalyst systems is called 'coordination polymerisation'.

The coordinated metal-carbon bond formed in the monomer-catalyst complex acts as the active centre from where propagation starts. The incoming monomer is incorporated at the active centre of the metal-carbon bond, from where the chain growth starts.

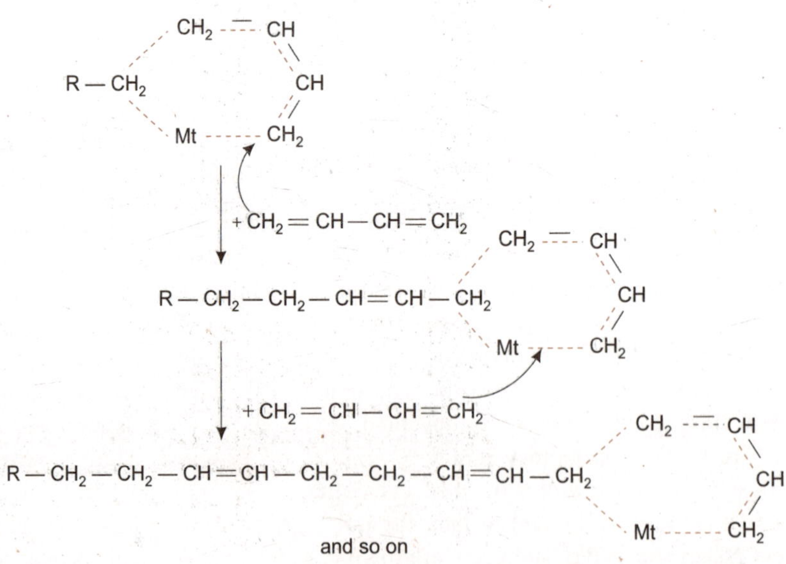

In coordination polymerisation, the catalyst-monomer complex is a heterogeneous system, with the metal ion in the solid phase and the carbanion of the alkyl group in the solvent phase. The monomer is inserted in between the metal ion and the carbanion with the result that the polymer chain formed is pushed out from the solid catalyst surface. For this reason, coordination polymerisation is also known as insertion polymerisation.

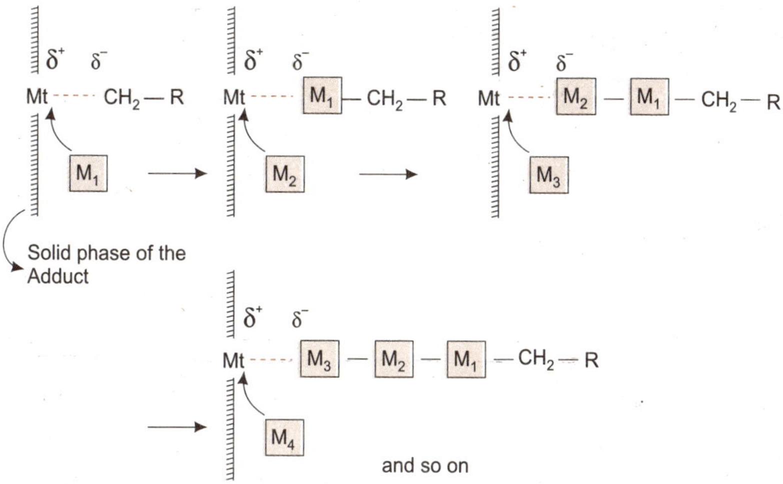

Here, M_1, M_2, M_3, and so on are the first, second, third (and so on) monomer units added to the growing chain.

Another interesting feature of coordination polymerisation is that, depending on the polarity of the metal-carbon bond and on that of the solvent medium, the metal counter-ion is placed in a particular spatial arrangement with respect to the anion. In certain catalyst systems, therefore, the spatial arrangement has a tremendous effect on the spatial orientation of the incoming monomer and also on the manner in which the monomer is inserted into the growing chain. The specific spatial arrangement of the monomeric unit inserted into the growing chain imparts stereo-regularity* to the polymer formed. By choosing a proper catalyst/solvent system, the coordination mechanism can be used to formulate highly stereo-regular polymers. The well-known Ziegler–Natta catalysts belong to this category of the polymerisation system.

Ziegler–Natta Catalysts

These are a special type of coordination catalysts, comprising two components as against single-component organo-metallic compounds that we came across earlier. The two components are, generally, referred to as the catalyst and the co-catalyst. The catalyst component consists of halides of IV–VIII group elements having transition valence and the co-catalysts are organo-metallic compounds such as alkyls, aryls and hydrides of group I–IV metals. Although there are hundreds of

*For 'stereo-regular polymers' see Section 5.3.3.

such catalyst—co-catalyst systems, those based on the organo-aluminium compounds such as triethyl aluminium ($AlEt_3$) or diethyl aluminium chloride ($AlEt_2Cl$) in combination with titanium chlorides—both tri and tetra ($TiCl_3$ and $TiCl_4$)—are, by far, the most commonly used. Aluminium alkyls act as the electron acceptor and the electron donor is titanium halides and the combination, therefore, readily forms coordination complexes. The complex formed is insoluble in the solvent and is, hence; heterogeneous in nature. Many structures have been proposed for these complexes by various authors. A few are shown here:

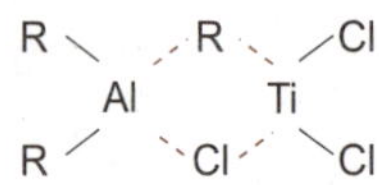

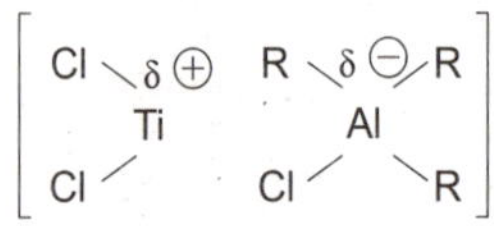

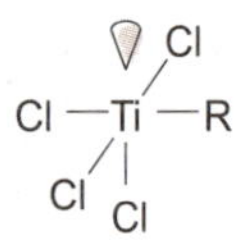

The active centres, from where the polymer chain growth propagates, are formed at the surface of the solid phase of the catalyst complex, and the monomer is complexed with the metal ion of the active centre before its insertion into the growing chain. The actual mechanism by which the propagation takes place and the factors governing the formation of stereo-regular polymers formed a topic of great debate at one time. Among the several mechanisms proposed, the bimetallic mechanism of Natta and the monometallic mechanism of Cossee had received much attention. Cossee's mechanism, however, is the one that is widely accepted presently. We shall, however, look into both these mechanisms. Natta's bimetallic mechanism stipulates the following: When the catalyst and co-catalyst components are mixed, there occurs a chemisorption of the aluminium alkyl (electropositive in nature) on the titanium chloride solid surface, resulting in the formation of an electron-deficient bridge complex of the structure:

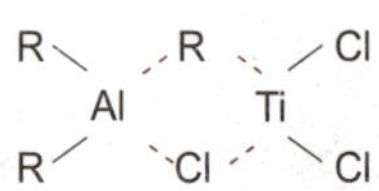

This complex now acts as the active centre. The monomer is then attracted towards the Ti—C bond (C from the alkyl group R) in the active centre, when it forms a π complex with the Ti ion:

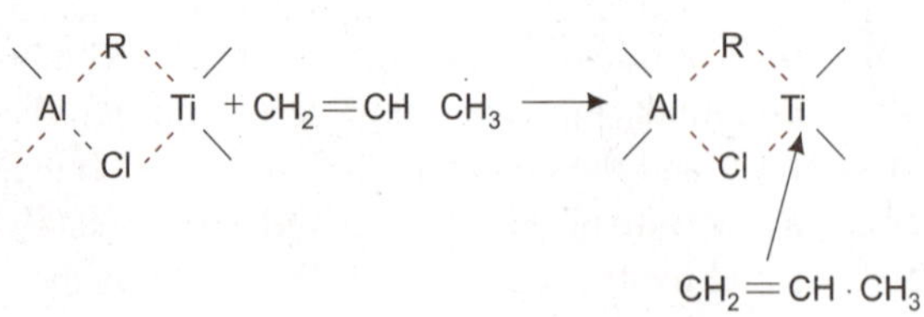

The bond between R and Ti opens up, producing an electron-deficient Ti and a carbanion at R:

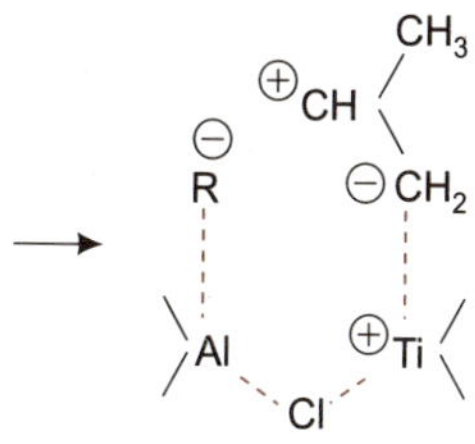

The Ti ion attracts the π electron pair of the monomer and forms a σ bond, while the counter-ion attracts the electron-deficient centre of the monomer. The monomer, in this way, is 'inserted' into a transition state ring structure:

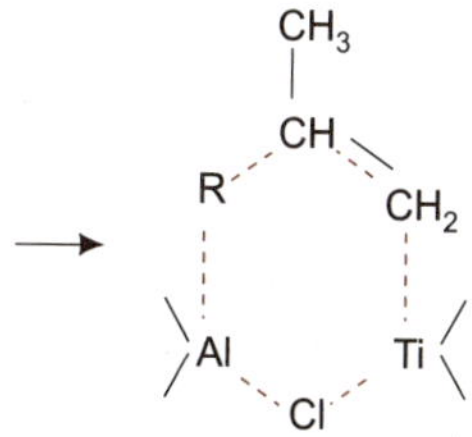

This transitional state now gives rise to the chain growth at the metal-carbon bond, regenerating the active centre:

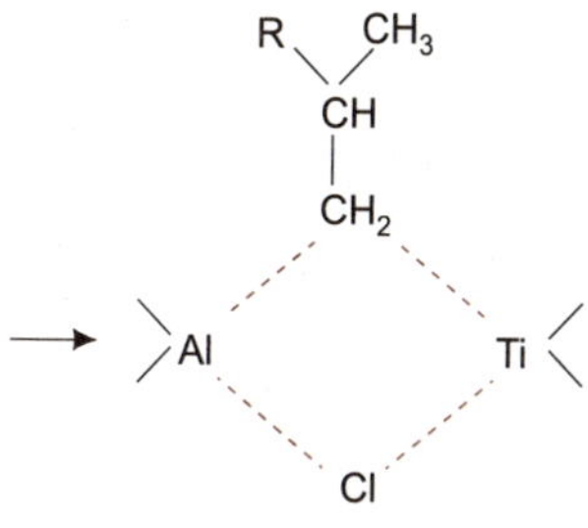

Repeating the whole sequence, with the addition of a second monomer molecule, we will get the structure of the resultant chain growth as:

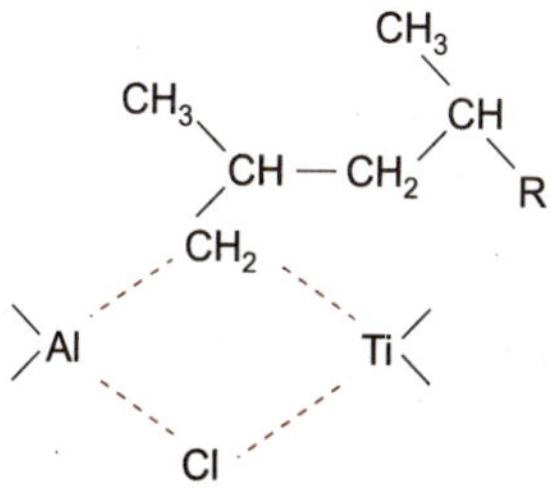

The monomer insertion is repeated in this manner and the orientation of the substituent group of the monomer is always taken from the metal-ion end, resulting in a stereo-regulated polymer.

On the other hand, the monometallic mechanism proposed by Cossee assumes that the active centre is at the Ti—R part of the catalyst, while the aluminium alkyl performs only the function of alkylating the $TiCl_3$. When the catalyst and co-catalyst are mixed, the aluminium is chemisorbed on the solid $TiCl_3$. The five-coordinated titanium ion on the surface of the catalyst is assumed to have a vacant *d*-orbital, as shown at 6 in the following structure:

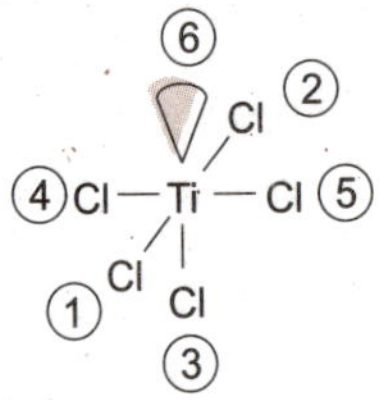

The Cl at 1 and 2 are assumed to be attached to another titanium in the crystal lattice of $TiCl_3$, and, except the Cl at 5 which can be replaced by an alkyl group, all the other four Cl are non-exchangeable. Immediately after the chemisorption of the aluminium alkyl on the $TiCl_3$ crystal, Ti^{3+} is alkylated by an exchange mechanism, as follows:

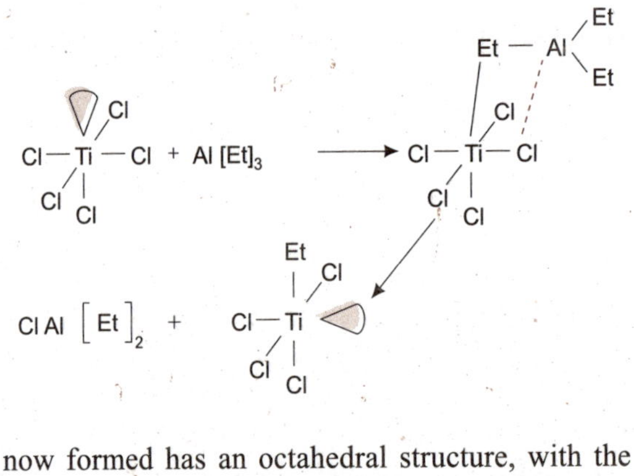

The active catalyst now formed has an octahedral structure, with the four Cl attached to the lattice and the ethyl group attached to the Ti through a σ bond, thus leaving a vacant orbital at the position where originally the fifth chlorine was attached. Once the active catalyst is formed, the monomer is attracted towards the vacant *d*-orbital which then forms a transition π complex with the Ti:

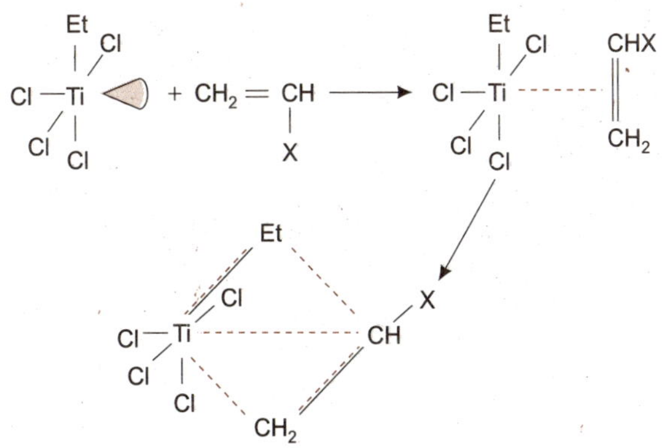

(Transition state)

The transition state quickly gives rise to the growth of the polymer chain by the monomeric insertion at the Ti—Et bond. While the monomeric group is inserted and the chain grows, the vacant *d*-orbital can either be regenerated at the same position or at the position where the ethyl group was originally attached:

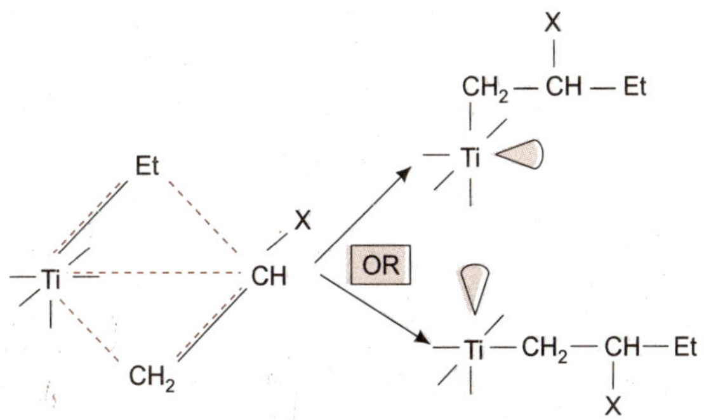

When the vacant *d*-orbital is regenerated at the same position all the time, the incoming monomeric units will be inserted with the same spatial arrangement, resulting in the formation of an isotactic polymer. With the insertion of each monomeric unit, if the *d*-orbital migrates from one position to the other alternatively, the monomers get inserted with alternating spatial arrangement, resulting in the formation of a syndiotactic polymer. Whether the vacant site is regenerated at the same position or migrates to the other site alternatively would depend on the interactive forces existing between (*i*) the substituent group of the incoming monomeric unit and the already inserted monomeric unit and (*ii*) the substituent group of the monomer and the chlorine atoms of the active catalyst.

Factors such as the type of catalyst system, catalyst crystal structure, molar ratios of the catalyst components and temperature of the catalyst complex influence the shift or migration of the vacant *d*-orbital and, hence, the stereo-regularity of the polymer formed.

Whatever be the structure of these active centres and mechanism of the active centre formation, the polymerisation is characterised by the initiation, propagation and termination reactions as follows:

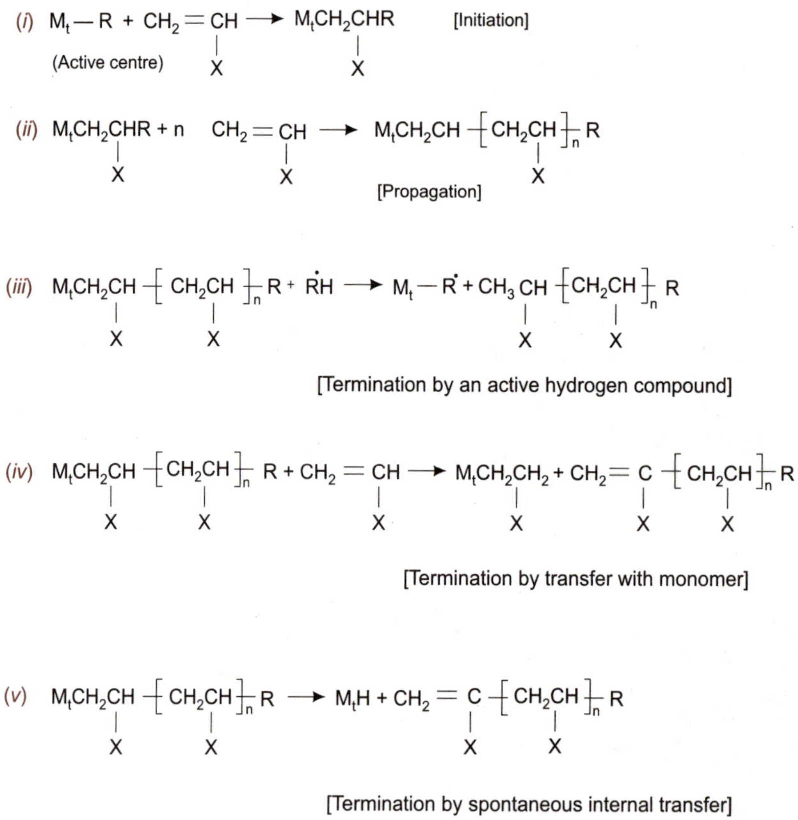

Here, Mt denotes transition metals such as Ti, Mo, Cr, V, Ni or Rh.

2.3 STEP POLYMERISATION

In step polymerisation, the polymer build-up proceeds through a reaction between functional groups of the monomers. The reaction takes place in a step-wise manner (*i.e.*, one after another), and the polymer build-up is, therefore, slow (unlike in chain polymerisation where the chain growth is very rapid). Although many known reactions with organic functional groups can be made use of in step polymerisation, condensation, addition, ring-opening, amidation and ester-interchange reactions are the most commonly used ones. Step polymerisation reactions are mostly accompanied by the elimination of small molecules. Whether the elemental composition of the polymer formed and that of the monomers are the same or not depends on whether the reaction is accompanied by the elimination of small molecules or not. With this brief background, we shall, in the forthcoming sections, examine some of the step polymerisation reactions.

2.3.1 Polycondensation

Polycondensation is brought about by monomers containing two or more reactive functional groups (such as hydroxyl, carboxyl and amino) condensing with each other. Basic reactions here are the same as between various functional groups in low molecular weight organic compounds. For example, we are familiar with the reaction between a hydroxyl group and a carboxylic group, giving an ester and a water molecule:

$$-OH + HO-\underset{\underset{O}{\|}}{C}- \longrightarrow -O-\underset{\underset{O}{\|}}{C}- + H_2O$$

If we take a monohydric alcohol, say, methanol, and a monocarboxylic acid, say, acetic acid, we get the following reaction:

$$CH_3OH + HOOC-CH_3 \rightarrow CH_3COOCH_3 + H_2O$$

It should, however, be noted that in this methyl acetate formation, the reactive functional groups (*viz.*, hydroxyl and carboxyl) are consumed, giving an unreactive functional group (*viz.*, ester). There is no more reactive functional group left with the product and, hence, it cannot react further with any other reactant molecules in order to form a different product.

Now, let us see the difference. Take ethylene glycol (EG), instead of methanol, and make it react with acetic acid. Observe what happens. First, one molecule of EG reacts with one molecule of acetic acid as follows:

$$\underset{(EG)}{HOCH_2CH_2OH} + HOOCCH_3 \longrightarrow HOCH_2CH_2OOCCH_3 + H_2O$$

In this case, one of the hydroxyl groups in EG has been consumed by one molecule of acetic acid, resulting in the formation of monoacetic ester of EG. This product, however, still contains another hydroxyl group which can react with another molecule of acetic acid to form the ethylene glycol di-acetate:

$$CH_3COOCH_2CH_2OH + HOOCCH_3 \longrightarrow CH_3COOCH_2CH_2OOCCH_3 + H_2O$$

The product now contains no more reactive functional group and, hence, cannot react further to give any other product. In the reaction between methanol and acetic acid, we have an instance of one molecule of each of the reactants reacting to give a single molecule of the product and one water molecule. The product contains one ester group in its molecule. In the case of reaction between EG and acetic acid, on the other hand, three reactant molecules are involved (but only two reactants) to give again a single molecule of the product. But in this case, two water molecules are liberated, and the product contains two ester groups in its molecule.

Let us experiment further. Let us replace acetic acid with adipic acid and take two molecules each of EG and adipic acid and see what happens. As expected, one molecule of EG reacts with one molecule of adipic acid to form a monoester product:

$$HOCH_2CH_2OH + HOOC \left(CH_2 \right)_4 COOH \longrightarrow HOCH_2CH_2OOC \left(CH_2 \right)_4 COOH + H_2O$$

Apart from a non-reactive ester group, this product still contains two reactive groups, and each (*i.e.,* one hydroxyl and one carboxyl) can merrily react with one more molecule of EG or adipic acid. It may be recalled that, we started off with two molecules of each of the reactants. We have only consumed one of each and are, therefore, still left with one molecule each of EG and adipic acid. The reaction will proceed further as follows:

$$HOOC \left(CH_2 \right)_4 COOH + HOCH_2CH_2OOC \left(CH_2 \right)_4 COOH + HOCH_2CH_2OH$$

$$\longrightarrow HOOC \left(CH_2 \right)_4 COOCH_2CH_2OOC \left(CH_2 \right)_4 COOCH_2CH_2OH + 2\ H_2O$$

The resultant molecule contains at its ends one carboxyl and one hydroxyl group, respectively, and, hence, if more quantities of EG and adipic acid are made available, it is capable of reacting further, resulting in an even bigger molecule.

In the foregoing case, four molecules of the reactants react to give a single product molecule (containing three ester groups) and three water molecules. But as against the previous two cases, here, the resultant product molecule is still capable of reacting further to form a bigger molecule, because each of the species of the starting reactants contains two functional groups in its molecule.

Now, the product formed from two molecules of EG and two molecules of adipic acid can also be written as:

$$H \left[OOC \left(CH_2 \right)_4 COOCH_2CH_2 \right]_2 OH$$

and the overall reaction as:

$$2\ HOCH_2CH_2OH + 2\ HOOC \left(CH_2 \right)_4 COOH \longrightarrow HO \left[CH_2CH_2OOC \left(CH_2 \right)_4 COO \right]_2 H + 3H_2O$$

The time has come to generalise this reaction. Say, there are *n* molecules of adipic acid and *n* molecules of EG. We will get the overall reaction as:

$$n\ HOCH_2CH_2OH + n\ HOOC \left(CH_2 \right)_4 COOH \longrightarrow HO \left[CH_2CH_2OOC \left(CH_2 \right)_4 COO \right]_n H + (2n - 1)\ H_2O$$

From this illustration, the following conclusions can be drawn regarding polycondensation (or for that matter, any step polymerisation):

(*i*) That monomers should have two reactive functional groups for polymerisation to proceed.

(*ii*) That polymerisation proceeds by step-wise reaction between reactive functional groups.

(*iii*) That only one type of reaction (*i.e.*, condensation reaction, in this case) between two functional groups is involved in polymer formation.

(*iv*) That the polymer formed still contains both the reactive functional groups at its chain ends (as end groups) and, hence, is 'active' and not 'dead', as in chain polymerisation.

In the above illustration, two different types of reactive functional groups, *viz.*,—OH and —COOH, were present on two different monomers. In some cases, both the reactive functional groups could be present on the same monomer, in which case, it can lead to a self-polycondensation reaction. A typical example is the polycondensation of amino enanthic acid to form polyenanthoamide:

$$n\,H_2N\text{-}(CH_2)_6\text{-}COOH \longrightarrow H\text{-}[HN\text{-}(CH_2)_6\text{-}CO]_n\text{-}OH + (n-1)\,H_2O$$

Apart from amino acids, hydroxy acids and alkylene chlorohydrines are also capable of undergoing self-polycondensation.

When a pair of bifunctional monomers (such as dicarboxylic acid/dialcohol, dicarboxylic acid/diamine or dialcohol/dihalide) undergoes polycondensation, this is usually described as an AA—BB type polycondensation and the reaction is depicted as follows:

$$n\,A\text{—}A + n\,B\text{—}B \longrightarrow A\text{-}[AB]_{2n-1}\text{-}B + \text{by-product}$$

When a single bifunctional monomer undergoes self-condensation, it is described as an A—B type polycondensation. The reaction, then, can be depicted as follows:

$$n\,A\text{—}B \longrightarrow B\text{-}[AB]_{n-1}\text{-}A + \text{by-product}$$

If in both the AA-BB and A-B types of polycondensation reactions intramolecular condensation occurs in preference to intermolecular condensation, one will end up with cyclic products of low molecular weight instead of, or in addition to, straight chain polymer molecules. During the preparation of polyethylene terephthalate (PETP), for example, cyclic products have also been isolated in addition to the required polymer, as shown:

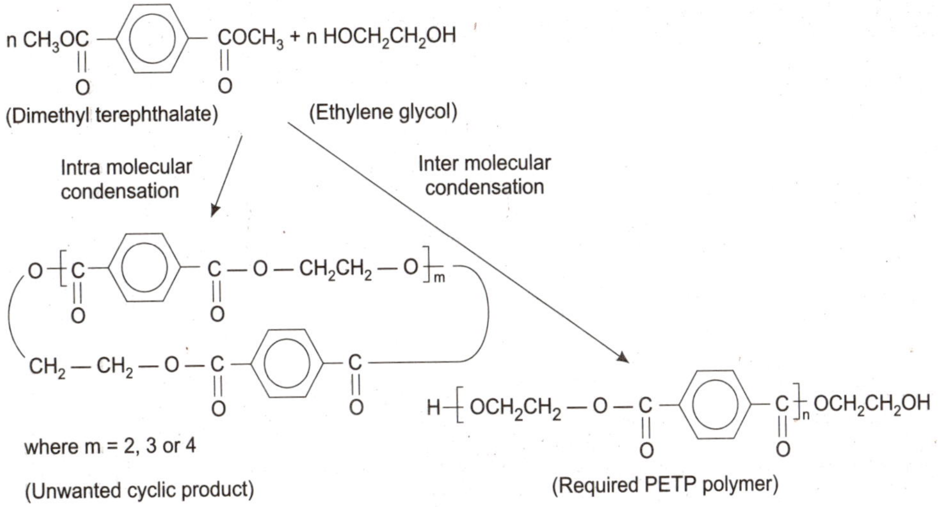

If in the AA-BB type of polycondensation reaction, one of the monomers is trifunctional, the polymer develops a three-dimensional network as against a linear chain structure that a difunctional monomer gives. For example, glycerol reacts with a dibasic acid to give a complex structure of the intermediate molecule, even at the early stage of polymerisation. For every glycerol molecule reacted, the functionality in the resultant intermediate product increases by one. This means that the number of possible sites where the polymer build-up can take place increases so rapidly and stupendously that several branching sites are formed on the growing polymer molecule. When many such branched-growths containing reactive functional end groups condense further, they lead to an enormously large, three-dimensional molecule of almost infinite molecular weight. When the three-dimensional network is just about to be formed, the whole polymer becomes an insoluble and infusible mass and is called 'gel'. The precise stage of the reaction at which gelation occurs depends on the ratio of the bifunctional monomer to the trifunctional one. For instance, when an equimolar ratio of glycerol and a dibasic acid is used, the gel point is reached when the extent of reaction* reaches a value of 0.765.

Some generic polycondensation reactions leading to different types of polymers are now enumerated.

(*i*) Dihydric alcohols resulting in the formation of polyethers:

$$n\,HO-R-OH \longrightarrow HO\left[RO\right]_n H + (n-1)\,H_2O$$

(*ii*) Dihydric alcohols and dicarboxylic acids forming polyesters:

$$n\,HO-R-OH + n\,HOOC-R'-COOH \longrightarrow HO\left[R-OOC-R'-COO\right]_n H + (2n-1)\,H_2O$$

(*iii*) Dicarboxylic acids and diamines giving polyamides (nylons):

$$n\,HOOC-R-COOH + n\,H_2N-R'-NH_2 \longrightarrow HO\left[OC-R-CONH-R'-NH\right]_n H + (2n-1)\,H_2O$$

(*iv*) Dihydric alcohols and methyl ester of dicarboxylic acids yielding polyesters:

$$n\,HO-R-OH + n\,CH_3OOC-R'-COOCH_3 \longrightarrow$$

$$HO\left[R-OOC-R'-COO\right]_n CH_3 + (2n-1)\,CH_3OH$$

(*v*) Amino acids undergoing self-condensation giving polyamides:

$$n\,H_2N-R-COOH \longrightarrow H\left[HN-R-CO\right]_n OH + (n-1)\,H_2O$$

(*vi*) ω-hydroxy carboxylic acids giving polyesters:

$$n\,HO-R-COOH \longrightarrow HO\left[R-COO\right]_n H + (n-1)\,H_2O$$

Examples for these types of polymers can be found in Chapter 9, where we shall be dealing with individual polymers.

---

*For 'extent of reaction', see Section 4.5.1.

2.3.2 Polyaddition Polymerisation

Polyaddition polymerisation is brought about by the migration of atoms from one monomer molecule to another, or to the intermediate product. Vinyl monomers as well as monomer pairs with reactive functional groups can undergo polyaddition polymerisation. Styrene, for example, can be polymerised in the presence of perchloric acid by this method. The process involves the following mechanism:

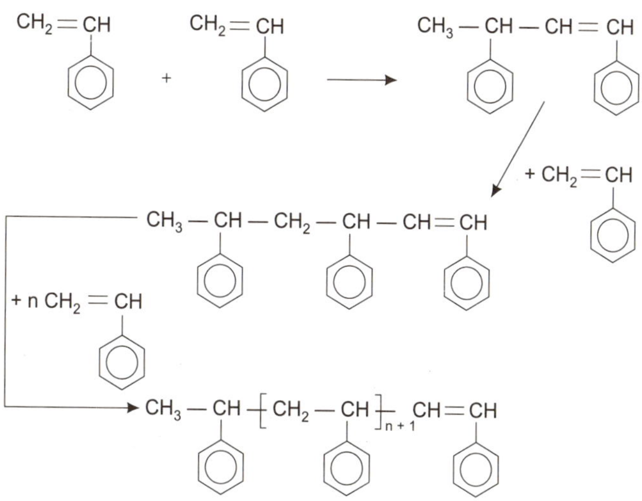

Notice the interesting manner in which the polymer is formed by a simple addition at the double bond, without any free-radical, carbonium ion or carbanion being formed at all. However, like in chain polymerisation, monomer units are simply added one after another to form a polymer without eliminating any small molecules. Since the polymer is formed by the migration of a hydrogen atom from the monomer and its addition at the double bond—a process involving a higher activation energy as compared to the free-radical or ionic process—the polymer growth is rather gradual and is brought about step by step, as against the rapid chain growth in chain polymerisation.

From the above reaction, it is noticed that intermediate products also have a double bond between the ultimate and the penultimate carbon atoms and hence, have the same structural similarity and reactivity as the starting monomer. The intermediate products are quite stable and can be isolated from the reaction mixture, whereas the growing chain in the case of chain polymerisation cannot be isolated.

Now, let us see how a pair of monomers having reactive functional groups can undergo polyaddition. It should, however, be recalled that the polycondensation reaction between two reactive functional groups involves the elimination of simple molecules and the polymer formed has quite a different elemental composition from that of the monomers. As against this, in the polyaddition reaction between reactive functional groups, no elimination of simple molecules is involved and the resultant polymer has the same elemental composition as that of the monomers. As an example, polyurethane polymer is formed from a diisocyanate and a diol by a polyaddition reaction, in which, the hydrogen from the hydroxyl group of the diol migrates and adds on to the nitrogen of the isocyanate group:

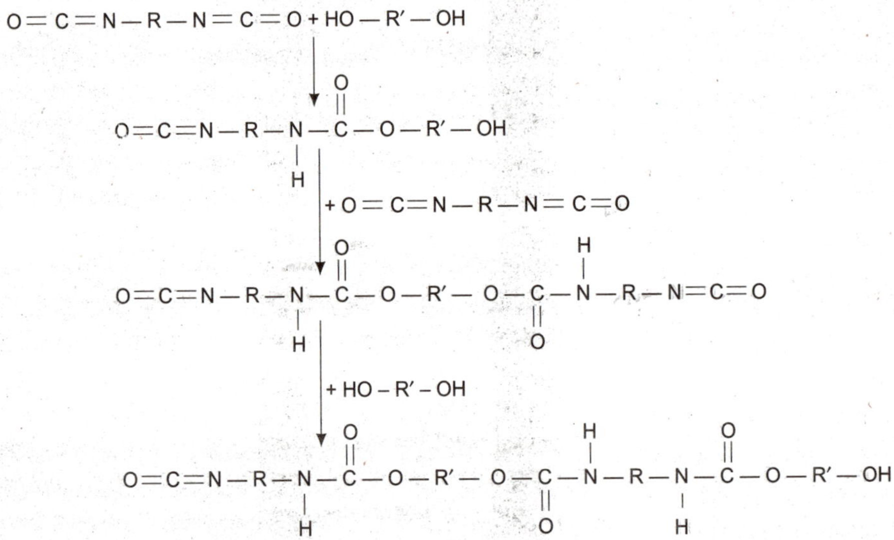

and so on

As in the polycondensation reaction, here also, the reactive functional groups are used up while forming the polymer, which, again, is made-up of repeat units comprising new non-reactive functional groups:

Polycondensation

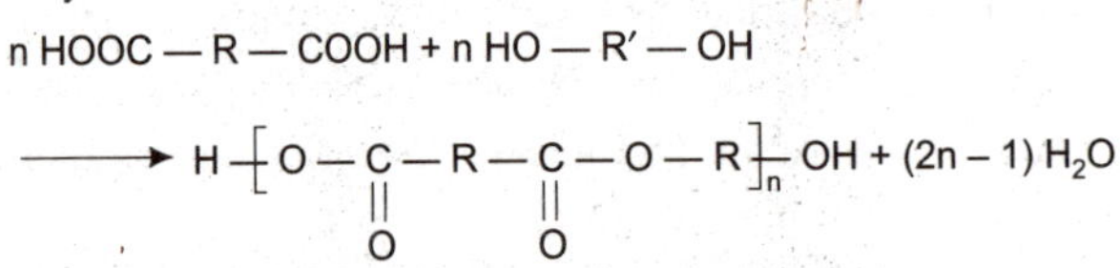

Polyaddition

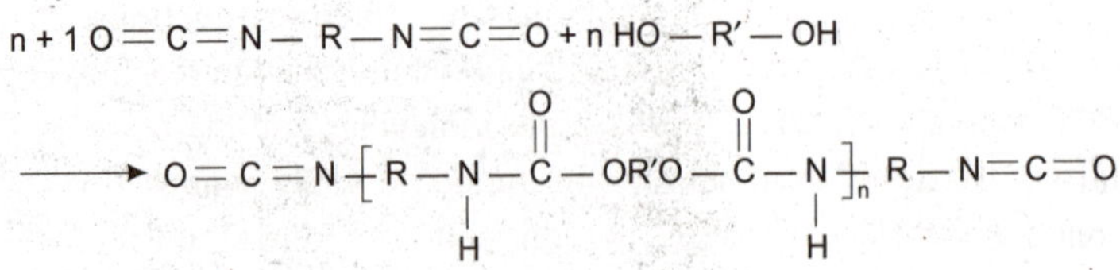

We have seen that both vinyl monomers as well as monomer pairs with reactive functional groups can undergo polyaddition polymerisation. However, since the hydrogen atom in a methylene group of the vinyl monomer is not that mobile, it is not so easy to get really high molecular weight polymers from vinyl monomers by this technique.

Oligomers (*i.e.,* liquid polymers of molecular weight of the order of 2,000 or so) of vinyl monomers are, nevertheless, made by this technique. On the other hand, the hydrogen atom in a hydroxyl or in an amino group is highly mobile and, hence, monomers such as polyhydric alcohols and polyamines can easily undergo polyaddition with polyisocyanates, yielding polyurethanes and polyureas of a very high molecular weight:

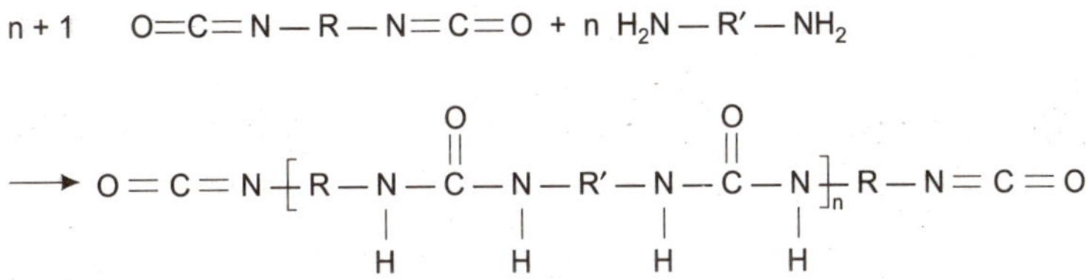

(Polyurea)

2.3.3 Ring-Opening Polymerisation

Monomers having a ring structure can be opened and polymerised, if conditions are favourable. We need hardly add that this 'opening up' is the opposite of the 'condensing' process.

Condensing of bifunctional low molecular weight compounds can give cyclic or ring compounds having a strain-free configuration:

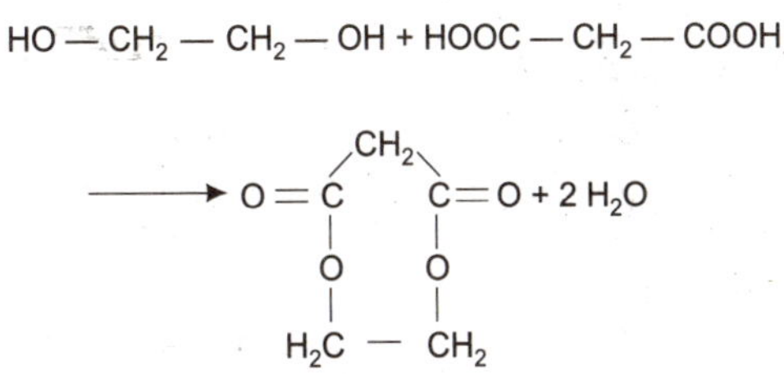

Conversely, when a ring structure molecule has a strained configuration, the same can be opened up and polymerised. Thus, cyclic compounds such as lactams and oxiranes can be opened up by aqueous alkali or acids to form an intermediate with active ends capable of undergoing further polymerisation. It may be noted that 4-and 8-membered lactones, possessing a strained ring structure, can be polymerised by ring opening. On the other hand, while the 5-membered lactone (γ-valerolactone which is a relatively strain-free molecule) is rather difficult to polymerise, the 6-membered δ-valerolactone can be polymerised only with a poor yield.

The mechanism of ring-opening polymerisation can be explained by taking the example of ethylene oxide which is a heterocyclic oxirane ring compound. Sodium methoxide (containing traces of methanol) dissolved in dioxane attacks the oxirane ring and cleaves the same to form an anion as shown in Fig. 2.1. This step can be conceived as the 'initiation' of the polymerisation. The 'propagation' step has to follow. This is effected by the anion attacking a fresh ethylene oxide ring which will be opened up and, in turn, added on with the simultaneous formation of an anion; and this process goes on and on. After 'propagation', is the 'termination' step, and this can be effected when the anion is caged by adding a proton, say, in the form of HCl. This polymerisation is similar to anionic chain polymerisation except that, in the latter, the attack by the anion is on the π electron pair of the C=C, while, in the present case, it is on the activated ring.

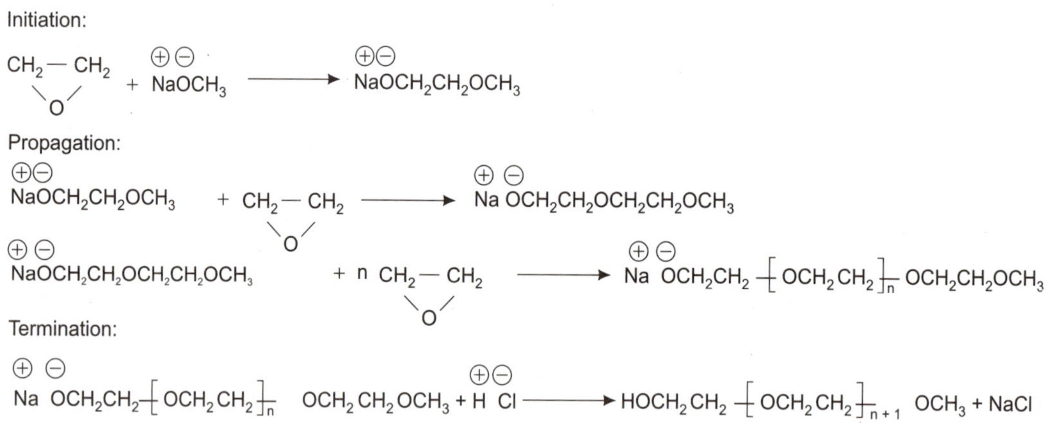

Fig. 2.1 Ring-opening polymerisation of ethylene oxide by sodium methoxide.

Another interesting example of ring-opening polymerisation by anionic mechanism is the polymerisation of caprolactam to form nylon 6, as shown in Fig. 2.2, In this case, the attacking anionic species is the caprolactam anion (I) introduced in the form of sodium caprolactam and N-acetyl caprolactam (1%). The anion attacks a caprolactam molecule and abstracts a proton, generating a new caprolactam anion (II). This new anion opens up the ring in the dimer (III) formed between the sodium caprolactam and N-acetyl caprolactam and gets itself attached, forming a trimer having, again, an anion (IV). The anion on the trimer now carries on the attack on a fresh caprolactam molecule and abstracts a proton, generating a new monomeric anion. Propagation of the polymer chain continues by a repeated and successive opening up of the activated rings by the monomeric anion, which, in turn, is followed by the proton abstraction.

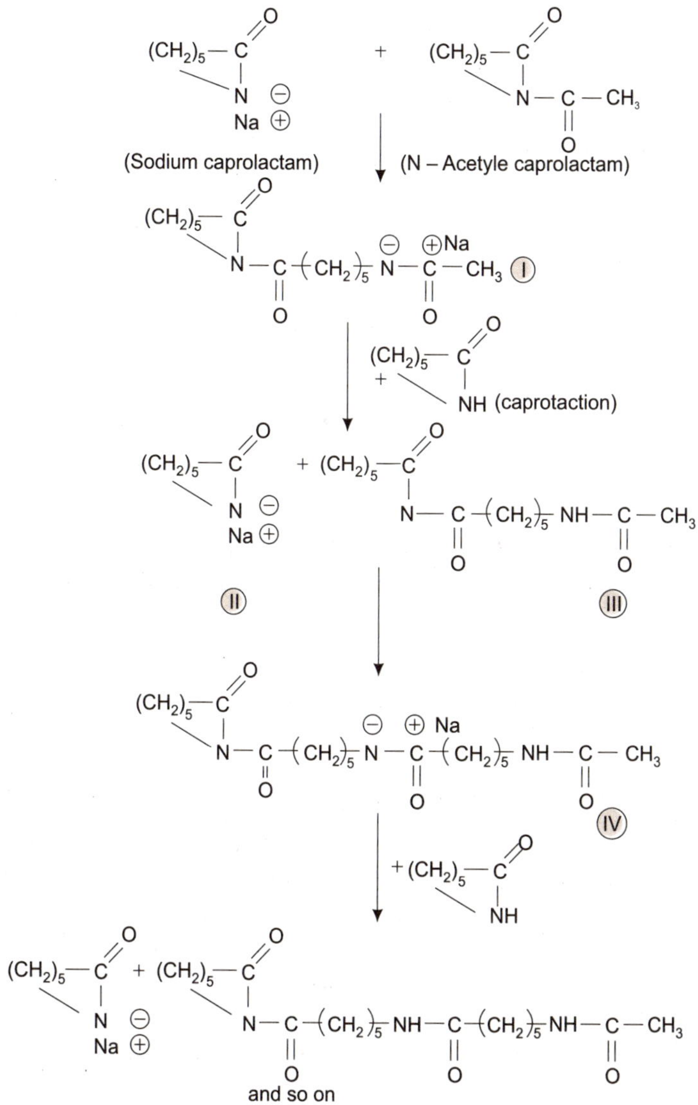

Fig. 2.2 Ring-opening polymerisation of caprolactam.

Another type of ring-opening polymerisation (of the lactams, for instance) brought about by the ring opening by catalytic quantities of water to form amino acid followed by the amide interchange reaction or polycondensation of the amino acid:

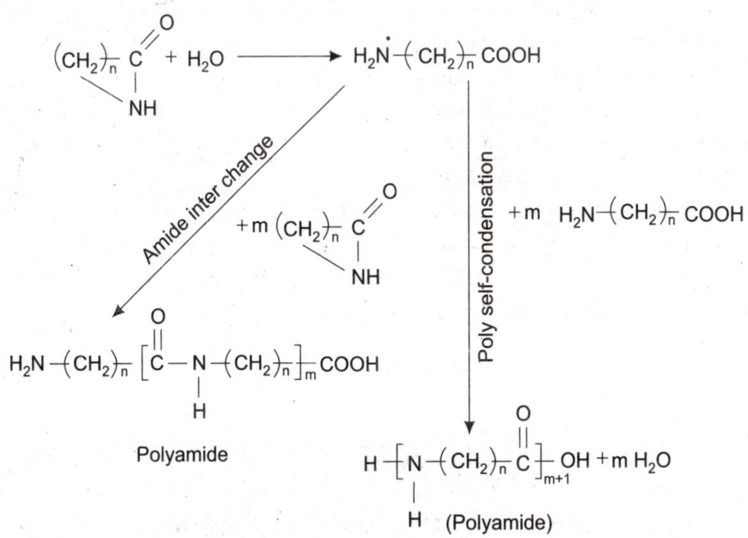

2.4 MISCELLANEOUS POLYMERISATION REACTIONS

There are some polymerisation reactions in which the reaction mechanism may conform to one or the other type we have seen so far, but need special mention because the initiating species are formed under special conditions or from specific catalyst systems.

2.4.1 Electrochemical Polymerisation

The initiating species in electrochemical polymerisation are produced at the electrolytic cell containing a monomer and other suitable solvents or additives which make up the electrolyte. These species are formed by the electrode processes, involving either the monomers themselves or the additive used for the purpose. The initiating species can be a free-radical, a cation or an anion, produced either at the anode or at the cathode. They are categorised as follows:

(i) $RCOO^{\ominus} \longrightarrow RCOO^{\cdot} + e^{\ominus}$ (Free-radical produced at the anode)

(ii) $R^{\oplus} + e^{\ominus} \longrightarrow R^{\cdot}$ (Free-radical produced at the cathode)

(iii) $M + e^{\ominus} \longrightarrow M^{\ominus}$ (Anion produced at the cathode)

(iv) $M \longrightarrow M^{\oplus} + e^{\ominus}$ (Cation produced at the anode)

$e^{\ominus}$ = electron M = monomer

Once the initiating species is formed at the electrode, the initiation and propagation are similar to those taking place under the free-radical, anionic or cationic polymerisation. The site for the chain growth can be either the electrode surface itself or the bulk of the electrolyte. The system is, therefore, a heterogeneous one and, hence, the interface boundaries play an important role. Variables such as the chemical nature and physical structure of the electrodes, electrode overpotential, half-cell reaction potential, current density and electrolyte concentration are very important in controlling the yield of the polymer, its molecular weight and molecular weight distribution.* In some cases, where polymer growth takes place exclusively at the electrode surface, stereo-regular polymers are obtainable. Typical examples of electrochemical polymerisation are now given.

1. Anodic free-radical initiation of methyl methacrylate, using an acetate and acetic acid electrolyte:

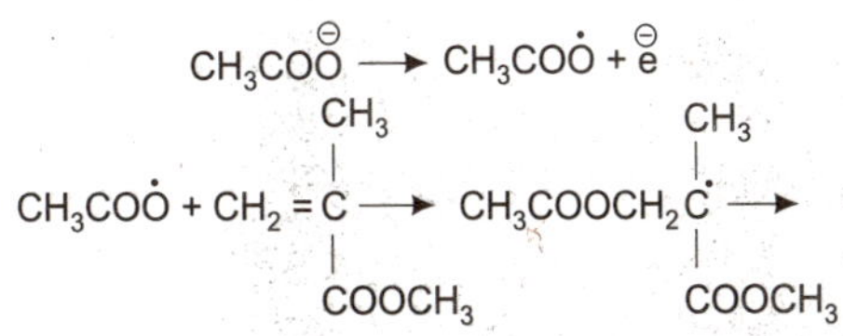

and so on.

2. Anionic initiation of molten caprolactam containing sodium benzoate and *N*-acetyl caprolactam at the cathode:

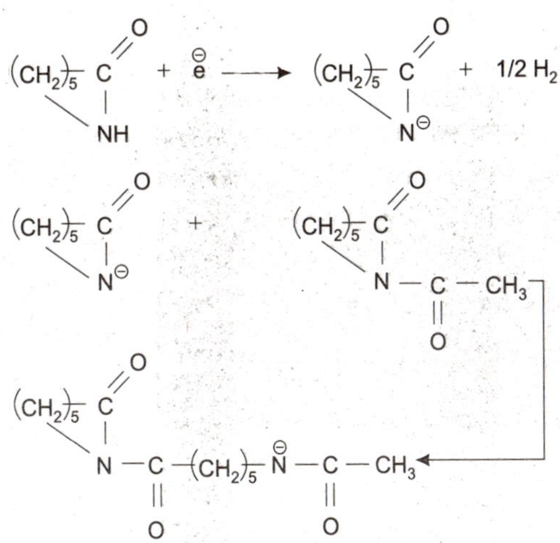

*For 'molecular weight distribution', see Section 3.5.

(Further steps are as given in Section 2.3.3).

3. Cationic initiation of ethylene oxide at the anode:

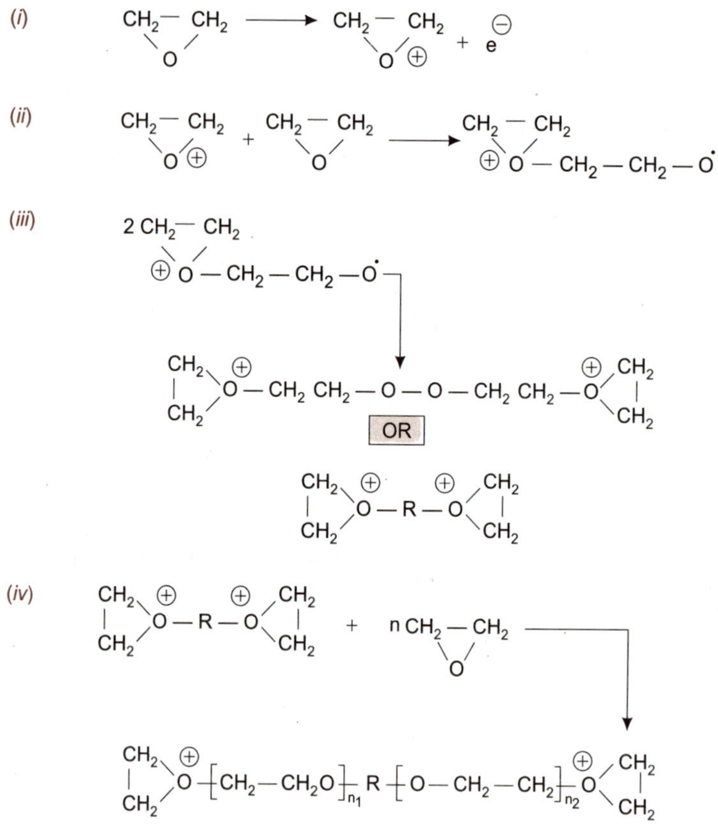

where $n_1 + n_2 = n$.

2.4.2 Metathetical Polymerisation

The olefine metathesis reaction can be used for the synthesis of polymer molecules. Olefine metathesis is basically an exchange reaction wherein the alkylidene moieties of the reactant molecules are redistributed through the cleavage and re-forming of olefinic double bonds, to form new product molecules. For example, at an elevated temperature and pressure, propylene could be converted into ethylene and butene-2 by employing a special heterogeneous catalyst system consisting of $Mo(CO)_6$ supported on alumina:

$$2CH_3-CH=CH_2 \rightleftharpoons CH_2=CH_2 + CH_3-CH=CH-CH_3$$

The product contains an equilibrium mixture of ethylene : propylene : butene in the mole ratio of 1 : 2 : 1. The composition of the equilibrium mixture fulfils the statistical requirement for a random exchange or redistribution of the alkylidene moieties according to the generalised scheme as shown:

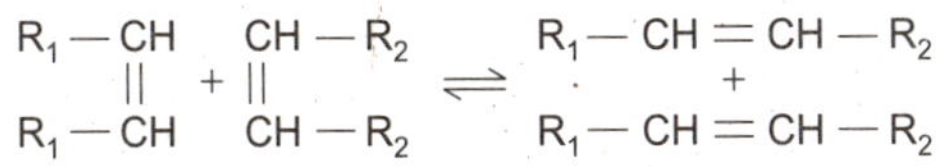

The olefine metathesis reaction is catalysed by transition metal compounds, especially by those of tungsten, molybdenum and rhenium. The catalyst system can further be activated by the addition of an organo-metallic co-catalyst and an oxygen-containing organic compound. A typical catalyst system can be formed by the combination of WCl_6, $C_2H_5AlCl_2$ and C_2H_5OH.

The mechanism of olefine metathesis is believed to involve the rearrangement of a complex containing a metal-carbene species (I) and a coordinated olefine, proceeding through a four-membered metallocyclic transition state (II):

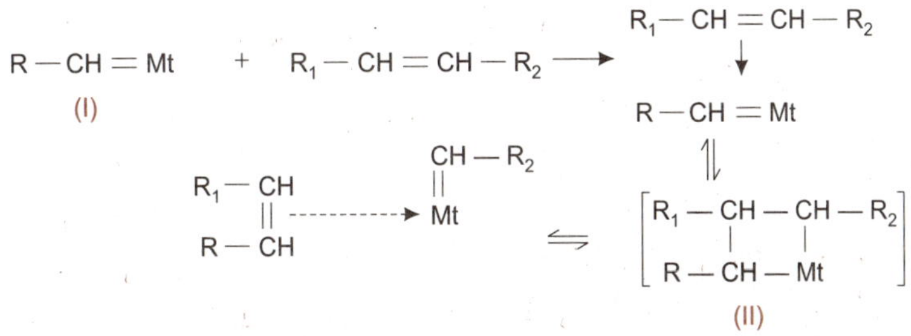

Cycloolefines can be polymerised using metathesis catalysts to yield polymers known as polyalkenamers. Cyclopentene, for example, can be polymerised to yield a rubbery polymer called polypentenamer:

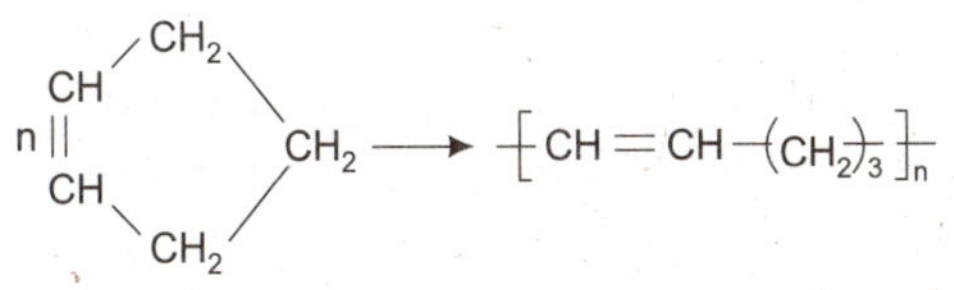

It may be noted that the double bond in the monomer does not take part in polymerisation. It was earlier thought that polymerisation proceeds through a simple opening up of the monomeric rings across the C—C bond adjacent to the double bond followed by a head-to-tail fusion. It has now been confirmed that the double bond does undergo a rupture followed by the transfer of the alkylidene moieties and re-formation of the double bond. The final product is an equilibrium mixture of linear and cyclic macromolecules containing all the double bonds originally present in the monomer molecules.

When cycloolefines are mixed with acyclic olefines, the random incorporation of the monoalkylidene units obtained from the acyclic monomer imparts a control on the chain length of the polymer. The degree of polymerisation is then decided by the ratio of the cyclic to acyclic monomers:

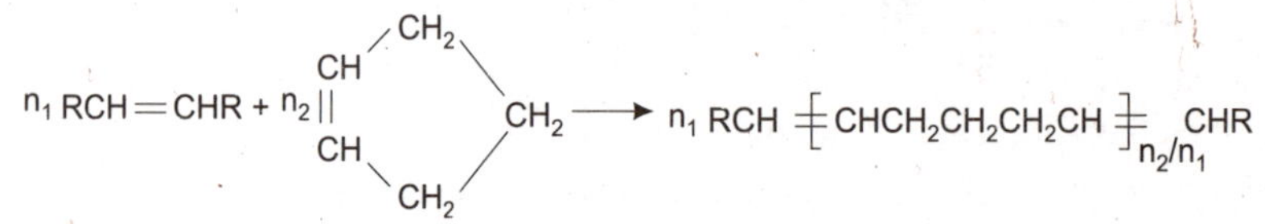

In the industrial preparation of the polypentenamer, the acyclic olefine is thus used as the chain modifier* to control the molecular weight. Cyclodienes can also undergo metathetical polymerisation. For example, metathetical polymerisation of methyl-1, 5, cyclooctadiene, yields an alternating copolymer comprising butadiene and isoprene repeat units:

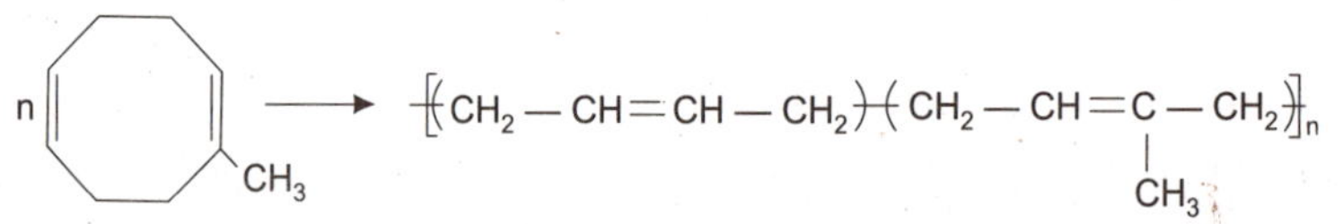

2.4.3 Group Transfer Polymerisation

Group transfer polymerisation (GTP) is a relatively new process, *i.e.*, developed during 1983 at Dupont, Wilmington. By this process, using an organo-silicon initiator and a bifluoride catalyst, α, β-unsaturated esters, ketones, nitriles and carboxamides can be polymerised to provide excellent yields of polymers of narrow molecular weight distribution. At Dupont, it has been successfully demonstrated that polymethyl methacrylate (PMMA) of $\overline{M}_n$ around 4300 with a dispersity** of 1.24 ($\overline{M}_w/\overline{M}_n = 1.24$) could be prepared even at room temperature by just adding the monomer methyl-methacrylate to a previously made solution of tris-(dimethylamino) sulfonium bifluoride (catalyst) and 1-methoxy-l-(trimethylsiloxy)-2-methyl-l-propene (group transfer initiator) in tetra-hydrofuran (THF) solvent. The polymerisation is exothermic and when no more heat is evolved, a non-solvent like methanol may be added to precipitate the polymer from the THF solution. Many other compounds such as ethyl trimethyl silyl acetate and trimethyl silyl cyanide also function as group transfer catalysts. The polymerisation is believed to take place as per the following scheme:

---

*For 'chain modifier', see Section 4.2.3.

**For 'polydispersity', see Section 3.5.

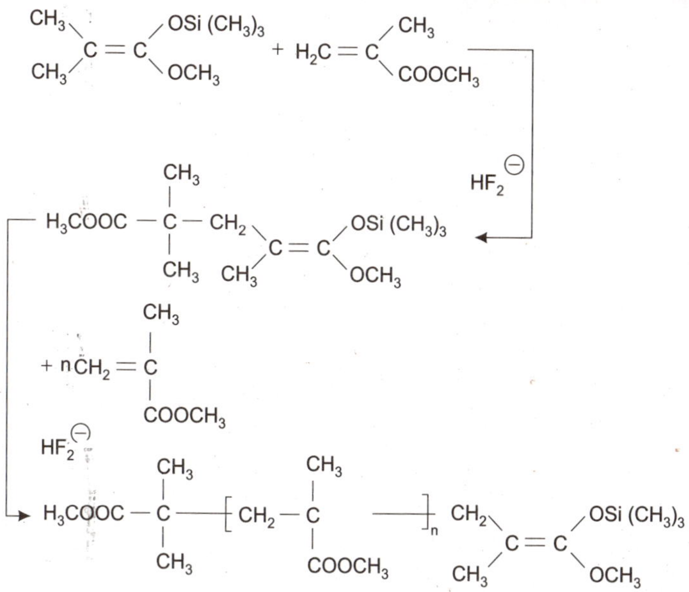

As can be seen from these reactions, the name GTP derives from the fact that the trimethyl silyl group gets transferred from the initiator molecule or from a growing chain end to an incoming monomer and thus enables the polymer molecule to grow to its full size. The transfer of the trimethyl silyl group takes place to the carbonyl oxygen of the new methacrylate group, as shown:

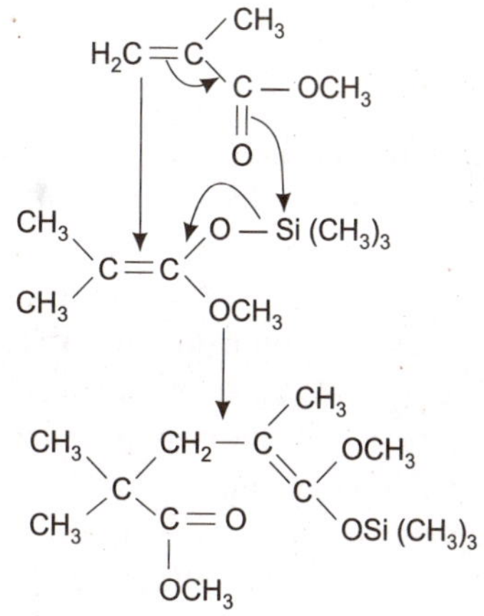

It may also be noted that the polymer chain having the trimethyl siloxy end group is a living polymer. This living polymer can be isolated and block co-polymers prepared by the sequential addition of monomers to it. So far, Dupont had successfully prepared homopolymers from N, N-dimethyl acrylamide, methyl vinyl ketone and 5-trimethyl siloxy 1, 2-dihydrofuran and random co-polymers of methyl and *n*-butyl methacrylates and of methyl, *n*-butyl and glycidyl methacrylates.

Block co-polymers of methacrylate and ethyl acrylate have also been prepared at Dupont using the GTP technique.

Group transfer polymerisation has some advantages: For instance, unlike anionic polymerisation, GTP does not require a very low temperature to bring about polymerisation. Many of the examples cited earlier occur at room temperature. Another major advantage is the perfect control that is achieved in the generation of polymer end groups. In many of the polymers prepared by the anionic polymerisation technique, only around 90% of chain ends have the required functional groups (such as the hydroxyl or carboxyl group) at both ends. In certain critical applications, all the chains must have functional groups at both ends. The group transfer polymerisation process is found to impart perfect control in producing polymer chains with terminal functional groups.

2.5 POLYMERISATION TECHNIQUES

We have studied so far the chemical aspects of polymerisation reactions. The terminology included terms such as polymerisation system, catalyst system and solvent system. These refer to the physical aspects of polymerisation reactions which decide whether the monomer is polymerised in its condensed or gaseous state, whether it is polymerised as such or along with other inert components such as solvents and non-solvents. Factors such as the nature of the monomer, the type of polymerisation mechanism chosen, the required physical form of the polymer and the viability of the process for industrial production are some of those dictating the physical conditions under which polymerisation is to be carried out.

These varying physical conditions lead to different polymerisation techniques, as enumerated in the following sections.

2.5.1 Bulk Polymerisation

The monomer is taken in the liquid state and the initiator is dissolved in the monomer. The chain transfer agent, whenever used to control the molecular weight, is also dissolved in the monomer itself. The whole system is, therefore, in a homogeneous phase. The reaction mass is heated or exposed to a radiation source for initiating the polymerisation and is kept under agitation for proper mass and heat transfers. As the polymerisation proceeds, the viscosity of the medium increases and mixing becomes progressively difficult, leading to products with very broad molecular weight distribution. Another disadvantage of bulk polymerisation is that as the medium gets viscous, the diffusibility of the growing polymer chains becomes restricted, the probability of chain collision becomes less, termination becomes difficult, active radical sites accumulate and the rate of polymerisation increases enormously. This whole phenomenon is called ‘autoacceleration’ and, sometimes, the uncontrolled exothermic reaction can lead to an explosion.

Bulk polymerisation, however, is quite simple and the product obtained has a high purity since, except the initiator and the chain transfer agent, no other additive that could contaminate the product is used. The polymer obtained can also be used as such since no isolation from other components is involved.

Finally, the bulk polymerisation technique is used in the free-radical polymerisation of methyl methacrylate or styrene to get transparent moulding powders and cast sheetings and also of vinyl chloride to get PVC resin.

2.5.2 Solution Polymerisation

In solution polymerisation, the monomer is dissolved in a suitable inert solvent along with the chain transfer agent, whenever used. The free-radical initiator is also dissolved in the solvent medium, while the ionic and coordination catalysts, can either be dissolved or suspended. The presence of the inert solvent medium helps to control viscosity increase and promote a proper heat transfer.

The major disadvantage of the solution polymerisation technique is that, however, inert the selected solvent may be, chain transfer to the solvent cannot be completely ruled out and, hence, it is difficult to get very high molecular weight products. The polymer formed will also have to be isolated from the solution either by evaporation of the solvent or by precipitation in a non-solvent, and removal of their final traces is always extremely difficult. Solution polymerisation techniques, nevertheless, can advantageously be used where the polymer is used in its solution form, as in the case of certain adhesives and coating compositions, or in systems where the polymer formed is insoluble in its monomer or solvent and precipitates out as a slurry and is amenable for easy isolation.

Coming to the practical applications, industrial production of polyacrylonitrile by free-radical polymerisation and that of polyisobutylene by cationic polymerisation, use the solution technique. Block co-polymers are also made exclusively by this technique.

2.5.3 Suspension Polymerisation

Only water-insoluble monomers can be polymerised by this technique. The monomer is suspended in water, in the form of fine droplets, which are stabilised and prevented from coalescing by using suitable water-soluble protective colloids, surface active agents and by stirring. The size of the monomer droplets formed depends on the monomer-to-water ratio, the type and concentration of the stabilising agents and also on the type and speed of agitation employed.

The initiators are monomer soluble. Since each monomer droplet is isolated and independent of the other droplets, it can be visualised to act as an independent bulk polymerisation nucleus. The continuous aqueous phase separating the monomer droplets acts as an efficient heat transfer medium and, hence, the exothermicity is well controlled. Since water is used as the heat transfer medium, the process is also economical as compared to solution polymerisation. As the entire bulk of the monomer is divided into innumerable tiny droplets, control on the kinetic chain length* of the polymer formed is also quite good, resulting in a fairly narrow molecular weight distribution of the product. Polymerisation proceeds to 100% conversion and the product is obtained as spherical beads or pearls. (For this reason, this technique is also known as bead or pearl polymerisation). Isolation of the product becomes easy as this involves only filtration of the beads and removal of the surface active agents and protective colloids by water washing. The water-washed and dried product can be used as such for moulding purposes or can be dissolved in a suitable medium for use as adhesives and coatings.

---

*For a discussion on 'kinetic chain length', see Section 4.2.1.

Expandable polystyrene beads (from which polystyrene foams are made), styrene-divinyl benzene co-polymer beads (for the preparation of ion-exchange resins) and polyvinyl acetate beads (for further conversion into polyvinyl alcohol) are produced by the suspension technique using free-radical initiators.

2.5.4 Emulsion Polymerisation

As in the case of suspension polymerisation, in emulsion polymerisation too, the monomer is dispersed in the aqueous phase not as discrete droplets, but as a uniform emulsion. The emulsion is stabilised by surface active agents (surfactants), protective colloids and also by certain buffers. The surfactants can be anionic (alkali salts of fatty acids and of aryl and alkyl sulfonic acids), cationic (alkyl amine hydrochlorides and alkyl ammonium halides) or non-ionic (alkyl glycosides and saccharose esters of fatty acids). Surfactants serve the purpose of lowering the surface tension at the monomer-water interface and facilitate emulsification of the monomer in water. Owing to their low solubility, surfactants get fully dissolved or molecularly dispersed only at low concentrations. Beyond a particular concentration, the excess quantity does not get molecularly dispersed, but forms molecular aggregates known as 'micelles', and an equilibrium is set up between the dissolved surfactant molecules and the aggregated ones. The highest concentration, wherein all the molecules are in a dispersed state, or the concentration beyond which only micelle formation is possible, is known as the 'critical micelle concentration' (CMC). Typical examples of surfactants and their CMC values are given in Table 2.6. We know that emulsifier molecules are made of two parts: a long non-polar hydrocarbon chain to which is attached a polar group such as—COONa, —SO_3Na, —NH_2HCI or —NBr. In micelle formation, the emulsifier molecules aggregate in such a way that the polar end

Table 2.6 List of surfactants and their critical miscelle concentration (CMC)*

Surfactant	CMC (g/l)	Temp (°C)
Anionic		
$CH_3(CH_2)_6COONa$	6.5×10^1	20
$CH_3(CH_2)_{10}COONa$	5.6	20–70
$CH_3(CH_2)_7OSO_3Na$	3.0×10^1	25–50
$CH_3(CH_2)_{11}OSO_3Na$	2.6	25–60
$CH_3(CH_2)_5C_6H_4SO_3Na$	9.8	75
$CH_3(CH_2)_{11}C_6H_4SO_3Na$	4.0×10^{-1}	50–75
Cationic		
$CH_3(CH_2)_9NH_2{\cdot}HCl$	8.5	25
$CH_3(CH_2)_{11}NH_2HCl$	2.7	30–50
$CH_3(CH_2)_7N(CH_3)_3Br$	7.8×10^1	25
$CH_3(CH_2)_{11}N(CH_3)_3Br$	5.4	25
Non-ionic		
$CH_3(CH_2)_7C_6H_{11}O_6$	7.3	25
$C_{12}H_{20}O_9(C_{16}H_{31}O_2)_2$	1.1×10^{-2}	20
$CH_3(CH_2)_{10}COOC_{12}H_{21}O_{10}$	7.1×10^{-3}	50

*Data from *Polymer Handbook*, Eds. Brandrup J. and Immergut E.H. John Wiley & Sons, New York, 1975.

of the molecules align themselves outward and the hydrocarbon ends come close to each other at the interior (Fig. 2.3).

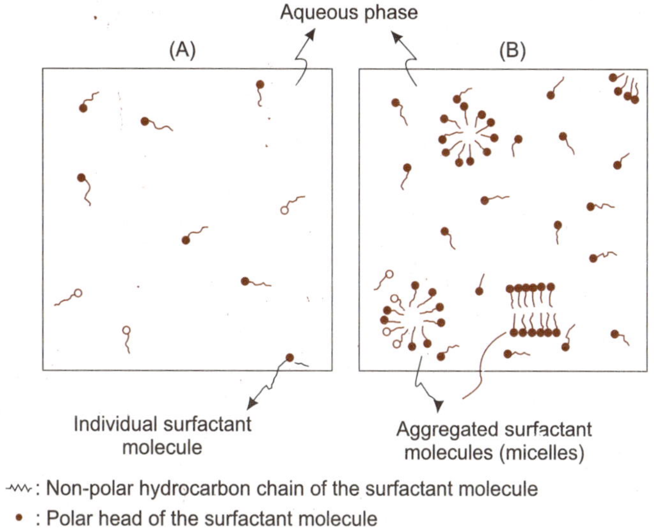

Fig. 2.3 Schematic representation of surfactant molecules distributed in water: (A) completely dissolved at low concentration and (B) dissolved as well as aggregated beyond critical micelle concentration (CMC).

Due to the close proximity of the hydrocarbon ends of all emulsifier molecules, the interior of the micelle acts as a hydrocarbon phase where the monomer can be solubilised. Now, when the monomer is added and agitated, emulsification takes place. The resultant emulsion is a complex system: A molecular solution of the emulsifier in water is the continuous phase wherein the monomer droplets and micelles (having the solubilised monomer at their interior) are uniformly dispersed (see Fig. 2.4). If the monomer is slightly soluble in water, then the aqueous emulsifier solution phase will also contain a part of the monomer dissolved in it.

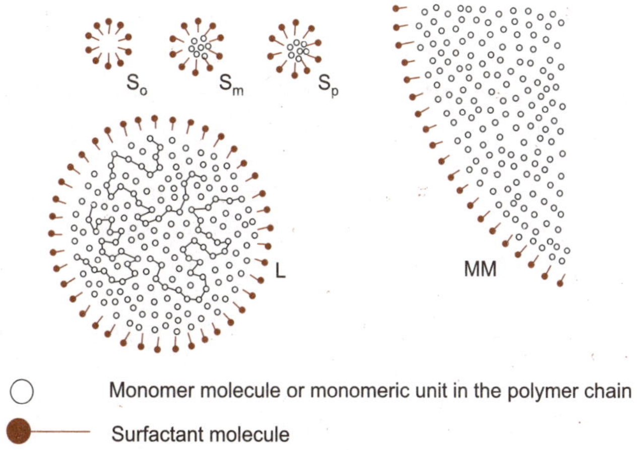

Fig. 2.4 Schematic details of the various components present in an emulsion polymerisation system. S_o, S_m, S_p, *MM* and *L* represent respectively the empty micelle, monomer-filled micelle, micelle with growing polymer chain, monomer droplets and latex particles containing monomer as well as growing polymer.

CHAPTER 2

The difference between the suspension system and the emulsion system is that, while both have monomer droplets, the emulsion system has, in addition, micelles in which the monomer is solubilised. If the quantity of the emulsifier added is less than the CMC, the system becomes one of suspension polymerisation and only when there is enough emulsifier to form the micelles, can we get an emulsion system. Emulsion systems generally utilise-water soluble initiators such as persulfates or hydrogen peroxide. Redox initiators are also widely used. In the emulsion polymerisation technique, we have a system where, as already mentioned, the monomer is present as suspended droplets as well as solubilised at the interior of the micelle. When the initiator is water soluble and monomer insoluble, there is hardly any initiator in the monomer droplet and, hence, no appreciable polymerisation occurs inside the monomer droplet. At the micelle, however, we have a very favourable condition for the polymerisation to occur. The surface layer of the micelle formed by the polar ends of the emulsifier molecules is highly hydrophilic and, hence, an appreciable concentration of the initiator is found at this place. At the same time, the interior of the micelle is filled with the solubilised monomer. Polymerisation, therefore, starts at the surface layer of the micelle and proceeds inwards. As and when the monomer inside the micelle is consumed, more monomer diffuses into the micelle from the monomer droplets present in the aqueous phase. The chain growth continues till another radical species enters and arrests the chain growth. With more and more polymer formation, the polymer chains agglomerate into fine particles and get surrounded and then stabilised by the emulsifier layer of the micelle. As the polymer particle concentration increases, it absorbs the monomer and swells in size, which, in turn, enlarges the size of the micelle. At a particular stage, however, the micelle bursts and loses its identity; all the emulsifier molecules originally forming the micelle now form an absorption layer around the polymer particles. Further supply of the monomer to the growing chain inside the swollen polymer particle comes, again, from the monomer droplets. Monomer droplets in this emulsion system, therefore, act only as a reservoir where from the monomer diffuses into micelles or polymer particles, as the case may be.

At the end of the polymerisation, we have fine particles of the polymer, stabilised by the emulsifier layer and dispersed uniformly in the aqueous phase. This milky white dispersion is often called 'latex'. The latex can either be used as such for making adhesives, water-soluble emulsion paints, etc., or the polymer can be isolated from the latex by destabilising the emulsion (using some electrolytes), by spray drying or by freezing.

The picture given above of the polymer formation is only representative of a system wherein the monomer is completely insoluble in water and the initiator is completely insoluble in the monomer. The actual mechanism is quite complicated and varies from system to system. Depending on whether the monomer is slightly soluble or completely insoluble in water, whether the initiator is soluble in water or monomer or in both, and the monomer/water and emulsifier/water ratios, the polymerisation can occur at one or more of the following spots: (*a*) at the dissolved phase of the monomer in water; (*b*) at the interface between the monomer droplet and water; (*c*) inside the monomer droplet; (*d*) at the surface layer of the micelle; (*e*) at the hydrocarbon phase of the micelle; (*f*) at the surface of the polymer particle; or (*g*) at the interior of the swollen mass. The ultimate result, however, is the same in that the monomer droplets as well as the micelles lose their identity and the polymer is formed as fine particles and is then stabilised into an emulsion.

Emulsion polymerisation is the most widely used industrial technique and monomers such as vinyl chloride, butadiene, chloroprene, vinyl acetate, acrylates and methacrylates are polymerised by this technique. Heat transfer is very efficient and the viscosity build-up of the polymerisation mass is quite low as compared to that in the bulk and solution techniques. Since the polymerisation proceeds at the micelles, where the surface/volume ratio is very high, the rate of polymerisation is quite high. Also, since the number of radicals in a micelle is very much limited and since there is a continuous supply of monomer to the growing chain, a very high molecular weight product is achieved by this technique.

2.5.5 Melt Polycondensation

This technique is used for the polymerisation of monomers which have at least one solid component and do not decompose around their melting points. The temperatures involved in melt polycondensation are rather high and, hence, the reaction has to be carried out in an inert atmosphere of N_2 or CO_2 to avoid such side reactions as can lead to oxidation, decarboxylation, degradation*, etc. In some cases, the reaction is carried out under reduced pressure to facilitate removal of the by-product, which becomes essential if a high molecular weight product is aimed at. Removal of the by-product becomes extremely difficult at later stages of the reaction as there is considerable increase in the viscosity of the medium, like in bulk polymerisation. The polymer formed is in the molten state at the reaction temperature and is discharged from the reactor while hot, as otherwise, it will solidify inside the reactor and removal will become extremely difficult. In most of the cases, the hot melt from the reactor is passed on directly to the processing equipments for extrusion, casting or spinning. Production of polyethylene terephthalate from dimethyl terephthalate and ethylene glycol is carried out by the melt condensation technique. Preparation of nylon 6, 6 is another example, illustrative of the melt condensation technique.

2.5.6 Solution Polycondensation

In the solution polycondensation technique, the reactants are taken as a solution in a suitable inert solvent. The reaction can be carried out at comparatively lower temperatures during which heat and mass transfer processes are easier than in the melt technique. The solvent can also serve as an entrapping agent for the by-product formed and, hence, the removal of the by-product becomes easy. Owing to the presence of the solvent phase, however, the kinetic probability of chain growth is low and this leads to a reduced rate and a low degree of polymerisation.

Many of the liquid polyester resins based on glycols and unsaturated dicarboxylic acid are prepared by this technique using high-boiling aromatic hydrocarbons as solvent with which the water (by-product) forms an azeotrope and hence, can be removed easily.

2.5.7 Interfacial Condensation

In this technique, polymerisation is allowed to proceed at the interface between an aqueous and an organic medium. Reactants having highly reactive functional groups, which can readily react at ambient temperatures to form condensation products, are suited to this technique. Preparation of fully aromatic polyamides from terephtholoyl chloride and paraphenylene diamine is a typical example:

---

*For a detailed discussion on 'degradation', see Chapter 10.

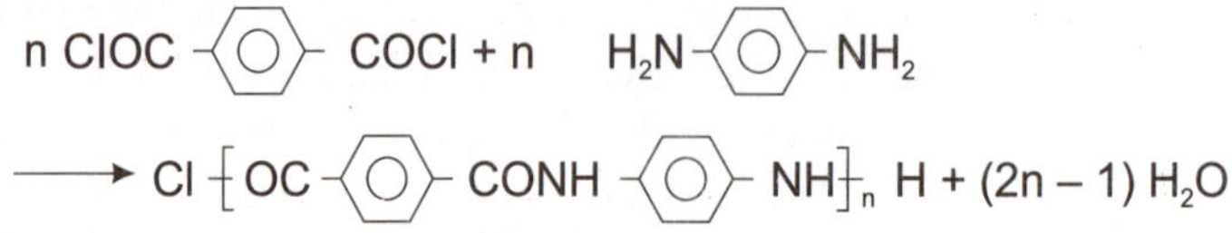

The diamine is dissolved in water and the acid chloride in chloroform or carbon tetrachloride. When the two solutions are brought in contact with each other, at the interface, the diamine molecules diffuse into the organic phase and react with the acid chloride, resulting in the formation of a polymer which precipitates out immediately. The by-product formed (HCl) diffuses back into the aqueous phase and gets dissolved. The precipitate is formed at the interface in the form of a thin film and, when removed from the system, exposes a fresh surface of the acid chloride to the organic phase, resulting in the formation of the fresh quantity of the polymer. Since the polymer formation at the interface is a diffusion-controlled process, very high molecular weight products can be achieved by this technique. In actual production, the two solutions are very thoroughly agitated so as to form an emulsion wherein the interface surface/volume ratio is increased tremendously and, hence, both the rate and degree of polymerisation become very high. As the polymer is formed, it precipitates out and, under the influence of high-speed agitation, forms a slurry. The polymer is isolated from the slurry and washed free of adherent reactants.

2.5.8 Solid and Gas Phase Polymerisation

In all the above mentioned techniques, the polymerisation is carried out in a liquid phase—either as pure liquid or as solution in suitable solvents or as liquid/liquid suspension and emulsion. Although polymerisations in liquid systems are very popular and very widely used for industrial production, the same can also be carried out in solid and gaseous phases.

Solid phase polymerisation is mostly restricted to chain polymerisation. Also, since such polymerisation is restricted to low temperatures, thermal activation of the reaction is quite difficult and hence, the photo or radiation-activation technique is resorted to. Since molecular mobility in the solid state is completely restricted, the polymer-building process becomes quite slow. A good example of solid-state polymerisation is the preparation of polyformaldehyde by the radiation polymerisation of solid trioxane. Single crystals of trioxane, on irradiation, polymerises to form highly crystalline and fibrous polyformaldehyde:

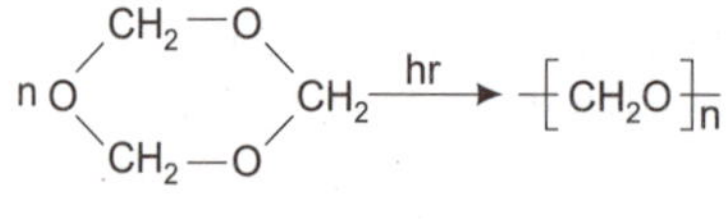

(Trioxane) (Polyformaldehyde)

Solid-state polymerisation of trioxane can also be brought about by the conventional cationic initiators such as BF_3. In this case, the monomer crystal is kept dipped in the catalyst medium and the polymerisation proceeds with the diffusion of the catalyst and the counter-ion into the monomer crystal.

Photopolymerisation of diacetylene derivatives is another example of solid-state polymerisation. The polymer formed is highly crystalline and also conducts electricity:

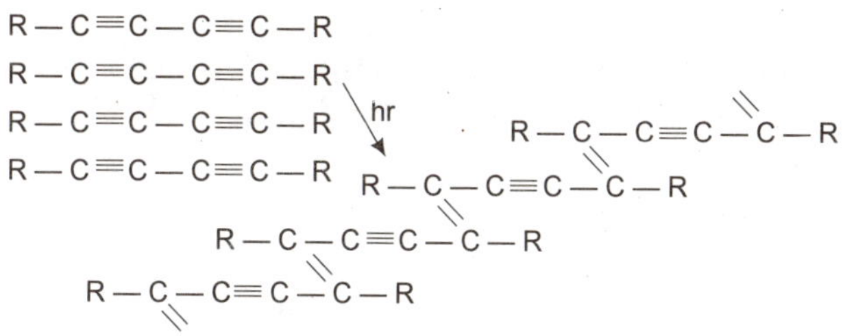

Gas phase polymerisation is known in the case of very few olefinic monomers. The disadvantage of the system is that it has a very poor heat transfer. The two methods used in the gas phase polymerisation are: (*i*) Spraying the catalyst (usually Ziegler–Natta type) into the gaseous monomer and (*ii*) feeding the gaseous monomer into a fluidised bed made-up of the catalyst particles. In both the cases, the polymer is obtained as a free-flowing powder. Polymerisation of ethylene and *p*-xylene can be cited as examples for gas phase polymerisation. In the first case, ethylene gas is passed through a fluidised bed column filled with the Ziegler catalyst (made by mixing titanium chloride and alkyl aluminium in pentane medium, spray drying and powdering). The gaseous monomer is passed at the room temperature at a pressure of 4–5 atmospheres and the polymer formed is collected as a free-flowing powder. The polymerisation of *p*-xylene is done in a different mode. First *p*-xylene is dehydrogenated to form its dimer (di-*p*-xylylene) using superheated steam at 950°C. The dimer, which is a crystalline solid at room temperature, is then heated to 600°C at one mm pressure, when it sublimes forming an equilibrium mixture of a diradical and a quinonoid. When this vapour is quenched to 50°C over metal surfaces, a linear polymer called poly-*p*-xylylene is formed on the metal surface:

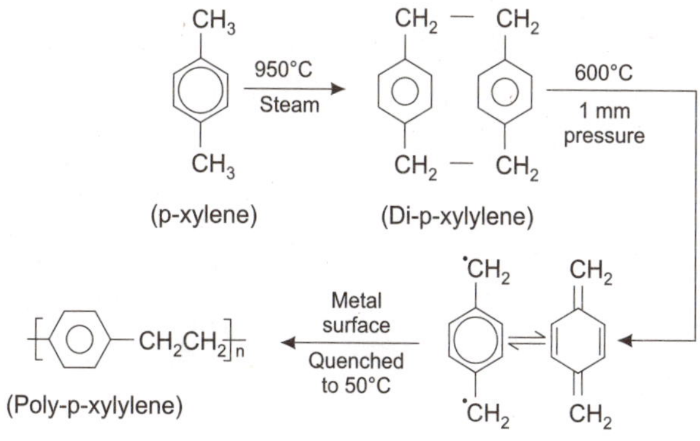

Salient features of different polymerisation techniques are summarised in Table 2.7.

Table 2.7 Salient features of different polymerisation techniques.

Bulk polymerisation (1)	Solution polymerisation (2)	Suspension polymerisation (3)	Emulsion polymerisation (4)	Melt polycondensation (5)	Solution polycondensation (6)
Homogeneous system. In certain cases where the polymer is insoluble in the monomer, the polymer precipitates out as polymerisation sets in, and the system becomes heterogeneous.	Homogeneous system. In certain cases, where the polymer is insoluble in the monomer/solvent, the polymer precipitates out as the polymerisation sets in and the system becomes heterogeneous.	Heterogeneous system. Proper control on agitation and stabiliser needed, to keep the suspension stable.	Heterogeneous system. Agitation and proper control on the emulsifier/stabiliser needed to keep the emulsion stable.	Heterogeneous system.	Homogeneous system. If the polymer is insoluble in the solvent the system becomes heterogeneous.
Viscosity of the medium increases very rapidly and at high conversions, proper mass transfer becomes difficult.	Viscosity of the medium increases slowly, but not difficult to maintain proper mass transfer even at high conversions.	Viscosity of the medium does not increase appreciably even at high conversions.	Viscosity of the medium does not increase thoughout the polymerisation.	Viscosity of the medium increases very rapidly and at high conversions proper mass transfer becomes difficult. Removal of by-product also becomes difficult at high conversions.	Viscosity of the medium increases slowly. Removal of by-product is comparatively easier if a suitable solvent is chosen as entrainer.
Exotherm dissipation is difficult. May lead to autoacceleration.	Exotherm dissipation is easy.	Exotherm dissipation is easy.	Extherm dissipation is the best.	No exotherm. But proper heat transfer within the medium becomes difficult at high conversions.	No exotherm. Heat transfer is uniform.
Requires longer duration for very high conversions.	Requires longer duration for very high conversions.	Requires longer duration for very high conversions.	High conversions (almost 100%) achieved in shorter-duration.	Requires longer duration for very high conversions.	Requires longer duration for very high conversions.
Suitable for producing medium to high molecular weight products.	Suitable for producing low molecular weight products.	Suitable for producing medium to high molecular weight products.	Suitable for producing very high molecular weight products.	Suitable for producing medium to high molecular weight products.	Suitable for producing medium to high molecular weight products.
Most widely used in industrial processess involving free-radical, polycondensation and polyaddition polymerisations.	Suitable for free-radical, cationic, anionic and coordination polymerisations.	Suitable for free-radical and polycondensation polymerisations.	Suitable for free-radical and polycondensation polymerisation.	Suitable for polycondensation polymerisation.	Suitable for polycondensation polymerisation.

Contd.

Bulk polymerisation (1)	Solution polymerisation (2)	Suspension polymerisation (3)	Emulsion polymerisation (4)	Emulsion polycondensation (5)	Solution polycondensation (6)
Product is minimum contaminated. easty to isolate and can be directly used for moulding purposes.	Isolation becomes difficult. Product contaminated with solvent and solvent related impurities.	Isolation is easy as the product is obtained as powder/beads and can be directly used for moulding porpose. Product contaminated stabiliser.	Emulsion can be directly used for paints adhesives and so on. Solid product can be isolated by breaking the emulsion. Product contaminated with emulsifier, stabiliser and so on.	Product minimum contaminated. Freeing from monomers and by-product difficult since the polymer solidifies at ambient temperature.	Product usually contaminated with solvent.

In all these and other polymerisation techniques, the ultimate properties of the polymer such as molecular weight, molecular weight distribution, molecular structure and chemical homogeneity are very much dependent on factors such as (*i*) effective removal of heat of polymerisation, (*ii*) provision of proper residence time, and (*iii*) effective mixing to ensure uniform temperature profile and reactant distribution throughout the system.

Books for Further Reading

Bawn, C.E.H., *The Chemistry of High Polymers*, Interscience, New York, 1948.

Burnett, G.M., *Mechanism of Polymer Reactions*, Wiley-Interscience, New York, 1954.

Bovey, F.A., Kolthaff, I.M., Medalia, A.L. and Meehan, E.J., *Emulsion Polymerisation*, Wiley-Interscience, New York, 1955.

Schildknecht, C.E., Ed., *Polymer Processes*, Interscienc, New York, 1956.

Bevington, J.C., *Radical Polymerisation*, Academic, London, 1961.

Vander Hoff, M.B.E., *Polymerisation and Polycondensation Processes*, American Chemical Society, Washington, 1962.

Plesch, P.H., *Chemistry of Cationic Polymerisation*, Pergamon, Oxford, 1963.

Morgan, P.W., *Condensation Polymers by Interfacial and Solution Methods*, Interscience, New York, 1965.

Natta, G. and Damusso, F., Eds., *Stereoregular Polymers and Stereospecific Polymerisations*, Vols. 1 and 2, Pergamon, Oxford, 1967.

Ham, G.E., Ed., *Vinyl Polymerisation*, Dekker, New York, 1967.

Szwarc, M., *Carbanions, Living Polymers and Electron Transfer Processes*, Wiley-Interscience, New York, 1968.

Frisch, K.C. and Reagen, S.L., Eds., *Ring-opening Polymerisation*, Dekker, New York, 1969.

Tsuruta, T. and O'driscoll, K.F., Eds., *Structure and Mechanism in Vinyl Polymerisation*, Dekker, New York, 1969.

Odian, G., *Principles of Polymerisation*, McGraw-Hill, New York, 1970.

Braun, D., Cherdon, H. and Kern, W., *Techniques of Polymer Synthesis and Characterisation*, Wiley-Interscience, New York, 1971.

Sandler, S.R. and Karo, W., *Polymer Synthesis*, Vol. 1, Academic, New York, 1974.

Blackley, D.C., *Emulsion Polymerisation: Theory and Practice*, Halsted, New York, 1975.

3

MOLECULAR WEIGHT AND SIZE

3.1 'AVERAGE' MOLECULAR WEIGHT

The molecular weight can vary from a single-digit figure for a simple chemical entity to a number in millions for a complex polymer. For example, the molecular weight of hydrogen is just two, while that of circular DNA in *Escherichia coli* is around 2 billion! One then wonders, if the term molecular weight conveys the same meaning in the two cases. Well, let us examine the matter in detail.

A simple compound has a fixed molecular weight which, for acetone, for instance, is 58 regardless of how it is made. If this 58 becomes 60, the compound is no longer acetone but may be acetic acid. Moreover, in any given sample of acetone, for example, each molecule has the same molecular weight. This is true for all low molecular weight compounds. There is thus some definiteness about the molecular weight of simple chemical compounds. In contrast, a polymer comprises molecules of different molecular weights and, hence, its molecular weight is expressed in terms of an 'average' value. There are many 'averages' as we will presently see.

Ethylene gas, for instance, is a low molecular weight, simple chemical compound; each of its molecules has the same chemical structure (*viz.*, $CH_2{=}CH_2$) and hence, a fixed molecular weight of 28. But we polymerise ethylene to make polyethylene and we encounter an indefinite chemical structure of

$$-\!\!\left[CH_2 - CH_2\right]_n\!\!-$$

wherein n can change its value from one polyethylene molecule to another present in the same polymer sample. This is so because when ethylene is polymerised to form polyethylene, a number of polymer chains start growing at any instant, but all of them do not get terminated after growing to the same size. The chain termination is a random process and, hence, each polymer molecule formed can have a different number of monomer units and thus different molecular weights.

The following three structures, for example represent polyethylene molecules, but each has a different molecular weight:

$$R\!-\!\!\left[CH_2-CH_2\right]_{500}\!\!-\!R \quad \text{(Molecular weight} \simeq 14{,}000)$$

$$R\!-\!\!\left[CH_2-CH_2\right]_{550}\!\!-\!R \quad \text{(Molecular weight} \simeq 15{,}400)$$

$$R\!-\!\!\left[CH_2-CH_2\right]_{600}\!\!-\!R \quad \text{(Molecular weight} \simeq 16{,}800)$$

We can see that the three molecules have different sizes, their molecular weights are different, and yet, they are all very much 'polyethylenes'. A polymer sample can, therefore, be thought of as a mixture of molecules of the same chemical type, but of different molecular weights. In this situation, the molecular weight of the polymer sample can only be viewed statistically and expressed as some average of the molecular weights contributed by the individual molecules that make the sample. By contrast, the molecular weight of a simple compound such as ethylene or acetone is constant.

3.2 NUMBER-AVERAGE AND WEIGHT-AVERAGE MOLECULAR WEIGHTS

We have to now consider ways and means of averaging out the molecular weight of the polymer. The two most common and experimentally verifiable methods of averaging are called 'number-average' and 'weight-average' methods. The following sections will clarify the concept.

3.2.1 Number-Average Concept

Consider a basket containing four vegetables: onions, brinjals, cabbages and cauliflowers. Just for ease of understanding, let each onion of the onion lot weigh the same and so also each of the other vegetables. Assume that the number of each vegetable and its weight are as tabulated.

Vegetable entity	*Number of units in each entity, n*	*Weight of each unit, M(g)*	*Total weight of each entity W = nM(g)*
Onions	2	10	20
Brinjals	4	20	80
Cabbages	6	100	600
Cauliflowers	3	250	750
	15	—	1450

Now, we have to find out the average weight of the vegetables present in the basket. To work this out, one might assume that the individual vegetable entity contributes to the average weight in the proportion of its numbers. What we then get is the number-average weight, arrived at as follows:

Total number of vegetables contained in the basket = 15

Number of onions present in the basket = 2

Therefore, Number fraction of onions

(to the total number of vegetables) = 2/15

Similarly,

$$\text{Number fraction of brinjals} = 4/15$$
$$\text{Number fraction of cabbages} = 6/15$$
$$\text{Number fraction of cauliflowers} = 3/15$$

Contribution made by 2 onions towards average weight of vegetables in the basket:

Number fraction of onions × weight of each onion = (2/15) × 10 = 1.33 g

Similarly, contribution made by 4 brinjals to the average weight of vegetables in the basket

$$= (4/15) \times 20 = 5.33 \text{ g}$$

Similarly, contribution made by 6 cabbages to the average weight of vegetables in the basket

$$= (6/15) \times 100 = 40.00 \text{ g}$$

Similarly, contribution made by 3 cauliflowers to the average weight of vegetables in the basket

$$= (3/15) \times 250 = 50.00 \text{ g}$$

Summing up the contributions made by each vegetable variety, we get the 'number-average' weight of the total vegetables as

$$1.33 + 5.33 + 40 + 50 = 96.66 \text{ g}$$

3.2.2 Weight-Average Concept

The other method of calculating the average weight is based on the assumption that the individual vegetable variety contributes to the total weight in the proportion not of its number but its weight. What we then get is the 'weight-average', arrrived at as follows:

$$\text{Total weight of all vegetables in the basket} = 1450 \text{ g}$$
$$\text{Weight of onions present in the basket} = 20 \text{ g}$$

Therefore, $\quad \text{Weight fraction of onions} = 20/1450$

Similarly, weight fractions of brinjals, cabbages and cauliflowers are 80/1450, 600/1450, and 750/1450, respectively.

Next, contribution made by onions towards average weight of vegetables in the basket = Weight fraction of onions × Average weight of onions = (20/1450) × 10 = 0.14 g

Similarly,

$$\text{Corresponding contribution by brinjals} = (80/1450) \times 20 = 1.10 \text{ g}$$
$$\text{Corresponding contribution by cabbages} = (600/1450) \times 100 = 41.38 \text{ g}$$
$$\text{Corresponding contribution by cauliflowers} = (750/1450) \times 250 = 129.31 \text{ g}$$

Summing up the contributions made by each vegetable variety, we get the weight-average weight of the total vegetables as

$$0.14 + 1.10 + 41.38 + 129.31 = 171.93 \text{ g.}$$

3.2.3 Generalisation of the Foregoing Concepts

In computing the molecular weight of a polymer we can also use either the number fraction or the weight fraction of the molecules present in the polymer to get either the number-average molecular weight (designated as $\overline{M}_n$) or the weight-average molecular weight (designated as $\overline{M}_w$), respectively.

The method of working out $\overline{M}_n$ and $\overline{M}_w$ can now be easily generalised, using *simple* mathematics. Assume that there are n number of molecules in a polymer sample and n_1 of them have M_1 molecular weight; n_2 have M_2 molecular weight and so on till we get n_i having M_i molecular weight (see Fig. 3.1).

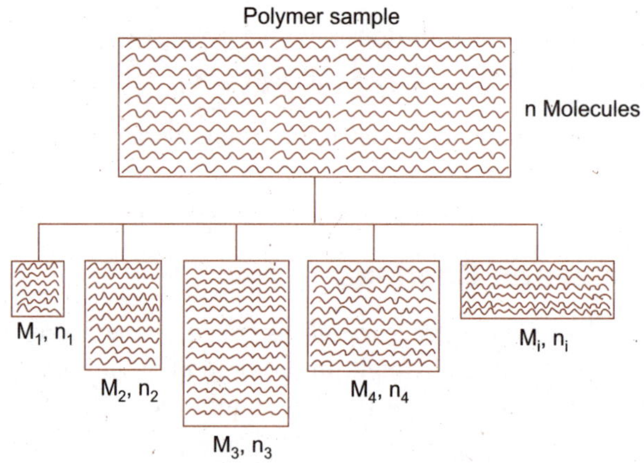

Fig. 3.1 Conceptual representation of a polymer sample containing n number of molecules of different molecular weights, of which n_1 have a molecular weight of M_1, n_2 have M_2 and so on.

Now, we have a total number of molecules (n) given by

$$n = n_1 + n_2 + n_3 + \cdots + n_i = \Sigma n_i$$

$$\text{Number of molecules in fraction } 1 = n_1$$

$$\text{Number fraction of fraction } 1 = \frac{n_1}{n} = \frac{n_1}{\Sigma n_i}$$

$$\text{Molecular weight contribution by fraction } 1 = \frac{n_1 M_1}{\Sigma n_i}$$

Similarly, molecular weight contribution by other fractions will be as follows:

$$\frac{n_2 M_2}{\Sigma n_i}, \frac{n_3 M_3}{\Sigma n_i}, \ldots, \frac{n_i M_i}{\Sigma n_i}$$

Number-average molecular weight of the whole polymer will then be given by

$$\frac{n_1 M_1}{\Sigma n_i} + \frac{n_2 M_2}{\Sigma n_i} + \cdots + \frac{n_i M_i}{\Sigma n_i} = \frac{\Sigma n_i M_i}{\Sigma n_i} = \overline{M}_n \tag{3.1}$$

Similarly, total weight of the polymer = $W = \Sigma n_i M_i$.

$$\text{Weight of fraction 1} = W_1 = n_1 M_1.$$

$$\text{Weight fraction of fraction 1} = \frac{n_1 M_1}{W} = \frac{n_1 M_1}{\Sigma n_i M_i}.$$

Molecular weight contribution by fraction 1 is given by

$$\frac{n_1 M_1 M_1}{\Sigma n_i M_i} = \frac{n_1 M_1^2}{\Sigma n_i M_i}$$

Similarly, the molecular weight contribution by the other fractions will be

$$\frac{n_2 M_2^2}{\Sigma n_i M_i}, \frac{n_3 M_3^2}{\Sigma n_i M_i}, \ldots, \frac{n_i M_i^2}{\Sigma n_i M_i}$$

The weight-average molecular weight of the whole polymer will then be

$$\frac{n_1 M_1^2}{\Sigma n_i M_i} + \frac{n_2 M_2^2}{\Sigma n_i M_i} + \cdots + \frac{n_i M_i^2}{\Sigma n_i M_i} = \frac{\Sigma n_i M_i^2}{\Sigma n_i M_i} = \overline{M}_w \tag{3.2}$$

For all synthetic polymers, $\overline{M}_w$ is greater than $\overline{M}_n$. If they were to be equal, the polymer sample may be considered as perfectly homogeneous (*i.e.*, each molecule has the same molecular weight); but this does not happen.

3.3 SEDIMENTATION AND VISCOSITY-AVERAGE MOLECULAR WEIGHTS

Apart from $\overline{M}_n$ and $\overline{M}_w$, there are two other ways of expressing the molecular weight based on the sedimentation and flow behaviour of the polymer in solution. They are: *z*-average molecular weight ($\overline{M}_z$) and viscosity-average molecular weight ($\overline{M}_v$), expressed as

$$\overline{M}_z = \frac{\Sigma n_i M_i^3}{\Sigma n_i M_i^2} \tag{3.3}$$

and

$$\overline{M}_v = \left(\frac{\Sigma n_i M_i^{1+a}}{\Sigma n_i M_i}\right)^{1/a} \tag{3.4}$$

CHAPTER 3

In these two equations n_i is, as before, the number of molecules having the molecular weight of M_i present in the sample and a is a variable (its value ranging from 0.5 to 1) in the Mark–Houwink equation which relates intrinsic viscosity $[\eta]$ with the viscosity average molecular weight ($\overline{M}_v$). (see Section 12.6).

We will deal with experimental techniques used for determining molecular weights in Chapter 13. It will suffice here to know that light scattering, osmotic pressure, viscosity and ultracentrifugation methods are used to get $\overline{M}_w$, $\overline{M}_n$, $\overline{M}_v$ and $\overline{M}_g$, respectively.

3.4 MOLECULAR WEIGHT AND DEGREE OF POLYMERISATION

We have briefly remarked in Section 1.3 that the size of a polymer molecule depends on the number of repeat units it contains and that this number represents the 'degree of polymerisation'. For instance, if there are 1000 repeat units in a polymer molecule, the degree of polymerisation (DP) is 1000. DP provides another way of expressing the molecular weight as follows:

$$M = D_p \cdot m \tag{3.5}$$

where M is the molecular weight of the polymer, D_p its degree of polymerisation and m the molecular weight of the monomer or the repeat unit.

Both DP and molecular weight are related to the molecular size. Hence, like the molecular weight, DP can also be averaged over the size of the sample. The 'number-average' and 'weight-average' degree of polymerisation can then be defined in a manner similar to that in which $\overline{M}_n$, $\overline{M}_w$, are defined earlier, *viz.*,

$$(\overline{D}_p)_n = \frac{\Sigma n_i (D_p)_i}{\Sigma n_i} \tag{3.6}$$

and

$$(\overline{D}_p)_w = \frac{\Sigma n_i (D_p)_i^2}{\Sigma n_i (D_p)_i} \tag{3.7}$$

Each of these averages can be related to the corresponding molecular weight averages by the following two equations:

$$\overline{M}_n = (\overline{D}_p)_n \cdot m \tag{3.8}$$

$$\overline{M}_w = (\overline{D}_p)_w \cdot m \tag{3.9}$$

3.5 POLYDISPERSITY AND MOLECULAR WEIGHT DISTRIBUTION IN POLYMERS

It may be recalled that a simple chemical compound contains molecules, each of which has the same molecular weight (a 'monodispersed' system), whereas a polymer contains molecules, each of which can have different molecular weights (a 'polydispersed' system). Monodispersity and polydispersity can be represented as shown in Fig. 3.2.

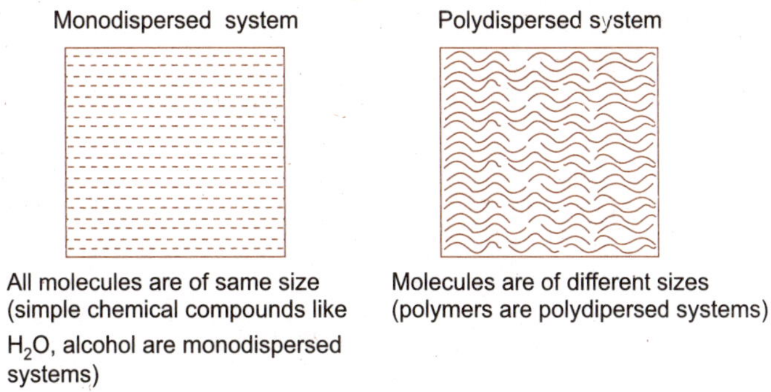

Fig. 3.2 Conceptual representation of monodispersed and polydispersed systems (dispersity is with respect to molecular weight).

While the polydisperse nature of the polymer is the basis of the concept of 'average' molecular weight, this 'average' by itself conveys nothing on the dispersity pattern in a given polymer sample. Take a polymer sample of, say, 40,000 average molecular weight. Now, this may mean that the molecules have molecular weights ranging from 20,000 to 80,000 or from 500 to 100,000—the figure is uncertain. This is the reason why two polymer samples of the same 40,000 $\overline{M}_n$ can display similar properties in some respects but not in some others. To know a polymer properly, we must have a knowledge of both the average molecular weight as well as its dispersion pattern. This dispersity with respect to the lowest to the highest molecular weight homologues is expressed by a simple molecular weight distribution curve. Such a curve for a polymer sample is computed by plotting the number fraction (n_i) of molecules having a particular molecular weight (M_i) against the corresponding molecular weight (Fig. 3.3). While the details of computing a molecular weight distribution curve are given in Chapter 13, we may observe here that such a curve reveals the pattern of the different molecular species present in a polymer sample. The curve shown in Fig. 3.3 brings out the relative positions of various average molecular weights in a polydispersed polymer sample. It may be noted that $\overline{M}_w$ is greater than $\overline{M}_n$ and that $\overline{M}_v$ is closer to $\overline{M}_w$ than to $\overline{M}_n$. Figure 3.4 shows molecular weight distribution in two samples having the same number-average molecular

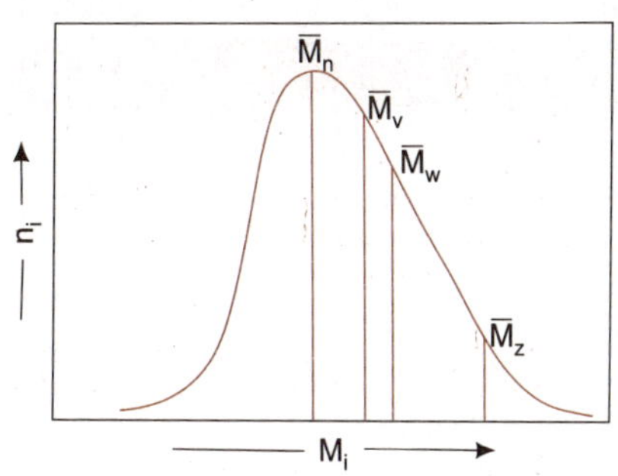

Fig. 3.3 A molecular weight distribution curve for a hypothetical polydispersed polymer sample (for details see text).

weight but different polydispersities. Sample 1 obviously has a narrower dispersion pattern and, hence, a lower polydispersity than sample 2.

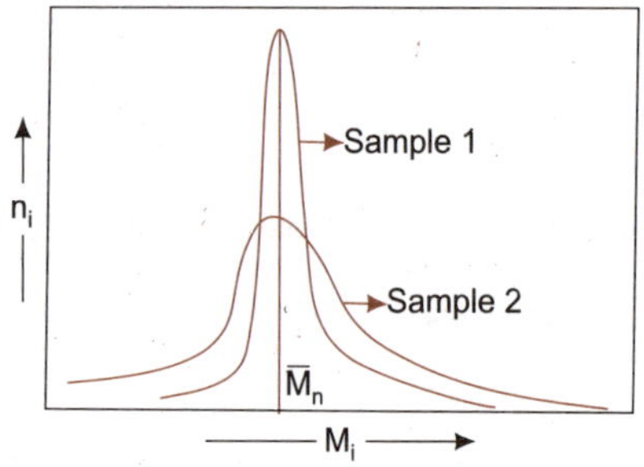

Fig. 3.4 Molecular weight distribution curves of two hypothetical polymer samples with the same number-average molecular weight but different polydispersities.

Polydispersity is a very important parameter. It gives an idea of the lowest and the highest molecular weight species as well as the distribution pattern of the intermediate molecular weight species. To quickly ascertain the degree of polydispersity, we can determine $\overline{M}_w$ and $\overline{M}_n$ by two different experimental methods and then compute $\overline{M}_w/\overline{M}_n$. This ratio is indicative of the extent of polydispersity. For all synthetic polymers, this ratio is higher than 1. As the molecular weight distribution becomes broader (that is to say, as the dispersity increases), the value of $\overline{M}_w/\overline{M}_n$ increases. The three curves in Fig. 3.5 show the dispersity ratio increasing for the three samples with the same $\overline{M}_n$, but a different molecular weight distribution.

We have mentioned earlier that polydispersity arises owing to a variation in the degree of polymerisation attained by different molecules during the polymerisation process. One would, therefore, expect polymers obtained by different polymerisation techniques to show different polydispersities, and this is indeed the case. Table 3.1 gives $\overline{M}_w/\overline{M}_n$ values for synthetic polymers obtained by different methods.

Table 3.1 $\overline{M}_w/\overline{M}_n$ values for synthetic polymers obtained by different polymerisation techniques

Polymerisation type	$\overline{M}_w/\overline{M}_n$
Free-radical chain polymerisation:	
(*i*) with precise temperature control (solution, suspension and emulsion systems)	1.5–2
(*ii*) with moderate temperate control (bulk system)	2–5
(*iii*) without temperature control (with autoacceleration)	8–10
Cationic/anionic chain polymerisation:	
(*i*) using homogeneous catalyst system	< 1.5
(*ii*) using heterogeneous catalyst system	> 10
Polycondensation, polyaddition and ring-opening polymerisation	2–3
Coordination polymerisation	> 10
Polymerisation systems leading to chain-branching	> 20

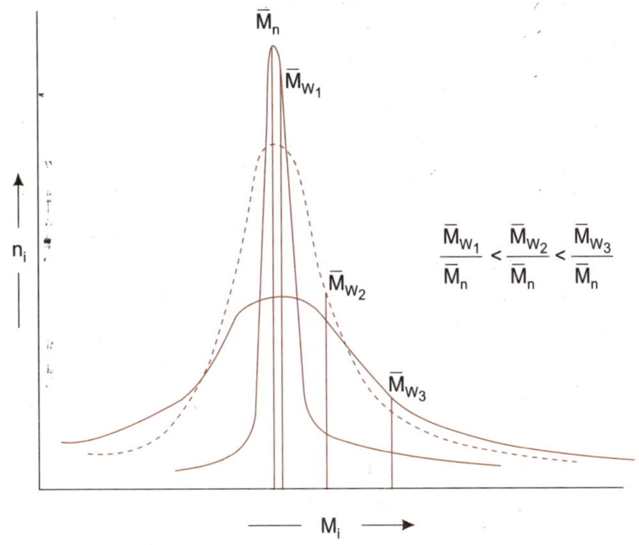

Fig. 3.5 Molecular weight distribution curves of three hypothetical samples with different polydispersities (one of the curves is shown as a dashed line for the purpose of clarity).

3.6 THE PRACTICAL SIGNIFICANCE OF POLYMER MOLECULAR WEIGHT

One might ask as to what is the practical use of the knowledge of molecular weights. Let us look at this aspect.

Many commercially useful polymers are selected on the basis of their properties such as melt viscosity, impact strength or tensile strength. These properties are directly dependent on the molecular weight of the polymer or its degree of polymerisation. For example, tensile and impact strengths increase with molecular weight. The increase is up to a point and then it levels off (Fig. 3.6). The melt viscosity of the polymer, however, shows a different trend. At very high molecular weights, the melt viscosity rises more steeply than at low molecular weights (Fig. 3.7). A commercially useful polymer should have a low melt viscosity to permit ease of processing, but, at the same time, should exhibit good strength. The patterns shown in Figs. 3.6 and 3.7 help to select the useful range of molecular weights for polymers.

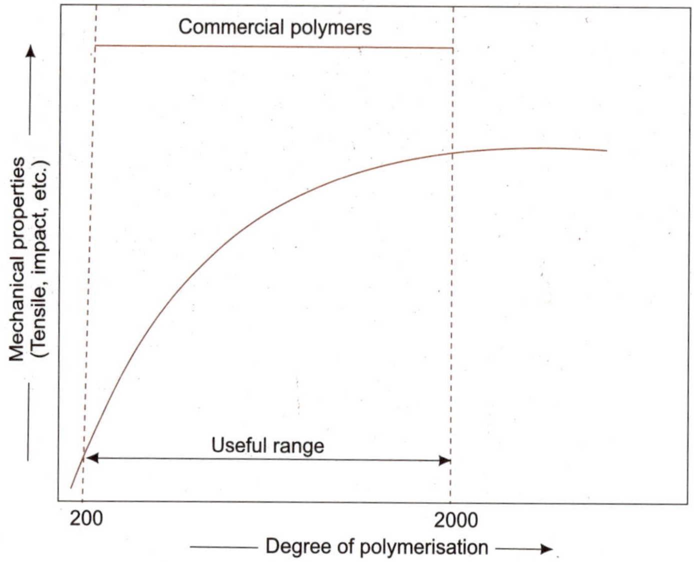

Fig. 3.6 Trend showing dependence of tensile or impact strength of a polymer on degree of polymerisation.

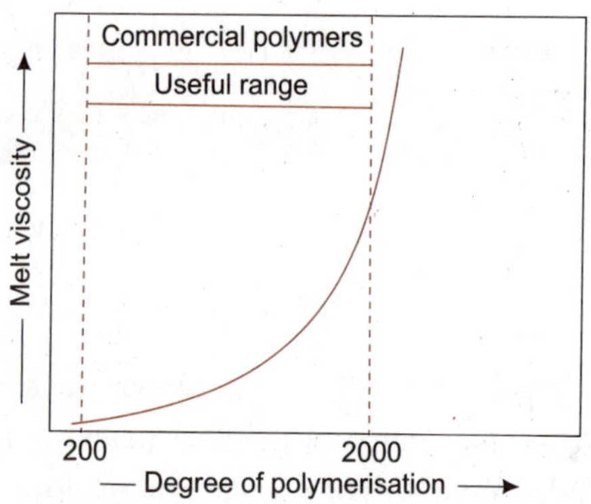

Fig. 3.7 Trend showing the dependence of melt viscosity of a polymer on degree of polymerisation.

To bring into sharper focus the effect of molecular weight on physical properties, a more generalised form of representation is shown in Fig. 3.8. Here, mechanical strength is plotted against DP. Qualitatively, the nature of the curve is valid for all polymers, differing from one polymer to another only with respect to the numerical value. Every polymer has a threshold value (TV) for its DP below which the polymer does not possess any strength and exists as friable powder or as liquid resin. The threshold DP value varies from one polymer to another. For instance, cellulose has a threshold DP value of 60, and a good number of vinyl polymers are found to have a value of around

100. Similarly, the knee portion of the curve (indicated as *X* in Fig. 3.8), at which DP value, the polymer attains more or less full strength, also varies from one polymer to another. For instance, this value is 400 for vinyl type polymers, 250 for cellulose and roughly 150 for polyamides. Based on a wide range of data collected, it has been established that polymers do not exhibit strength for DP values of lower than 30 and the optimum strength of most of the polymers is obtained at a DP value of around 600. The useful range of DP is from 200 to 2,000 which corresponds to molecular weights of roughly 20,000 to 200,000.

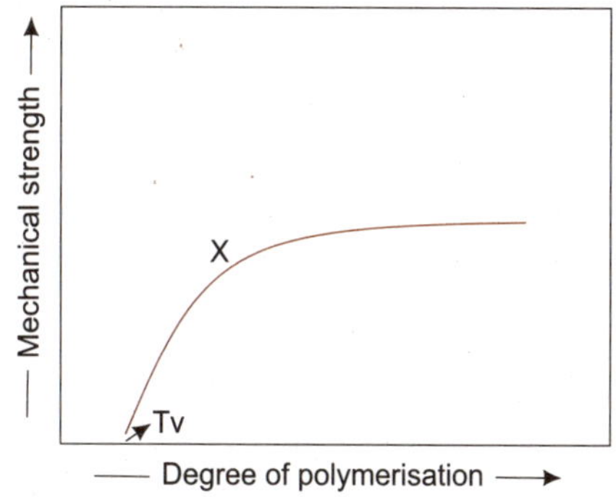

Fig. 3.8 General trend showing dependence of mechanical strength of commercial polymers on degree of polymerisation (for TV and X see text).

3.7 SIZE OF POLYMER MOLECULES

3.7.1 Freely Jointed Chain Model

For the sake of simplicity, let us imagine that a polymer molecule is made up of n segments, each having a length of l joined together freely end to end. Also, let us assume the joints at each segment are such that (say, as in the case of the ball-and-socket joint) the segments can freely rotate at these joints and that there is no restriction on the angle formed between any two segments. This simplified version of an imaginary polymer chain is called a 'freely jointed chain'. Figure 3.9 (A) shows a simple representation of a freely jointed chain.

Now, how are we to express the size of this chain? One way of expressing it would be by its total length which is equal to $n \times l$. This length is known as the 'contour length'. Another way of expressing the size of the chain would be by the distance between the two chain ends. Let us call this distance as chain end-to-end distance, and denote it by r. But how do we comprehend this distance r? Let us first assume that in a particular model of a freely jointed chain, the joints are not of the ball-and-socket type but of the hinge type. The result would be that the segment cannot undergo free rotation at each joint but can only have free movement in one plane without any restriction on the angle formed between them. All the segments of this chain will, therefore, be on the same plane and the shape of the whole chain will be coplanar. Because of the unrestricted closing or opening of any segment at its joint with respect to its neighbouring one, the chain as a whole can get folded or stretched to different extents. The two extreme cases will be (*i*) the completely folded state and

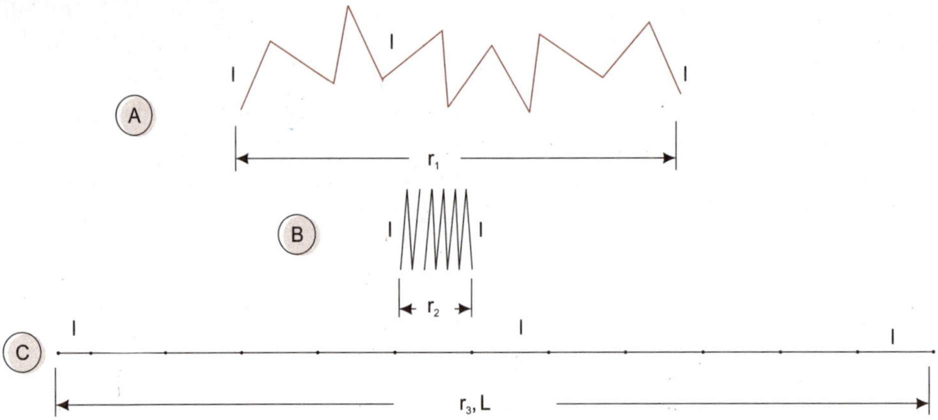

Fig. 3.9 Illustration of a freely jointed chain model: (A) simple representation, (B) completely folded form, and (C) fully extended form (l = length of each segment, L = extended chain length, r_1, r_2 and r_3 are chain end-to-end distances and $r_3 = L$).

(*ii*) the fully stretched or extended state as shown in Figs. 3.9 (B) and (C). The distance between the chain ends will be the shortest at its fully folded state, and the longest at its fully extended state. In the fully extended state, the chain will take the shape of a straight line and r will be equal to $l \times n$. This distance between the two chain ends, when the chain is fully extended, is known as the 'extended chain length'. Let us denote it by L. (Extended chain length should not be confused with contour length. In the case of the freely jointed chain model, these two lengths are the same. But for an actual polymer molecule, they will not be so. We shall come across this later in Section 3.7.2.) In between the completely folded and fully extended states, the chain can take up innumerable shapes. The majority of the shapes are, however, decided by what is known as the 'random-walk' probability.

According to the random-walk probability, the shape of the chain is comparable to the path covered in n steps by a drunken man walking randomly without any control over his direction, but with a fixed distance of l between each step. (A few of the random-walk shapes are shown in Fig. 3.10.) The chain end-to-end distance in these cases can be denoted by r_w, the subscript w indicating the random-walk mode. Statistically, it has been shown that for shapes assumed on random-walk mode, the values of r_w are much smaller than L.

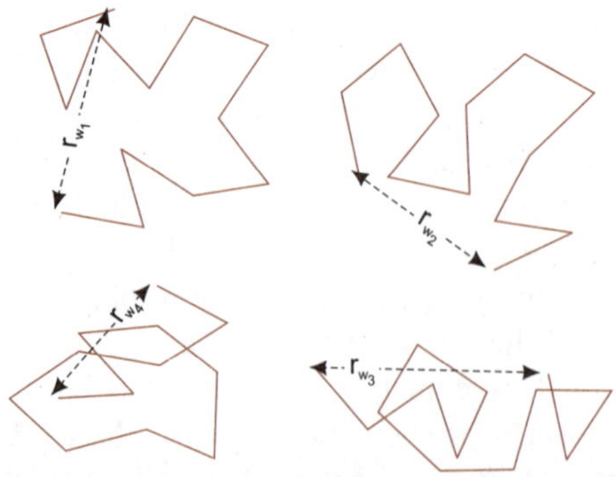

Fig. 3.10 A few of the innumerable shapes that a freely jointed chain model can assume according to random-walk probability. r_{w1}, r_{w2}, r_{w3} and r_{w4} are chain end-to-end distances.

Now, let us take the case of a freely jointed chain model where there is no restriction either on the free rotation at the joints or on the angle formed between any two segments. Because of the freedom both in respect of rotation and intersegment angle, the segments can occupy positions in any plane irrespective of the plane occupied by the neighbouring ones. Hence, the chain as a whole can take a three-dimensional shape. This chain can also assume innumerable shapes, ranging from a compactly wound ball made of the freely jointed chain (*i.e.*, the most coiled shape) to a fully stretched out long thread (*i.e.*, fully extended shape). The value of r in the fully stretched mode will again be equal to L. The majority of the shapes assumed by such a chain is given by what is known as 'random-flight' probability. According to this probability, the shape of the chain is comparable to the path traced in space by a free particle undergoing n number of successive displacements with the conditions that the distance of each displacement is the same (equal to l) and the direction of each displacement is at random and is independent of that of all the preceding ones. The distance between the chain ends for these shapes can be denoted by r_f, the subscript f indicating the random-flight mode. The magnitude of r_f is much smaller than L. The overall shapes of the chain attained in the random-flight mode are loosely wound spheres or ellipsoids and the distance r_f is usually used to indicate the end-to-end chain distance.

Let us now assume that a freely jointed chain changes its shape continuously. How do we comprehend its shape over a period of time? At this stage, we draw an analogy with men exercising in a gymnasium. Our objective is to study the various postures of an exercise. But this is difficult, as the movements are fast and rhythmic. As all the men are performing the same exercise, to identify the various poses, we have two options. One is that we take a series of photographs of one man, till he completes the cycle and comes back to the original pose. Another method is to take an instant photograph of several people in different postures of the same exercise. The two situations are depicted in Fig. 3.11.

Fig. 3.11 The two ways of arriving at the various postures adopted by an individual doing an exercise.

We have made the above mentioned analogy because of the similarity of situations when it comes to describing the shapes assumed by polymer chains at different instants of time. If we could take an instant photograph of several freely jointed chains, we get an idea of the various coiling patterns assumed by these chains. This imaginary photograph could reveal, say, n_1 chains with end-to-end distance r_{f1}, n_2 chains with end-to-end distance r_{f_2} and so on. We can then write an equation for an average end-to-end distance as

$$\bar{r}_f = \frac{n_1 r_{f1} + n_2 r_{f2} + \cdots + n_i r_{fi}}{n_1 + n_2 + \cdots + n_i} \tag{3.10}$$

To get a better representation of the average, however, a term $\langle \bar{r}_f^2 \rangle^{1/2}$, called the root-mean-square end-to-end distance, is used as given by the following expression:

$$\langle \bar{r}_f^2 \rangle^{1/2} = \left[\frac{n_1 r_{f1}^2 + n_2 r_{f2}^2 + \cdots + n_i r_{fi}^2}{n_1 + n_2 + \cdots + n_i} \right]^{1/2}$$

or

$$\langle \bar{r}_f^2 \rangle^{1/2} = \left[\frac{\Sigma n_i r_{fi}^2}{\Sigma n_i} \right]^{1/2} \tag{3.11}$$

According to the random-flight calculations, the root mean square end-to-end distance is related to the chain segments by the equation

$$\langle \bar{r}_f^2 \rangle^{1/2} = l(n)^{1/2} \tag{3.12}$$

where n is the number of segments in the polymer chain and l is the segment length.

3.7.2 Actual Polymer Molecules

Actual polymer molecules are often made of many segments linked together in a linear fashion. The difference between the freely jointed chain model and the actual polymer chains is that, in the latter case, the bond angles between the adjacent links are fixed and individual segments do not rotate that freely, owing to steric and other factors.

Assume a linear polymer molecule made of n number of segments of length l each and that the segments are made up completely of the C—C bonds. The fully extended form of such a chain is represented in Fig. 3.12. The segment length, l, for such a —C—C— chain is 1.54 Å and the bond angle is 109°28′. Assuming a value of 500 for n, we can show by calculation that the fully extended length of the chain is 629 Å and the contour length (*i.e.*, actual length of the chain following all the zig-zags) is 1.54 × 500 = 770 Å.

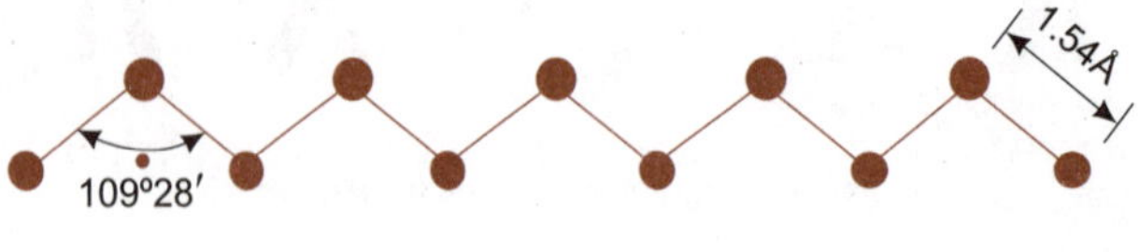

Fig. 3.12 Fully extended form of a polymer chain made of C—C backbone comprising ten C—C bonds.

A real polymer molecule can, and, in fact, does, assume different shapes over a period of time, governed again by the random-flight probability. Since the bond angle of the actual polymer molecule is fixed and there is hindrance to free rotation, the end-to-end distance of the polymer molecule will always be higher than that obtained for the random-flight model of the freely jointed chain of the same contour length. The dimension of a polymer molecule and that of the freely jointed chain (both containing the same number of segments or equal segment lengths) are related by the equation

$$\left\langle \bar{r}_0^2 \right\rangle^{1/2} = N\left\langle \bar{r}_f^2 \right\rangle^{1/2} \tag{3.13}$$

where $\left\langle \bar{r}_0^2 \right\rangle^{1/2}$ is the root mean square end-to-end distance of the polymer molecule and N is a number whose value is always more than unity. The value of N for polyethylene, for instance, is 2.6. Here, $\left\langle \bar{r}_0^2 \right\rangle^{1/2}$ is called the unperturbed dimension of the polymer molecule. The word 'unperturbed' is used to denote the fact that apart from the fixed bond angle and hindrance to rotation, there are no other attractive or repulsive forces acting upon the segments of the chain that can alter its shape.

Polymer molecules keep on changing their shape continuously and as a consequence, the end-to-end distance of the molecule also keeps on varying. The change in shape can be from the fully extended form at one end of the spectrum to a compactly coiled up form at the other. The fully extended form will have the highest value for $\left\langle \bar{r}_0^2 \right\rangle^{1/2}$, whereas the compactly coiled up form will have the least. According to the random-flight probability, neither the fully extended form nor the compactly coiled up form is the favoured shape, but randomly coiled forms are more common. Calculations have shown that the value of $\left\langle \bar{r}_0^2 \right\rangle^{1/2}$ for a polymer chain is quite small as compared to its extended length. The value for a chain of 500 —C—C— segments would be as low as 45 Å against its extended length of 629 Å.

It may be added that the values for the root mean square end-to-end distances assumed by a polymer molecule in solution, denoted by the term $\left\langle \bar{r}_f^2 \right\rangle^{1/2}$, are larger compared to its unperturbed dimension $\left\langle \bar{r}_0^2 \right\rangle^{1/2}$. This is due to the effect of solvent molecules exerting certain attractive or repulsive forces upon the segments of the polymer chain (*i.e.*, solvent-solute interaction), the details of which are given in Chapter 12 (Section 12.5).

Books for Further Reading

Volkenstein, M.V., *Configurational Statistics of Polymeric Chains*, trans. by Timasheff, S.N. and Timashaff, M.J., John Wiley & Sons, New York, 1963.

Boishtein, T.M. and Ptitsyn, O.B., *Conformations of Macromolecules*, Inter-science, New York, 1966.

Flory, P.J., *Statistical Mechanics of Chain Molecules*, Wiley-Interscience, New York 1969.

Bovey, F.A., *Polymer Conformation and Configuration*, Academic, New York, 1969.

Peebles, L.H., *Molecular Weight Distribution in Polymers*, Wiley-Interscience, New York, 1971.

4

KINETICS OF POLYMERISATION

4.1 INTRODUCTION

Having dealt with various polymerisation reactions in Chapter 2, the present chapter covers the kinetic aspects of some of those reactions. Kinetics has a bearing on the speed of a reaction under a given set of conditions and on how these conditions will influence, among other things, the polymer molecular weight, molecular weight distribution, chemical composition and indeed, the extent of polymer conversion. Let us look into some of these aspects.

4.2 FREE-RADICAL CHAIN POLYMERISATION

To understand the underlying kinetics of radical polymerisation, we have to recall the three steps mentioned earlier, *viz.*, initiation, propagation and termination, in a polymerisation reaction. To initiate the reaction, there must be free-radicals available, which are usually generated by the homolytic decomposition of an initiator as shown

$$\mathrm{I} \xrightarrow{k_d} \dot{\mathrm{R}} + \dot{\mathrm{R}} \tag{4.1}$$

Here I is the initiator, $\dot{\mathrm{R}}$ the free-radicals formed and k_d the initiator decomposition rate constant. The rate of decomposition (R_d) of the initiator is related to the concentration of the initiator (I) by the following equation:

$$\frac{-d[\mathrm{I}]}{dt} = R_d = 2k_d\,[\mathrm{I}] \tag{4.2}$$

A factor of 2 is used in Eqn. (4.2) as a pair of radicals is produced by decomposing one molecule of the initiator. From the Arrhenius equation, we know that the rate constant is related to temperature as follows:

$$k_d = A\,e^{-E/RT} \tag{4.3}$$

where A is a constant called the frequency factor, E the activation energy term, R the gas constant and T the temperature in the absolute scale. Substituting k_d from Eqn. (4.3). Equation (4.2) can be rewritten, as

$$R_d = 2A\, e^{-E/RT}[\text{I}] \tag{4.4}$$

This equation tells us that the rate of decomposition of an initiator increases with the temperature and concentration of the initiator.

The free-radicals generated by the decomposition of the initiator, as shown in Eqn. (4.1), can attack the monomer to give a new free-radical as shown

$$\dot{\text{R}} + \text{M} \xrightarrow{k_i} \text{R}\dot{\text{M}} \tag{4.5}$$

where M is the monomer and k_i is the initiation rate constant. The rate of initiation denoted by R_i may then be, written as

$$\frac{-d[\text{M}]}{dt} = R_i = k_i\,[\dot{\text{R}}][\text{M}] \tag{4.6}$$

The free-radicals have such a high reactivity that as soon as they are formed by the decomposition of the initiator, they attack the monomer and start the initiation process. We can assume, therefore, that the rate of formation of free-radicals is equal to the rate of the disappearance of the same according to Eq. (4.5). We can, therefore, visualise that R_i is equal to R_d. R_i can now be written as

$$R_i = 2k_d\,[\text{I}] \tag{4.7}$$

This equation holds good when all the free-radicals produced are effective in initiating the polymer growth. In actual practice, some of them are not effective since they are lost as side products by way of processes such as recombination. If we denote by f, the fraction of the free-radicals produced that is effective in initiating the chain growth, then R_i can be further modified as

$$R_i = 2fk_d\,[\text{I}] \tag{4.8}$$

The value of f, which is usually between 0.6 and 1.0, gives us the efficiency of the initiator. A value of 0.6 indicates that only 60% of the radicals generated have effectively initiated chain growth, the remaining 40% being lost as side products.

Initiation is followed by a propagation step which can be represented as:

$$\text{R}\dot{\text{M}} + \text{M} \xrightarrow{k_p} \text{RM}\dot{\text{M}} \tag{4.9a}$$

$$\text{RM}\dot{\text{M}} + \text{M} \xrightarrow{k_p} \text{RMM}\dot{\text{M}} \tag{4.9b}$$

$$\text{RMM}\dot{\text{M}} + n\text{M} \xrightarrow{k_p} \text{RM}_{n+2}\dot{\text{M}} \tag{4.9c}$$

where k_p is the propagation rate constant.

The rate of propagation (R_p) is represented by

$$\frac{-d[\text{M}]}{dt} = R_p = k_p[\dot{\text{M}}][\text{M}] \tag{4.10}$$

From Eqn. (4.10), we understand that the propagation rate is proportional to $[\dot{\text{M}}]$, the concentration of reactive free-radical sites (*i.e.*, concentration of growing chains) as well as [M], the concentration of the monomer.

The final step in polymerisation is termination. If termination occurs by the combination of two growing chains (i.e. coupling), it can be represented as

$$\mathrm{RM}_{n1}\dot{\mathrm{M}} + \mathrm{RM}_{n2}\dot{\mathrm{M}} \xrightarrow{k_t} \mathrm{RM}_{n1}\mathrm{MM}\dot{\mathrm{M}}_{n2}\mathrm{R} \text{ or } \mathrm{RM}_{n3}\mathrm{R} \quad (4.11)$$

where k_t is the termination rate constant.

The corresponding rate equation for the termination process is given by

$$\frac{-d[\dot{\mathrm{M}}]}{dt} = R_t = 2k_t[\dot{\mathrm{M}}]^2 \quad (4.12)$$

However, when the steady state prevails (*i.e.*, when the number of chain growths initiated equals the number of chain growths arrested), the rate of initiation is equal to the rate of termination, *i.e.*,

$$R_i = R_t \quad (4.13)$$

From Eqns. (4.8) and (4.12), we can write

$$2k_d f[\mathrm{I}] = 2k_t[\dot{\mathrm{M}}]^2$$

$$[\dot{\mathrm{M}}]^2 = \frac{k_d f[\mathrm{I}]}{k_t}$$

$$[\dot{\mathrm{M}}] = \left(\frac{k_d f[\mathrm{I}]}{k_t}\right)^{1/2} \quad (4.14)$$

From Eqn. (4.10), we get the value or [M] as

$$[\dot{\mathrm{M}}] = \frac{R_p}{k_p[\mathrm{M}]} \quad (4.15)$$

On substituting this value of $[\dot{\mathrm{M}}]$ in Eqn. (4.14), we get

$$\frac{R_p}{k_p[\mathrm{M}]} = \left(\frac{k_d f[\mathrm{I}]}{k_t}\right)^{1/2}$$

or

$$R_p = k_p \frac{k_d^{1/2}}{k_t^{1/2}} (f[\mathrm{I}])^{1/2}[\mathrm{M}] \quad (4.16)$$

This equation states that the rate of polymer formation is proportional to the first power of the monomer concentration and also to the square-root of the initiator concentration.

4.2.1 Equation for Kinetic Chain Length

The kinetic chain length (ν) can be defined as the average number of monomer molecules consumed by each effective free-radical generated by the initiator. It, thus, expresses the number of monomer molecules added to a growing chain from its initiation up to termination. The kinetic chain length can be expressed as

$$\nu = \frac{\text{rate of propagation}}{\text{rate of initiation}} = \frac{\text{rate of propagation}}{\text{rate of termination}}$$

or $$\nu = \frac{R_p}{R_i} = \frac{R_p}{R_t} \tag{4.17}$$

Substituting the values of R_p and R_i from Eqns. (4.16) and (4.8) in Eqn. (4.17), we get

$$\left[\nu = k_p \frac{k_d^{1/2}}{k_t^{1/2}} (f[\text{I}])^{1/2}[\text{M}] \frac{1}{2k_d f[\text{I}]} = \frac{k_p[M]}{2(k_d k_t f[\text{I}])^{1/2}} \right] \tag{4.18}$$

4.2.2 Degree of Polymerisation

As the kinetic chain length gives the average number of monomer molecules present in a growing chain at the time of termination, the average degree of polymerisation, $\overline{D}_p$, can be correlated as follows:

(*i*) $\overline{D}_p = 2\nu$, when termination occurs by coupling;

(*ii*) $\overline{D}_p = \nu$, when termination occurs by disproportionation; and

(*iii*) $\overline{D}_p = \nu/N$ where N has a value of 0.5 for termination by coupling and 1 for termination by disproportionation.

From an overall study of the kinetics, we venture to provide the following generalised statements regarding radical polymerisation:

(*a*) The degree of polymerisation decreases with an increase of temperature and of initiator concentration and with a decrease of monomer concentration.

(*b*) The extent of conversion (*i.e.*, the amount of monomer converted into polymer) increases with an increase of (*i*) temperature, (*ii*) polymerisation time and (*iii*) initiator and monomer concentrations.

The foregoing generalisation is graphically depicted in Fig. 4.1.

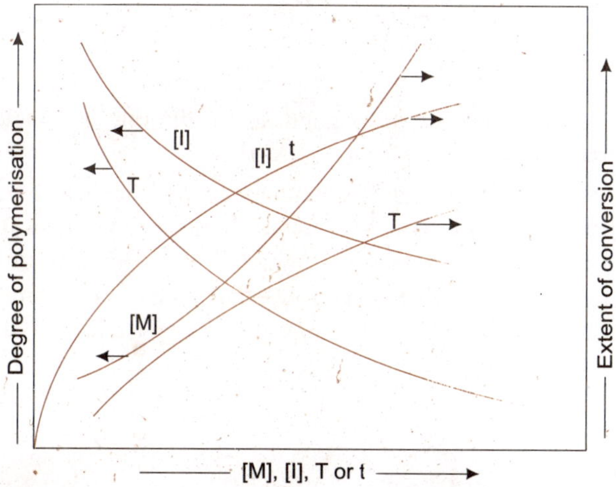

Fig. 4.1 Dependence of degree of polymerisation and extent of conversion on: (i) monomer concentration, [*M*]; (ii) initiator concentration, [*I*]: (iii) temperature of polymerisation, *T*: or (iv) polymerisation time, *t*–in radical polymerisation.

4.2.3 Chain Transfer Reactions

We have seen that the termination of a growing polymer chain can take place also through chain transfer. The chain transfer reaction could be represented as

$$\sim\sim\sim M^{\bullet} + RH \xrightarrow{k_{tr}} \sim\sim\sim MH + R^{\bullet} \quad (4.19)$$

where k_{tr} is the chain transfer rate constant and RH the chain transfer agent. Equation 4.19 informs us that in the chain transfer process, the total number of active centres will not change before and after the reaction even though a growing chain is terminated. Many initiators, such as hydroperoxides, can bring about the chain transfer reaction, as shown

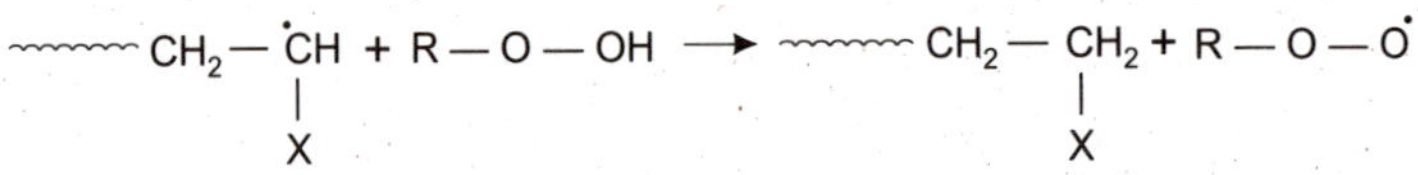

All initiators, however, do not encourage the chain transfer reaction. (Azo bis isobutyronitrile is a case in point).

Generally, in free-radical polymerisation, the initiator concentration is very low and the growing chain interaction with the initiator may also be quite small. The chain transfer to the initiator is, therefore, not of great significance. The monomer concentration, on the other hand, is considerably high, especially in the initial stages of polymerisation and hence, chain transfer to the monomer molecules is more probable. But, usually, hydrogen or other atoms in many monomer molecules are not that easily removable and hence, in a good majority of cases, the chain transfer to monomers is not very significant despite their high concentrations.

Of greater significance are chain transfer reactions with the solvents. A term, *i.e.*, chain transfer constant, is used to assess the ability of different solvents (and for that matter, of any substance) to act as chain transfer agents, and is defined as

$$C = \frac{k_{tr}}{k_p} \quad (4.20)$$

where C is the chain transfer constant and k_{tr} and k_p are the rate constants for the transfer and propagation reactions, respectively.

The chain transfer constant depends on the temperature and the nature of the growing polymer radical which brings about the transfer reaction. As will be seen from the examples in Table 4.1, the chain transfer constant is very high for carbon tetrachloride, which, in turn, is attributed to the presence of the easily removable chlorine atom in the molecule. Thus, a growing polymer chain of styrene abstracts a chlorine atom from CCl_4 in a transfer reaction as follows; the radical CCl_3 formed can attack a monomer and start a new chain.

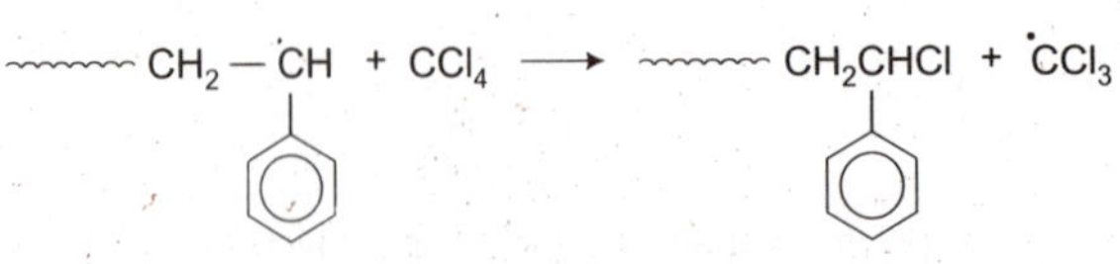

Table 4.1 Chain transfer constants for different solvent/radical systems*

Solvent	*Transfer constant $C_s \times 10^5$*				
		$\sim CH_2-\dot{C}H(C_6H_5)$	$\sim CH_2-\dot{C}(CH_3)(COOCH_3)$		$\sim CH_2-\dot{C}H(OCOCH_3)$
	60°C	80°C	100°C	80°C	60°C
Benzene	0.18	0.61	1.84	0.75	29.6
Cyclohexane	0.24	0.66	1.6	1.0	65.9
Toluene	1.25	2.98	6.45	5.25	208.9
Ethylbenzene	6.7	10.7	16.2	13.5	551.5
Cumene	8.2	13.1	20.0	19.0	899.0
Triphenylmethane	35.0		80		
Butylchloride	0.4		3.7		
Dichloromethane (methylene chloride)	1.5		118		
Carbon tetrachloride	920	1330	1850	23.9	

*Data from '*Polymer Handbook*', Eds. Brandrup J. and Immergut E.H., John Wiley & Sons, New York,1975.

The chain transfer with CCl_4 is confirmed by the fact that polystyrene obtained by polymerising styrene in CCl_4 invariably contains chlorine at the chain terminal. It should be noted, however, that a solvent which acts as a very good chain transfer agent for one particular monomeric radical may not be so for another. CCl_4, for instance, is not that effective in butadiene polymerisation.

At times, especially in industrial polymerisation, specific chain transfer agents are intentionally added while carrying out the polymerisation reaction in order to control the molecular weight of the polymer. A chain transfer agent used for the purpose of controlling molecular weight is called a 'chain modifier'. Diisopropyl xanthate disulfide is one of the modifiers used in the industrial production of styrene-butadiene rubber. The chain transfer mechanism, with this modifier, is as follows:

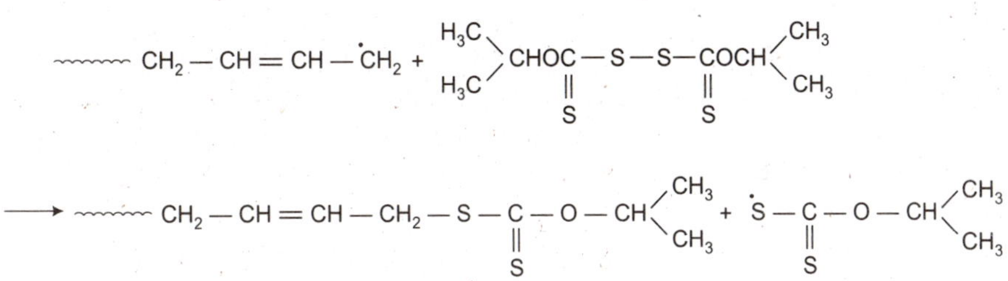

The modifier thus prevents the uncontrolled growth of a polymer chain. It terminates the growing chains and promotes fresh chains in such a way that the average molecular weight of the polymer remains within a close range. Mercaptans are also very commonly employed in industry as chain modifiers, and their interaction with a growing chain may be represented as follows:

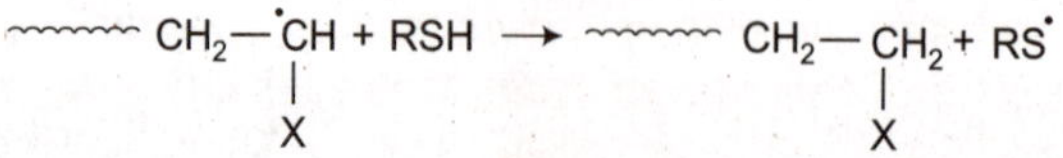

Dodecyl mercaptan is one of the most widely used modifiers in industry.

Solvents with very high chain transfer ability are used for producing low molecular weight liquid polymers. A solvent having a chemical structure similar to that of the monomer is selected so that the terminal end groups formed by the solvent molecules do not alter the chemical structure of the polymer formed. This method of polymerisation, utilising a solvent of a high chain transfer ability and having a chemical structure similar to that of the monomer to form liquid polymers, is known as 'telomerisation'. The solvent used in such cases is known as the 'telogen'. Polymerisation of bromotrifluoro ethylene, to obtain liquid polymers, is carried out by this technique using bromotrifluoro methane as the telogen:

(i) $\sim\sim CF_2-\dot{C}FBr + CF_3Br \longrightarrow \sim\sim CF_2CFBr_2 + \dot{C}F_3$

(ii) $\dot{C}F_3 + n\,CF_2{=}CFBr \longrightarrow F_3C \sim\sim CF_2-\dot{C}FBr$

(iii) $CF_3 \sim\sim CF_2-\dot{C}FBr + CF_3Br \longrightarrow CF_3 \sim\sim CF_2CFBr_2 + \dot{C}F_3$ and so on.

Chain transfer to an already formed dead polymer is also possible. As the number of dead polymer molecules in the system increases the chances of the chain transfer reaction taking place through the polymer also increase. When chain transfer occurs in this way, only branching of the polymer chain takes place and no new polymer molecule is produced:

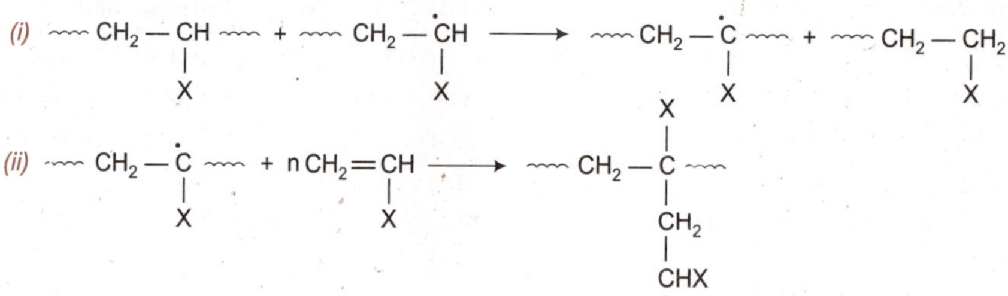

(Chain branching)

While deriving Eqn. (4.18) for the kinetic chain length, we tacitly assume that the polymerisation proceeds without a chain transfer reaction. A more realistic equation for the kinetic chain length may be written by incorporating the chain transfer reaction also into the kinetic scheme of free-radical polymerisation:

$$\sim\sim\dot{M} + ZX \xrightarrow{k_{tr}} \sim\sim MX + \dot{Z} \tag{4.21}$$

$$\dot{Z} + M \xrightarrow{k_i} Z\dot{M} \tag{4.22}$$

where ZX is a chain transfer agent.

Now, $\nu = R_p/R_t$. Since the chain transfer involves the termination of a growing chain, its degree of polymerisation (denoted by D_p) will be equal to its kinetic chain length at the time of transfer.

Therefore,

$$\overline{D}_P = \nu = \frac{R_p}{R_t} = \frac{R_p}{R_{tr}} = \frac{k_p[\text{M}][\text{M}^\cdot]}{k_{tr}[\text{ZX}][\text{M}^\cdot]} \tag{4.23}$$

or

$$\frac{1}{\overline{D}_p} = \frac{k_{tr}[\text{ZX}]}{k_p[\text{M}]} \tag{4.24}$$

Combining Eqns. (4.20) and (4.24), we get

$$\frac{1}{\overline{D}_p} = C_{zx}\frac{[\text{ZX}]}{[\text{M}]} \tag{4.25}$$

where C_{zx} is the transfer constant of the compound ZX. When polymerisation takes place in the presence of a solvent (S), initiator (I), monomer (M), and the chain modifier '(ZX), and if chain termination takes place by the chain transfer reaction to solvent, monomer and initiator, in addition to coupling as well as disproportionation, Eqn. (4.25) will assume the following form:

$$\frac{1}{\overline{D}_p} = C_{zx}\frac{[\text{ZX}]}{[\text{M}]} + C_\text{S}\frac{[\text{S}]}{[\text{M}]} + C_\text{I}\frac{[\text{I}]}{[\text{M}]} + C_\text{M}\frac{[\text{M}]}{[\text{M}]} + N\frac{R_i}{R_p} \tag{4.26}$$

The last term in Eqn. (4.26) governs termination by coupling and/or by disproportionation. C_S, C_I and C_M are the chain transfer constants for the solvent, the initiator and the monomer, respectively.

4.2.4 The Mayo Equation and Evaluation of the Chain Transfer Constant

The chain transfer constant of a solvent can be evaluated by carrying out the polymerisation under appropriate conditions so that transfers to the initiator, monomer and polymer are essentially avoided and transfer only to the solvent is predominant. Under such conditions, Eqn. (4.26) will assume a simpler form:

$$\frac{1}{\overline{D}_p} = \frac{1}{\overline{D}_{p0}} + C_\text{S}\frac{[\text{S}]}{[\text{M}]} \tag{4.27}$$

Here $\overline{D}_{p0}$ and $\overline{D}_p$ represent the degree of polymerisation in the absence of any chain transfer and in the presence of the same, respectively. Naturally, $\overline{D}_{p0}$ will always be greater than $\overline{D}_p$. This simple expression, known as the Mayo equation, is widely used for the determination of the chain transfer constant for solvents. The synthesis of polystyrene by the thermal polymerisation of styrene in various solvents offers a good example. Figure 4.2 shows plot of $1/D_p$ versus [S]/[M] in the case of styrene for different solvents at the reaction temperature of 100°C. The ordinate intercept of the lines gives $1/D_{p0}$ and the chain transfer constant C_S for a particular solvent is obtained from the slope of the corresponding straight line.

CHAPTER 4

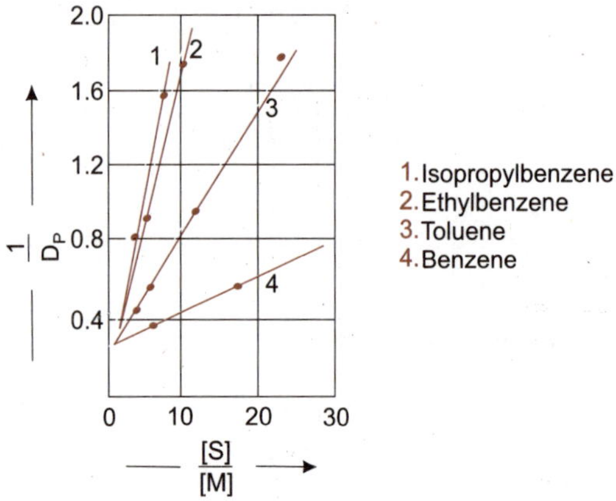

Fig. 4.2 Dependence of degree of polymerisation on solvent-to-monomer molar ratio in free-radical polymerisation of styrene in different solvents at 100°C.

4.2.5 Ceiling Temperature

We know that many chemical reactions are reversible. The propagation reaction, too, can be considered as reversible and is written as

$$RM + M \underset{k_{dp}}{\overset{k_p}{\rightleftharpoons}} RMM \qquad (4.28)$$

The reverse reaction in Eqn. (4.28) is called 'depropagation'. At the normal polymerisation temperature, the rate of depropagation is quite low, compared to the rate of propagation and hence, polymerisation proceeds. As the temperature is increased, the rates of both forward as well as backward reactions increase. However, the increase in the rate of the backward reaction becomes so much larger than that of the forward reaction, that, at a particular temperature, the two rates become equal, and as much of polymer formed depropagates back to the monomer; hence, practically no polymerisation occurs. This temperature, at which the rates of propagation as well as depropagation are equal, is called the 'ceiling temperature'. It should be noted that, polymerisation is possible only below the ceiling temperature. At the ceiling temperature, the monomer and polymer exist in equilibrium as given by

$$nM \circ M_n$$

4.3 CATIONIC POLYMERISATION

The general scheme for cationic polymerisation can be written as follows:

$$A + BH \rightleftharpoons \overset{\oplus}{H}\overset{\ominus}{BA} \qquad \text{(ion-pair formation)} \qquad (4.29)$$

$$\overset{\oplus}{H}\ \overset{\ominus}{BA} + M \xrightarrow{k_i} \overset{\oplus}{HM}\overset{\ominus}{BA} \qquad \text{(initiation)} \qquad (4.30)$$

$$H\overset{\oplus}{M}\overset{\ominus}{BA} + M \xrightarrow{k_p} HM\overset{\oplus}{M}\overset{\ominus}{BA} \qquad \text{(propagation)} \qquad (4.31)$$

$$HM_n\overset{\oplus}{M}\overset{\ominus}{BA} \xrightarrow{k_t} HM_nM + \overset{\oplus}{H}\overset{\ominus}{BA} \qquad \text{(termination)} \qquad (4.32)$$

$$HM_n\overset{\oplus}{M}\overset{\ominus}{BA} \xrightarrow{k_{tr}} HM_nM + \overset{\oplus}{M}\overset{\ominus}{BA} \qquad \text{(chain transfer to monomer)} \qquad (4.33)$$

In this scheme, A is the catalyst, BH the co-catalyst and M the monomer.

The rate equations for the initiation and termination steps are:

$$R_i = k_i[M]\,[\overset{\oplus}{H}] \qquad (4.34)$$

$$R_t = k_t\,[\overset{\oplus}{M}] \qquad (4.35)$$

At this point, it is interesting to note that unlike radical polymerisation [see Eqn. (4.12)], termination follows the first-order kinetics.

On applying the steady-state approximation:

$$k_i[M]\,[\overset{\oplus}{H}] = k_t[\overset{\oplus}{M}]$$

$$[\overset{\oplus}{M}] = \frac{k_i}{k_t}[M]\,[\overset{\oplus}{H}] \qquad (4.36)$$

and the propagation rate may be written as

$$R_p = \frac{-\,d[M]}{dt} = k_p[M]\,[\overset{\oplus}{M}] = \frac{k_pk_i}{k_t}[M]^2\,[\overset{\oplus}{H}] \qquad (4.37)$$

The overall polymerisation rate, unlike in the case of radical polymerisation [see Eqn. (4.16)], is second order with respect to the monomer concentration. Thus, in the propagation rate equation, the term $[M]^2$ appears rather than [M].

The degree of polymerisation (which is the same as the kinetic chain length in case of cationic polymerisation) may be expressed by two equations

$$\bar{D}_P = \frac{R_P}{R_t} = \frac{K_P[M]\,[\overset{\oplus}{M}]}{K_t\,[\overset{\oplus}{M}]} = \frac{k_P}{k_t}[M] \qquad (4.38)$$

$$\bar{D}_P = \frac{R_P}{R_{tr}} = \frac{K_P[M]\,[\overset{\oplus}{M}]}{K_{tr}[M]\,[\overset{\oplus}{M}]} = \frac{k_P}{k_{tr}} \qquad (4.39)$$

depending on whether simple termination [Eqn. (4.38)] or chain transfer occurs [Eqn. (4.39)].

If both termination as well as chain transfer occur, then, we have

$$\bar{D}_P = \frac{R_P}{R_t + R_{tr}} = \frac{K_p[M]\,[\overset{\oplus}{M}]}{K_t[\overset{\oplus}{M}] + K_{tr}[M]\,[\overset{\oplus}{M}]} = \frac{K_p[M]}{K_t + K_{tr}[M]} \tag{4.40}$$

Equations (4.38), (4.39) and (4.40) indicate that the degree of polymerisation is independent of initiator concentration, and when chain transfer to monomer predominates, the degree of polymerisation is independent of monomer concentration also. This degree, however, depends on the temperature at which the reaction is carried out. Thus, for instance, the molecular weight decreases as the temperature increases, as shown in the case of isobutylene polymerisation (Table 4.2).

Table 4.2 Effect of polymerisation temperature on molecular weight of polyisobutylene*

Temperature (°C)	Viscosity-average molecular weigh
−10	10,000
−25	13,000
−45	25,000
−80	74,000
−90	120,000

*Data (unpublished) from Chemicals Group, Vikram Sarabhai Space Centre, Trivandrum.

This trend is explained on the basis of the fact that the activation energy for propagation is less than the activation energy for chain 'killing' (by termination or transfer). Consequently, as the temperature of polymerisation increases, more chain 'killing' will take place and, hence, the molecular weight of the polymer will decrease.

The nature of the solvent medium has a telling effect on the rate of the polymerisation reaction. As the polarity of the medium increases, the rate of polymerisation also increases. This is confirmed by experimental results, as in the case of methyl styrene polymerisation (see Table 4.3). It is quite evident from these data that the dielectric constant and, hence, the polarity of the solvent affect the rate of polymerisation to a large extent. This is explained on the basis of the fact that an increase in the value of the dielectric constant of the solvent medium would favour separation of ions and hence, enhance the rate of initiation and retard the rate of termination. Also, a decrease in the rate of termination will increase the degree of polymerisation.

Table 4.3 Effect of solvent polarity on the rate of cationic polymerisation of α-methyl styrene*

Solvent	Dielectric constant	Polymerisationc ratec (mole/minute)
Cyclohexane	1.9	1.25
Dichloroethane	10.0	3.30
Nitroethane	28.0	20.40
Nitrobenzene	36.0	150.00

*Data from *A First Course in Polymer Chemistry*, trans, by David Sobolev, Mir Publishers, Moscow, 1971.

4.4 ANIONIC POLYMERISATION

A generalised scheme for the anionic polymerisation may be written as:

$$\mathrm{IA} \xrightarrow{k_d} \overset{\oplus}{\mathrm{I}}\,\overset{\ominus}{\mathrm{A}} \qquad \text{(ion-pair formation)} \qquad (4.41)$$

$$\overset{\ominus}{\mathrm{A}}\,\overset{\oplus}{\mathrm{I}} + \mathrm{M} \xrightarrow{k_i} \mathrm{A}\overset{\ominus}{\mathrm{M}}\;\overset{\oplus}{\mathrm{I}} \qquad \text{(initiation)} \qquad (4.42)$$

$$\mathrm{A}\overset{\ominus}{\mathrm{M}}\,\overset{\oplus}{\mathrm{I}} + \mathrm{M} \xrightarrow{k_p} \mathrm{AM}\overset{\ominus}{\mathrm{M}}\;\overset{\oplus}{\mathrm{I}} \qquad \text{(propagation)} \qquad (4.43)$$

$$\mathrm{AM}_n\overset{\ominus}{\mathrm{M}}\;\overset{\oplus}{\mathrm{I}} + \mathrm{HA} \xrightarrow{k_t} \mathrm{AM}_n\mathrm{MH} + \overset{\ominus}{\mathrm{A}}\,\overset{\oplus}{\mathrm{I}} \qquad \text{(termination)} \qquad (4.44)$$

where IA is the initiator, HA is a protonating agent and k_d, k_i, k_p and k_t are the dissociation, initiation, propagation and termination rate constants, respectively.

The propagation rate, as per the foregoing scheme, will be given by

$$R_p = \frac{k_i k_p}{k_t} \frac{[\overset{\ominus}{\mathrm{A}}]\,[\mathrm{M}]^2}{[\mathrm{HA}]} \qquad (4.45)$$

The degree of polymerisation is governed by the following equation:

$$\bar{D}_p = \frac{K_p}{K_t} \frac{[\mathrm{M}]}{[\mathrm{HA}]} \qquad (4.46a)$$

Equation (4.46a) is applicable for anionic polymerisation systems involving termination, as shown in the generalised scheme. For systems in which no termination is involved, as in the case of 'living polymers', an expression for the degree of polymerisation can be found on the basis of the following assumptions:

(*i*) the entire quantity of the initiator dissociates at the same time;

(*ii*) all the anions formed by the dissociation initiate chain growth essentially simultaneously; and

(*iii*) all the anions compete equally for the monomer and the monomer is consumed more or less equally by all the anions.

Further, the expression for the degree of polymerisation depends on whether the initiating species is a monoanion (*e.g.*, butyl lithium) or a dianion (*e.g.*, sodium complex of naphthalene). If it is a monoanion (where the chain growth takes place at one end only as shown in Fig. 4.3), the expression for D_p takes the form

$$\bar{D}_p = \frac{[\mathrm{M}]}{[\mathrm{IA}]} \qquad (4.46b)$$

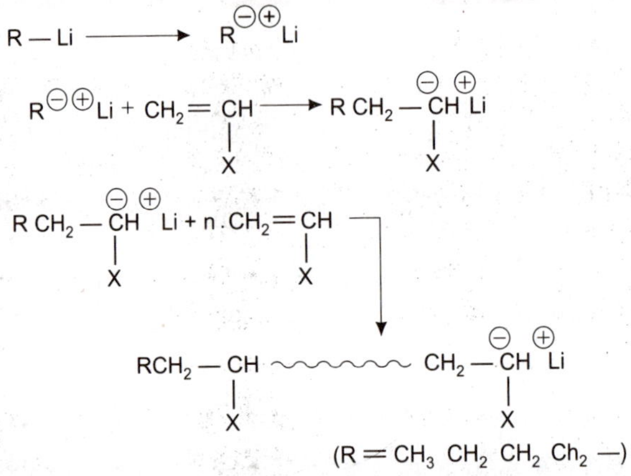

Fig. 4.3 Anionic chain growth by a monoanion. The initiator here is butyl lithium.

If, on the other hand, it is a dianion (where the chain growth takes place at both the ends as shown in Fig. 4.4), the D_p will be governed by the equation

$$\overline{D}_p = 2\frac{[M]}{[IA]} \tag{4.46c}$$

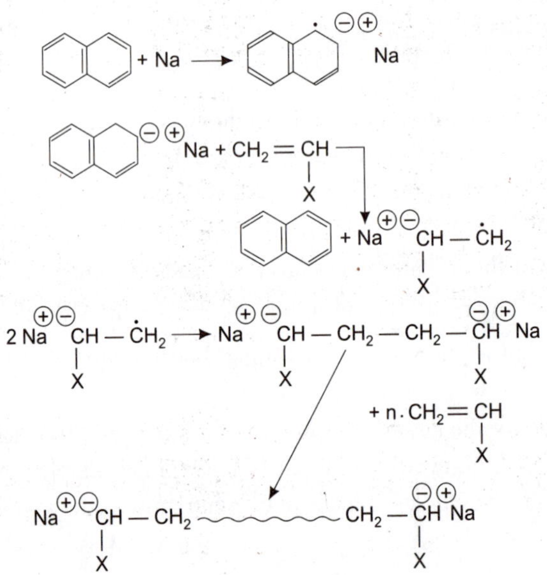

Fig. 4.4 Anionic chain growth by a dianion. The initiator here is sodium complex of naphthalene.

4.5 POLYCONDENSATION

In so far as the kinetics of condensation reaction is concerned, it is relatively simple, once we introduce the assumptions proposed by Flory. According to Flory, so long as the molecules are not quite small, the intrinsic reactivity of the functional groups present in them are independent of the molecular size. Similarly, the effect of viscosity of the reaction medium and factors such as diffusion rate contribute negligibly to the overall kinetics of a condensation reaction. With these assumptions, we shall see the kinetic aspects of polycondensation reactions of the type:

$$n\,A—A + n\,B—B \longrightarrow A\!\left(AB\right)_{2n-1}\!B + \text{by-product}$$

4.5.1 Non-catalysed Polycondensation

To derive a mathematical expression, let us take the synthesis of a polyester from a dicarboxylic acid and a diol. This type of esterification reaction is usually catalysed by acids. When, however, no strong acid is added to catalyse this reaction, the dicarboxylic acid has a dual function, *i.e.*, as a reactant and as a catalyst. Hence, the polyesterification reaction rate will be proportional to the square of the concentration of the carboxylic acid group and to the first power of the hydroxyl group concentration. Mathematically, this means

$$\frac{-d[\mathrm{OH}]}{dt} = k[\mathrm{COOH}]^2[\mathrm{OH}] \qquad (4.47)$$

where k is the esterification rate constant.

When the concentration of —OH and —COOH groups are equal, *i.e.*, when we take stoichiometric quantities of the functional groups (FG), we can represent that at a given time t:

$$[\mathrm{COOH}] = [\mathrm{OH}] = [\mathrm{FG}]$$

Then Eqn. (4.47) can be rewritten as:

$$\frac{-d[\mathrm{FG}]}{dt} = k[\mathrm{FG}]^2[\mathrm{FG}] = k[\mathrm{FG}]^3 \qquad (4.48)$$

Integrating Eqn. (4.48) and rearranging the terms, we get

$$2\,kt = \frac{1}{[\mathrm{FG}]^2} - \frac{1}{[\mathrm{FG}]_0^2} \qquad (4.49)$$

where $[\mathrm{FG}]_0$ represents the initial concentration of —COOH or —OH. If p represents the fraction of the functional groups which has undergone the esterification reaction (*i.e.*, the extent of reaction) at a time t, then [FG] could be related to $[\mathrm{FG}]_0$ through the following equations:

$$\frac{[\mathrm{FG}]_0 - [\mathrm{FG}]}{[\mathrm{FG}]_0} = p \qquad (4.50a)$$

$$\frac{[\mathrm{FG}]}{[\mathrm{FG}]_0} = (1 - p) \qquad (4.50b)$$

and $$[FG] = [FG]_0(1 - p) \tag{4.51}$$

On substituting this value of [FG] in Eqn. (4.49), we get

$$2kt = \frac{1}{[FG]_0^2(1-p)^2} - \frac{1}{[FG]_0^2}$$

or $$2k[FG]_0^2 t = \frac{1}{(1-p)^2} - 1 \tag{4.52}$$

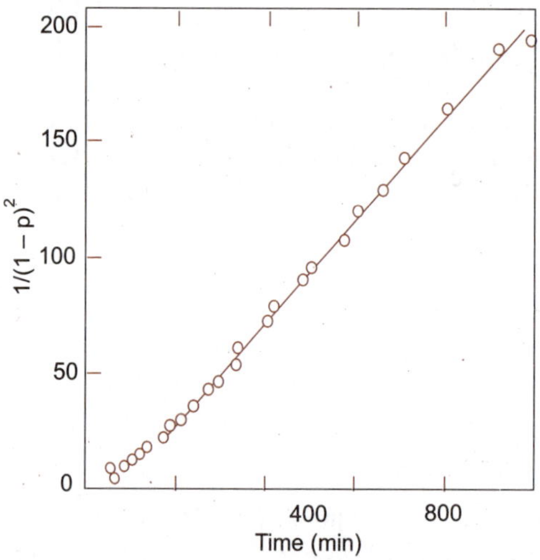

Fig. 4.5 Plot showing reaction kinetics for condensation of adipic acid with diethylene glycol (in absence of any added catalyst) at 166°C.

This implies that a graphical plot of $1/(1 - p)^2$ versus t must be linear. Experimentally, this has been found to be the case; Fig. 4.5 represents a plot for the reaction between adipic acid and diethylene glycol at 166°C.

For a simple esterification reaction, which proceeds without any side reaction, the number-average degree of polymerisation $\overline{D}_{pn}$ may be related to p, the fraction of the functional groups which has undergone the esterification at a time t, by the following equation:

$$\overline{D}_{pn} = \frac{1}{(1-p)} = \frac{[FG]_0}{[FG]} \tag{4.53}$$

From Eqns. (4.51), (4.52) and (4.53), it can be seen that

$$\overline{D}_{pn}^{\,2} = 2k[FG]_0^2 t + 1 \tag{4.54}$$

In Eqn. (4.54), the $\overline{D}_{pn}$ appears as the second power and the time as the first power and, hence, the polymer molecular weight will increase only slowly.

4.5.2 Acid-catalysed Polycondensation

We have said earlier that the condensation reactions are enhanced by the presence of strong acids. The acid-catalysed polycondensation reaction becomes governed by the following equation:

$$\frac{-d[\mathrm{FG}]}{dt} = k_x[\mathrm{COOH}][\mathrm{OH}] \tag{4.55}$$

where $k_x = k[X]$; $[X]$ being the catalyst concentration. It must be noted that throughout the reaction the catalyst concentration remains the same. Since the whole pattern of the reaction remains the same, an equation similar to Eqn. (4.52) could be derived for the acid-catalysed polycondensation as:

$$k_x[\mathrm{FG}]_0 t = \frac{1}{(1-p)} - 1 \tag{4.56}$$

An actual experimental plot of $1/(1 - p)$ versus time, in the case of adipic acid-diethylene glycol system catalysed by 0.4 mole per cent of toluene sulphonic acid, gives a straight line as in Fig. 4.6. Similarly, we can also derive an equation

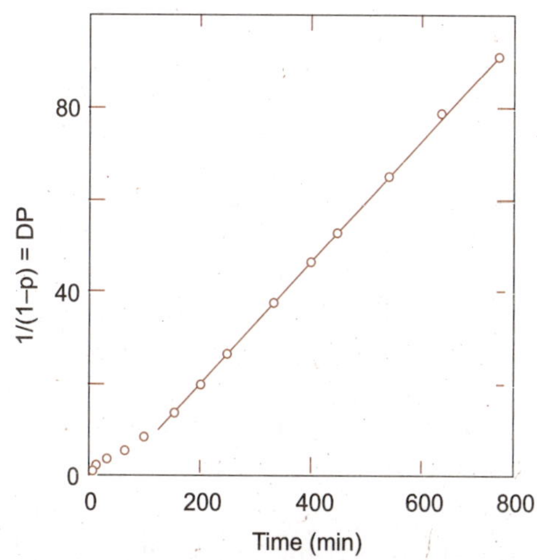

Fig. 4.6 Plot showing reaction kinetics for the condensation of adipic acid with diethylene glycol (catalysed by 0.4 mole % of toluene sulphonic acid) at 109°C.

$$\overline{D}_{pn} = k_x[\mathrm{FG}]_0 t + 1 \tag{4.57}$$

indicating that the molecular weight built up in this case will be faster than in the polycondensation reaction without acid catalyst.

4.5.3 Molecular Weight Distribution

Let us recall our assumption that equimolar quantities of reactants AA and BB react and that the reactivity of groups A or B remains the same at all stages of conversion. We know that the degree of polymerisation varies with the extent of polymerisation that has taken place. Now, let us see how the probability factor can give an idea about the molecular weight distribution at a particular conversion p. Here, p is the extent of reaction, which also gives the probability of reaction occurring between two functional groups to give a repeat unit of —AB— as per the reaction

$$n\text{A–A} + n\text{B–B} \longrightarrow \text{A–[–AB–]}_{2n+1}\text{B} + \text{by-product}$$

The fraction of unreacted groups is $1 - p$. Now, assume that a tetramer is formed as a result of the reaction between 4 reactant molecules, 2 of A—A and 2 of B—B, wherein 3 pairs of reactive groups have reacted. The probability of a tetramer formation is, therefore, p^3. For that matter, the probability of occurrence of an n-mer, *i.e.*, a polymer molecule containing n number of repeat units, will be p^{n-1}. Assuming that at a conversion of p, the total number of polymer molecules formed is N and that of the n-mers formed is N_n, then N and N_n can be related as follows:

$$N_n = Np^{n-1}(1 - p) \tag{4.58}$$

If N_0 is the number of reactant molecules present to start with, then

$$N = N_0(1 - p) \tag{4.59}$$

Now, combining Eqns. (4.58) and (4.59), we get

$$N_n = N_0(1 - p)p^{n-1}(1 - p)$$

or

$$N_n = N_0(1 - p)^2 p^{n-1} \tag{4.60}$$

N_n in this equation gives the number fraction of n-mer in the total number of polymer molecules present, N. The weight fraction W_n, similarly, would be given by

$$W_n = \frac{N_n \cdot n}{N_0} = n(1 - p)^2 p^{n-1} \tag{4.61}$$

For different values of p, the N_n and W_n distribution functions are shown in Figs. 4.7 and 4.8, respectively. The curves in Fig. 4.7 and Fig. 4.8 show that if we consider the number fraction, an appreciable quantity of very low molecular weight species will be present at all stages of conversion, as compared to high molecular weight species. On the other hand, if we consider the weight fraction, the very low molecular weight species will be present in a negligible quantity, as compared to high molecular weight species.

Equation 4.61 which relates the weight-fraction distribution with the extent of reaction can also be used in determining the extent of reaction that should be attained if we want to get a maximum yield of a particular molecular weight species. Thus, from Eqn. (4.61) we can derive:

$$p_n = \frac{n-1}{n+1} \tag{4.62}$$

where p_n represents the extent of reaction required to obtain the n-mer. Using Eqn. (4.62), we can calculate the extent of reaction necessary in order to get a polymer with a maximum yield of, say, 9-mers as follows:

$$p_9 = \frac{9-1}{9+1} = \frac{8}{10} = 0.8$$

That is, at 80% conversion, the polymer formed will have a maximum percentage of 9-mers.

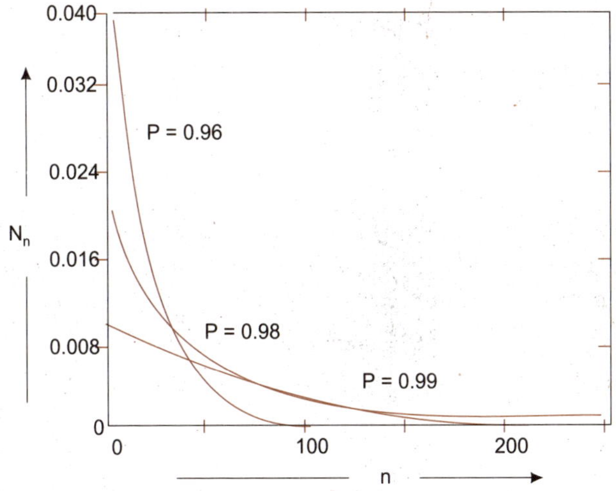

Fig. 4.7 Number-fraction distribution of chain molecules at different extents of reaction in polycondensation.

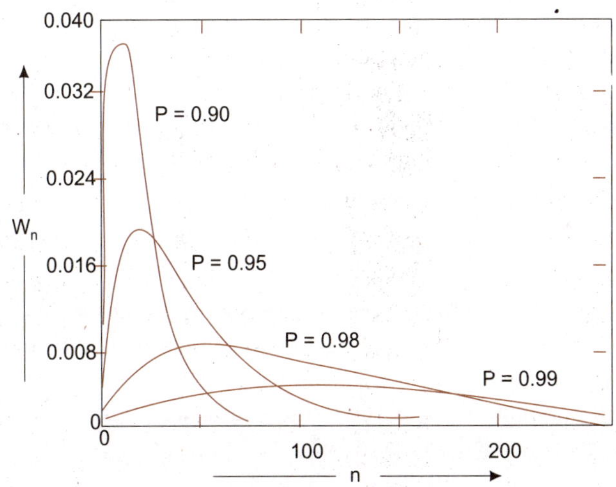

Fig. 4.8 Weight-fraction distribution of chain molecules at different extents of reaction in polycondensation.

4.5.4 Ratio of Weight-average Molecular Weight to Number-average Molecular Weight

The breadth of the molecular weight distribution curve may be obtained as follows. From Eqn. (4.53), we know that

$$\overline{D}_{pn} = \frac{1}{1-p}$$

Hence, $\overline{M}_n$ (the number-average molecular weight) can be expressed as

$$\overline{M}_n = \frac{M_0}{1-p} \tag{4.63}$$

where M_0 is the molecular weight of the repeat unit.

On the basis of similar arguments, we can get an expression for $\overline{M}_w$ the weight-average molecular weight in the case of a condensation polymerisation as

$$\overline{M}_w = M_0\frac{1+p}{1-p} \tag{4.64}$$

Hence, the ratio

$$\frac{\overline{M}_w}{\overline{M}_n} = M_0\frac{1+p}{1-p} \times \frac{1-p}{M_0} = 1 + p \tag{4.65}$$

It will be seen from the foregoing that at very high conversions (that is, when p approaches unity), $\overline{M}_w/\overline{M}_n$ approaches a value of 2.

4.5.5 Extent of Reaction and Degree of Polymerisation

We know that

$$\overline{D}_{pn} = \frac{1}{1-p}$$

Let us put different values of p and get the corresponding values of $\overline{D}_{pn}$, which are listed in Table 4.4. It may be noted that in order to get a relatively high degree of polymerisation or a high molecular weight polymer, the extent of conversion has to be very high. We can see from Table 4.4 that even for a degree of polymerisation of 100 (which, in fact, corresponds to low molecular weight polymers), conversion has to be as high as 99%.

Table 4.4 Effect of extent of conversion on degree of polymerisation for polycondensation reaction

Values of p	*Extent of conversion (%)*	*Degree of polymerisation, $\overline{D}_{pn} = \frac{1}{1-p}$*
0.50	50	2
0.90	90	10
0.95	95	20
0.99	99	100
0.999	99.9	1,000
0.99999	99.999	100,000

4.5.6 Stoichiometry of Reactants and Degree of Polymerisation

We have been dealing so far with polycondensation reactions wherein the two reactants are present in stoichiometric equivalents. What will happen if there is a stoichiometric imbalance—that is, if one of the reactants is present in excess quantity? The imbalance will show up on the degree of polymerisation. Let us see how.

Assume that N_A moles of reactant A—A are reacted with N_B moles of reactant B—B and that $N_A < N_B$. The stoichiometric ratio of A—A to B—B will be less than 1. Let

$$\frac{N_A}{N_B} = r \tag{4.66}$$

Number of reactive groups A to start with = $2N_A$.

Number of reactive groups B to start with = $2N_B$.

Total number of reactive groups to start with

$$= 2N_A + 2N_B = 2N_A\,[1 + (1/r)].$$

When the extent of reaction is p, the fraction of groups A consumed will also be p.

Fraction of groups B consumed = rp.

Fraction of groups A remaining unreacted = $1 - p$.

Fraction of groups B remaining unreacted = $1 - rp$.

Total number of groups remaining unreacted

$$= 2N_A\,(1 - p) + 2N_B\,(1 - rp).$$

Now, the degree of polymerisation is given by

$$\overline{D}_{pn} = \frac{[\mathrm{FG}]_0}{[\mathrm{FG}]} = \frac{2N_A\left(1+\dfrac{1}{r}\right)}{2N_A(1-p)+2N_B(1-rp)}$$

$$= \frac{r+1}{r+1-2rp} \tag{4.67}$$

When there is no stoichiometric imbalance (*i.e.*, when both the reactants are present in stoichiometric equivalents), $r = 1$ and, hence,

$$\overline{D}_{pn} = \frac{1}{1-p}$$

Similarly, when the extent of conversion is 1 (*i.e.*, when polymerisation goes to 100% completion), $p = 1$ and, hence,

$$\overline{D}_{pn} = \frac{1+r}{1-r} \tag{4.68}$$

Table 4.5 Salient features of some polymerisation reactions

Free-radical polymerisation (1)	*Cationic polymerisation (2)*	*Anionic polymerisation (3)*	*Polycondensation (4)*
Monomers have double bonds and the polymerisation proceeds by an attack on the π electrons of the double bond by a free radical.	Monomers have double bonds and the polymerisation proceeds by an attack on the π electrons of the double bond by a cation (usually a carbonium ion).	Monomers have double bonds and the polymerisation proceeds by an attack on the π electrons of the double bond by an anion (usually a carbanion).	Monomers contain reactive functional groups and the polymerisation proceeds by condensation reaction between the functional groups.
Polymer chain growth is brought about by a rapid addition of monomer units in quick succession, one after another, and hence, is very fast.	Polymer chain growth is brought about by a rapid addition of monomer units in quick succession, one after another, and hence, is very fast.	Polymer chain growth is brought about by a rapid addition of monomer units in quick succession, one after another, and hence, is very fast.	Polymer built up is by the step-wise reaction between functional groups and hence, is slow.
No elimination of small molecules involved during polymer formation. The elemental composition of the polymer is, therefore, the same as that of the monomer.	No elimination of small molecules involved during polymer formation. The elemental composition of the polymer is, therefore, the same as that of the monomer.	No elimination of small molecules involved during polymer formation. The elemental composition of the polymer is, therefore, the same as that of the monomer.	Due to condensation reaction, small molecules are eliminated during polymer formation. The elemental composition of the polymer is, therefore, different from that of the monomers.
As the polymerisation proceeds, the monomer concentration steadily decreases and the polymer concentration simultaneously increases. The concentration of the growing chains, however, remains almost constant throughout. The reaction mixture at any instant contains unreacted monomer, growing polymer chains and the dead polymer.	As the polymerisation proceeds, the monomer concentration steadily decreases and the polymer concentration simultaneously increases. The concentration of the growing chains, however, remains almost constant throughout. The reaction mixture at any instant contains unreacted monomer, growing polymer chains and the dead polymer.	As the polymerisation proceeds, the monomer concentration steadily decreases and the polymer concentration simultaneously increases. The concentration of the growing chains, however, remains almost constant throughout. The reaction mixture at any instant contains unreacted monomer, growing polymer chains and the dead polymer. In the case of living polymers, where no termination occurs, the growing chains consume the entire monomer and the polymer formed is not 'dead', but still 'living'.	Almost the entire quantity of the monomer is consumed at an early stage (less than 1% of the monomer remains when the DP is as low as 10). Polymers of sufficiently high molecular weights are formed only at very high conversions (say, more than 99% conversion). At intermediate stages, molecules of varying molecular weights are found in appreciable distribution.
The process is exothermic due to the release of energy associated with the cascading of the electron pair from the π level to the sigma level.	The process is exothermic due to the release of energy associated with the cascading of the electron pair from the π level to the sigma level.	The process is exothermic due to the release of energy associated with the cascading of the electron pair from the π level to the sigma level.	The process is endothermic and in the neighbourhood of the equilibrium condition associated with the condensation reactions, energy is to be supplied to continuously remove the small molecules eliminated so as to force the reaction in the forward direction.

Contd.

Free-radical polymerisation (1)	*Cationic polymerisation (2)*	*Anionic polymerisation (3)*	*Polycondensation (4)*
Molecular weight of the polymer depends on monomer and initiator concentrations and temperature, but it is not much dependent on the extent of conversion. Solvent medium also contributes to the molecular weight by affecting the initiator decomposition rate and through chain transfer.	Molecular weight depends on monomer concentration and temperature but is independent of initiator concentration and extent of conversion. Polarity of the solvent medium, however, has a profound effect on the molecular weight.	Molecular weight depends on monomer concentration, temperature and solvent medium. In the case of living polymers, molecular weight depends on initiator concentration and extent of reaction also.	Molecular weight depends on the mole ratio of the monomers. Extent of conversion and hence, the reaction time, have a significant effect on the molecular weight.
Polymer has a broad molecular weight distribution. $\overline{M}_w/\overline{M}_n$ ranges from 1.5 to 5.	Polymer has a medium molecular weight distribution. $\overline{M}_w/\overline{M}_n$ ranges from 1.5 to 2.	Polymer has a very low molecular weight distribution. $\overline{M}_w/\overline{M}_n$ of less than 1.5 is possible.	Polymer has a medium molecular weight distribution. $\overline{M}_w/\overline{M}_n$ approximately from 2–3.
Chain branching is favoured due to chain transfer to already formed polymer molecules.	Chain branching is negligible.	Chain branching is negligible.	Chain branching is possible only with trifunctional monomers.
Product formed is usually atactic.	Product formed is usually atactic.	Product formed is usually atactic. With special heterogeneous catalysts, stereoregular polymers can be obtained.	
The reaction is simple and no stringent care or precision is required to carry out polymerisation. Hence, most widely used in industrial processes.	Reaction is sensitive to moisture and other ionic impurities, hence the system should be anhydrous and free from other impurities. Has limited application in industrial processes.	Reaction is sensitive to moisture and other ionic impurities, hence, the system should be anhydrous and free from other impurities. Has only limited application in industrial processes.	The reaction is simple and no stringent care (except for very high purity of the difunctional monomers) or precision required to carry out polymerisation. Hence, most widely used in industrial processes.
Suitable for bulk, solution, suspension and emulsion techniques.	Suitable for solution technique. Due to moisture sensitivity, unsuitable for suspension and emulsion techniques.	Suitable for solution technique. Due to moisture sensitivity, unsuitable for suspension and emulsion techniques.	Suitable for bulk, solution, emulsion and melt techniques.

The effect of the stoichiometric ratio of the reactants on the degree of polymerisation at 100% conversion is depicted in Fig. 4.9. As expected from Eqn. 4.68, at a stoichiometric ratio of 1, the degree of polymerisation would tend to approach a value of infinity. An imbalance on the stoichiometry by 0.02 will, however, bring down the degree of polymerisation value from infinity to as low a value as 100.

4.5.7 Monofunctional Reactants and Degree of Polymerisation

The availability of bifunctional reactants in 100% purity poses a practical problem. Bifunctional reactants, often, contain some percentage of monofunctional impurities. What will be the effect of

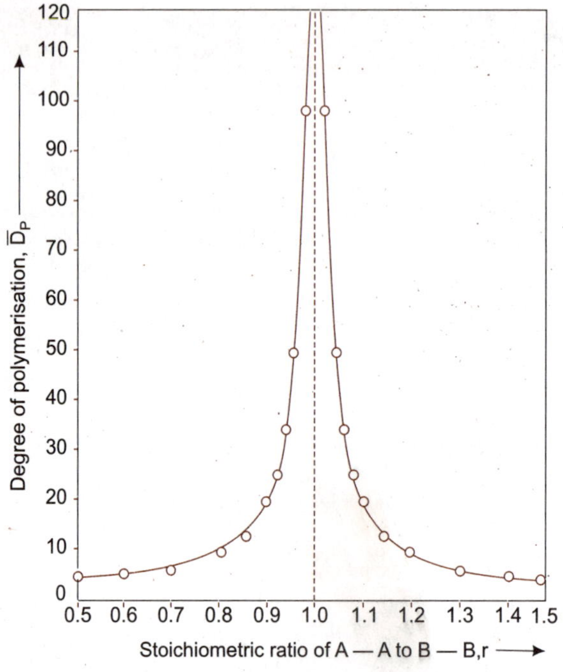

Fig. 4.9 Effect of stoichiometric ratio of bifunctional reactants on degree of polymerisation in polycondensation.

such impurities on the degree of polymerisation? Monofunctional impurities upset the stoichiometric balance and bring about a decrease in the degree of polymerisation in a manner similar to what we have seen in the previous section. Assume that N_A moles of A—A, N_B moles of B—B and N_M moles of a monofunctional impurity of the type B are reacted. The stoichiometric imbalance will be represented by

$$r = \frac{N_A}{N_B + N_M} \tag{4.69}$$

Some generalisations can now be made regarding polycondensation as follows:

(*i*) A high degree of polymerisation can be achieved only at very high conversions.

(*ii*) It is essential to start with exactly equimolar quantities of the reactants to achieve a high degree of polymerisation.

(*iii*) The presence of monofunctional impurities drastically reduces the degree of polymerisation.

Salient features of free-radical, cationic, anionic and polycondensation polymerisation reactions are summarised in Table 4.5.

Books for Further Reading

Bamford, C.H., Barb, W.G., Jenkins, A.D. and Onyon, P.F., *The Kinetics of Vinyl Polymerisation by Radical Mechanisms*, Butterworths, London, 1958.

Ham, G.E., Ed., *Kinetics and Mechanics of Polymerisation*, Vols. 1, 2 and 3, Dekker, New York, 1969.

Jenkins, A.D. and Ledwith, A., Eds., *Reactivity, Mechanisms and Structure in Polymer Chemistry*, John Wiley & Sons, New York, 1974.

Keii, T., *Kinetics of Ziegler–Natta Polymerisation*, Halsted, New York, 1973.

5

CHEMICAL AND GEOMETRICAL STRUCTURE OF POLYMER MOLECULES

5.1 GENERAL REMARKS ON POLYMER MICROSTRUCTURE

We have stated in Chapter 3 that many of the polymer properties depend on (*i*) average molecular weight and (*ii*) molecular weight distribution of the polymer (see Sections 3.5 and 3.6). But even these two parameters do not fully characterise a polymer. What other factors are there to completely define the polymer? Let us now take up this question.

Two polymer samples can have the same chemical structure and almost similar molecular weight distribution, but may have different properties—one sample may be amorphous, while the other crystalline; one soft and flexible and the other one hard and rigid; one dissolving in a particular solvent, the other not; one melting on heat, the other only charring. And we can go on and on. All such properties depend on intermolecular and intramolecular interactions of polymer chains in the aggregated state.

Such interactions are caused by forces, such as dispersion force, induction force, dipole interaction and hydrogen bonding, acting within various parts of the same molecule as well as between the neighbouring molecules. The nature and the magnitude of such interaction forces depend, apart from the chemical characteristics of the monomer, on the manner in which these monomeric units are linked with each other in the polymer. Whereas the number and chemical nature of monomeric units forming the macromolecule gives an overall picture of the polymer chain, the manner in which they are linked gives the finer aspects of the picture. That is to say, the number of monomeric units depicts the macrostructure, and the mode of their interlinking indicates the microstructure of the polymer molecule. It would be interesting to look into microstructures based on both the chemical structure and the geometrical structure. We will now examine these two broad categories.

5.2 MICROSTRUCTURES BASED ON THE CHEMICAL STRUCTURE

5.2.1 Organic and Inorganic Polymers

It may be recalled (cf. Section 1.4.2) that polymers can be classified as 'organic' or 'inorganic', depending on whether the chain backbone is made essentially of carbon-carbon (C—C) links or not. This classification is based on the chemical structure. Polygermane is an inorganic polymer analogous to the well-known organic polymer polyethylene (Fig. 5.1).

5.2.2 Homochain and Heterochain Polymers

In some polymers, the chain backbone is made entirely of a single type of atoms, whereas, in some other polymers, the backbone has different types of atoms. A polymer with the chain backbone made entirely of the same type of atoms is called a 'homochain polymer'. Examples are polyethylene (the backbone made entirely of C) and polygermanes (the backbone made entirely of Ge). In contrast, a polymer with the chain backbone made of different types of atoms is called a 'heterochain polymer'. Examples are polyethylene adipate (the backbone made of C and O) and poly (dimethyl siloxane) type silicone polymer (the backbone made of Si and O). Structural formulae illustrating these examples are given in Fig. 5.1.

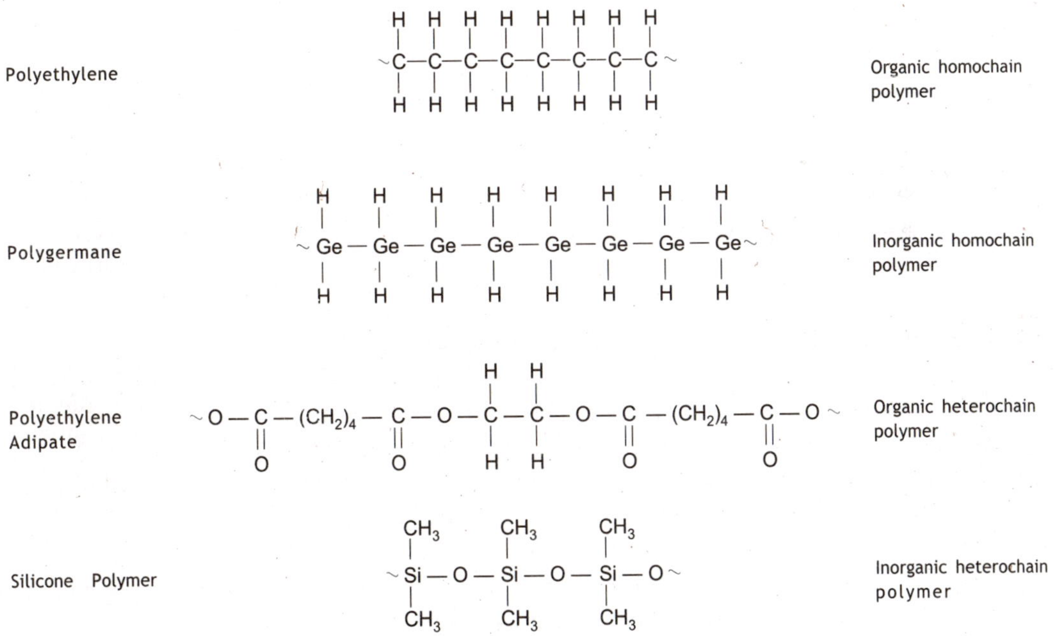

Fig. 5.1 Examples of organic, inorganic, homochain and heterochain polymers.

5.2.3 Homopolymers and Copolymers

We know that the ethylene monomer can be polymerised to give polyethylene. Similarly, two other monomers, vinyl chloride and vinyl acetate, can be separately polymerised to give polyvinyl chloride (PVC) and polyvinyl acetate (PVAc), respectively. Structural formulae of these two monomers and polymers are given in. Fig. 5.2A. Now, we can straightaway observe that PVC is made of a single repeat unit —CH_2—CH.Cl— and PVAc is made of another single repeat unit —CH_2—CH—O.CO.CH_3—. Polyethylene, PVC and PVAc are examples of polymers where the entire polymer chain is made of one single repeat unit. These are called 'homopolymers'.

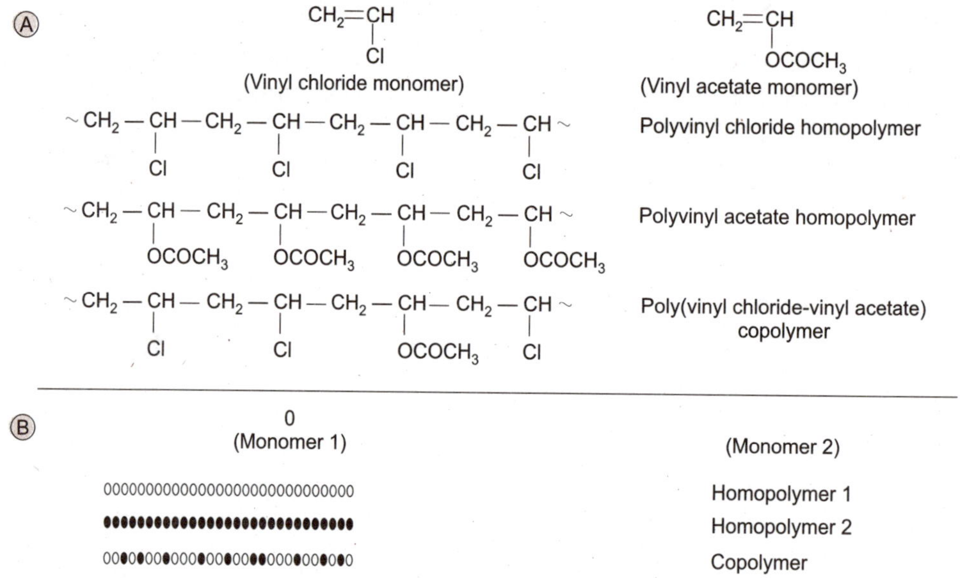

Fig. 5.2 (A) Examples and (B) conceptual representation of homopolymers and copolymers.

Now, the two monomers—vinyl chloride and vinyl acetate—can be mixed to give a (vinyl chloride-vinyl acetate) mixture and then the mix polymerised to give a poly (vinyl chloride-vinyl acetate). The chain backbone is obviously made of the two repeat units shown in the brackets. Polymers like poly (vinyl chloride-vinyl acetate) having more than one type of repeat unit are called 'copolymers' and the monomers from which the copolymer is made are called 'comonomers'.

In general, if A and B represent two monomers, the structure of the homopolymers and the copolymer obtainable from them can be represented as follows:

—A—A—A—A—A—A—A—A—A—A—A— } Homopolymers
—B—B—B—B—B—B—B—B—B—B—B—

—A—A—B—A—B—B—B—A—A—B—A—A—A—A— Copolymer

Figure 5.2B shows a schematic representation of two monomers and two homopolymers and a copolymer made of them.

5.3 MICROSTRUCTURES BASED ON THE GEOMETRICAL STRUCTURE

While the chemical structure of a macromolecule depends on the chemical nature of the monomeric units, the geometrical structure depends on the spatial arrangement of the monomeric units with respect to each other. Polymers having the same chemical structure (*i.e.*, made-up of the same monomeric units) can have different geometrical structures.

While discussing the spatial arrangement of monomeric units in a polymer chain, two terminologies are common: configuration and conformation. A configuration is an arrangement fixed by chemical bonding between adjacent monomeric units and between atoms of individual monomeric units. The configuration remains unaltered as long as chemical bonds are not broken or re-formed. A polymeric chain cannot shift from one configuration to another without breaking or re-forming the chemical bonds. A conformation, on the other hand, is an arrangement resulting from the rotation of chain segments (or adjacent monomeric units) around single bonds. It will be noticed that this kind of rotation does not involve any breaking or re-forming of the chemical bonds; a polymeric chain of a given configuration can, over a period of time, assume innumerable shapes or conformations, depending on such external forces as thermal energy, solvent interaction and applied stress.

The conformations taken by a polymer molecule depend on whether the polymer is of a flexible-chain type or a rigid-chain type. A flexible-chain polymer is one in which the chain segments can rotate with respect to each other with a sufficient degree of freedom. Polymers composed of non-polar segments or segments with a very low polarity are of the flexible-chain type. Polyethylene, polystyrene and rubber are examples. By contrast, a rigid chain polymer is one in which the rotation of the chain segments is hindered due to steric factors (such as the presence of bulky side groups or of aromatic groups in the chain backbone) or due to strong forces of attraction (such as dipole or hydrogen bonding) between the neighbouring chains. Polyamides, aromatic polyesters and cellulose esters belong to this group. The conformations possible with polymer molecules vary from a linear rigid rod-like structure to that of a flexible random coil. Polymer molecules in the solid state have conformations of interpenetrating random coils (in amorphous polymers), orderly folded chains (in crystalline polymers) and spiralled or helical chains (in biological polymers such as proteins). In dilute solutions, molecules of flexible-chain polymers exist as isolated random coils and those of rigid-chain polymers exist as rigid rods or helixes. In highly concentrated solutions or swollen gels, they form interconnected network or interconnected cell structure (Fig. 5.3). Details of these conformations will be given in appropriate chapters. In the forthcoming sections, we deal mainly with various configurations.

5.3.1 Linear, Branched and Cross-linked Polymers

Homopolymers and copolymers can exist in three different types of chain configurations, *viz.*, linear, branched or cross-linked. Consider a PVC homopolymer and a PVC-PVAc copolymer. Their

Interpenetrating random coils
(Solid Polymers Amorphous)

Folded chains
(Solid Polymers Crystalline)

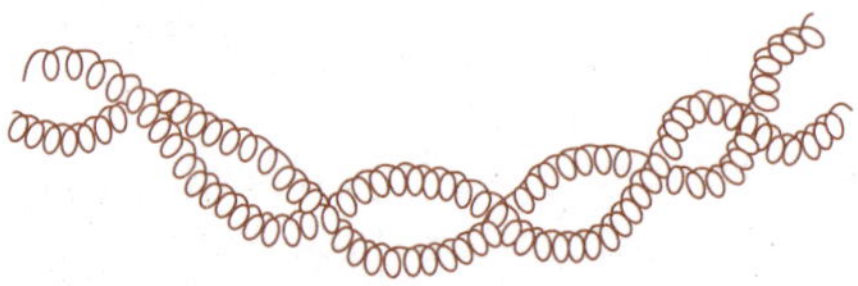

Spiralled or-helical chains
(Polypeptides or Proteins)

(Linear)
Isolated random coils
(Dilute Polymer Solution)

(Branched)

Interconnected network structure
(Cross-linked Polymers)

Fig. 5.3 Representative conformations of polymer molecules under different situations.

structures, shown in Fig. 5.2, are of a linear chain type. Here, each monomeric unit is linked with two other monomeric units on either side, forming a continuous straight chain. Monomeric units are added on to each other, like railway bogies; forming a long train. A linear polymer can be schematically represented by a single line.

In some cases, however, while the polymer chain is growing in a linear manner, some side growth also takes place from the main chain. In such cases, while most of the monomeric units are linked with two others on either side, some monomeric units may also be linked with a third monomeric unit. Such polymer molecules will have an overwhelmingly linear chain, with a few short side chains or branches attached at random points. These are called 'branched polymer molecules', the shape of which can be represented by a line with a few branches.

During polymerisation, a large number of branched polymer molecules may be formed and, on further polymerisation, the side chains from one molecule could possibly interact with those from the neighbouring ones. Since there are many molecules present with side chains attached at random positions, the process of the interuniting of side chains of neighbouring molecules results in a single molecule with a criss-crossed network of chain segments in all the three dimensions. Such a polymer is called a cross-linked or network polymer.

Schematic representations of linear, branched and network polymer molecules are shown in Fig. 5.4.

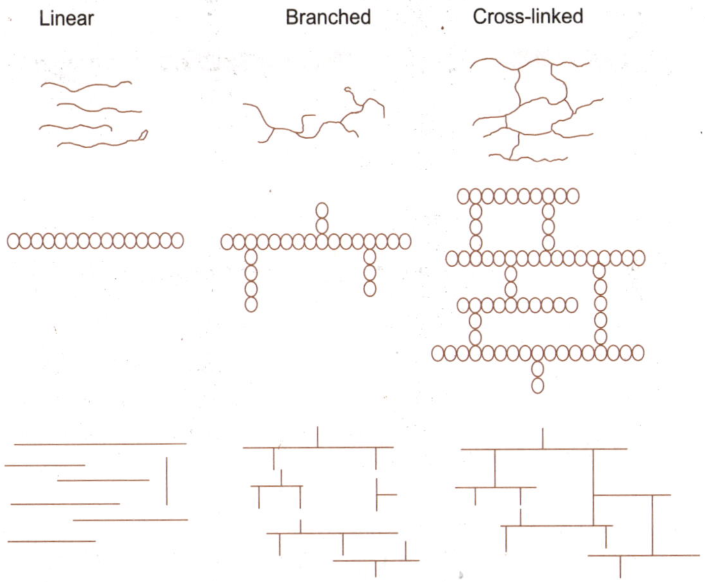

Fig. 5.4 Schematic representation of linear, branched and cross-linked polymer molecules.

5.3.2 Random, Alternating Block and Graft Copolymers

When two different repeat units in a copolymer are distributed at random throughout the chain, the polymer is called a 'random copolymer'. In certain copolymers, a more ordered sequence of repeat units is seen and then these copolymers are termed alternating, block or graft copolymers.

When two repeat units are distributed alternately throughout the chain, the polymer is termed an 'alternating copolymer'.

When a sequence or block of one repeat unit is followed by a block of another repeat unit, which, in turn, is followed by a block of the first repeat unit and so on, the polymer is called a 'block copolymer'. Block copolymers are, therefore, linear molecules in which both repeat units are consecutively distributed in fairly long sequences.

Graft copolymers, on the other hand, are branched molecules where the main chain is made entirely of one repeat unit, while the branch-chains are made of yet another repeat unit.

Figure 5.5 shows schematic representations of various copolymers.

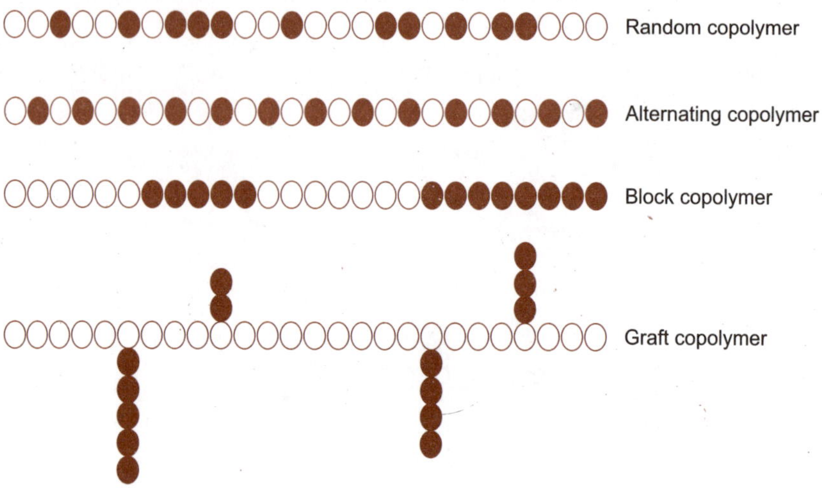

Fig. 5.5 Schematic representation of different types of copolymer structures.

5.3.3 Stereo-regular Polymers

In stereo-regular polymers, each monomer segment is in a regular configuration, giving a definite structural regularity to the polymer molecule as a whole. This structural regularity, in a polymer may arise due to what are termed as "optical" and "geometric" isomerisms of the main chain atoms or substituents in the polymer molecule. The meaning of these two isomerisms is now explained.

Optical Isomerism

Optical isomerism has its origin in the way different substituents sit on an asymmetric carbon atom in a polymer molecule. We will give an example. A polyethylene molecule has fully saturated carbon atoms with the following chemical formula:

$$—CH_2—CH_2—CH_2—CH_2—CH_2—CH_2—$$

We can write this formula in a linear structural way, showing how various atoms are attached (but without bothering about the bond angles and molecular plane) as follows:

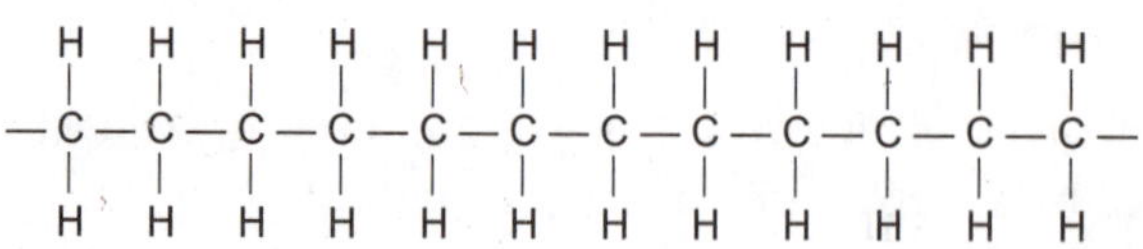

Let us now assume that the carbon-carbon bonds are all on the plane of this paper and that the bonds between the carbon atom and its substituent atoms are either below or above the plane of the paper. The structure of polyethylene in a planar zig-zag configuration is then as shown in Fig. 5.6. In doing this, of course, we have not bothered about the free rotation around each carbon-carbon single bond and the various resultant spatial positions of the polymer segment. It is seen from Fig. 5.6 that all carbon atoms are attached to each other through single bonds which are arranged in a zig-zag manner considering their bond angles. All carbon atoms lie in the plane of the paper but hydrogen atoms attached to the carbon atoms protrude away, up or down, from the plane of the paper. The hydrogen atoms above or below the plane of the paper are interchangeable without altering the overall polyethylene structure. This is possible because in polyethylene all hydrogen atoms attached to carbon atoms are indistinguishable.

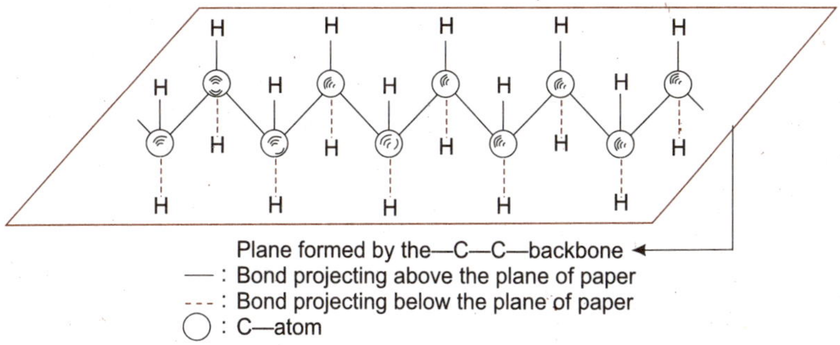

Fig. 5.6 Planar zig-zag structure of polyethylene molecule.

But imagine that one of the hydrogen atoms in all ethylene units of this polymer is substituted by a substituent R (such as CH_3, Cl or CN). The polymer will then have the chemical formula —CH_2—CHR—CH_2—CHR—CH_2—CHR— and its structural formula would be as follows:

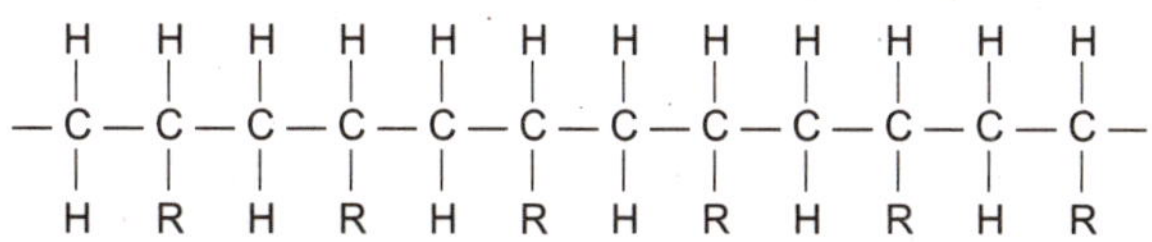

In such a polymer chain, every alternate carbon atom can be considered to be asymmetric. This means that the alternate carbon atom, which we designate as C*, carries four different substituents —H, R and two polymer chain segments of chain length *m* and *n* as shown

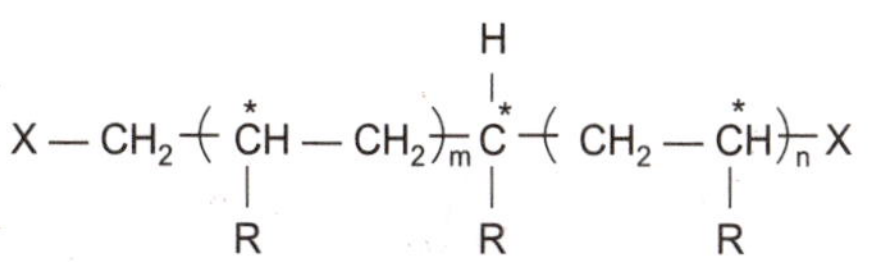

Here, X denotes end groups.

Each of these C* atoms provides a site for optical isomerism. Each such site can exhibit either *d*- or *l*-type isomerism, depending on whether the R group is located below the plane of carbon-carbon chain or above. The regularity or the order in which the successive asymmetric carbon sites, C*, exhibit their *d*- or *l*-form gives rise to three different types of isomeric structure in the polymer molecule. The structures shown in Fig. 5.7 will clarify the concept. In structure I, we have all the R groups located on one side of the plane of the carbon-carbon chain. In structure II, we have the R groups located alternatively above and below the plane of the carbon-carbon chain and in structure III, we have them located randomly. Polymers having structure I are called isotactic; those having structure II are called syndiotactic; while the third type are known as atactic or heterotactic. These three types of polymers have the same chemical structure, but exhibit entirely different properties because of their differing configurations and the resulting geometrical structure. Atactic polymers are, for instance, generally low melting and easily soluble, while isotactic and syndiotactic polymers are high melting and less soluble.

Geometric Isomerism

In optical isomerism, while single bond C—C atoms are involved, in geometric isomerism double bond C=C atoms are involved. Geometric isomerism arises from different configurations of the substituents on a carbon-carbon double bond and depends on how the substituent groups are positioned. For example, let us consider 1, 3-butadiene polymerisation. The butadiene monomer has two double bonds in its structure:

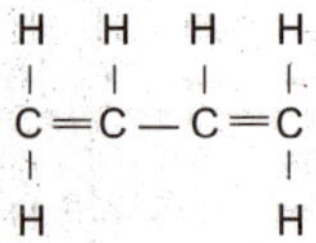

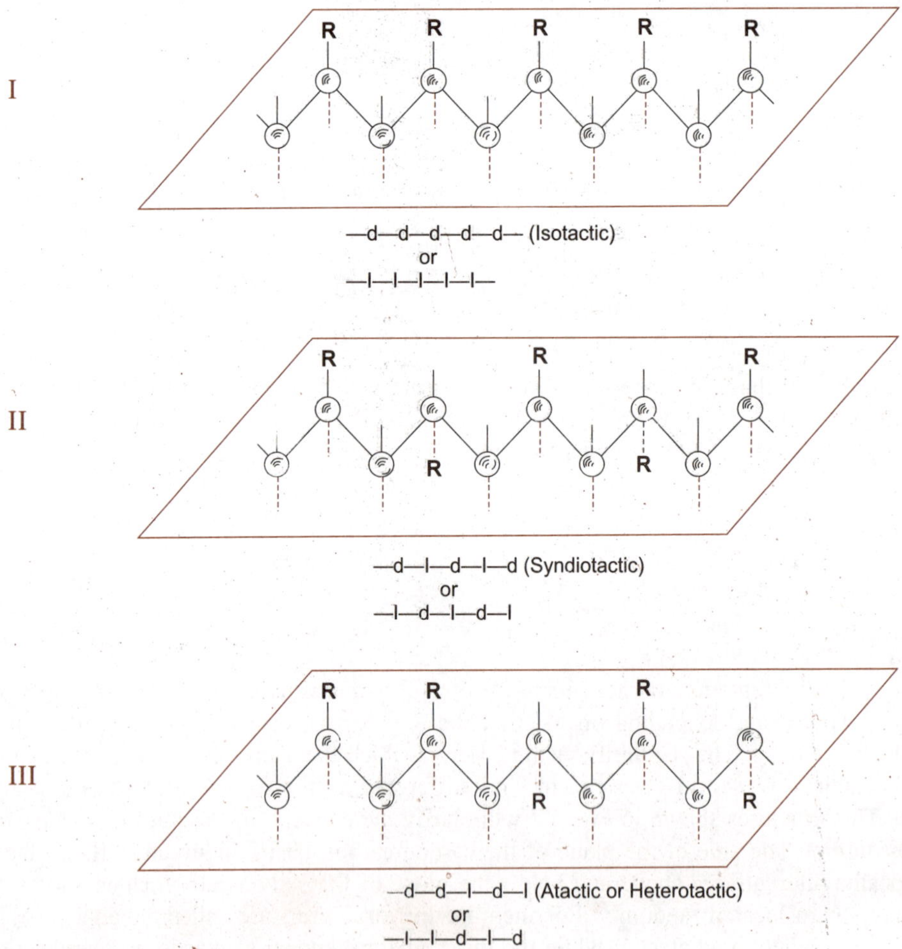

Fig. 5.7 Planar zig-zag structure of polymer molecules showing: (I) isotactic, (II) syndiotactic, and (III) heterotactic configurations. (Hydrogen atoms are not shown for the purpose of clarity).

It can polymerise as shown

$$n\,CH_2{=}CH-CH{=}CH_2 \longrightarrow \left(CH_2-CH{=}CH-CH_2\right)_n$$

The resultant polymer structure has a double bond in each repeat unit of

$$\left(Ch_2-CH=CH-CH_2\right)$$

Each of these double bonds provides a site for a steric isomerism, depending on whether the CH_2 groups attached to the carbon atoms on either side of the double bond are close to or away from each other. The two possible configurations (Fig. 5.8) are: (*a*) where both the —CH_2 groups are on the same side of the double bond close to each other (called cis-configuration), and (*b*) where they are on the opposite sides and quite apart (called trans-configuration).

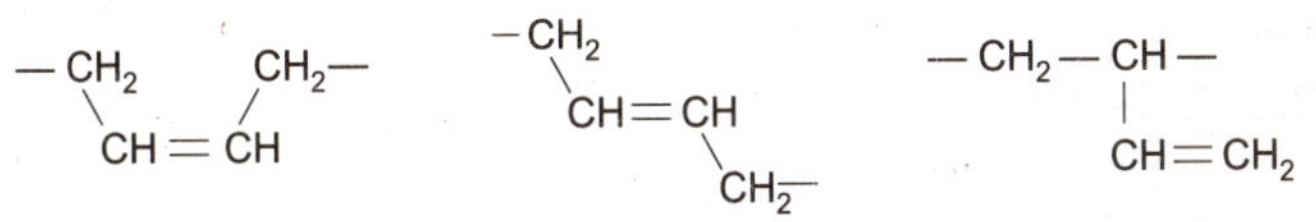

Cis-configuration   Trans-configuration   1, 2 vinyl configuration

Fig. 5.8 Cis-, trans- and 1, 2 vinyl configurations of butadiene units in polybutadiene molecule.

A third configuration, called 1, 2 vinyl configuration, is also possible. (see Fig. 5.8). In the case of 1, 2 vinyl configuration, however, the three types of optical isomerism, *i.e.*, atactic, isotactic and syndiotactic configurations, are possible. It may also be seen from this figure that in cis-configuration, there is a bending back of the carbon-carbon chain, whereas, in trans-configuration, there is a straightening out of the carbon-carbon chain. During the polymerisation of 1, 3-butadiene, if all repeat units take cis-configuration, we get a 100% cis-polybutadiene and due to bending back of all the successive carbon-carbon chain segments, the molecule as a whole assumes a shape of a spring (Fig. 5.9) and exhibits good elongation. On the other hand, during polymerisation, if all the repeat units take trans-configuration, the resultant polymer is a 100% trans-polybutadiene and due to the straightening out of all the successive carbon-carbon chain segments, the molecule assumes a straight and stiffened rod-like structure (Fig. 5.9) and exhibits lower elongation.

In practice, however, polymerisation of butadiene to 100% cis or 100% trans is extremely difficult, and we usually obtain mixtures of cis- and trans-configurations, randomly distributed throughout the chain length (presence of 1, 2-vinyl structure is also possible) (Fig. 5.9). Depending on the ratio of cis- to trans-chain segments present, the polymer can exhibit either high or low elongation.

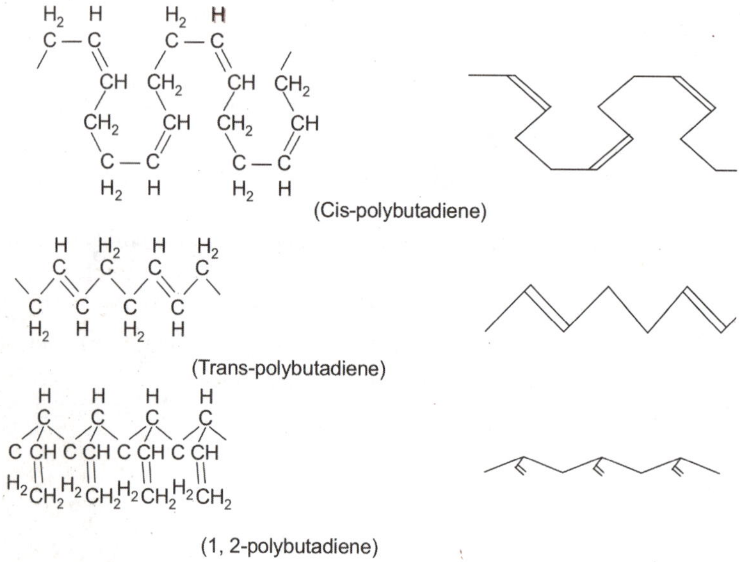

Fig. 5.9 Configurations of 100% cis-, 100% trans- and 100% 1, 2-polybutadiene molecules.

Books for Further Reading

Natta, G. and Damusso, F., *Stereoregular Polymers and Stereospecific Polymerisations,* Pergamon, Oxford, 1967.

Bovey, F.A., *Polymer Conformation and Configuration,* Academic, New York, 1969.

Koening, J.L., *Chemical Microstructure of Polymer Chain,* John Wiley & Sons, New York, 1980.

6

GLASS TRANSITION TEMPERATURE

6.1 WHAT IS GLASS TRANSITION TEMPERATURE?

An ordinary rubber ball if cooled below –70°C becomes so hard and brittle that it will break into pieces like a glass ball falling on a hard surface! Why does a rubber ball become like glass below –70°C? This is because there is a 'temperature boundary' for almost all amorphous polymers (and many crystalline polymers) only above which the substance remains soft, flexible and rubbery and below which it becomes hard, brittle and glassy. This temperature, below which a polymer is hard and above which it is soft, is called the 'glass transition temperature'. The hard, brittle state is known as the glassy state and the soft flexible state as the rubbery or viscoelastic state. On further heating, the polymer (if it is uncross-linked) becomes a highly viscous liquid and starts flowing; this state is termed viscofluid state, the transition taking place at its flow temperature T_f (Fig. 6.1).

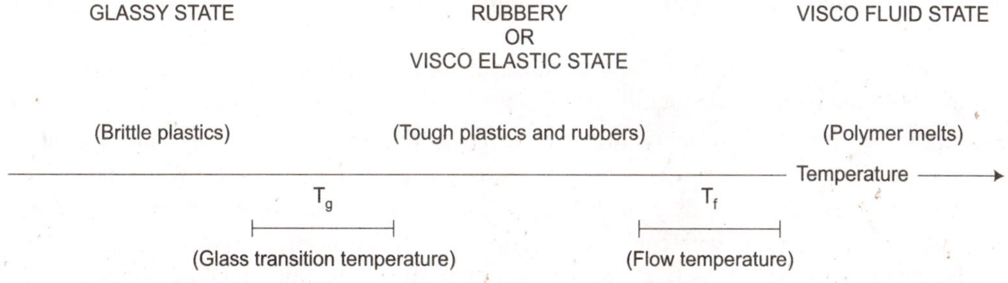

Fig. 6.1 Change of slate with temperature in polymeric materials.

6.2 GLASSY SOLIDS AND GLASS TRANSITION

To understand the concepts of glassy solids and glass transition, the states of aggregation and of phases of low and high molecular weight substances are now briefly reviewed.

6.2.1 States of Aggregation

Low molecular weight compounds can exist in three states of aggregates: solid, liquid and gas. Polymers, however, exist only as solid or liquid but never as gas as they decompose before reaching their boiling point.

Each individual molecule in a crystalline, low molecular weight solid substance sits in the crystal lattice in a certain equilibrium position from which it moves about only a little, in a vibrational mode. There is no molecular translational motion or 'Brownian' movement in a crystalline solid. The molecules sit in their respective positions within the crystal lattice almost like a well-disciplined audience in an auditorium. The audience is certainly not motionless, but the individuals do not leave their seats and loiter about. Their limited movement in their respective seats corresponds to vibrational motion of the molecules within the crystal lattice; the 'non-loitering' of the audience can be compared to non-translational motion or the absence of Brownian movement on the part of the molecules. The molecules occupy definite positions within the crystal lattice and this well-defined arrangement is repeated several times in all the three dimensions of the crystal to give rise to, what is called, the long-range order (lro) of the crystalline solid.

You might wonder as to what would happen to this system when an external force is applied? Well, individual molecules in a solid are in a rigid state and, naturally, the planes formed by these molecules within the solid are also rigid and inflexible. This rigidity gives solids a capability of transferring external stresses. A low molecular weight crystalline solid can thus, be defined essentially in terms of the (*i*) position and conditions of individual molecules inside the crystal, (*ii*) geometry of the crystal lattice and (*iii*) response to an external force.

Now what would happen to such a system in a high temperature environment? As we increase the temperature, more and more kinetic energy is added to the system and individual molecules start vibrating more and more vigorously. A stage is eventually reached when the vibrational motion increases sufficiently to cross the energy barrier which holds individual molecules in the equilibrium position. When this happens, the molecules start moving out of their fixed positions, activating Brownian movement. At higher temperatures, the Brownian movement becomes progressively more vigorous. The molecules diffuse all over randomly, like a relaxed audience during a tea-break. No well-defined pattern of the molecular arrangement is possible and the long-range order is lost. If now an external force is applied, the system can no longer transfer the stress, but will only yield and flow. This state is called the liquid state.

So, with low molecular weight substances in the condensed state, we can have two states of affairs: (*i*) the first without Brownian movement, but with long-range order when stress transfer phenomenon is possible (the state is called 'solid') and (*ii*) the second with Brownian movement, without long-range order and with energy dissipation (the state is called 'liquid'). The temperature at which the change of state occurs (*viz*., from solid to liquid) is called the 'melting point' (T_m).

Now, let us consider a polymer, say, polyethylene. At room temperature, polyethylene is solid, exhibiting all characteristics of a low molecular weight substance. At high temperatures, however, the characteristic difference between high and low molecular weight substances can be seen. What is this difference? Depending on the temperature, the molecules of a low molecular weight substance either move apart as a whole or do not move at all, *i.e.* there is a definite temperature (T_m) below which the molecules do not move and above which they do move. On the other hand, with polymers, as we increase the temperature, localised units (chain segments) within the long chain molecule are first mobilised before the whole molecule starts moving. In some parts within the molecule, there is a considerable localised motion, but not in other parts of the same molecule. You might compare the situation with a section of an audience where individuals stretch out their hands and legs, lean backward and forward, turn this way or that and even stand up and sit down, but do not go away from their seats. The perfect order of a thoroughly disciplined audience is certainly not there, but there is some modicum of order.

Similarly, within the long chain of the polymer molecule, some segments have a certain freedom of movement, whereas others do not. The molecule as a whole does not move although some of its segments do. (The situation is similar to that of a nailed snake which can wriggle violently but is not able to move away).

At this stage we have to differentiate between the two types of motions exhibited within a polymeric material, namely, (*a*) the localised mobility of the segments, called segmental motion, and (*b*) the total mobility of the molecule as a whole, called molecular motion. In other words, we come across two types of Brownian movement (BM) in a polymeric material: (*a*) the 'internal' or 'micro' Brownian movement (IBM), describing the segmental motion, and (*b*) the 'external' or 'macro' Brownian movement (EBM), describing the molecular motion. As we increase the temperature of the polymeric material, the internal Brownian movement would be first activated, and only at a higher temperature, would the external Brownian movement also come into play. At high temperatures, when both Brownian movements are activated, the individual polymer chains start moving apart and there will be a flow of material instead of stress transfer. This state compares with the liquid state of low molecular weight substances; the only difference is that polymeric materials on account of their large molecular size are very highly viscous liquids at such temperatures. So, with polymers, we have something corresponding to the solid state of low molecular weight substances at one side of the temperature scale, where the two Brownian movements are frozen; we also have something corresponding to the liquid state at the other side of the temperature scale, where both Brownian movements are activated. But what happens in between? The very fact that there are two temperature ranges where the two BMs are both frozen or activated respectively, suggests that there should be an intermediate range where one BM be activated and the other frozen.

In the case of polymers, there is indeed such an intermediate stage where the localised mobility is activated but the overall mobility is not. The local segments, where mobility is already activated, correspond to the liquid state, while the molecule as a whole, where mobility is forbidden, is in the solid state. This state, which is really a combination of liquid and solid, is called the rubbery state. In this rubbery state, the polymeric materials exist as liquid with superimposed solids at fixed points or as solid with localised segments of a liquid nature. Under the influence of an applied stress, it exhibits properties or a viscous fluid as well as an elastic solid and undergoes what is called, viscoelastic deformation.

6.2.2 States of Phase

A phase is simply a part of a system separated from the other parts of the same system by an interface. It differs from the other parts with respect to thermodynamic properties, and is also separable from them. From the structural point of view, different phases within the same system differ with regard to the order of arrangement of their molecules.

Low molecular weight substances can exist in three phase states, *viz*., crystalline, liquid and gaseous. Since the gaseous state of phase is not pertinent to polymeric substances, we shall compare only solid and liquid phases of low and high molecular weight substances.

Let us first consider a low molecular weight substance. It may be recalled that a crystalline solid is qualified by the existence of a well-defined long-range order in all the three dimensions of the crystal lattice. The individual atoms or molecules sit in their equilibrium positions and a definite and repetitive molecular pattern is observed over distances which are hundreds and thousands of times the size of individual molecules. These molecules, nevertheless, do not exhibit molecular mobility but only vibrate about their fixed centres of equilibrium. This is the characteristic of a crystalline phase state.

The liquid phase state, on the other hand, is characterised by the absence of any long-range order. This means that there is no repetitive pattern or orderliness of the positions occupied by the molecules over long distances; molecules in the near vicinity of any given molecule may exhibit some orderly arrangement (short-range order) due to aggregation or association, but even this limited orderliness ceases to exist at a small distance from the given molecule.

Two situations are possible in the liquid phase: first where molecular mobility is frozen, and second where molecular mobility is activated. In both the situations, no long-range order is exhibited. Let us consider the first situation, *i.e.* where molecular mobility is frozen. Since the substance does not possess any molecular mobility, it exists as a physical solid, but, at the same time, as it does not possess any long-range order, it is unable to crystallise. Such solids, which are unable to crystallise are called amorphous solids. Since soda glass is a typical solid substance that does not have any crystalline lattice, amorphous solids are compared to soda glass and the amorphous solid state is termed as the glassy state. When such an amorphous solid or glass is heated, the energy of motion (as in the case of crystalline solids) increases and beyond a certain level, the Brownian movement is activated and there is molecular mobility. Now, in the second situation, we have the substance in a liquid phase possessing molecular mobility. Here, due to the onset of molecular mobility, the solid has become liquid. There is no change in the internal order of the molecular arrangement, but there is a change only in the energy of motion. Hence, transition from an amorphous solid state to a liquid state occurs gradually over a temperature range.

What can we conclude from the foregoing description of phase states of a substance and their comparison with the states of aggregates? Three major observations are: (*i*) both crystalline and amorphous (or glassy) solids are in the solid state of aggregation, whereas only crystalline solid exist in the crystalline state phase; (*ii*) amorphous (glassy) solids, although in solid physical (aggregate) state, really exist in a liquid phase state; and (*iii*) both amorphous and crystalline substances can exist in a molten liquid state.

A crystalline substance can thus exist as a 'crystalline' solid (solid phase state) or as a molten liquid (liquid phase state), with solid-to-liquid transition occurring at a temperature known as the melting point (T_m). An amorphous substance cannot exhibit crystallinity (*i.e.* cannot exist in the crystalline phase state) but can exist as a 'glassy' solid (liquid phase state) or as a molten liquid (liquid phase state), and the transition from glassy solid to molten liquid occurs at a temperature; known as the glass transition temperature (T_g). When a crystalline substance passes from the molten liquid state to the solid crystalline state, there is not only a physical change (or a change in the state of aggregation), but also a phase change, involving transition from an equilibrium state of short-range order to another equilibrium state of long-range order. In other words, it is a real phase transition. On the other hand, when an amorphous substance changes from a molten liquid into a glassy solid, there is no change occurring in its internal order. The absence of a long-range order, which is characteristic of the liquid phase, is still retained. However, owing to a decrease in the energy of thermal motion, the molecules lose their mobility and the substance becomes physically solid. The glassy solid is, therefore, considered to be a supercooled liquid only. The transition from the molten liquid to the glassy solid, therefore, involves only a physical change and no phase change.

Now, let us consider polymeric substances and for the sake of simplicity, the uncross-linked ones. Depending on whether a polymer possesses long-range order or not, it can be in the solid (glassy or crystalline), rubbery or liquid molten state. A polymer which does not possess long-range order cannot exist in the crystalline state. It is non-crystallisable or amorphous and exists in the glassy, rubbery or liquid state. At low temperatures, where it does not possess either segmental or molecular mobility, it exists as a glassy solid. On heating beyond a certain temperature, segmental mobility comes into play and it becomes rubbery. This temperature at which the amorphous polymer passes from the glassy state to a rubbery state is the glass transition temperature (T_g). On further heating the rubbery polymer much above its T_g, a temperature is reached beyond which molecular mobility also sets in and the polymer starts flowing. The temperature at which the molecular mobility sets in and the polymer passes from the rubbery state to the liquid state is called its flow temperature or flow point (T_f). It may be noted that due to the absence of a crystal lattice in an amorphous polymer, no true melting takes place, and hence, there is nothing called a melting point (T_m) associated with it.

A polymer capable of exhibiting long-range order, on the other hand, is called a crystalline polymer.* At low temperatures, when molecular mobility is arrested, it exists in the crystalline solid state. The segments in a crystalline polymer are firmly held in the crystallites (see Section 7.6) by intermolecular forces and, hence, do not exhibit any segmental mobility, even at fairly high temperatures. The energy required to overcome such forces forming the crystallites is quite high and almost equal to the energy required to initiate molecular mobility. This means that when a crystalline polymer is heated, the possibility of segmental mobility alone setting in without molecular mobility almost does not exist. The crystalline polymer, therefore, does not pass from solid state to rubbery state and does not possess any glass transition temperature. On the other hand, at a sufficiently high temperature, molecular mobility (associated with segmental mobility) sets in almost abruptly, making the long-range order lose simultaneously. The polymer now passes from the crystalline state directly to the liquid state. The temperature at which this transition takes place is called the melting temperature (T_m). A crystalline polymer can, nevertheless, be made to exist in a glassy state by rapidly cooling its melt to a very low temperature without allowing enough time for the molecules to orient and form crystallites.** This is called 'quenching'.

Most of the polymers in practical use are, however, neither fully crystalline nor completely amorphous. They are partially crystalline and consist of both amorphous as well as crystalline regions. Partially crystalline polymers possess both the glass transition temperature as well as the

*The term 'crystalline polymer', at this juncture, is used to represent a polymer which is 100% crystalline, although in practice, one does not get a polymer with 100% crystallinity, except when grown as single crystal.

**For a discussion on 'crystallites', see Section 7.6.

melting temperature. Below T_g, the amorphous regions of the polymer exist in the glassy state and the crystalline regions in the crystalline state. At T_g, the amorphous regions pass on to the rubbery state, while the crystalline regions continue to be in the crystalline state. At T_m, the crystalline regions melt and pass on to the liquid state. Beyond T_m, the amorphous regions and the crystalline regions become one and the polymer as whole is in a liquid state. The values of T_g and T_m of some commercially important polymers are given in Table 6.1.

Table 6.1 Glass transition and melting temperature of selected polymers.*

Polymer	T_g°C	T_m°C
Cellulose triacetate	49–478	306
Cellulose trinitrate	53	700
Ethyl cellulose	43	–
Natural rubber	–73	36
Nylon 6	50	250
Nylon 6, 6	50	270
Polyacrylonitrile	97	341
Polybutadiene		
cis–	–102	6
trans–	–58	100
Polybutyl acrylate	–54	47
Polycarbonate (Bisphenol-A)	145	–
Polyethyl acrylate	–24	–
Polyethylene (high density)	–125	146
Polyethylene terephthalate	69	264
Polyisobutylene	–73	44
Polymethyl acrylate	10	–
Polymethyl methacrylate		
atactic	105	–
isotactic	38	160
syndiotactic	105	>200
Poly α-methyl styrene	170	–
Polypropylene		
atactic	–13	–
isotactic	–8	208
Polystyrene	100	250
Polyvinyl acetate	32	–
Polyvinyl alcohol		
atactic	85	–
isotactic	–	212
syndiotactic	–	267
Polyvinyl carbazole	150	> 320
Polyvinyl chloride	81	310
Polyvinylidene chloride	–18	190

*Data from *Polymer Handbook*, Eds. Brandrup J. and Immergut, E.H., John Wiley & Sons, New York, 1975.

The whole concept of solid-liquid transition in low molecular weight compounds and solid-rubber-polymer melt transition in polymeric materials can be represented as shown in Fig. 6.2.

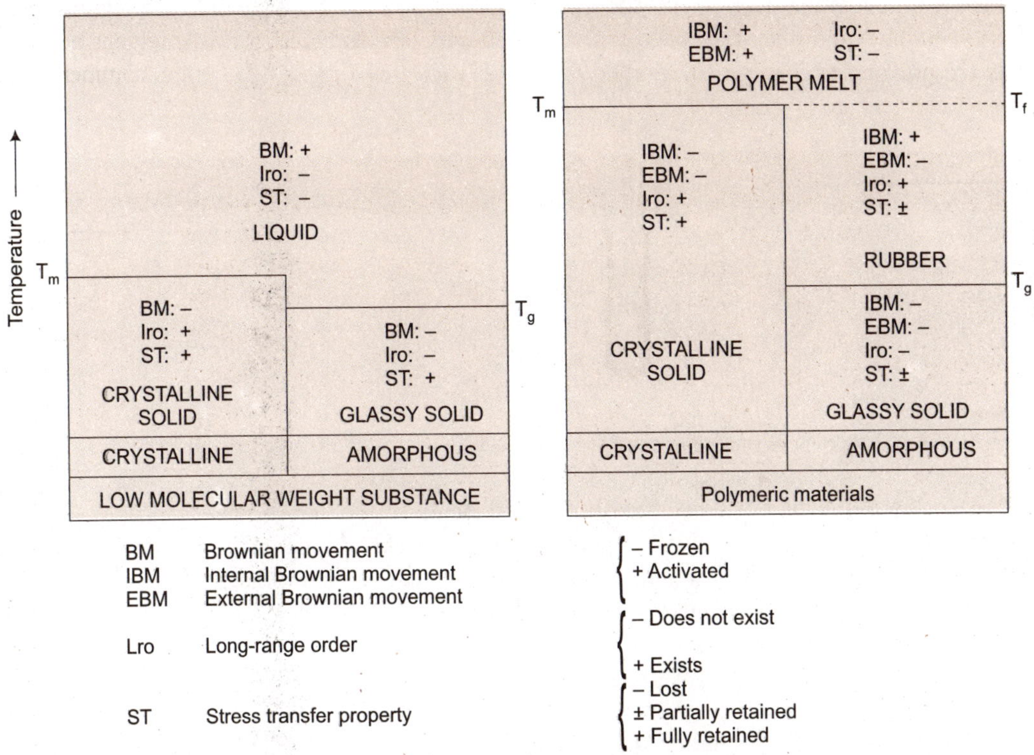

Fig. 6.2 Concept of various transitions of states associated with low molecular weight compounds and polymeric materials.

6.3 TRANSITIONS AND ASSOCIATED PROPERTIES

When a polymer passes from one state to another, there is a gradual change in many of its physical properties. Among them, specific volume is perhaps the most significant property which reveals changes during the transition and is frequently utilised to determine the transition temperatures of the polymer. (Of course, other properties such as density, specific heat and dielectric constant also show gradual changes at the transition points.) The changes in specific volume of a polymeric sample with temperature are measured using a diatometer (Fig. 6.3). The bulb of the instrument containing the sample along with a confined liquid can be heated or cooled to measure the change in volume of the sample with temperature. When the sample expands or contracts with temperature, a corresponding amount of the liquid is displaced and the quantity displaced is measured in a graduated capillary tube.

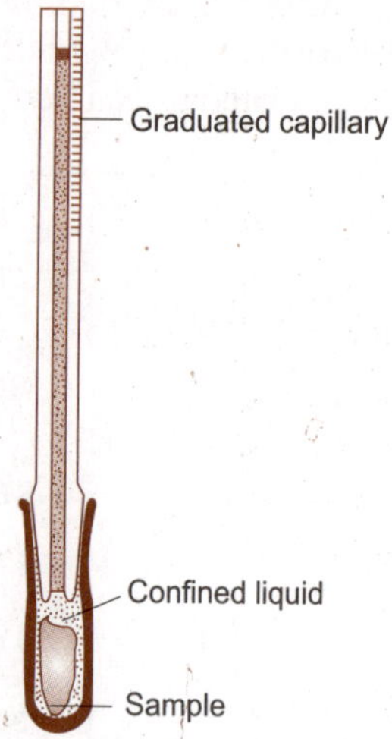

Fig. 6.3 Dilatometer showing polymer sample and confined liquid.

When a low molecular weight crystalline solid melts into a liquid, the change of state is characterised by a discontinuity in its thermodynamic variables. One such variable is specific volume. If we take a sample of a crystalline solid in a dilatometer and measure its volume change with temperature, we will get a plot as shown in Fig. 6.4. It can be seen from the plot that at the

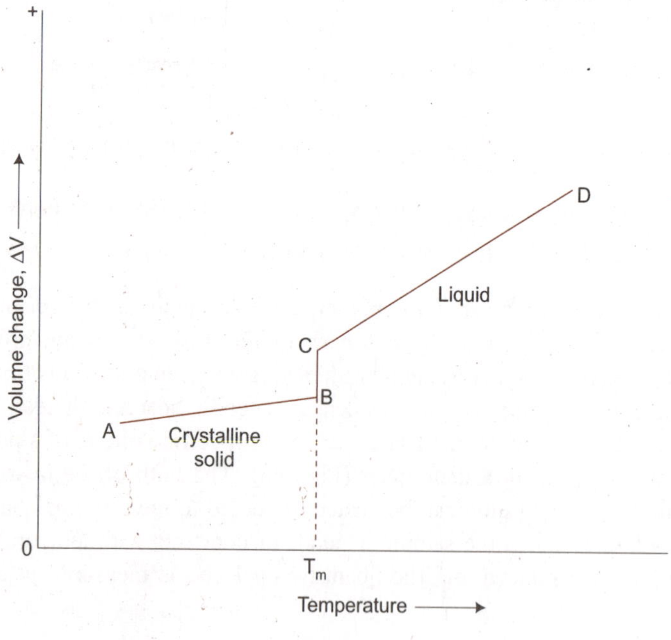

Fig. 6.4 Plot showing trend of change in volume as a function of temperature in the case of low molecular weight crystalline substances.

melting point (T_m), there is an abrupt jump in the volume change of the material, indicating an abrupt change in its specific volume at T_m. When the solid is heated slowly, there is a small but steady increase in its volume, as indicated by the line *AB* (the slope giving the volume coefficient of expansion for the solid), until it reaches T_m, when a sharp increase in its volume is noticed, as indicated by the line *BC*. After the entire sample has melted and as the temperature is increased further, there is yet another steady increase in the volume of the liquid material, as shown by the straight line *CD*. Since the volume coefficient of expansion is higher for a liquid than for a solid, the slope of *CD* is higher than that of *AB*.

Now, let us see what will be the nature of the curve showing specific volume versus temperature (*V* versus *T*) for polymeric substances. As expected, the nature will depend on whether the polymer is crystalline or not. Let us assume that we have three polymer samples of the same chemical structure, but different crystallinity. Assuming sample 1 as highly crystalline,* sample 2 amorphous and sample 3 partially crystalline, we can represent the *V* versus *T* behaviour of these samples by curves as shown in Figs. 6.5 *A*, *B* and *C*. It can be seen from curve *A* that as the highly crystalline polymer is heated, there is a steady increase in volume and, as the crystallites melt at T_m, there is an abrupt increase in its volume. The concept of melting in the case of highly crystalline polymers is almost the same as in low molecular weight crystals, except that, due to the giant size of the polymer molecule and the polydispersity of molecular weight within the sample, there is some deviation from the ideal melting behaviour. Although there are strong intermolecular forces holding the chain segments in the crystallite, at sufficiently high temperatures, they do not completely prevent the onset of segmental motion without the associated molecular motion and there can always be the onset of some segmental mobility just prior to the onset of molecular motion. Because of this and also because of the molecules of different molecular weights melting at somewhat different temperatures, the melting process in a crystalline polymer occurs not at one sharp temperature, but over a finite temperature range. This is shown by a gradual bend, and not by a sharp bend, in the plot just before T_m. The process of melting starts at a temperature corresponding to the start of the upward bend in the curve. Molecules of lower chain lengths start melting first and those of higher chain lengths melt at higher temperatures. As more and more crystallites melt and pass from the crystalline state to the liquid state, the crystallinity of the sample gradually decreases until, at T_m the whole sample melts into a liquid. This range depends on the degree of polydispersity—the broader the molecular weight distribution, the wider the range.

Now, let us take sample 2 which is amorphous. As we start from the glassy state of the polymer and heat it slowly, there is a steady rise in its volume, as indicated by the straight portion *MN* of curve *B* (Fig. 6.5B). The slope of the curve, however, gradually changes near the glass transition temperature (T_g) and then, all of a sudden, the slope increases steeply, as shown by the straight portion *NO*. It can be seen from the curve that the point *N* is obtained by the extrapolation of the straight line portions of the curves representing the glassy state and the rubbery state. This point *N* represents the T_g which is computed as an average of the temperature range in which the volume expansion coefficient gradually increases from a low value for the glass to a high value for the rubber. It can also be seen from the curve that the volume expansion coefficient of the rubbery state is just the same as that of the liquid state of the crystalline polymer, as indicated by the same slope of

*It is difficult to get polymers that are 100% crystalline, except when grown as single crystals; hence Sample 1 can only be assumed as 'highly crystalline' and not 'completely crystalline'.

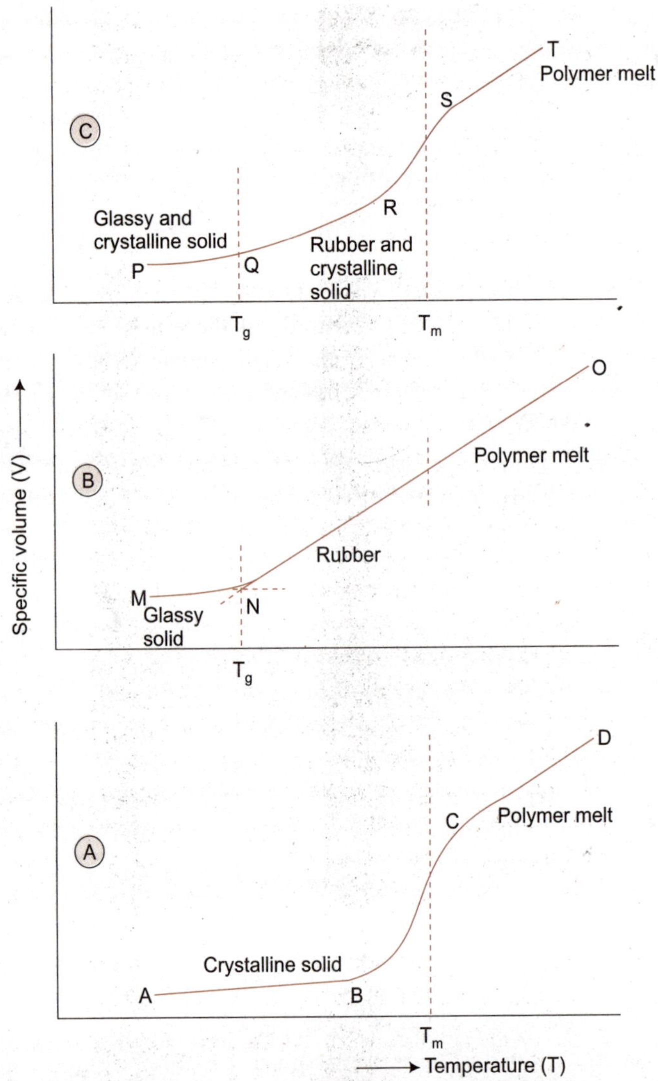

Fig. 6.5 Plots showing variation of specific volume with temperature for (A) highly crystalline, (B) amorphous and (C) partially crystalline, polymers.

the straight line representing these two stages. In fact, if these two curves *A* and *B* are superimposed, they almost overlap beyond T_m, indicating that beyond the temperature at which both segmental and molecular mobility start operating, the polymer of a particular chemical structure will be in its liquid state of aggregate; this is irrespective of whether it is crystallisable or not. It can be seen that beyond T_g, curve *B* shows no change in its slope even at temperatures as high as the T_m in curve *A*, indicating that the change in specific volume while moving from the rubbery to the molten state is imperceptible. An amorphous polymer, in fact, does not undergo true melting.

In the case of sample 3, which has both crystalline and amorphous regions, we see a mixed behaviour. The V versus T plot (curve C, Fig. 6.5C) shows a change of slope at two temperature ranges, the lower temperature range representing the T_g region and the higher one the T_m region.

6.4 FACTORS INFLUENCING THE GLASS TRANSITION TEMPERATURE

The presence or absence of segmental and molecular motions decides whether a polymer is in a solid, rubbery or molten state. The nature and the magnitude of these motions depend on the size and geometry of the polymeric chain, flexibility of the chain segments and the type of molecular aggregates formed. The following paragraphs list some of the pertinent factors influencing the same.

If molecular geometry permits formation of a definite molecular orientation leading to a long-range three-dimensional order, the polymer has a greater crystallisability. That is why polymers with a symmetrical or stereo-regular structure are crystalline. Polymers with irregular chain backbone or randomly placed side groups are non-crystallisable. Segmental and chain mobilities are easier in non-crystallisable or amorphous polymers than in crystalline polymers. Flexibility of the chain segments is determined by the degree of freedom with which different segments along the chain backbone can rotate around the covalent bonds. Linear polymer chains made of C—C, C—O or C—N single bonds have a high degree of freedom for rotation. The presence of aromatic or cyclic structure in the chain backbone or of bulky side groups on the backbone C atoms hinders the freedom for rotation. The higher the freedom to rotate, the more flexible are the chain segments, and hence, the higher their segmental mobility. Similarly, the intermolecular forces determine the magnitude of the molecular aggregates. In the case of hydrocarbon polymers, only van der Waals' forces act on neighbouring chains and hence, molecular aggregates are not that strong. Chain segments can slip past each other easily. Polymer chains containing polar groups, on the other hand, are held together more strongly by neighbouring dipoles as well as intermolecular hydrogen bonding, and are unable to move that easily.

Hindrance to segmental rotation and intermolecular cohesive forces restrict the molecular and segmental motions and the chains are held down to positions of minimum potential energy. The thermal energy of a polymer (which naturally increases with temperature), however, activates thermal vibrations of the chain elements. As the temperature is increased, these invigorated thermal vibrations are able to get over the intermolecular cohesion and the hindrance to segmental rotation. When these restrictions are overcome at appropriate temperatures, segmental and molecular motions set in. The onset of these motions determines the T_g as well as the T_m of the polymer. Hence, it can be seen that the glass transition temperature (as also the melting temperature) of a polymer depends on such parameters as chain geometry, chain flexibility and molecular aggregates.

We now venture to provide the following generalisations: Highly crystalline polymers possessing a regular chain geometry show a high glass transition temperature. A bulky side group is, if we may draw an analogy, like the heavy garment of a swimmer, affecting his or her movements. Truly, bulky groups come in the way of segmental motion and increase the T_g of the polymer.

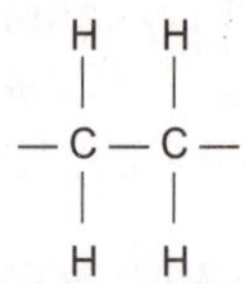

Polyethylene, which is a hydrocarbon polymer and made up of repeating units of ethylene, has a T_g of –125°C. In this case, T_g is quite low because (*i*) strong intermolecular cohesive forces are absent and (*ii*) the substituent group on C atoms is only hydrogen, which is not bulky at all. On the other hand, nylon 6, a polyamide, has a high glass transition temperature (50°C) primarily because of the presence of a large number of polar groups in the molecules, leading to strong intermolecular hydrogen bonding (see Fig. 6.6). It should be noted that intermolecular cohesive forces restrict segmental motion below 50°C in the case of nylon 6.

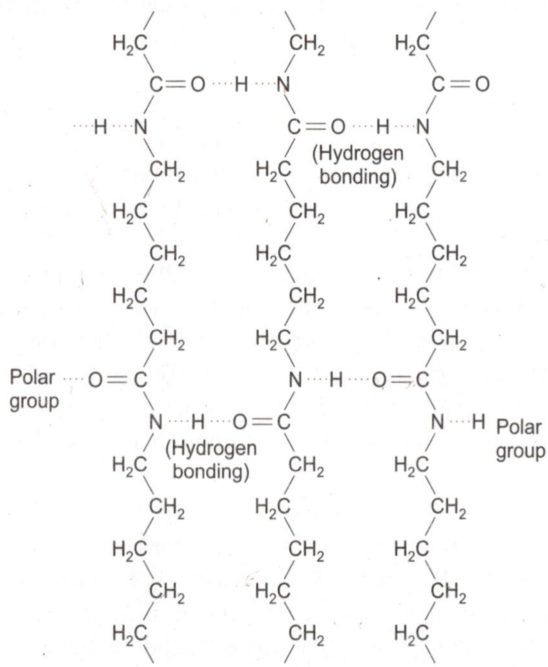

Fig. 6.6 Chemical structure of nylon 6 is showing hydrogen bonding between adjacent chains through polar groups.

The effect of the side group can be appreciated by comparing the T_g of polystyrene with that of polymethylstyrene and the T_g of polymethylacrylate with that of polymethylmethacrylate. The presence of —CH_3 groups in poly α-methylstyrene and polymethylmethacrylate, comes in the way of free rotation around C—C bond of the chain backbone and, hence, hinders the chain mobility, resulting in an increase of around 70°C and 100°C in their T_g values over polystyrene and polymethylacrylate, respectively.

A different trend in the effect of substituent side chains on T_g is seen if we examine the T_g values of methyl, ethyl and butyl polyacrylates. In this series of polyacrylates, the T_g values get reduced as the length of the side chain increases (*e.g.* 10°C, –24°C and –54°C for methyl, ethyl and butyl, respectively). Here also, the presence of the side chain, no doubt, hinders the free rotation of the C—C bond of the main chain, but this effect is almost the same in all the three acrylates. The side chains, nevertheless, due to the presence of C—C and C—O linkages, are themselves flexible. As the side chain length increases from methyl to butyl, their freedom of flexibility increases and they also tend to push the neighbouring main chains further apart, increasing thereby the free volume and, hence, the chain mobility.

A T_g value of 150°C for polyvinyl carbazole is understandably high, as it contains bulky side groups as shown:

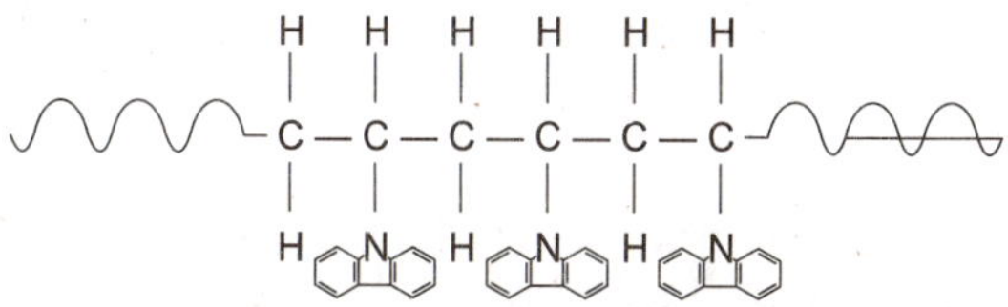

(Polyvinyl carbazole)

In the case of polyethylene terephthalate, represented as

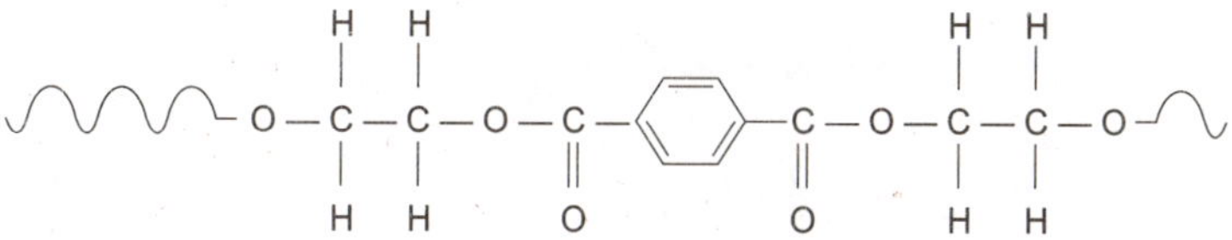

(Polyethylene terephthalate)

the presence of the aromatic rings in the chain backbone increases the inflexibility of the chain and the T_g value is high (69°C). The higher T_g value for the derivatives of cellulose, such as cellulose nitrate, is attributable to the rigid ring structures in the macromolecular chain:

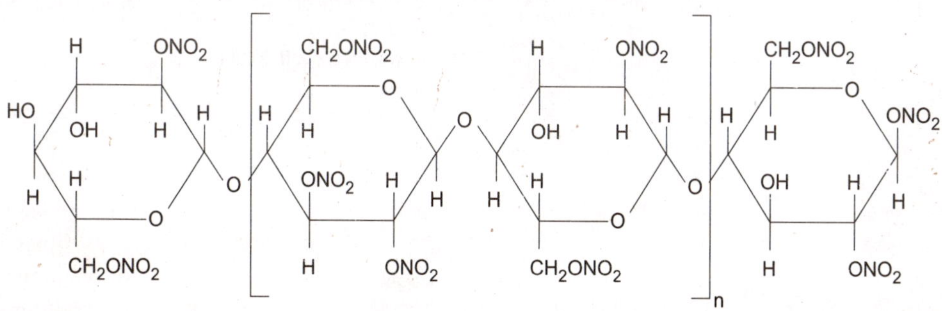

Cellulose nitrate

The phenomenon of ultimate chain stiffening devoid of any segmental rotation may be seen in 'ladder polymers' (name derived from their structural resemblance to a ladder):

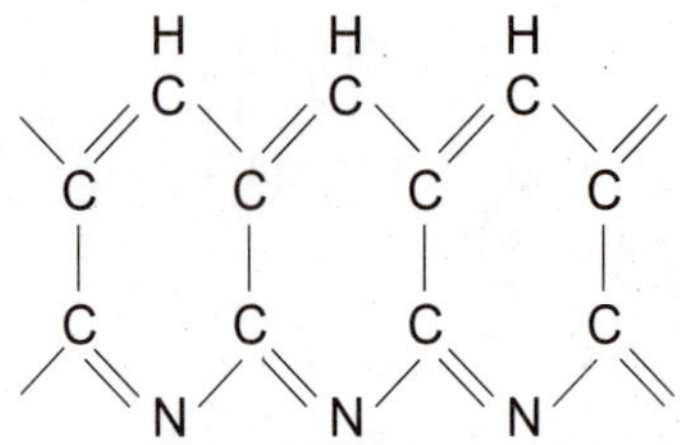

Ladder structure

The ladder polymer has been found to be very stable thermally. Even upon direct exposure to a flame, the material shows a remarkable stability.

6.5 GLASS TRANSITION TEMPERATURE AND MOLECULAR WEIGHT

The glass transition temperature of a polymer is influenced by its molecular weight, at least up to around a value of 20,000; beyond this, the effect of the molecular weight is not pronounced and this can be seen from Fig. 6.7. Many attempts have been made to compute mathematical relationships, among which the following two are frequently used:

$$T_g = T_g^{\infty} - \frac{K}{M_n} \tag{6.1}$$

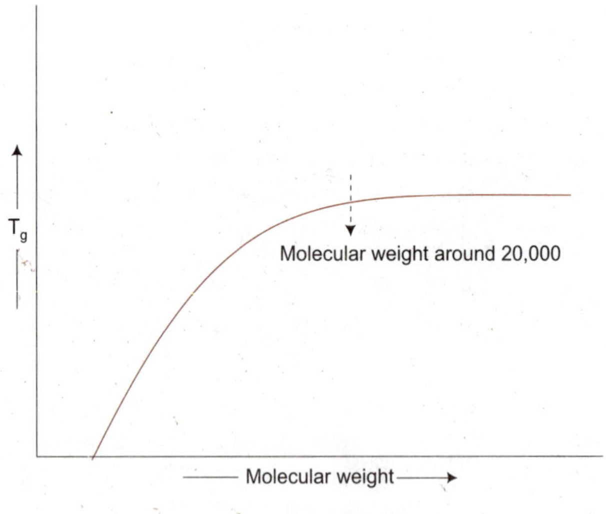

Fig. 6.7 Plot showing effect of molecular weight on glass transition temperature of polymers.

$$\frac{1}{T_g} = \frac{1}{T_g^{\infty}} + \frac{A}{M_n} \tag{6.2}$$

where T_g^{∞} is the glass transition temperature at infinite molecular weight and K and A are arbitrary constants.

Now, let us see how the molecular weight can affect the T_g values. We know that polymer chains made of hundreds of monomeric units comprise several segments, each made of a few tens of monomeric units. We also know from earlier discussions that the mobility of chain segments influences the glass transition temperature. The segments that are away from the chain ends, although part of the main chain, can be thought of as connected at the two ends to the main chain (see Fig. 6.8) and hence, their mobility is rather restricted. The two end segments containing the chain ends, on the other hand, are connected to the main chain at one of their ends only and hence, have more freedom for motion. For a given weight of the polymer, a low molecular weight sample will have more chain end segments than a high molecular weight sample. The larger the number of chain end segments, the larger will be the effective segmental motion. Thus, the T_g value will be lower for low molecular weight polymers.

Fig. 6.8 Schematic representation of a polymer chain showing chain ends and chain segments.

6.6 GLASS TRANSITION TEMPERATURE AND PLASTICISERS

Plasticisers are low molecular weight non-volatile substances (mostly liquids), which when added to a polymer, improve its flexibility, processibility and hence, utility. The plasticiser substantially reduces the brittleness of many amorphous polymers because its addition even in small quantities markedly reduces the T_g of the polymer. This effect is due to a reduction in cohesive forces of attraction between polymer chains. Plasticiser molecules (being relatively small in size compared to polymer molecules) penetrate into the polymer matrix and establish polar attractive forces between it and the chain segments. These attractive forces reduce the cohesive forces between the polymer chains and increase the segmental mobility, thereby reducing the T_g value.

The reduction in T_g with the addition of different plasticisers to a polystyrene sample can be seen in Fig. 6.9. It would appear that although all plasticisers reduce T_g considerably, the extent of reduction (*i.e.*, the 'plasticiser efficiency') is different, and depends on factors such as solubility parameter, polarity and density.

Many plasticisers such as dibutyl phthalate, dioctyl sebacate, diisooctyl phthalate and tricresyl phosphate are used to decrease the T_g value of PVC. Selecting a plasticiser for commercial application is obviously a matter of compromise, after having considered various factors such as compatibility, efficiency, migration effect and cost.

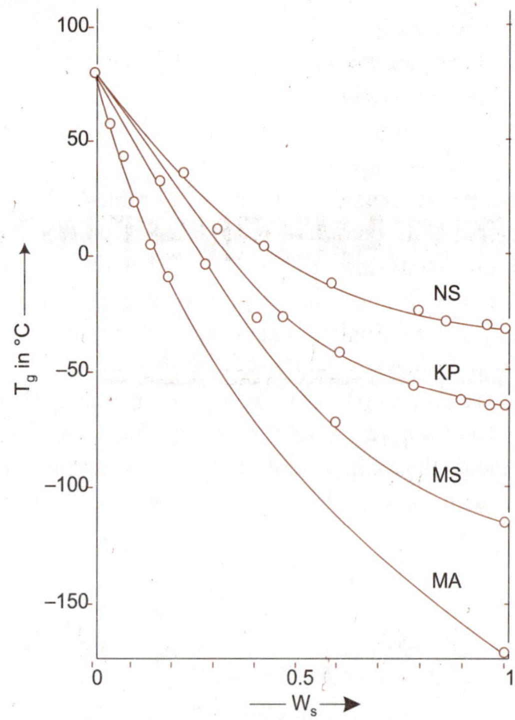

Fig. 6.9 Plot showing lowering of glass transition temperature of a polystyrene sample by different weight fractions (W_s) of various plasticisers. (NS = naphthyl salicylate, KP = tricresyl phosphate, MS = methyl salicylate and MA = methyl acetate).

6.7 GLASS TRANSITION TEMPERATURE OF COPOLYMERS

We will take the case of a random copolymer. The net effect of the interaction between two monomeric components of a copolymer on the T_g value is as though the copolymer is a mixture of two homopolymers. The T_g value of random copolymers, therefore, depends on the T_g value of the individual homopolymers and on the ratio of the two monomeric components. This leads to the formulation of an empirical relation:

$$\frac{1}{T_{g(AB)}} = \frac{W_A}{T_{g(A)}} + \frac{W_B}{T_{g(B)}} \tag{6.3}$$

where $T_{g(A)}$, $T_{g(B)}$ and $T_{g(AB)}$ are the glass transition temperatures of homopolymers A and B and the copolymer AB, respectively, and W_A and W_B are the weight fractions of the respective monomer components in the copolymer. It can be seen from this equation that the glass transition temperature of a copolymer will be inbetween those of the respective homopolymers. The glass transition temperature of many homopolymers could thus be lowered by copolymerisation with small quantities of another suitable monomer. This process is generally referred to as 'internal plasticisation', which is in contrast to 'external plasticisation', wherein T_g is lowered by the physical addition of a plasticiser.

6.8 GLASS TRANSITION TEMPERATURE AND MELTING POINT

Many attempts have been made to study the interrelationship between the glass transition temperature and melting point. We could anticipate, on basic considerations, that factors affecting T_g will also affect T_m. Based on experimental observations, T_g and T_m (in degrees Kelvin), have been shown to be interconnected as follows:

$$T_g = \frac{1}{2} T_m \quad \text{(for symmetrical polymers)} \tag{6.4}$$

$$T_g = \frac{2}{3} T_m \quad \text{(for unsymmetrical polymers)} \tag{6.5}$$

However, these are not universal relations and deviations have been reported in the case of many polymers. A combined version of these two equations, in which the effect of molecular symmetry is not taken into account, is also in vogue, as now shown:

$$\frac{1}{2} < \frac{T_g}{T_m} < \frac{2}{3} \tag{6.6}$$

Since this relation gives a range for T_g/T_m (rather than a sharp value), it may more realistically reflect the thermal behaviour of polymers.

6.9 IMPORTANCE OF GLASS TRANSITION TEMPERATURE

The glass transition temperature is an important parameter of a polymeric material. It is used as a measure for evaluating the flexibility of a polymer molecule and the type of response the polymeric material would exhibit to mechanical stress. The T_g value of a polymer, therefore, decides whether a polymer at the 'use temperature' will behave like rubber or plastic. We will come across more details of this aspect in Chapter 14. At present, it would suffice to know that polymers above their T_g will be soft and flexible and exhibit a delayed elastic response (viscoelasticity), while those below their T_g will be hard and brittle and will possess dimensional stability.

The T_g value along with the T_m value gives an indication of the temperature region at which a polymeric material transforms from a rigid solid to a soft viscous state. This helps in choosing the right processing temperature, *i.e.*, the temperature region in which the material can be converted into finished products through different processing techniques such as moulding, calendering and extrusion, details of which are given in Chapter 15.

6.10 HEAT DISTORTION TEMPERATURE

Closely associated with T_g is the heat distortion temperature. While T_g gives a temperature range over which a hard polymeric material tends to become soft under static conditions, the heat distortion temperature indicates the temperature range over which the polymer begins to soften and gets deformed under the influence of a load. In the polymer industry, the heat distortion temperature is, in fact, more often used (than T_g) as an empirical method of comparing the transition behaviour of different polymers from their hard solid state to the soft viscous state. The heat distortion temperature can be expressed as the narrow temperature range over which a standard test specimen at a standard rate of heating gets distorted to a fixed extent under a given load. It will be close to T_g for amorphous polymers and close to T_m for highly crystalline polymers.

Books for Further Reading

Shen, M.C., and Isenberg, A.E., *Glass Transition in Polymers*, Pergamon, New York, 1966.

Deanin, R.D., *Polymer Structure, Properties and Applications*, Cahners, Boston, 1972.

Haward, R.N., Ed., *The Physics of Glassy Polymers*, John Wiley & Sons, New York, 1973.

7

CRYSTALLINITY IN POLYMERS

7.1 CRYSTALLINE SOLIDS AND THEIR BEHAVIOUR TOWARDS X-RAYS

We all know that a definite shape is one main characteristic of solids and this distinguishes them from gases and liquids. We also know that solids are conventionally classified as crystalline or amorphous, depending on whether they possess a 'long-range order' or not (Section 6.2.1).

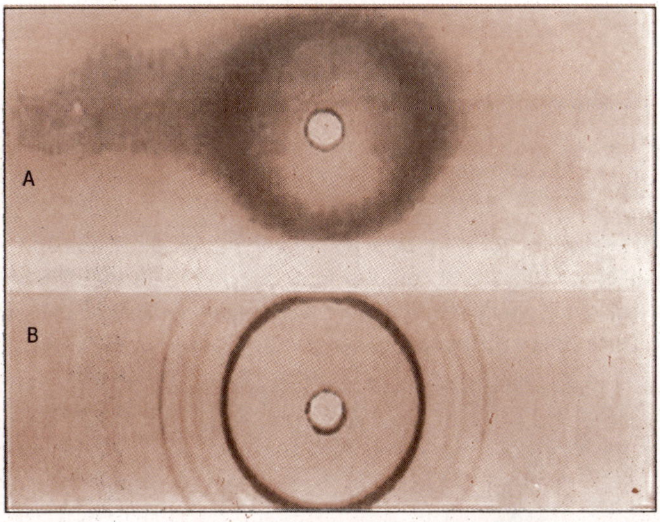

Fig. 7.1 X-ray diffraction patterns of low molecular weight solids: (A) crystalline and (B) amorphous

The presence or absence of a long-range order makes different solids behave differently when exposed to X-rays. While crystalline substances give sharp and well-defined X-ray diffraction patterns, non-crystalline (amorphous) substances give rather diffuse and broad patterns (Fig. 7.1).

We might recall at this stage as to why diffraction patterns are formed when X-rays fall on solids. X-rays, like visible light, exhibit a wave nature. The wavelength of light can be measured using a diffraction grating, which consists of a series of lines separated by a distance of roughly one wavelength (λ) of the monochromatic light. When a monochromatic beam of light falls on the grating, the light waves emerging from the grating interfere with one another, to either amplify or eliminate the light in a particular direction and, finally, give alternately dark and bright bands on the screen. Now, in a crystal lattice, the molecules are seated in a well-defined and repetitive pattern and can act as a diffraction grating. A crystalline substance, therefore, diffracts X-rays and gives patterns which are, indeed, the 'fingerprints' of the substance.

A scheme for the X-ray diffraction set-up is shown in Fig. 7.2.

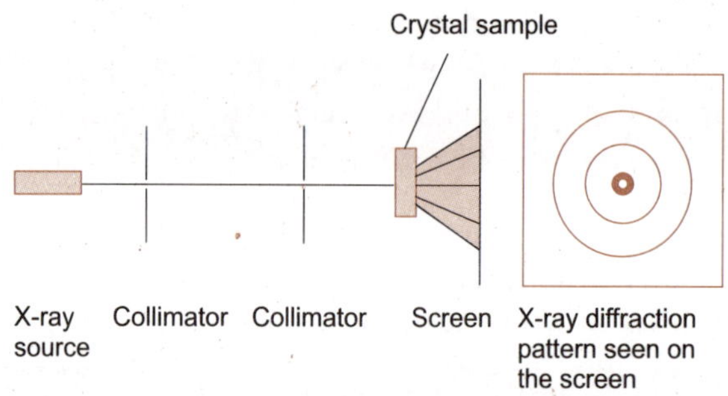

Fig. 7.2 Scheme for an X-ray diffraction set-up.

7.2 POLYMERS AND X-RAY DIFFRACTION

When X-ray crystallographic techniques were extended to polymeric solids, some interesting features of the internal structure of these substances came to light. A good majority of polymers diffract X-rays like any crystalline substance would. On the other hand, many behave like amorphous materials giving very broad and diffuse X-ray diffraction patterns. This is seen in Fig. 7.3.

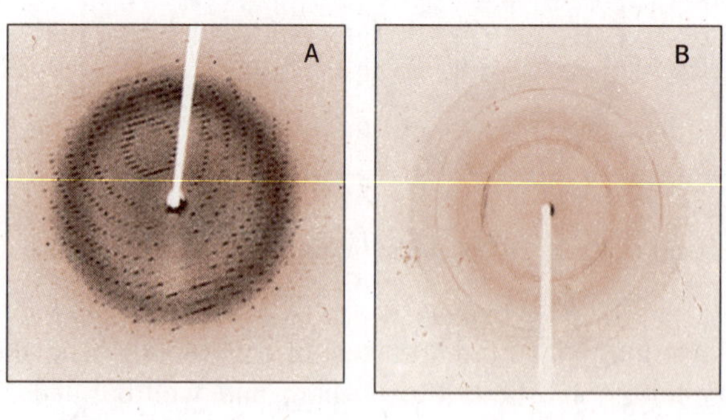

Fig. 7.3 X-ray diffraction patterns of (A) highly crystalline polymer and (B) amorphous polymer

Unlike simple inorganic compounds (such as NaCl or KCl), polymers do not have a perfectly ordered crystal lattice formation and are not completely crystalline. In fact, they contain both crystalline and amorphous regions. As a consequence, the X-ray diffractions from them are found to be a mixture of sharp as well as diffused patterns.

7.3 DEGREE OF CRYSTALLINITY

The crystalline and amorphous components influence polymer properties in as much as the molecular weight of a polymer is influenced by that of the various fractions which make the polymer. Actually, crystallinity of a polymer sample is expressed in terms of that fraction of the sample which is crystalline. This definition of the degree of crystallinity is, of course, based on the premise that crystalline and non-crystalline components of a polymer can (and indeed do) co-exist. It is for this reason that X-ray diffraction patterns of most polymers contain both sharp as well as broad and diffuse bands. The sharp bands correspond to crystalline (orderly) regions and the diffuse and broad ones refer to amorphous (non-orderly) regions. Figure 7.4 shows how the crystalline and amorphous regions can co-exist and how they are held together by long polymer chains running through them. The highest crystallinity is generally associated with polymers which have a simple unit structure and a relatively high degree of molecular order. (This will be explained later). We could, therefore, express the overall property (Q) of a partially crystalline polymer as a sum total of its two components as follows:

$$Q = Q_c + Q_a \tag{7.1}$$

Fig. 7.4 Arrangement of polymer chains forming crystalline and amorphous regions in polymers.

where Q_c and Q_a are the contributions of the crystalline and amorphous components of the sample, respectively. We know that the density of a material increases when it changes i's phase from liquid to crystal. This is also true of polymers. Amorphous components in a polymer are in the liquid phase and the crystalline components in the crystalline phase. The density of the crystalline component is, therefore, higher than that of the amorphous component. For a given polymer, a 100% crystalline sample will have the highest density and a 100% amorphous sample will have the lowest. Many of the actual polymers are only partially crystalline and their densities will naturally be in between those of their crystalline and amorphous components. Conversely, from the knowledge of the densities of the crystalline and amorphous components of a polymer, and that of the aggregate sample, the degree of crystallinity of the sample can be computed as follows:

$$X_v = \frac{d - d_a}{d_c - d_a} \tag{7.2}$$

where X_v is the degree of crystallinity by volume and d, d_a and d_c are the densities of the sample, the fully amorphous and fully crystalline components, respectively. Apart from density, specific volume (V) of the sample could also be used to measure the degree of crystallinity. The corresponding equation would be

$$X_m = \frac{V_a - V}{V_a - V_c} \tag{7.3}$$

where X_m is the degree of crystallinity by mass and V, V_a and V_c are the specific volumes of the sample, the fully amorphous and fully crystalline components, respectively.

While the degree of crystallinity of a partially crystalline sample can be experimentally determined by measuring the density or specific volume, the same can also be calculated by using enthalpy (H) as per the equation:

$$X_{cal} = \frac{H_a - H}{H_a - H_c} \tag{7.4}$$

where H_a, H_c and H are fully amorphous, fully crystalline and sample enthalpies, respectively.

7.4 CRYSTALLISABILITY

There is another term, namely, 'crystallisability' which is apt to be confused with crystallinity. Crystallisability is the maximum crystallinity that a polymer can achieve at a particular temperature, regardless of the other conditions of crystallisation. Crystallisability at a particular temperature depends on the chemical nature of the macromolecular chain, its geometrical structure, molecular weight and molecular weight distribution. On the other hand, the actual degree of 'order' or the extent of crystallinity attained at a particular temperature depends very much on such conditions as the rate of cooling, residence time, temperature of the molten polymer and heat dissipation, under which crystallisation takes place.

7.5 POLYMER CRYSTALLISATION

Any polymer at a temperature higher than its T_m or T_f is in the form of a viscous melt. The melt is in a liquid phase where polymer chains exist in a completely disordered manner. They have conformations of random coils interpenetrating each other and are constantly undergoing segmental and molecular motions due to thermal energy. The state of polymer molecules in such a polymer melt can be compared to a bunch of long worms kept in a warm bowl. Imagine, however, that the interior of the bowl is slippery and the worms are unable to get out of it. Since all of them are simultaneously trying to get out, they get entangled with each other, and the bulk of the worms looks like a soft mass exhibiting constant wavy motion.

The degree of chain coiling and chain entanglement in the melt, however, depends on the chemical nature and geometrical structure of the polymer chains. Polymers composed of non-polar flexible chains assume maximum chain coiling, and hence, are more entangled. Polar polymers and polymers with bulky side groups are less coiled and less entangled. Also, linear polymers are less entangled than branched polymers.

What happens to a polymer melt when it is cooled? It will solidify, of course. But the type of solid obtained will depend on the nature of the polymer. On cooling the melt, the thermal energy associated with the molecules decreases and exactly at the point of solidification the molecular mobility of the chains is arrested. The chains which had been continuously changing their conformations in the molten state are frozen to fixed conformations at the point of solidification. Depending on the conformations taken by the individual chains and also on the nature of the aggregation of nighbouring chains at the time of solidification, the polymer may either crystallise or vitrify, *i.e.*, the polymer will become either a crystalline or a glassy (amorphous) material.

The crystalline state is characterised by the existence of a long-range order with respect to molecular arrangement. It is, therefore, essential that if a polymer melt has to crystallise, the chains which were completely disordered in the melt state, take up conformations leading to a highly ordered and oriented arrangement of chain aggregates at the time of solidification (Fig. 7.5).

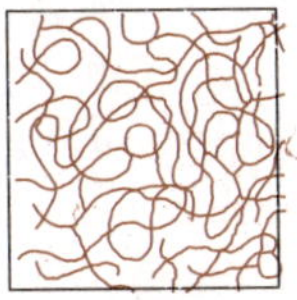

(A) Molten state

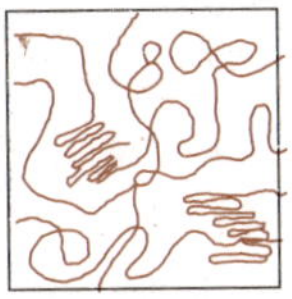

(B) Solidified state

Fig. 7.5 Polymer chains exist as disorderly and interpenetrating random coils in a polymer melt (A). During solidification of melt, these chains diffuse and form orderly arranged chain aggregates in crystalline regions (B).

What takes place first at the point of solidification is the formation of several nuclei for crystal growth. These nuclei are formed either by impurities present or by minute volume elements of chain segments spontaneously aggregating together in a sterically ordered manner. Once the nuclei are formed, the crystal growth around them starts due to chain segments from the melt regions diffusing on to the nuclei and taking orderly orientation around them. This process of polymer chain segments diffusing from a disorderly molten state to an orderly oriented crystalline state is accompanied by a decrease in the free energy, which actually provides the driving force for an increasing number of the chain segments to diffuse into the orderly crystalline region. The process does not, however, go on endlessly. The diffusion of chain segments continues smoothly in regions where there is no chain entanglement. At regions of chain entanglement, the chain segments lying between the already crystallised portion of the chain and the portion caught within the entanglement get strained. The work required to overcome the stress developed within the chain opposes the free energy lowering. In regions where chain entanglements are minimum, the work required to overcome the stress is not sufficient to completely offset the free energy lowering and the process of crystal growth continues. On the other hand, in regions of high chain entanglements, the work required to overcome the stress actually offsets the free energy lowering and hinders the process of chain diffusion and crystal growth. We can, therefore, expect crystallisation to be favoured in polymer melts where chain entanglement is minimum, and only such polymers with chemical and geometrical structures as would lend themselves to a highly ordered molecular arrangement can crystallise, *i.e.*, only certain types of polymers are crystallisable. Other polymers, whose chain structure and geometry would not lend their chains to an orderly arrangement cannot crystallise and can only form glassy solids.

In the case of crystallisable polymers too, it is very difficult to get a fully crystalline solid. When the melt is cooled, the viscosity increases rapidly and, at the point of crystallisation, it becomes extremely high. This high viscosity considerably slows down the process of chain diffusion from the disorderly state to a highly ordered state. While some chains settle in the orderly state, others are still in the entangled and disorderly state. Another condition wherein the process of crystal growth does not go to completion is seen when different portions of the same chain, spaced quite far apart, may simultaneously diffuse into more than one crystalline region and get involved in the crystal growth around more than one nucleus. As the process continues, at some stage, the chain segments lying between these different regions of crystal growth get strained and prevent further segmental diffusion. Within the same chain, therefore, only some segments get settled in an orderly state, while the remaining are still in a disorderly state. The process of crystallisation, therefore, never goes to completion. The resultant solid contains both crystalline as well as amorphous regions. The ratio of the crystalline region to the amorphous region depends upon the chemical nature of the polymer and physical parameters of crystallisation such as temperature and rate of cooling. Maximum crystallisation becomes possible if the polymer is annealed just at its T_m for a sufficiently long time so as to allow the maximum number of chains or chain segments to diffuse into an orderly arrangement. Crystallisation can, sometimes, be completely avoided by rapidly cooling down the polymer melt (quenching) much below its T_m without allowing any time for the process of chain orientation to take place. In this way, a polymer which can otherwise be crystallised is obtained as a completely glassy (amorphous) solid.

7.6 CRYSTALLITES

The crystalline regions in a polymer, unlike those of a low molecular weight substance, do not have a regular shape nor do they have a perfect lattice structure. The expression 'highly ordered state of chain molecules in the crystalline regions in a polymer' has been used to suggest that in these volume elements of the polymeric substance, the chain segments are so arranged as to form small orderly bundles or aggregates resembling, but not quite representing, the limited portion of a three-dimensional crystal lattice of low molecular crystals. These orderly regions in a polymeric substance are called 'crystallites' instead of 'crystals'. Crystallites can be visualised as regions composed of imperfect crystal, like chain aggregates. A partially crystalline polymeric material consists of several such crystallites co-existing with amorphous regions of disorderly placed chain segments. No sharp boundaries, however, exist between the crystallites and the amorphous regions. The former are connected to the latter by polymer chains running through both the regions (Fig. 7.4).

7.7 STRUCTURAL REGULARITY AND CRYSTALLISABILITY

The most important factor which determines whether a polymer can crystallise or not is its geometrical regularity. By 'geometrical' we mean the configuration of the chain. Stereo-regular polymers, *i.e.*, isotactic and syndiotactic are found to crystallise, whereas atactic ones are unable to do so. Similarly, linear polyethylene has a highly regular configuration and is, therefore, highly crystalline (crystallinity 90%). But this high crystallinity drops sharply to around 40% when there is branching. Branched polyethylenes find it more difficult to crystallise than linear polyethylenes do. Figure 7.6 shows the structural difference between linear and branched polyethylenes. We can see that branching imparts irregularity to the molecular structure and reduces the ability of the molecules to get themselves packed closely and hence, their ability to crystallise. The extent to which this ability is reduced depends on the pattern of branching.

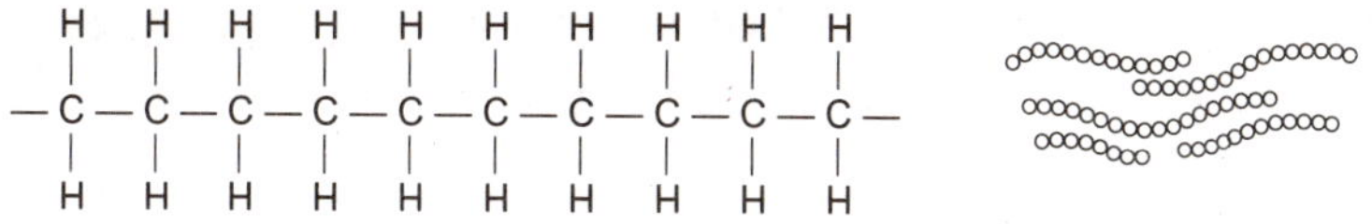

Linear polyethylene: crystallinity–90%

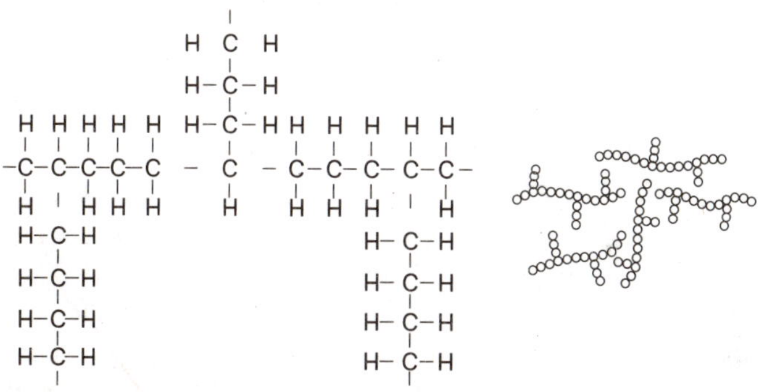

Branched polyethylene: crystallinity–40%

Fig. 7.6 Structure of linear and branched polyethylenes.

Natural rubber and gutta-percha provide yet another example to illustrate the effect of macromolecular geometry on crystallisability. Both are polyisoprenes with structures as shown in Fig. 7.7. In the cis-isomer, *i.e.*, natural rubber, there is a bending back of the successive isoprene units, giving the molecule a coiled structure. The trans-isomer, *i.e.*, gutta-percha, on the other hand, exhibits a straightening out of the successive isoprene units, giving a rod-like structure to the molecule. A rod-like structure enables gutta-percha molecules to orient themselves in an orderly manner and pack up closely, as compared to the coiled molecules of the natural rubber. Gutta-percha is, therefore, more crystalline.

The effect of structural regularity on crystallisability can also be explained by comparing a homopolymer with its copolymer. Linear polyethylene is highly crystalline. A random copolymer of ethylene and propylene, on the other hand, is completely non-crystalline. Random distribution of

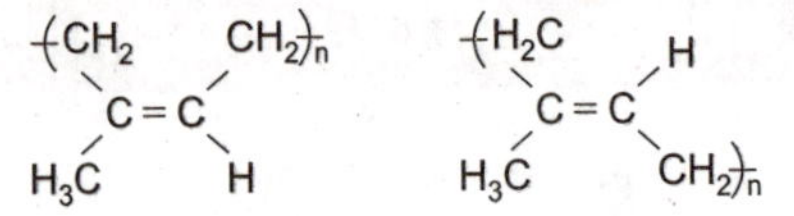

Fig. 7.7 Configuration of isoprene units in natural rubber and gutta-percha.

propylene repeat units in the copolymer disturbs the structural regularity of the chain molecule and curbs the tendency to crystallise. Consequently, random copolymers do not generally crystallise. Strictly alternating copolymers, where the two repeat units are arranged in regular alternation, however, retain the structural regularity and, hence, can show a tendency to crystallise.

7.8 OTHER FACTORS AFFECTING CRYSTALLISABILITY

Whether a polymer can crystallise or not depends, apart from its structural regularity, on many other factors. One such factor is polarity. The highly crystalline nature of nylon 6, for instance, is attributable to the polar groups in the molecule which leads to the formation of hydrogen bonds (see Fig. 6.6). It can be seen from Fig. 6.6 that the carbonyl oxygen atoms of one polymer chain come opposite to the NH groups of another chain in such a way that the formation of the hydrogen bond is favoured. The formation of hydrogen bonds between adjacent chains increases the interchain forces of attraction and facilitates tighter packing and perfect bonding of the chain elements with each other.

Let us now examine a few more polymers and see how their ability to crystallise is affected by different parameters. Take, for instance, polyvinyl carbazole, which has very bulky groups randomly projecting from the chain backbone as shown in Fig. 7.8. This polymer (unlike polyethylene, which has no bulky side groups) finds it very difficult to crystallise because of the bulky side groups which come in the way of a closer molecular packing. But if the substitute (side group) is relatively small, as in the case of polyvinyl alcohol (Fig. 7.9) or polyvinyl fluoride, the polymer can crystallise. Polyvinyl alcohol and polyvinyl fluoride do, in fact, crystallise.

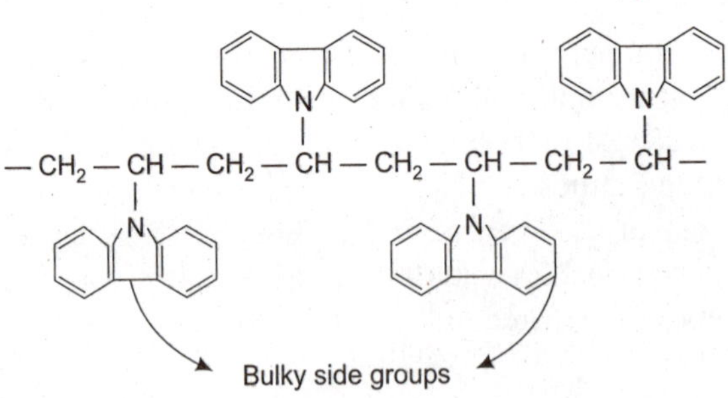

Fig. 7.8 Structure of polyvinyl carbazole.

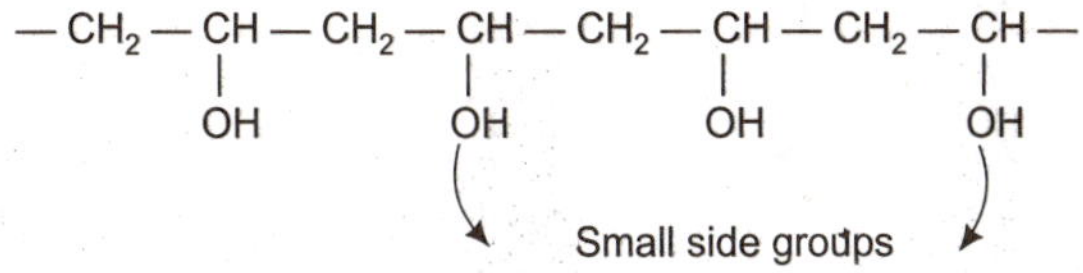

Fig. 7.9 Structure of polyvinyl alcohol.

7.9 HELIX STRUCTURES

To facilitate closer packing of molecules in the crystalline phase, many polymers tend to assume a helical structure. Isotactic vinyl polymers assume conformations as shown in Fig. 7.10. It can be noted from this figure that in all these conformations the carbon atoms (shown by smaller circles) forming the chain backbone are so arranged that an imaginary line drawn through all the successive substituent groups (shown by bigger circles) attached to the alternate C atoms of the chain backbone forms a helix. If three successive substituent groups complete one turn in the helix, the arrangement is called a 3_1-helix. It should be noted that a 7_2-helix and a 4_1-helix are also possible.

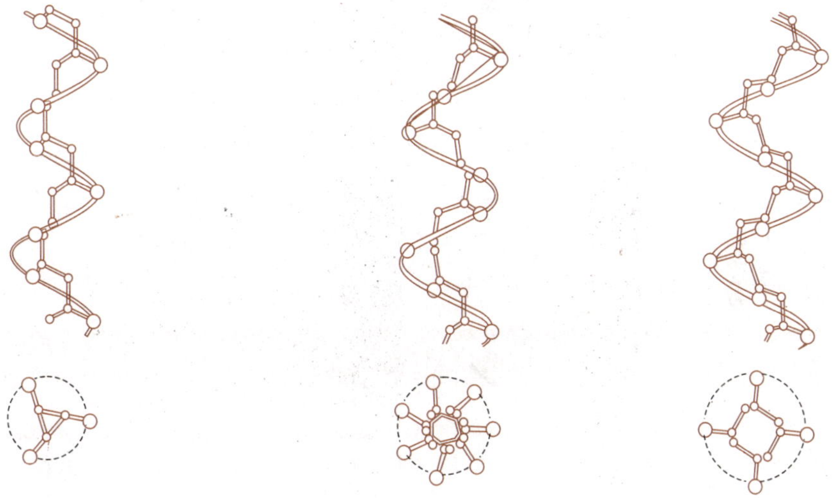

OR

(A) (B) (C)

R = $—CH_3$, $—C_2H_5$, $—CH{=}CH_2$
$—CH_2—CH_2—CH—(CH_3)_2$
$—O—CH_3$, $—O—CH_2—CH—(CH_3)_2$
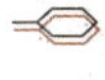

R = $—CH_2—CH—(CH_3)—C_2H_5$
$—CH_2—CH—(CH_3)_2$

R = $—CH—(CH_3)_2—C_2H_5$

Fig. 7.10 Helical structure of isotactic vinyl polymers. Smaller circles represent carbon atoms forming the chain backbone and bigger circles the substituent group R attached to alternate carbon atoms. Hydrogen atoms attached to carbon atoms are not shown for the purpose of clarity. (a = 3_1-helix; b = 7_2- helix; and c = 4_1-helix).

The helical structure has a special significance in polymers of biological origin. Many polypeptides have, in fact, a helical structure. The DNA structure, for instance, is helical (see Fig. 7.11). This was determined by Watson and Crick, for which they received the Nobel Prize. Hydrogen bonding plays an important role in the formation of the double helix of the DNA molecule.

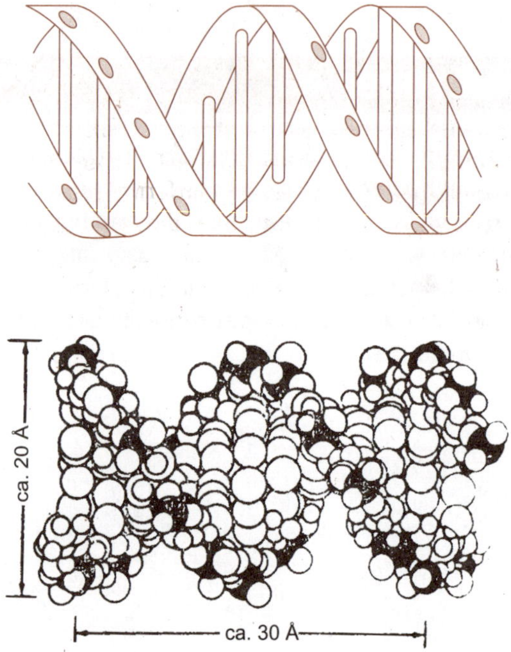

Fig. 7.11 Watson–Crick DNA model showing its helical structure.

7.10 SPHERULITES

Polymers that are crystallised from a melt reveal some interesting microstructures when observed under a polarising microscope. Studies show that these structures are very complex polycrystalline bodies. Figure 7.12 shows circular birefringent regions, as seen under a microscope. These birefringent regions are called spherulites. If a molten polymer such as polypropylene is made into a thin film between two hot glass plates (kept above 200°C) and cooled, it is seen that, from different nucleating centres, spherulites start developing. If left undisturbed, the spherulites would grow to attain a spherical structure with a very high geometrical symmetry. This geometrical symmetry of the spherulite would be lost when it comes in contact with other spherulites during its growth.

Transparency of crystalline polymers is more or less directly dependent on the size of spherulites. Spherulites can scatter light rays and, hence, the transparency of a given material decreases as the size of the spherulites increases. Some of the recent investigations have shown that the spherulites consist, in turn, of fibrils of crystalline nature having a thickness of 100 Å. Thus, a

Fig. 7.12 Circular birefringent regions called spherulites, as seen under a polarising microscope.

bunch of crystalline fibres constitutes a spherulite (Figs. 7.13 and 7.14). Based on the data available from various experiments to date, the macromolecular chains in the fibrils are seen to be perpendicular to the axis of fibril.

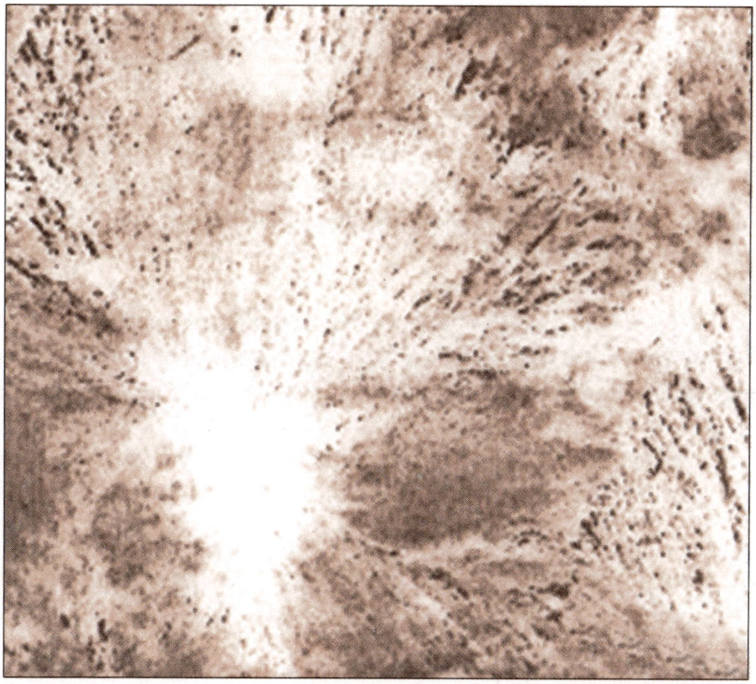

Fig. 7.13 Spherulites in crystalline polymers made-up of fibrils (as seen under electron microscope).

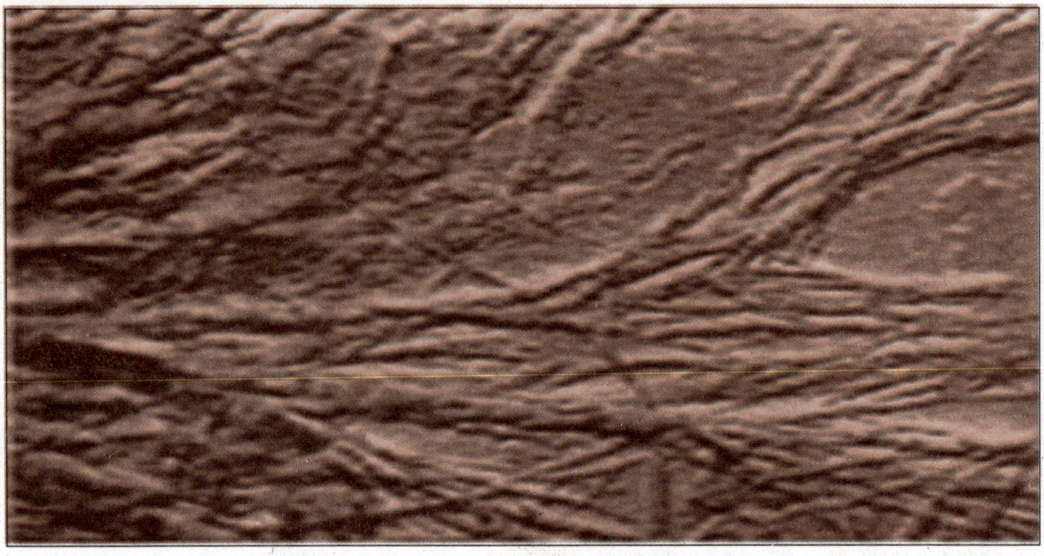

Fig. 7.14 Spherulites in crystalline polymers made of fibrils (as seen here under electron microscope).

7.11 POLYMER SINGLE CRYSTALS

Many simple, low molecular weight materials are thrown out in the form of crystals from a fairly saturated solution as it cools. These crystals, as pointed out earlier, possess a very high internal order. At one time, however, it was thought that polymers do not have any internal order. The argument then was that molecular entanglements of polymeric chains would come in the way of attaining internal order. The argument lost ground when Keller and Schlesinger showed that single crystals of many polymers could be produced by crystallisation from a very dilute solution, the concentration of which was around 0.01% by weight. Single crystals of various polymers appear similar when viewed under an electron microscope. For instance, Fig. 7.15 shows single crystals of two polymers.

A polymer single crystal looks as though it consists of many platelets (lamella) kept one over the other in a decreasing order of size. The usual lamellar thickness of the single crystal is around 100 Å. X-ray or electron diffraction experiments reveal that single crystals have a very high internal order. This is evident from the sharp and well-defined electron diffraction patterns (see Fig. 7.16) produced by them.

7.12 FOLDING OF CHAINS DURING CRYSTAL FORMATION

For a standard polymer, the lamellar thickness is around 100 Å and the molecular chain length is around 1,000 to 10,000 Å. The question is: How to accommodate the long chain into the narrow

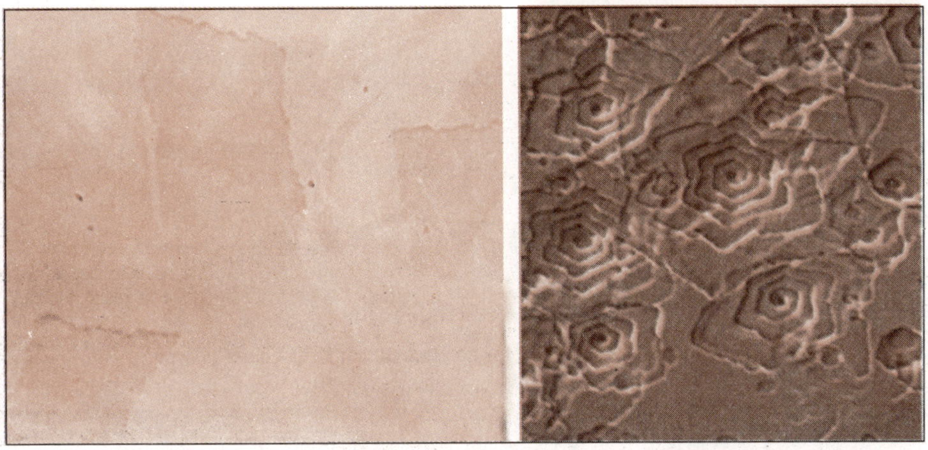

(A) (B)

Fig. 7.15 Single crystals of (A) polyethylene and (B) poly-4-methyl pentene-1 viewed under electron microscope.

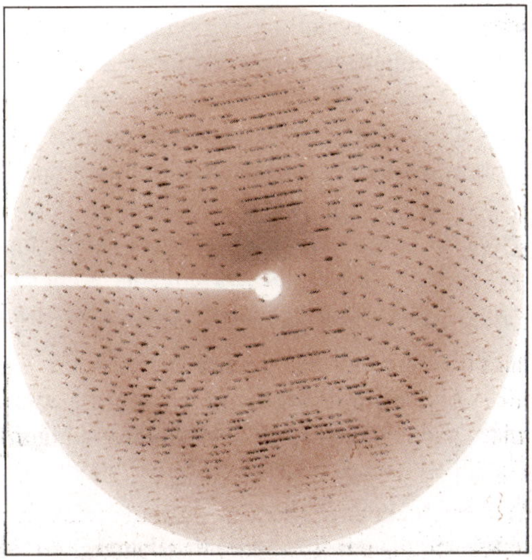

Fig. 7.16 Electron diffraction pattern of polymer single crystal.

lamella? This problem has been solved by assuming that chain folding takes place during the process of crystallisation, as shown in Fig. 7.17. Many experimental techniques (such as, electron diffraction) prove beyond any reasonable doubt that the chains in a single crystal are folded and oriented perpendicular to the plane of the polymer crystal lamella as shown in Fig. 7.17.

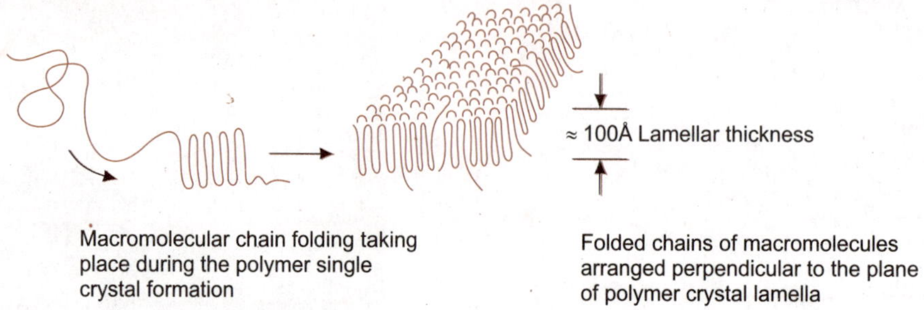

Fig. 7.17 Schematic representation of chain folding taking place during formation of polymer crystal lamella.

7.13 EFFECT OF CRYSTALLINITY ON THE PROPERTIES OF POLYMERS

The properties of a polymer such as density, modulus, hardness, permeability and heat capacity will be largely affected by its crystallinity. For a partially crystalline polymer, its crystalline and amorphous regions will exhibit different properties, even though both the regions are chemically the same. For instance, the density of the crystalline regions will be higher than that of the amorphous regions. Ultimately, many properties of the polymer depend on the percentage of crystalline material present in the bulk. To illustrate this point, let us take the case of natural rubber of different crystallinity values and plot Young's modulus values against crystallinity (Fig. 7.18). The graph clearly shows that the initial low value, characteristic of an amorphous polymer, steeply increases along with the amount of crystalline component in the sample. A comparison of properties such as density, hardness, tensile strength, and flexural modulus of two samples of polyethylene A and B (where sample B is more crystalline than A) brings out this point clearly (see Table 7.1). Evidently, sample B has significantly different physical properties as the other sample is less crystalline.

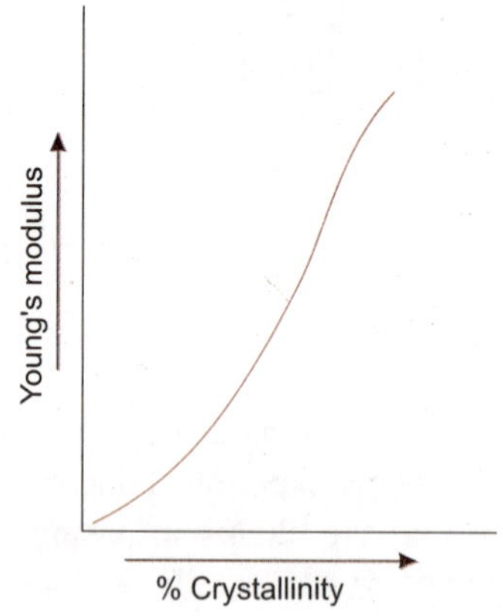

Fig. 7.18 Plot showing dependence of Young's modulus on crystallinity in a sample of natural rubber.

Table 7.1 Mechanical properties of two polyethylene samples of different crystallinity*

Property	Sample A (less crystalline)	Sample B (highly crystalline)
Density	0.917–0.932	0.952–0.965
Hardness, Shore D	44–50	66–73
Tensile strength (10^5 Pa)	83–314	221–310
Flexural modulus (10^6 Pa)	241–331	1000–1551

*Data from 1984–85 *Modern Plastics Encyclopedia*, McGraw-Hill, New York.

Molecular weight too has a profound effect on a polymer's physical properties. A combined effect of the molecular weight and crystallinity on the physical properties is shown in Fig. 7.19.

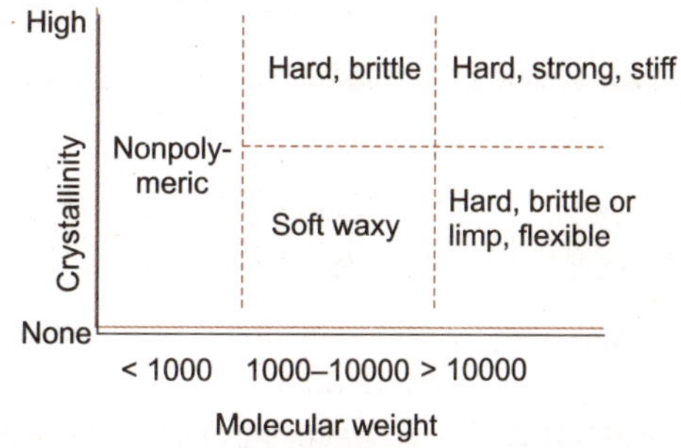

Fig. 7.19 Dependence of physical properties of polymers on their molecular weight and crystallinity.

Permeability is another property affected by crystallinity. Extensive studies have been carried out in this regard as permeability is a factor of great practical importance, especially in the package industry. Here again, crystallinity plays a major role because permeability depends on the extent and rate of penetration of liquid or vapour molecules through the polymer matrix which, in turn, depends on the physical structure of the polymer. It has been observed that the crystalline regions put up stiffer resistance to a penetrating molecule than the amorphous regions. In other words, crystalline polymers, in general, are far less permeable than the amorphous variety. Thus, we could say that as the polymer crystallinity increases the permeability decreases.

Permeability and hence, crystallinity have some effect on the chemical degradation of the polymer. The acid hydrolysis of cellulose, for example, takes place at the amorphous regions with more ease than at the crystalline regions. Similarly, in the whole mass of a polymer, such as polyethylene or polypropylene, the amorphous regions are more rapidly attacked by oxygen than their crystalline counterparts.

Books for Further Reading

Geil, P.H., *Polymer Single Crystals*, John Wiley & Sons, New York, 1963.

Mandelkern, L., *Crystallisation of Polymers*, McGraw-Hill, New York, 1964.

Sharples, A., *Introduction to Polymer Crystallisation*, Edward Arnold, London, 1966.

Alexander, L.E., *X-ray Diffraction Methods in Polymer Science*, John Wiley & Sons, New York, 1969.

Kakudo, M. and Kasai, N., *X-ray Diffraction by Polymers*, Elsevier, Amsterdam, 1972.

Deanin, R.D., *Polymer Structure, Properties and Applications*, Cahners, Boston, 1972.

8

COPOLYMERISATION

8.1 INTRODUCTION

The polymerisation reactions that we dealt with in Chapter 2 used either a single monomer or a single monomer pair, as will be seen from the tabulated matter.

Sl. No.	Monomer/monomer pair	Polymer formed
1.	Ethylene	Polyethylene
2.	Vinyl chloride	Polyvinyl chloride (PVC)
3.	Methyl methacrylate	Polymethyl methacrylate
4.	Isobutylene	Polyisobutylene
5.	Butadiene	Polybutadiene
6.	Propylene oxide	Polypropylene glycol
7.	Caprolactam	Nylon 6
8.	Hexamethylene diamine + adipic acid	Nylon 6, 6
9.	Ethylene glycol + toluene diisocyanate (TDI)	Polyurethane

In each of the first seven cases, there is only one monomer which gives the corresponding polymer. In the last two cases, we use a single monomer pair, which is (*a*) one diamine and one dicarboxylic acid for getting nylon 6, 6 and (*b*) one diol and one diisocyanate for getting polyurethane.

The process of polymerising a single monomer or a single monomer pair, is termed homopolymerisation and the polymer thus obtained is termed a homopolymer. In copolymerisation, on the other hand, two or more monomers/monomer pairs are polymerised simultaneously, and hence, the polymer is composed of more than one repeat unit. For example, styrene-butadiene rubber (SBR), the first commercially made synthetic rubber, is a copolymer obtained by the free-radical polymerisation of two monomers, styrene and butadiene. It is composed of both styrene as well as butadiene repeat units.

In general, copolymerisation can be represented as

$$xA + yB \rightarrow$$

—A—A—B—A—A—A—B—B—B—B—A—B—A—B—B—

The copolymer formed is composed of both the monomeric units A and B arranged at random. The A : B ratio and the degree of randomness depend on the quantity of the individual monomers taken, their amenability for copolymerisation and the polymerisation mechanism used.

Copolymerisation can be brought about by many types of polymerisation reactions, with which we are already familiar. The majority of the commercially important copolymers, however, are made by free-radical, ionic or polycondensation polymerisation. We shall, therefore, confine our discussion to these three copolymerisation methods.

8.2 FREE-RADICAL COPOLYMERISATION

Let two monomers, M_1 and M_2, be mixed together and polymerised using a free-radical initiator. Once the initiator decomposes, the free-radical formed can attack M_1 or M_2 as follows:

$$\dot{R} + M_1 \rightarrow R\dot{M}_1$$

$$\dot{R} + M_2 \rightarrow R\dot{M}_2$$

Two polymer chains can thus be initiated, one carrying a free-radical site on monomer unit M_1 and the other on monomer unit M_2. After initiation, comes the process of propagation. Here, the chain carrying a free-radical site on M_1 can either add another $\dot{M}_1$ or M_2. Similarly, the chain carrying a free-radical site on M_2 can either add another M_2 or an M_1, *i.e.*:

(*i*) $R\dot{M}_1 + M_1 \rightarrow RM_1\dot{M}_1$ [propagation reaction (PR) type 11].

(*ii*) $RM_1 + M_2 \rightarrow RM_1M_2$ (PR type 12).

(*iii*) $R\dot{M}_2 + M_1 \rightarrow RM_2\dot{M}_1$ (PR type 21).

(*iv*) $R\dot{M}_2 + M_2 \rightarrow RM_2\dot{M}_2$ (PR type 22).

Assuming that all the four types of addition take place, we still have the growing chains ending with $\dot{M}_1$ or $\dot{M}_2$ and the possibility of all the four types of propagation. Now, the rates of these four propagation reactions are as follows:

$$\dot{M}_1 + M_1 \xrightarrow{k_{11}} M_1\dot{M}_1\,;\quad R_{11} = k_{11}[\dot{M}_1][M_1] \tag{8.1}$$

$$\dot{M}_1 + M_2 \xrightarrow{k_{12}} M_1\dot{M}_2\,;\quad R_{12} = k_{12}[\dot{M}_1][M_2] \tag{8.2}$$

$$\dot{M}_2 + M_1 \xrightarrow{k_{21}} M_2\dot{M}_1\,;\quad R_{21} = k_{21}[\dot{M}_2][M_1] \tag{8.3}$$

$$\dot{M}_2 + M_2 \xrightarrow{k_{22}} M_2\dot{M}_2\,;\quad R_{22} = k_{22}[\dot{M}_2][M_2] \tag{8.4}$$

The basic assumption is that the reactivity of any growing chain depends only on the end monomer unit carrying the free-radical site and not on the number or type of monomer units already added to the chain. The rate at which the monomers M_1 and M_2 are consumed during the course of propagation can be expressed as follows:

$$\frac{-d[M_1]}{dt} = k_{11}[\dot{M}_1][M_1] + k_{21}[\dot{M}_2][M_1] \tag{8.5}$$

$$\frac{-d[M_2]}{dt} = k_{12}[\dot{M}_1][M_2] + k_{22}[\dot{M}_2][M_2] \tag{8.6}$$

Now, assuming a steady state, wherein the rate of a particular chain end (say, $\dot{M}_1$) disappearing is equal to the rate of formation of the same chain end, we can write

$$k_{12}[\dot{M}_1][M_2] = k_{21}[\dot{M}_2][M_1] \tag{8.7}$$

$$\frac{[\dot{M}_1]}{[\dot{M}_2]} = \frac{k_{21}[M_1]}{k_{12}[M_2]} \tag{8.8}$$

A combination of all the foregoing equations gives the 'copolymer equation':

$$\frac{d[M_1]}{d[M_2]} = \frac{[M_1]}{[M_2]} \bullet \frac{(k_{11}/k_{12})[M_1]+[M_2]}{(k_{22}/k_{21})[M_2]+[M_1]} \tag{8.9}$$

The terms k_{11}/k_{12} (denoted by r_1) and k_{22}/k_{21} (denoted by r_2) appearing in this equation are two important terms, known as the reactivity ratios for any given pair of monomers M_1 and M_2. These ratios (r_1 and r_2) indicate whether a growing chain carrying a free-radical on a particular monomer unit would prefer to add its own monomer species or the comonomer species. In other words, the composition of the copolymer formed at any given instant is dependent not only on the concentration of the monomer species present in the system at that instant but also on their reactivity ratios. The point to be noted is that r_1 and r_2 for any given pair of monomers are dependent purely on the nature of the two monomers and temperature and independent of other parameters such as the solvent, initiator and chain transfer agent. The latter, however, will have a pronounced influence on the molecular weight and the molecular weight distribution of the polymer formed. The chemical composition of the copolymer formed depends exclusively on the monomer concentrations and their reactivity ratios as per Eqn. (8.9), which can be rewritten in the following form:

$$\frac{[m_1]}{[m_2]} = \frac{[M_1]}{[M_2]} \cdot \frac{r_1[M_1]+[M_2]}{r_2[M_2]+[M_1]} \tag{8.10}$$

where $[m_1]/[m_2]$ gives the ratio of the monomers M_1 and M_2, entered into the copolymer formed. If we express the monomer components $[m_1]$ and $[m_2]$ in terms of mole fraction instead of molar concentration, the mole fraction of monomer M_1 in the copolymer can be given by

$$\frac{[m_1]}{[m_1]+[m_2]} = \frac{d[M_1]}{d[M_1]+d[M_2]} = n_1 = 1 - n_2 \tag{8.11}$$

where n_1 and n_2 are the mole fractions of monomers M_1 and M_2, respectively, in the copolymer formed. Similarly, the mole fraction of monomer M_1 in the monomer feed can be given by

$$\frac{[M_1]}{[M_1]+[M_2]} = N_1 = 1 - N_2 \tag{8.12}$$

where N_1 and N_2 are the mole fractions of the monomers M_1 and M_2, respectively, in the monomer feed.

Now, the 'copolymer equation' [Eqn. (8.9)] can be expressed as

$$n_1 = \frac{r_1(N_1)^2 + N_1N_2}{r_1(N_1)^2 + 2N_1N_2 + r_2(N_2)^2} \tag{8.13}$$

By knowing r_1 and r_2 and also the monomer feed ratio, we can easily predict the instantaneous molar composition of the copolymer formed. The r_1 and r_2 values for some important monomer pairs are given in Table 8.1.

CHAPTER 8

Table 8.1 Reactivity ratios of monomers*

Monomer pair		*Reactivity ratios (r_1 and r_2) at*									
		5°C		50°C		60°C		80°C		100°C	
M_1 1	M_2 2	r_1 3	r_2 4	r_1 5	r_2 6	r_1 7	r_2 8	r_1 9	r_2 10	r_1 11	r_2 12
Acrylic acid	Acrylonitrile			1.15	0.35			6.00	0.13		
	2-chloro allyl acetate									1.0	0.0
	Butyl acrylate			0.58	1.07						
	Butyl methacrylate			0.29	3.67						
	Styrene					0.25	0.15	0.45	0.25		
Acrylonitrile	Allyl chloride					5.50	0.0				
	Butadiene	0.02	0.28	0.05	0.35	0.25	0.33				
	Butyl methacrylate					0.31	1.08				
	Butyl vinyl ether					0.14	0.003				
	Diethyl fumerate					8.0	0.0				
	Ethyl acrylate			1.17	0.67			0.44	0.95		
	5-ethyl, 2-vinyl pyridine					0.02	0.43				
	Isobutylene			1.80	0.02	1.02	0.0				
	Maleic anhydride					6.0	0.0				
	Methyl acrylate			1.50	0.84	1.26	0.67	0.50	0.71		
	Methyl methacrylate					0.15	1.20	0.15	1.22	0.16	1.19
	Methyl vinyl ketone					0.61	1.78				
	Styrene			0.05	0.37	0.04	0.40	0.03	0.52		
	Vinyl acetate			4.20	0.05	4.05	0.06				
	Vinyl chloride			3.70	0.07	2.70	0.04				
Butadiene	Butyl acrylate	0.99	0.08								
	p-chloro styrene			1.07	0.42						
	Isoprene			3.38	0.47						

Contd.

(Table 8.1 Contd.)

Monomer pair		Reactivity ratios (r_1 and r_2) at									
		5°C		50°C		60°C		80°C		100°C	
M_1	M_2	r_1	r_2	r_1	r_2	r_1	r_2	r_1	r_2	r_1	r_2
1	2	3	4	5	6	7	8	9	10	11	12
	Methacrylonitrile	0.36	0.04								
	Methyl acrylate	0.76	0.05								
	Methyl methacrylate	0.53	0.06								
	Styrene	1.40	0.44	1.40	0.50	1.39	0.78				
	Vinyl chloride			8.80	0.04						
	Vinylidene chloride	1.90	0.05								
Butyl acrylate	Butyl methacrylate			0.30	2.20						
	Methyl vinyl ketone			0.65	1.6						
	Styrene					0.15	0.76				
	2-vinyl pyridine					2.51	0.10				
2-chloro allyl acetate	Methyl acrylate									0.0	0.70
	Methyl methacrylate			0.0	1.0						
	Styrene			0.0	4.10						
	Vinyl chloride									0.0	0.70
p-chloro styrene	Methacrylonitrile							0.45	0.73		
	Methyl methacrylate					0.89	0.42				
	p-methyl styrene					1.15	0.61				
	Styrene					1.03	0.74				
Chloro trifluoro ethylene	Methyl methacrylate					0.005	75.0				
	Styrene					0.001	7.0				
	Tetra fluoro ethylene					1.00	1.00				
	Vinyl acetate					0.01	0.60				
	Vinyl chloride					0.01	2.53				
Maleic anhydride	Methyl acrylate					0.0	2.5				
	Methyl methacrylate					0.03	3.5				
	Styrene					0.0	0.02	0.0	0.04		
	Vinylidene chloride					0.0	9.0				
Methacrylonitrile	Ethyl methacrylate							0.46	0.83		
	Methacrylic acid							0.59	1.63		
	Methyl methacrylate					0.65	0.67	0.70	0.74		
	α-methyl styrene							0.21	0.15		
	Styrene					0.16	0.30	0.28	0.43		
Methyl methacrylate	Diethyl maleate					20.0	0.0				
	Divinyl ether					10.0	0.006				

Contd.

(Table 8.1 Contd.)

Monomer pair		*Reactivity ratios (r_1 and r_2) at*									
		5°C		50°C		60°C		80°C		100°C	
M_1	M_2	r_1	r_2	r_1	r_2	r_1	r_2	r_1	r_2	r_1	r_2
1	2	3	4	5	6	7	8	9	10	11	12
	α-methyl styrene					0.50	0.14				
	Styrene					0.46	0.48				
	Trichloro ethylene					100.0	0.0				
	Vinyl acetate					20.0	0.02				
	Vinyl chloride					12.5	0.0				
	Vinylidene chloride					2.53	0.24				
	2-vinyl pyridine					0.44	0.77				
Styrene	2-chloro allyl acetate			4.10	0.0						
	2-chloro butadiene			0.0	6.30						
	Diethyl maleate					6.52	0.005				
	m-divinyl benzene					0.65	0.60				
	Divinyl ether					40.0	0.02				
	Ethyl acrylate			0.79	0.19			0.80	0.48		
	Ethyl methacrylate			0.67	0.26						
	Isoprene			1.38	2.05						
	Methacrylic acid					0.15	0.70				
	Methyl acrylate					0.75	0.18				
	Methyl vinyl ketone					0.29	0.35				
	2-methyl 5-vinyl pyridine					0.88	1.19				
	p-nitro styrene					0.19	1.15				
	Trichloro ethylene			16.5	0.0	16.0	0.0				
	Vinyl acetate					55.0	0.01				
	Vinyl chloride			35	0.07	17.0	0.02				
	Vinylidene chloride			2.10	0.15	2.0	0.14				
	Vinyl isocyanate					6.9	0.10				
	2-vinyl pyridine					0.55	1.14				
Tetra fluoro ethylene	Chloro trifluoro ethylene					1.0	1.0				
	Ethylene							0.85	0.15		
	Isobutylene							0.3	0.0		
Vinyl acetate	Allyl acetate					1.0	0.70				
	Chloro trifluoro ethylene					0.6	0.01				
	Methyl acrylate					0.1	9.0				
	Vinyl chloride					0.23	1.68				
	Vinylidene chloride					0.0	3.6				
	N-vinyl carbazole									0.152	3.02
Vinyl chloride	Isobutylene					2.05	0.08				
	Methyl acrylate			0.12	4.4	0.0	5.0				
	Vinyl isobutyl ether			2.0	0.02						
	Vinylidene chloride			0.2	4.5	0.3	3.2				

*Data from *Polymer Handbook*, Eds. Brandrup J. and Immergut E.H., John Wiley & Sons New York, 1966.

8.2.1 Determination of Reactivity Ratios

Reactivity ratios of a given pair of monomers can be experimentally determined by polymerising mixtures of two monomers in varying ratios to low conversion (usually less than 10%, so as to minimise the effect of the change in the monomer ratio as polymerisation progresses) and then analysing the resultant copolymer for its monomeric composition.

Let us denote the molar concentration ratio of the monomers in the feed, $[M_1]/[M_2]$, by M and the corresponding molar concentration ratio of the monomeric components in the polymer, $[m_1]/[m_2]$, by P. The reactivity ratios r_1 and r_2 can be computed by plotting the data according to the following two methods.

Mayo–Lewis Method

The copolymer equation [Eqn. (8.10)] can be rewritten in the form

$$r_2 = r_1 \frac{M^2}{P} + \left[\frac{M}{P} - M\right] \tag{8.14}$$

Now, for the assumed values of r_1 the corresponding values of r_2 are calculated for each set of values of M and P of the experimental data. If we plot r_1 and r_2 for each set of M and P, we get a series of straight lines intersecting almost at a point (Fig. 8.1). At this point, all the M and P values have the same r_1 and r_2 values which represent the actual r_1 and r_2 for the monomer pair.

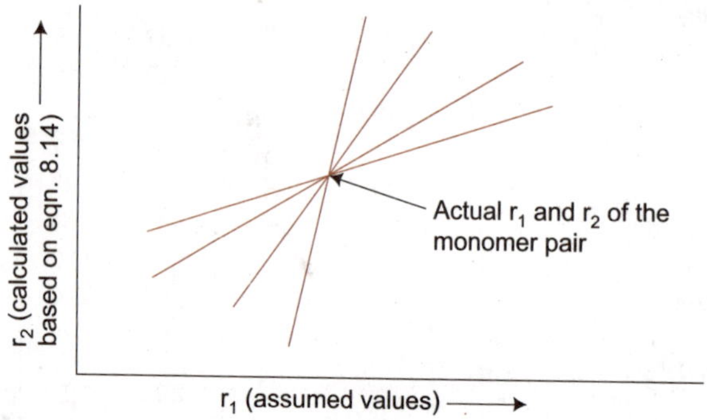

Fig. 8.1 Plots for computing r_1 and r_2 according to the Mayo–Lewis method.

Fineman–Ross Method

The Fineman–Ross method utilises the equation

$$M - \left(\frac{M}{P}\right) = -r_2 + r_1\left(\frac{M^2}{P}\right) \tag{8.15}$$

When we plot $M - (M/P)$ versus (M^2/P), we get a straight line, the ordinate intercept giving $-r_2$ and the slope of the line giving r_1 (Fig. 8.2).

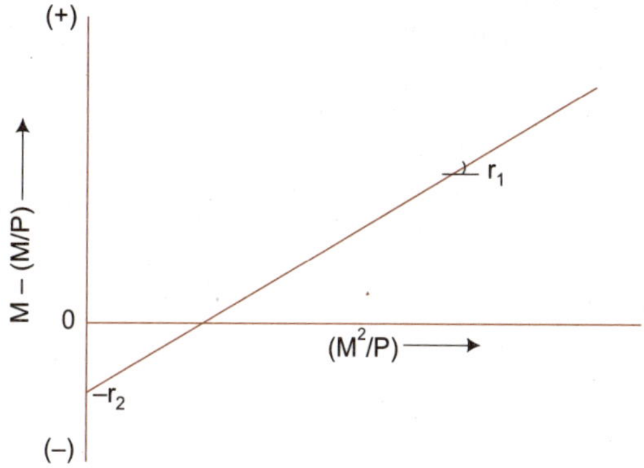

Fig. 8.2 Plot for computing r_1 and r_2 according to the Fineman–Ross Method.

8.2.2 Reactivity Ratios and Copolymerisation Behaviour

Let us recall that $r_1 = k_{11}/k_{12}$ and $r_2 = k_{22}/k_{21}$. When $r_1 = 1$, the propagation rate constants are the same for the reaction PR type 11 and for the PR type 12. Hence, a chain with a radical end M_1 can add either the M_1 or M_2 monomer molecule with equal probability. When r_1 is less than 1, the M_2 monomer will be added more readily than M_1, and when r_1 is more than 1, the reverse will occur. The same is true of r_2 also. Now, depending on the values of r_1 and r_2 various copolymerisation behaviours are possible, which are now discussed.

(*a*) When $r_1 = r_2 = 0$. In this case, the propagation reaction types 11 and 22 are not at all possible, and hence, a chain ending with M_1 will add only an M_2 and this M_2 will add only an M_1, which, in turn, will add only an M_2 and so on. In other words, the monomer feed ratio notwithstanding, the copolymer formed will have an equal number of M_1 and M_2 monomeric units arranged alternatively in the copolymer chain formed. We thus get a strictly alternating copolymer, irrespective of the monomer feed ratio. It may be interesting to note that the individual monomer falling under this category cannot form homopolymers, but when polymerised together, can form an alternating copolymer.

(*b*) When $r_1 = r_2 = 1$. In this case, all the four types of propagation reactions are equally possible. A chain ending with $\dot{M}_1$ can add either M_1 or M_2, and similarly, a chain ending with $\dot{M}_2$ can add either M_1 or M_2. The probability of M_1 or M_2 adding to the M_1 or M_2 depends purely on the monomer concentrations. The copolymer formed will have the same ratio of the monomeric components as that of the monomer feed ratio, with the two components arranged in the chain in a purely random sequence. This case can be called an 'ideal copolymerisation', since the copolymer composition can be ideally controlled by adjusting the initial monomer feed ratio. Since the feed compositions and the copolymer composition remain constant throughout the course of the polymerisation, this type of copolymerisation is also called the 'azeotropic copolymerisation'.

(*c*) When $r_1 > 1$ and $r_2 < 1$, in this case, the propagation reaction types 11 and 21 will be preferred to types 12 and 22, and hence, the probability of M_1 entering into the copolymer chain is higher as compared to M_2. The copolymer formed will, therefore, be richer in M_1. If r_1 and r_2 differ very widely, the copolymer formed will comprise almost entirely M_1 and it will be very difficult to incorporate any appreciable quantity of M_2 in the copolymer chain. Control on the copolymer composition, to some extent, can be achieved by adjusting the monomer feed ratios, *i.e.*, by keeping M_1/M_2 always less than 1.

(*d*) When $r_1 < 1$ and $r_2 > 1$. The behaviour in this case will just be the opposite of that for case (*c*). The copolymer formed will, therefore, be richer in M_2.

(*e*) When $r_1 < 1$ and $r_2 < 1$. In this case, the propagation reaction type 12 and 21 are preferred to types 11 and 22. In other words, while M_2 can add on to $\dot{M}_1$, M_1 can add on to $\dot{M}_2$ at the same time. The preference for one of these two will depend on the actual values of r_1 and r_2 and also on the monomer feed ratios. When r_1 and r_2 are equal, the copolymer formed will be richer in M_1 up to an N_1 value of 0.5, and then onwards, the copolymer will be richer in M_2. At $N_1 = 0.5$ (*i.e.*, when the monomer feed ratio is made-up of an equimolar quantity of M_1 and M_2), the copolymer formed will have an n_1 value of 0.5. (*i.e.*, the copolymer also will have an equal number of M_1 and M_2 components but arranged in a random sequence). Since this composition of the monomer feed results in the formation of a copolymer with the same monomeric composition, it is called 'azeotropic composition'. When r_1 is less than r_2, the azeotropic composition decreases from an N_1 value of 0.5 and when r_1 is higher than r_2, there is an increase in the same. In all the cases, however, there is an 'azeotropic composition' below which the copolymer formed will be richer in M_1 and above which it will be richer in M_2. The value of N_1 for the azeotropic composition can be arrived at from the following equation:

$$N_1 = \frac{1 - r_2}{(2 - r_1 - r_2)} \tag{8.16}$$

When $r_2 = r_1$, we get N_1 equal to 0.5. We can now make a comparison with the azeotropic copolymerisation, where both r_1 and r_2 are unity. In such a case, any monomer feed composition will act as an azeotropic composition, whereas in the present case (*i.e.*, where r_1 and r_2 are less than one), only one particular monomer feed composition will act as an azeotropic composition. As the values of r_1 and r_2 approach unity, the copolymers formed also tend to have their compositions close, though not identical, to the monomer feed ratio.

(*f*) When $r_1 > 1$ and $r_2 > 1$, such a system does not usually undergo random copolymerisation. It forms either a mixture of the two homopolymers or, under certain favourable conditions, block copolymers having long sequences of M_1 followed by a long sequence of M_2.

(*g*) When r_1 or $r_2 = 0$, such a system usually results in a copolymer with a particular azeotropic composition. If the mole fraction of the monomer with a zero reactivity ratio is below the azeotropic composition, the azeotropic copolymer along with the homopolymer of the comonomer will be formed. If the mole fraction is higher than the azeotropic composition, the azeotropic copolymerisation will proceed until the comonomer is consumed and the excess of the monomer with zero activity ratio remains unpolymerised.

The copolymerisation behaviour of the monomer pairs with various values of r_1 and r_2 as explained above, can be depicted as plots of mole fractions of monomers in the polymer (n_1) against the mole fraction of the monomer in the monomer feed (N_1), as shown in Fig. 8.3.

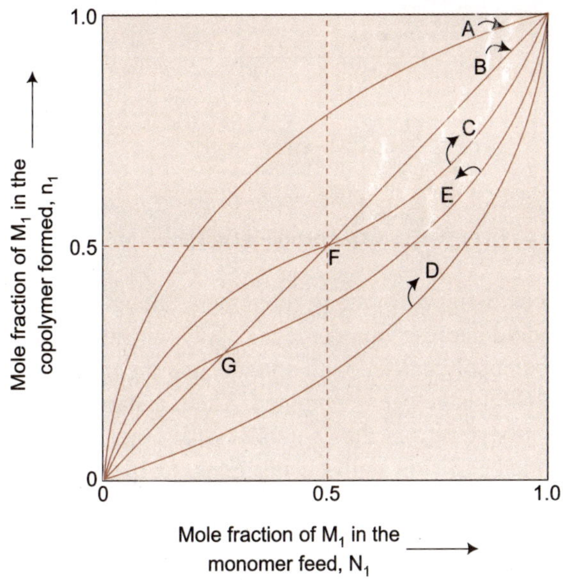

(A) $r_1 > 1, r_2 < 1$
(B) $r_1 = r_2 = 1$
(C) $r_1 = r_2 = < 1$
(D) $r_1 < 1, r_2 > 1$
(E) r_1 and $r_2 < 1$ and $r_1 < r_2$

Plot B showing azeotropic polymerisation behaviour. Point F and G give the azeotropic compositions for plots C and E respectively.

Fig. 8.3 Plots showing the copolymerisation behaviour of monomer pairs as a function of their reactivity ratios and monomer feed ratios.

8.2.3 The *Q*-*e* Scheme of Alfrey and Price

Alfrey and Price derived a semiquantitative relationship to compute the reactivity ratios of various monomers. The method deals with the resonance stabilisation and polarisation characteristics of a monomer and predicts its reactivity behaviour with reference to another monomer radical. Assuming that a monomer M_2 adds on to a radical $\dot{M}_1$, the rate constant for the propagation reaction can be written as

$$k_{12} = P_1 Q_2 \exp(-e_1 e_2) \tag{8.17}$$

where P_1 denotes a factor characterising the state of the radical $\dot{M}_1$ at the growing chain end and Q_2 indicates the resonance stability, and hence, the measure of the reactivity of the monomer M_2; e_1 and e_2 refer to the polarisation characteristics of monomers M_1 and M_2 and also to those of the radicals $\dot{M}_1$ and $\dot{M}_2$, respectively. Similarly, we have

$$k_{11} = P_1 Q_1 \exp(-e_1 e_1) \tag{8.18}$$

Now, we know that $r_1 = k_{11}/k_{12}$, which can be rewritten as

$$r_1 = \frac{P_1Q_1 \exp(-e_1e_1)}{P_1Q_2 \exp(-e_1e_2)}$$

or
$$r_1 = \frac{Q_1}{Q_2} \exp[-e_1(e_1 - e_2)] \tag{8.19}$$

and
$$r_2 = \frac{Q_2}{Q_1} \exp[-e_2(e_2 - e_1)] \tag{8.20}$$

Q and e values have been assigned to many monomers taking styrene as the standard. Styrene has been chosen as the standard because if undergoes copolymerisation with most of the monomers. Several monomers have been copolymerised with styrene, and the r_1 and r_2 values for the respective monomer (M_1) and styrene (M_2) have been computed. Then, a standard value of $Q = 1$ and that of $e = -8$ have been assigned to styrene and the Q-e values for other monomers have been calculated (from r_1 and r_2 values obtained experimentally) using Eqns. (8.19) and (8.20). Q-e values for some monomers are given in Table 8.2.

Table 8.2 Q-e values of monomers*

Monomer	Q	e
Acrylic acid	1.15	0.77
Acrylonitrile	0.60	1.20
Butadiene	2.39	−1.05
n-butyl acrylate	0.50	1.06
n-butyl vinyl ether	0.087	−1.21
p-chloro styrene	1.03	−0.33
Chloro trifluoro ethylene	0.02	1.48
Ethyl acrylate	0.52	0.22
Ethylene	0.015	−0.20
5-ethyl 2-vinyl pyridine	1.37	−0.74
Isobutylene	0.033	−0.96
Maleic anhydride	0.23	2.25
Methacrylo nitrile	1.12	0.81
Methyl acrylate	0.42	0.60
Methyl methacrylate	0.74	0.40
p-nitro styrene	1.63	0.39
Propylene	0.002	−0.78
Styrene (assumed standard)	1.0	−0.80
Tetrafluoro ethylene	0.049	1.22
Trichloro ethylene	0.019	1.86
Vinyl acetate	0.026	−0.22
Vinyl chloride	0.044	0.20
Vinylidene chloride	0.22	0.36
Vinyl isocyanate	0.16	−0.70
2-vinyl pyridine	1.30	−0.50

*Data from *Polymer Handbook*, Eds. Brandrup, J. and Immergut, E. H., John Wiley & Sons, New York, 1975.

Once Q-e values for different monomers are known, the r_1 and r_2 values for any chosen pair among those monomers can be computed from Eqns. (8.19) and (8.20) without the need for actual experimentation. The Q-e scheme suggests that: (*a*) monomers with widely differing Q values do not copolymerise; (*b*) monomers with widely differing e values have a tendency to form alternating copolymers; and (*c*) monomers having roughly equal Q values and almost identical e values will undergo azeotropic copolymerisation.

8.2.4 Copolymer Composition at Higher Conversions

The copolymer equation, from which we predict the composition of the copolymer formed, speaks only of the instantaneous molar composition of the copolymer formed from a given monomer feed with particular values of r_1 and r_2. We cannot predict the overall composition of the copolymer formed at higher conversions, say, 50–60% conversion, as encountered in most of the industrial processes. This is so because the moment copolymerisation begins (except in the case of azeotropic composition), the ratio of the unconsumed monomers continuously drifts away from the initial monomer feed ratio. The continuous change in the monomer ratio is the result of the relative rate of consumption of the two monomers not being the same, and hence, the two monomers do not enter into the copolymer chain in the same ratio as that of the initial feed. Usually, the monomer with a higher reactivity ratio is consumed sooner, and hence, the monomer mixture becomes progressively poorer in that monomer. As the composition of the monomer mixture is continuously changing, the composition of the copolymer formed also changes continuously, thereby, introducing a composition inhomogeneity as polymerisation progresses. The ultimate composition of the polymer formed will be the average composition of all the copolymer molecules formed at various stages of conversion from varying ratios of the monomer mixture. The composition distribution (or the extent of inhomogeneity) depends on the percentage conversion; the higher the conversion, the broader the distribution.

In commercial copolymerisation processes, where the aim is to achieve as narrow a composition distribution as possible, the following principles are adopted:

(*i*) Selection of the monomer pairs with r_1 and r_2 values close to unity, so that the monomer ratio is disturbed to the minimum throughout the polymerisation.

(*ii*) Working at or close to azeotropic compositions, wherever possible.

(*iii*) Restriction of the copolymerisation to low conversions only.

(*iv*) Maintenance of the monomer concentrations by controlled addition of the monomer which gets consumed faster.

8.3 IONIC COPOLYMERISATION

Monomer pairs, when polymerised by ionic initiators, undergo ionic copolymerisation to yield copolymers. Here, the basic principles involved are the same as in free-radical copolymerisation. The copolymerisation equation for free-radical copolymerisation is applicable in this case also. There is, however, a major difference when it comes to reactivity ratios. Although the concept of the reactivity ratios r_1 and r_2 of the monomer pair M_1 and M_2 is the same (as in free-radical copolymerisation), the dependence of the same on various polymerisation parameters is different in ionic copolymerisation. While, in free-radical copolymerisation, reactivity ratios are dependent only on the monomer pair and temperature of polymerisation, in ionic copolymerisation, they are dependent not only on the monomer pair and temperature but also on the solvent and the initiator systems used. Since the reactivity of an ion (which is the propagating species in the case of ionic polymerisation) as against that of a free-radical is very much dependent on the polarity of the polymerisation system as a whole (which, in turn, is influenced by the initiators and the solvents used), the reactivity ratios of the same monomer pair at the same temperature can vary with the initiator and the solvent system employed. Since the direction of the propagation (*i.e.*, whether the propagation is cationic or anionic) is also directed by the initiator and the solvent system, the same monomer pair can again have different reactivity rates, depending on whether they undergo cationic copolymerisation or anionic copolymerisation.

We notice, therefore, that since the same monomer pair will have different values of r_1 and r_2 depending on the mode of propagation, the same monomer pair, by different techniques, will give copolymers of entirely different monomeric compositions. The well-known example is the copolymerisation of the monomer pair methyl methacrylate-styrene, whose composition curves are given in Fig. 8.4. At an initial equimolar ratio of styrene to methyl methacrylate (shown by the vertical broken line in the figure), the free-radical polymerisation using benzoyl peroxide yields a copolymer containing roughly 50% of styrene; a cationic polymerisation initiated by borontrifluoride etherate gives a copolymer containing more than 95% of styrene, while the anionic polymerisation with triphenyl methyl sodium results in a copolymer having less than 5% styrene. Determining the reactivity ratios and the copolymer compositions obtained from a monomer pair using a new catalyst and comparing the same with the values obtained for the same pair with known initiator systems is, therefore, a very useful method of understanding the mechanism by which the new catalyst system brings about the copolymerisation.

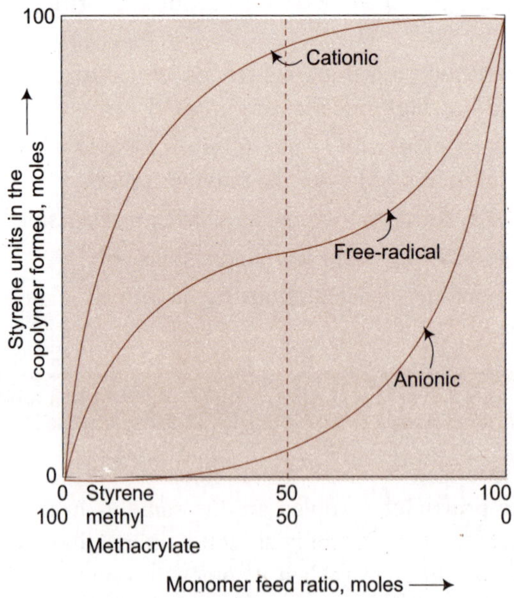

Fig. 8.4 Plots showing effect of polymerisation type on composition of copolymer formed from the same monomer pair.

8.4 COPOLYCONDENSATION

We are familiar with a polycondensation reaction, wherein, a pair of difunctional monomers is used to form a homopolymer. If the reactants in a polycondensation reaction are more than a single pair, there is copolycondensation resulting in a copolymer. The simple polycondensation and copolycondensation reactions can be represented as follows.

Simple polycondensation reaction:

$$\text{A—R—A} + \text{B—R}'\text{—B} \rightarrow \text{A—R—AB—R}'\text{—B} + \text{by-product}$$

Copolycondensation reactions:

$$\text{A—R—A} + \text{B—R}'\text{—B} + \text{A—R}''\text{—A} \rightarrow \text{A—R—AB—R}'\text{—BA—R}''\text{—A} + \text{by-product}$$

$$\text{A—R—A} + \text{B—R}'\text{—B} + \text{A—R}''\text{—A} + \text{B—R}'''\text{—B} \rightarrow \text{A—R—AB—R}'\text{—BA—R}''\text{—AB—R}'''\text{—B} + \text{by-product}$$

where A and B are functional groups capable of undergoing condensation reaction.

The copolymers obtained by copolycondensation are sometimes referred to as mixed polycondensates. Mixed polyamides and mixed polyesters are well-known instances in this regard.

We know, for example, the simple polyester-polyethylene terephthalate (PETP)—made from dimethyl terephthalate and ethylene glycol. PETP (I), due to the presence of a large number of para-linked aromatic rings in the chain backbone, is highly crystalline and rigid, and hence, difficult to process. A small percentage of dimethyl isophthalate, when introduced in the reaction mixture, results in a copolymer or a mixed polyester (II), wherein, the para-linkages are interposed at random with meta-linkages, which reduces the crystallinity and chain rigidity:

(I)

(II)

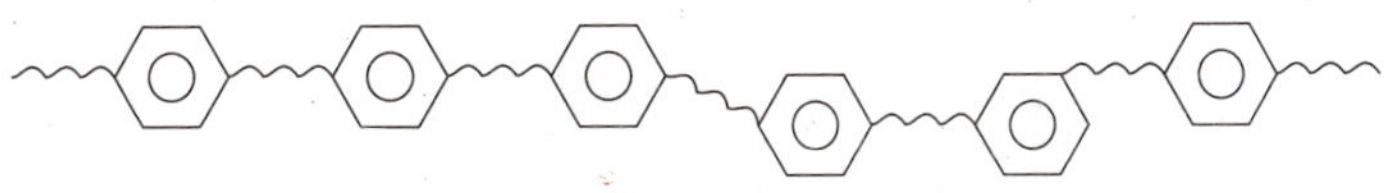

(〰 = —$COOCH_2CH_2OCO$—)

Similarly, nylon 6, 6 is a polyamide prepared by the polycondensation of hexamethylene diamine and adipic acid. Nylon 6, 6, a tough and fairly crystalline material, has a melting point of 270°C. If the melting point of this polyamide is to be increased, we can easily do so by the copolycondensation of hexamethylene diamine, adipic acid and terephthalic acid. What happens in this case is that the resultant polyamide, instead of containing purely aliphatic chains all through the chain backbone, contains an aromatic nucleus also; this results in an increase in its crystallinity, and hence, the melting point, depending on the ratio of aliphatic diacid to aromatic diacid:

$$n.H_2N-(CH_2)_6-NH_2 + n.HOOC-(CH_2)_4-COOH \longrightarrow \sim\!\sim[CO-(CH_2)_4-CONH-(CH_2)_6-NH]_n\sim\!\sim$$

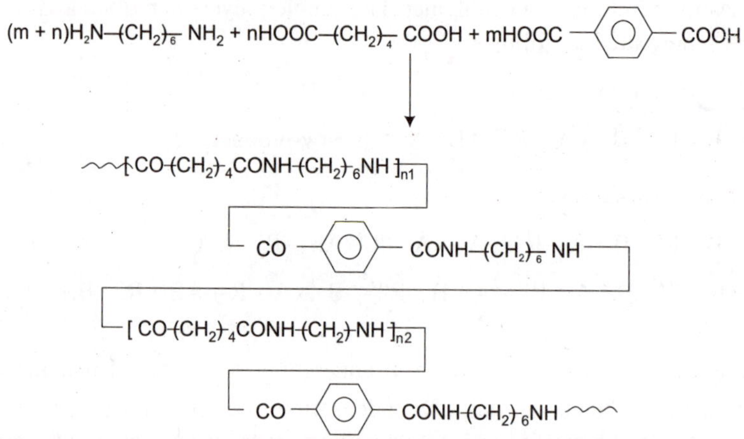

Books for Further Reading

Alfrey, T., Bohrer, J., and Mark, H.F., *Copolymerisation,* Interscience, New York, 1953.

Burlant, N.J. and Hoffmann, A.S., *Block and Graft Copolymers,* Reinhold, New York, 1960.

Ceresa, R.J., *Block and Graft Copolymers*, Butterworth, London, 1962.

Ham, G.E., (Ed.), *Copolymerisation*, Interscience, New York, 1964.

Battaerd, H.A.J. and Treglar, G.W., *Graft Copolymers,* Wiley-Interscience, New York, 1967.

9

INDIVIDUAL POLYMERS

9.1 POLYETHYLENE

Polyethylene is the simplest hydrocarbon polymer and has the following structure:

$$-\!\!\left[\!-CH_2 - CH_2-\!\right]_n\!\!-$$

It is a very commonly used polymer and was first produced in England by the Imperial Chemical Industries (ICI) in 1933 by polymerising the ethylene monomer.

Ethylene can be prepared either by the hydrogenation of acetylene or by the dehydration of ethanol. In some countries, it is produced from petroleum products by the process of cracking. In this process, saturated hydrocarbons such as ethane or propane undergo catalytic degradation and dehydrogenation, forming ethylene. Ethylene is a gas at room temperature. It has a boiling point of –104°C.

There are two varieties of polyethylenes, *viz.*, low density and high density. Low density polyethylene consists of molecules with branches, whereas the high density variety is essentially linear.

Low density polyethylene (LDPE) is produced by the high pressure polymerisation of ethylene, using oxygen as the initiator. The reaction occurs at pressures as high as 1.5×10^8 pascal (approximately 1500 atmospheres) and in the temperature range of 180–250°C. It is interesting to note that even an extremely small quantity of oxygen, say, 0.1%, is sufficient to initiate the polymerisation. Apart from oxygen, other initiators include peroxides, hydroperoxides and azo compounds. Ethylene can be polymerised by solution or bulk polymerisation techniques. For the solution method, the solvents employed are benzene, chlorobenzene and so on. The reaction conditions are such that the polymer as well as the monomer dissolve in the solvent and the system follows a true solution polymerisation technique.

Low density polyethylene melts at 110–125°C and is only around 40% crystalline. The density is around 0.91–0.92 g/cc. While practically no solvent dissolves it at room temperature, several solvents can do so at high temperatures. Some of the useful solvents for polyethylene at high temperatures are carbon tetrachloride, toluene, xylene, decaline and trichloroethylene. The dissolved polymer precipitates out as the solution cools down to room temperature.

As mentioned earlier, LDPE consists of molecules which are branched. The branching occurs during the polymerisation process either by intermolecular or intramolecular chain transfer reactions, as follows:

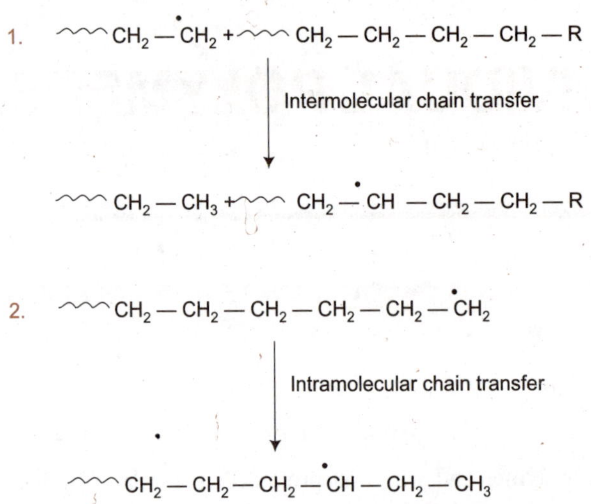

In the first case, the branches could be as long as the backbone chain itself, whereas, in the second case, the branches are much shorter in length.

High density polyethylene (HDPE) is produced principally by two methods. The first method is the coordination polymerisation of the monomer by triethyl chromium and titanium tetrachloride. In the second method, the polymerisation is done with the help of metal oxide catalysts such as chromium or molybdenum oxides, supported over alumina-silica bases. In both the methods, the polymerisation is carried out at relatively low pressures.

Further, high density polyethylene is 90% crystalline and possesses densities as high as 0.965 g/cc. It has a melting point in the range of 144–150°C. It is much stiffer than LDPE and has a higher tensile strength and hardness. Chemically, it is more resistant than the low density variety and has a significantly lower gas permeability.

LDPE films are mainly used for packing and wrapping frozen food, textile products and so on. The film uniquely combines in itself low density, high tear strength, extreme flexibility and chemical and moisture resistance. The flexibility of the thin film is the inherent property of the polymer without recourse to a plasticiser. LDPE's inertness to chemicals and resistance to breakage is made use of in 'squeeze bottles' and in many attractive containers. Pipes made of LDPE are used for both agricultural, irrigation and domestic water line connections. The non-polar nature of the polymer makes it ideal for providing insulations to electric cables.

HDPE finds use in the manufacture of toys and other household articles. Whenever high tensile strength and stiffness are required, high density polyethylene finds better use.

9.2 POLYPROPYLENE

The polypropylene polymer can be represented as follows:

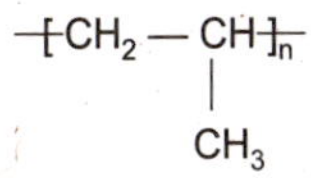

Polypropylene is perhaps the lightest known industrial polymer. It is produced by the polymerisation of propylene using the famous Ziegler–Natta catalysts. The monomer, propylene, is obtained as a by-product from gasoline refineries. The typical polymerisation catalysts are Al (iso-$C_4H_9)_3$ and $TiCl_3$; or $Al(C_2H_5)_3$ and $TiCl_3$.

Polypropylene can be prepared in isotactic, syndiotactic or atactic forms. The isotactic polymer melts at 208°C and is highly crystalline. The polymer molecules are essentially linear and take up configurations leading to a 3_1-helical structure as shown in Fig. 7.10. Being highly crystalline, polypropylene exhibits high stiffness, hardness and tensile strength. Its high strength-to-weight ratio makes it very useful industrially. The articles moulded out of this polymer can be sterilised since it possesses a melting point higher than 100°C. It is insoluble at room temperature in many of the known solvents. However, polypropylene, on heating above its melting point, can be dissolved in aromatic and chlorinated hydrocarbons. It is resistant to many chemicals such as acids, alkalis and oils, but less resistant to oxidation compared to polyethylene. While its stability towards heat and light is lesser than that of polyethylene, polypropylene has excellent mechanical and dielectric properties and it is quite comparable to polyethylene with regard to moisture resistance. Components made of polypropylene are used in appliances such as refrigerators, radios and TVs. It is also used for producing package films, pipes, storage tanks, seat covers, monofilaments and ropes.

9.3 POLYSTYRENE

Polystyrene, also known as polyvinyl benzene, has the following structure:

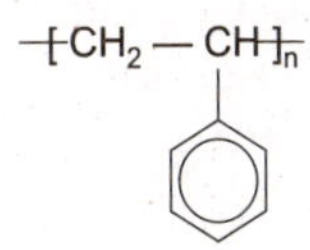

Polymerisation of styrene had been accidentally observed as early as 1839. However, its commercial potential was realised nearly a century later, around 1930.

Styrene is produced from ethylene and benzene. When ethylene is passed into benzene in the presence of aluminium trichloride catalyst, ethyl benzene is produced, which, on passing over a catalyst such as iron oxide or magnesium oxide at high temperatures, gets converted into styrene and hydrogen. Usually, the reaction proceeds at a reasonably good rate at 600°C:

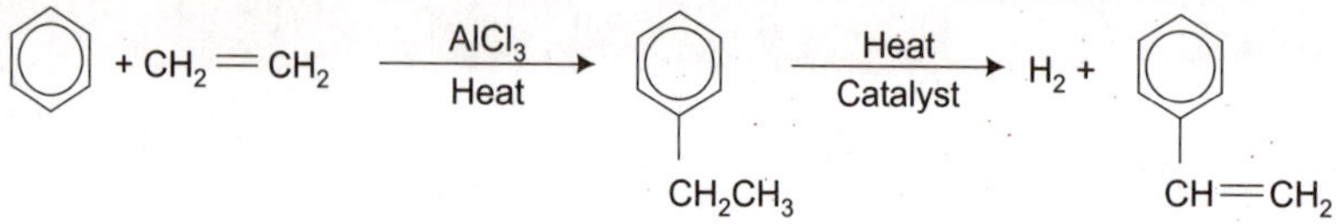

The styrene thus obtained is further purified by distillation. Styrene can be polymerised by radical, coordination, cationic or anionic method. It is industrially produced by free-radical polymerisation, using the suspension or bulk technique. The bulk method is amenable for a continuous process, while the suspension method is highly suitable for a batch process. Commercially available polystyrene is mostly of the atactic variety and hence, is amorphous in nature. Polystyrene consists generally of linear molecules and is chemically inert. Acids, alkalis, oxidising or reducing agents have little effect on it.

Thermal depolymerisation* of polystyrene is possible, giving back the styrene monomer along with a number of low molecular weight compounds. Many organic solvents dissolve polystyrene. Its major defects are poor weathering properties and brittleness. It has a low heat distortion temperature (85°C) and hence, articles made of it cannot be sterilised with steam. Another drawback is that outdoor exposure causes yellowing and crazing of this polymer. Despite all these drawbacks, polystyrene is very widely used in the manufacture of articles such as moulded containers, lids, jars, bottles, radio and television cabinets, toys, foamed plastics, and many other household items. Copolymers of styrene with acrylonitrile, vinyl carbazole, or diphenyl acrylamide are a few important ones. Introduction of a comonomer increases the heat and impact resistance of the polymer, without sacrificing the useful properties of the styrene homopolymer. Styrene-acrylonitrile (SAN) copolymer is a transparent plastic with a very good impact strength and is used for moulding crockery items and machine components. Acrylonitrile-butadiene-styrene (ABS) terpolymer is another commercially important plastic which possesses good strength and toughness. It is most suited for injection-moulded articles of household items and packing containers.

9.4 POLYACRYLONITRILE

Polyacrylonitrile (PAN), also known as polyvinyl cyanide, has the following structure:

$$\left[CH_2 - \underset{\displaystyle CN}{\underset{|}{CH}} \right]_n$$

It is produced from acrylonitrile by the radical polymerisation technique using peroxide initiators. Acrylonitrile monomer can be obtained from acetaldehyde and hydrogen cyanide:

$$CH_3CHO + HCN \longrightarrow CH_3 - \underset{\displaystyle OH}{\underset{|}{CH}} - CN \longrightarrow H_2O + CH_2{=}CH - CN$$

*For a discussion on 'depolymerisation', see Section 10.2.

It can also be produced by ammonoxidation of propylene:

$$CH_2{=}CH-CH_3 + NH_3 + \frac{3}{2}O_2 \xrightarrow[\text{Catalyst}]{500°C} 2CH_2{=}CH-CN + 3H_2O$$

PAN is soluble in dimethyl formamide, dimethyl sulphoxide, adipo nitrile, and so on. It has a remarkable resistance to heat up to around 220°C and exhibits very good mechanical properties. Polyacrylonitrile is used to produce what are known as PAN fibres. The copolymer of acrylonitrile with butadiene (nitrile rubber) is a material of great industrial importance.

9.5 POLYMETHYL METHACRYLATE

The polymethyl methacrylate polymer has the following structure:

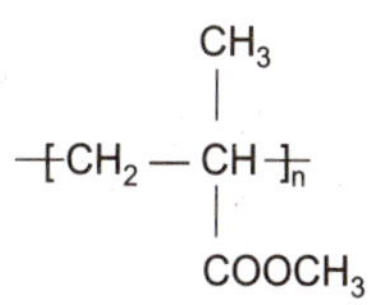

The monomer methyl methacrylate is produced from acetone as follows:

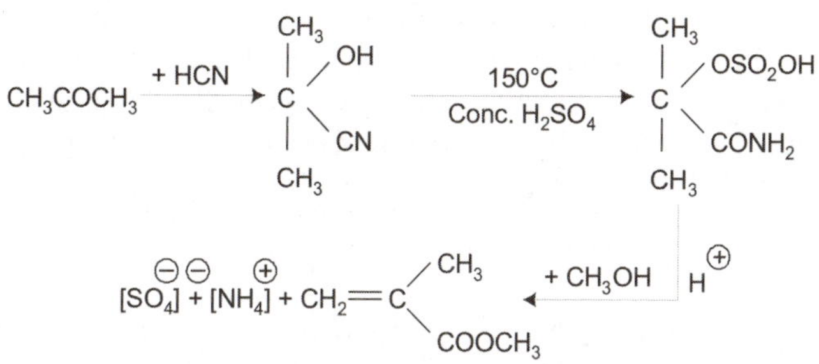

Polymethyl methacrylate (PMMA) is prepared by radical polymerisation in bulk or suspension. It is a colourless transparent plastic with an excellent outdoor life period and good strength. It is amorphous by nature, owing to the presence of bulky side groups in the molecules. It is resistant to many chemicals but soluble in organic solvents such as ketones, chlorinated hydrocarbons and esters. It can be thermally depolymerised to yield back the entire quantity of monomer. Optical clarity is the main feature of this plastic. In many applications, it is an excellent substitute for glass. It has good mechanical properties also. However, compared to glass, it has poor scratch resistance. PMMA is used to make attractive signboards and durable lenses for automobile lighting. It is also used in buildings for decorative purposes.

9.6 POLYESTERS

Polyesters have the structure with ester linkages as follows:

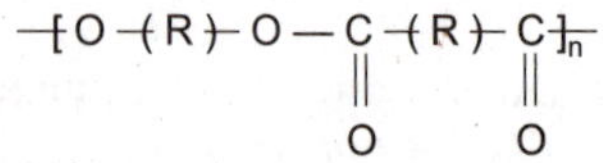

They are prepared by a polycondensation reaction between a dicarboxylic acid and a diol. If both R′ and R are aliphatic, the resultant polyester will be a fully aliphatic polyester.

For a long time aliphatic polyesters were not of much industrial importance, mainly because of their low melting points. For instance, the polyester with a degree of polymerisation comparable to that of commercial polyethylene melts somewhere in the range of 50 – 80°C, whereas, the melting point of polyethylene is around 120°C. This problem of a low melting point was overcome by introducing aromatic rings into the polyester chain. This is evident from the following examples:

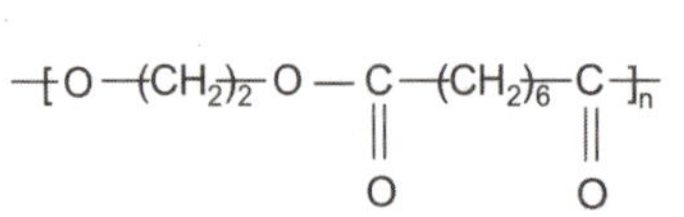

(Melting point 65°C)

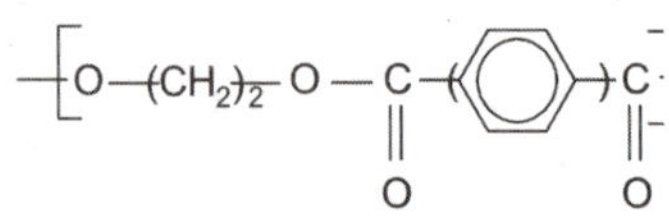

(Melting point 265°C)

The remarkable rise in the melting point of polyesters produced by the incorporation of the aromatic ring is due to the stiffening of the polymer backbone. Thus, a polyester such as polyethylene terephthalate (PETP) has a high melting point because of the presence of the aromatic ring and is commercially one of the most popular polymers marketed under the trade name of Terylene or Terene. The starting materials for PETP are ethylene glycol and terephthalic acid. In commercial practice, however, dimethyl terephthalate (DMT) is taken in place of terephthalic acid. DMT is melt condensed with ethylene glycol to give PETP:

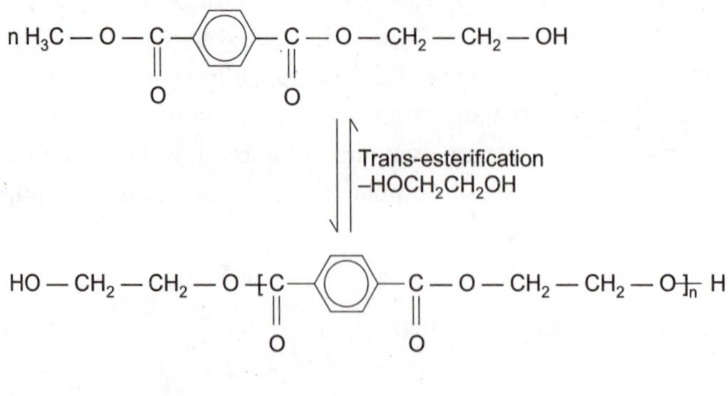

The first stage of the reaction (condensation) is carried out at the reflux temperature of ethylene glycol with a low vacuum, when methanol is recovered. The second stage (trans-esterification) is carried out at high temperatures around 200–250°C, under a very high vacuum, so that the ethylene glycol can be effectively removed to push the reaction in the forward direction.

Polyethylene terephthalate melts at around 265°C and is resistant to heat and moisture. This polyester is virtually unattacked by many chemicals. It is extensively used to make textile fibres. It has good mechanical strength up to 175°C. Garments made from its fibres resist the formation of wrinkles. PETP is also made into films, which are used in the manufacture of magnetic recording tapes, aluminised sheets for certain special applications and so on.

Aliphatic unsaturated polyesters, however, find some industrial use. Unsaturation is introduced in the resin molecule by taking an unsaturated dicarboxylic acid such as maleic acid. For example, with ethylene glycol, maleic acid will give a polyester of the type.

$$n.\ HOOC-CH{=}CH-COOH + n.\ HO-CH_2-CH_2OH$$

$$\downarrow$$

$$H{+}O-\underset{\underset{O}{\|}}{C}-CH{=}CH-\underset{\underset{O}{\|}}{C}-O-CH_2-CH_2{+}_n OH + (2n-1)H_2O$$

The water of condensation is to be continuously removed from the system, if the reaction should proceed further and give a high molecular weight resin. Commercial unsaturated polyesters are based on phthalic acid, maleic acid, ethylene glycol and butanediol. The double bond—present in the backbone of the chain—is utilised to cross-link the resin with styrene using free-radical catalysts (see Section 11.9.3). The unsaturated polyester-styrene combination is used as the resin matrix in fibre-reinforced plastics (FRP) structures. These resins also find use as decorative coatings.

9.7 POLYFORMALDEHYDE

Polyformaldehyde is composed of $—CH_2—O—$ repeat units and can be obtained by the cationic polymerisation of either formaldehyde or trioxane,

$$n.\ HCHO \xrightarrow[\text{(Moisture)}]{BF_3} HO{+}CH_2-O{+}_n H$$

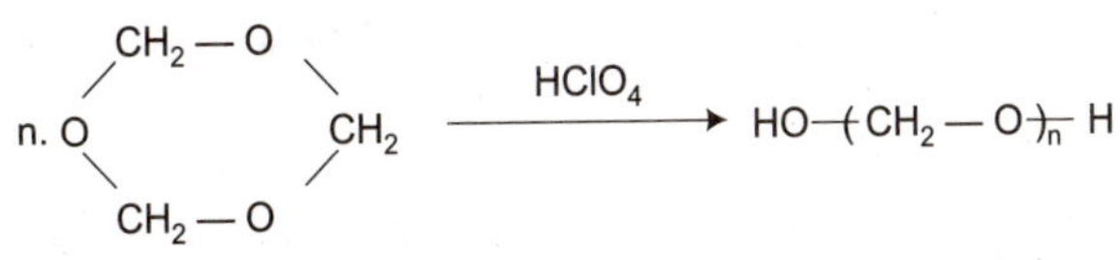

The polymer formed, however, is prone to degradation by unzipping,* on storage and hence, is usually stabilised by esterifying the hydroxyl end groups with acetic acid and pyridine. Commercial grade polyformaldehyde is a fibrous polymer with 50–80% crystallinity. It has a melting point of around 185°C. It has properties such as a high degree of hardness and rigidity, which make it useful as an engineering plastic.

9.8 POLYPARAPHENYLENE

This polymer is composed entirely of aromatic rings forming the chain backbone:

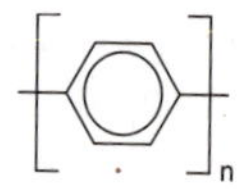

The starting material is paradibromobenzene. This compound, when treated with activated copper powder at high temperatures, yields parapolyphenylene as follows:

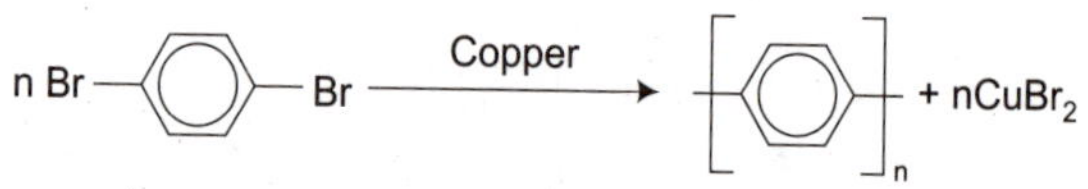

This polymer can also be prepared by the cationic polymerisation of benzene with aluminium trichloride and cupric chloride.

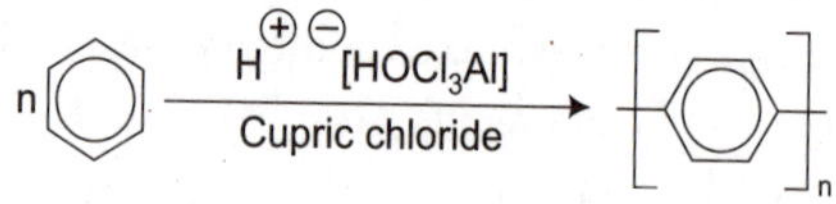

This polymer is very brittle and insoluble, owing to the rigid nature of the polymer chain, caused by the presence of aromatic rings linked together through parapositions. It can withstand temperature up to 560°C. This property is derived from the presence of resonance-stabilised conjugated double bonds in the aromatic rings.

9.9 POLYCARBONATES

Polycarbonates are polyesters of phenols and carbonic acid. They can be prepared by condensing diphenoxymethylene derivatives with diphenylcarbonate. Bisphenol—A with diphenylcarbonate gives a polycarbonate as follows:

---

*For 'unzipping', see Section 10.2.

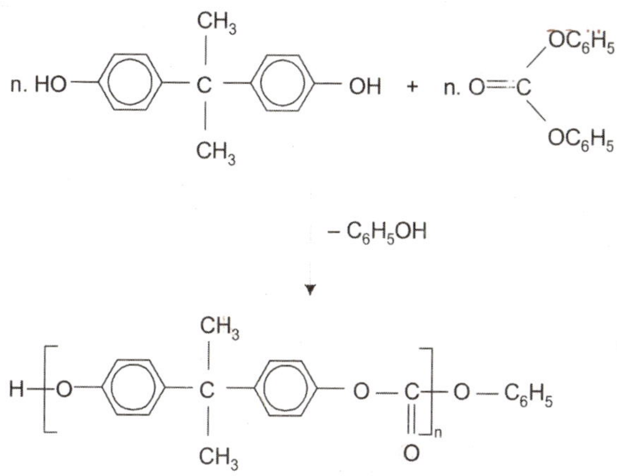

Like PETP, the polycarbonate reaction is also carried out in two stages, *i.e.*, the first stage at 200°C with low vacuum when the oligomers are obtained and the second stage at 300°C under high vacuum when the solid polymer is formed.

Polycarbonate melts at around 265°C and has a very high impact strength. It is resistant to water and many organic compounds but alkalies will slowly hydrolyse it. It is a transparent plastic. Many useful articles such as safety goggles, safety shields, telephone parts and machinery housings can be made from this plastic.

Interfacial condensation of sodium salt of bisphenol—A and phosgene also yields the same polymer:

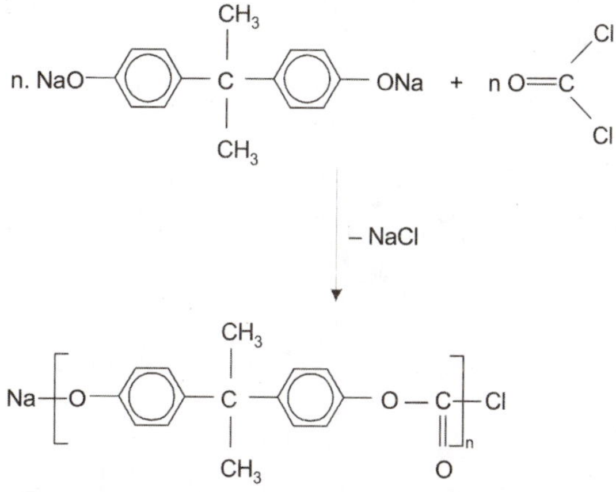

9.10 POLYSULPHONE

Polysulphone is a condensation product of bisphenol—A and dichlorodiphenyl sulphone:

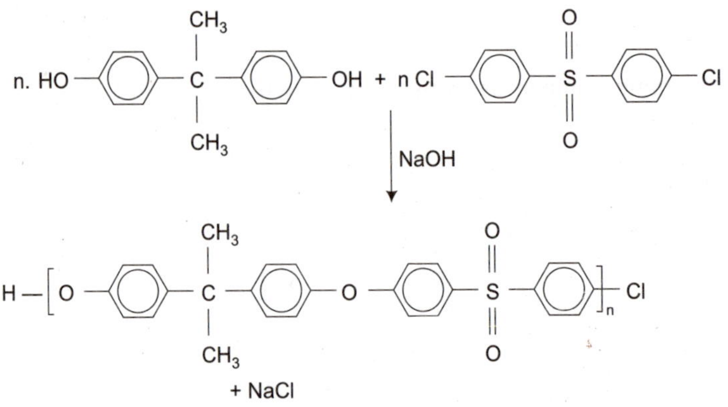

It is a thermoplastic with a very high thermal stability and good mechanical strength over a wide range of temperature (from –180°C to + 140°C). The extraordinary thermal stability arises out of the resonance stabilisation due to aromatic rings near—SO_2^- groups. Polysulphone is used for making injection-moulded items and films.

9.11 POLYIMIDES

Polycondensation of pyromellitic anhydride and p, p′-diamino diphenyl ether results in the formation of polyimides. The reaction, however, cannot be carried out in a single stage. To start with, the reaction is carried out in suitable solvents, such as DMF, at around 50°C, when a polyaddition reaction takes place with the formation of polyamic acid:

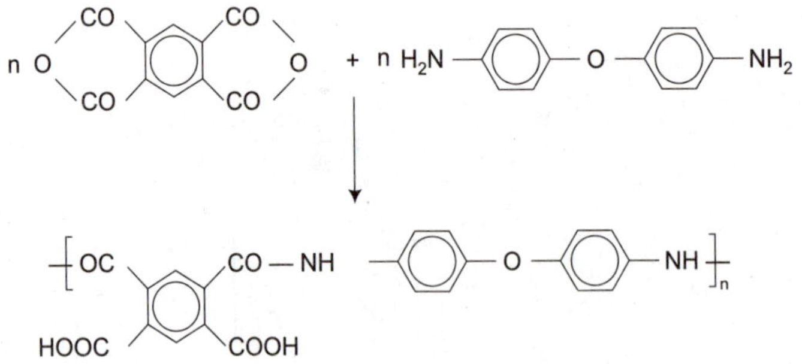

During the second stage, the polyamic acid is cast as a film, the solvent is evaporated and the film is baked at 300°C in a nitrogen atmosphere, when the condensation reaction takes place as follows:

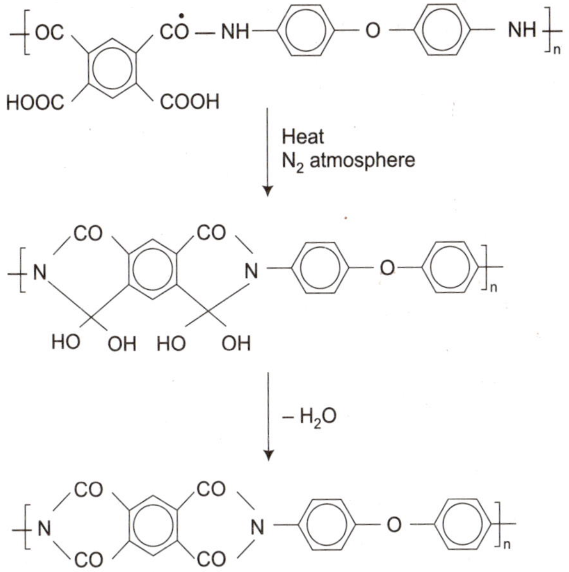

The product formed during the second stage from polyamic acid is insoluble and infusible and has to be formed in the final required shape and form.

Polyimide finds extensive application in electrical industry as the insulation coating of electromagnetic wirings. In practice, the wire is coated with polyamic acid and baked at 300°C under an inert atmosphere when the polyimide is deposited on the surface of the wire.

Polyimide can withstand temperatures up to 425°C for short time exposures without undergoing any degradation or deformation. For this reason, the polymer finds application as surface coating in supersonic aircraft.

9.12 POLYAMIDES

Polyamides are prepared by the melt polycondensation between dicarboxylic acids and diamines. They have the general structure as follows:

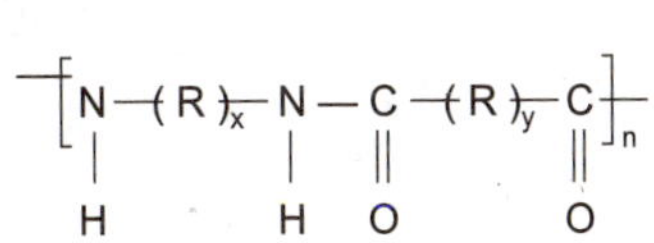

where R = $-CH_2-$ or 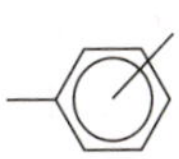

The aliphatic polyamides are generally known as nylons. There are different types of nylons and are usually indicated by a numbering system. This number gives the number of carbon atoms present in the monomer molecules. The nylons obtained from dibasic acids and diamines are usually represented by two numbers, the first one indicating the number of C atoms in the diamine and the second that in the dicarboxylic acid. Thus, a polyamide made from hexamethylene diamine and adipic acid is written as nylon 6, 6. Commercially, nylon 6, 6 and nylon 6, 10 are extremely important. Nylons made by the self-polycondensation of an amino acid or by the ring-opening polymerisation of lactams are represented only by a single number, as in the case of nylon 6, which is made from caprolactam.

Nylon 6, 6 is used as a plastic as well as a fibre. It has good tensile strength, abrasion resistance and toughness up to 150°C. Also, it offers resistance to many solvents. However, formic acid, cresols, and phenols dissolve this polymer. A large quantity of nylon 6, 6 is used to produce tyre cord. Nylon 6, 6 is also used to make monofilaments and ropes. Another fraction of the polymer manufactured goes to make textile fibres for use in dresses, undergarments and the like. Nylon 6, 6, being a tough plastic, is used as a good substitute for metals in gears and bearings. Nylon 6, 10 is not much used as a textile fibre, but is used to manufacture articles like brushes and bristles. Nylon 11 (made by the self-condensation of ω-amino undecanoic acid), is used as a textile fibre. It is less affected by water, compared to other nylons. Aliphatic nylons have melting points in the range of 250–300°C. Recently, fibres having high tensile strength have been produced from polyterephthalamides. These polymers are produced by the interfacial polycondensation of terephthalic acid dichloride with aliphatic diamines as follows:

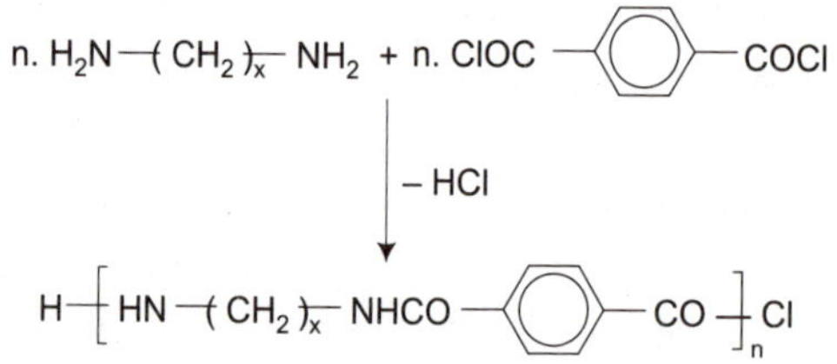

They have high melting points in the range of 435–450°C. Fully aromatic polyamides (or aramides), prepared either by the self-condensation of aromatic amino acids or by polycondensation between aromatic diacid chlorides and aromatic diamines, are also gaining industrial importance in recent years:

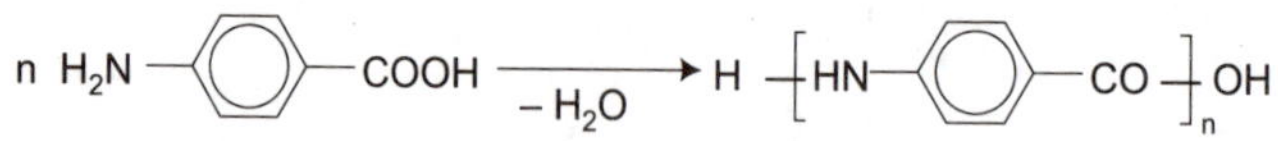

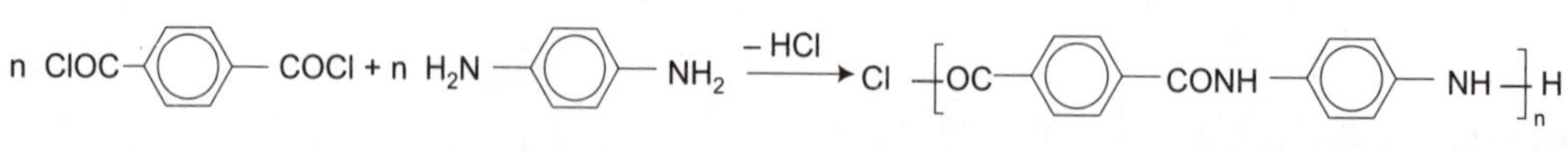

Polycondensations in these cases are carried out by melt, solution or interfacial techniques. Aramides have a very high melting point (>500°C). Fibres made of them have extremely high modulus and find application in reinforced plastics with unusually high strength-to-weight ratios. Fibres spun from poly (paraphenylene terephthalamide), commercially known as 'Kevlar' fibres, fall in this category.

9.13 POLYURETHANES

Polyurethane polymers are characterised by the presence of urethane linkages in their repeat units:

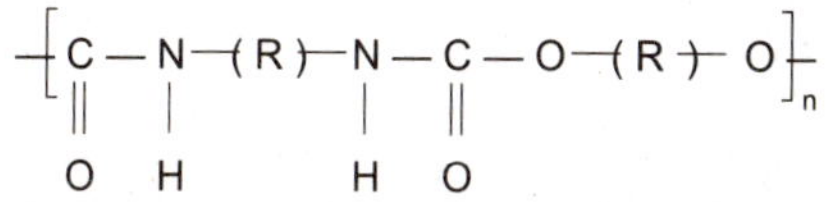

The structure has some resemblance to polyamides because both of them contain —CONH— groups. The principal linkage in polyurethane, however, is —NHCOO—. The presence of additional oxygen in the chain increases its flexibility. This is the reason why the melting point of polyurethane is much less than that of the corresponding polyamide as shown:

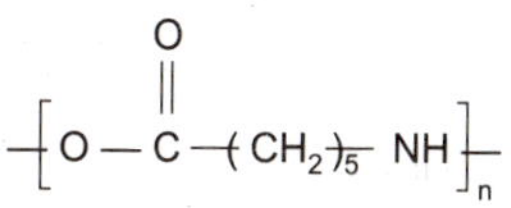

Polyurethane m.p. = 150°C

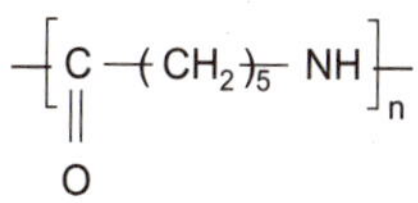

Nylon 6 (Polyamide) m.p. = 250°C

The polyaddition reaction between a diisocyanate with a diol or triol is very frequently used to prepare polyurethanes commercially:

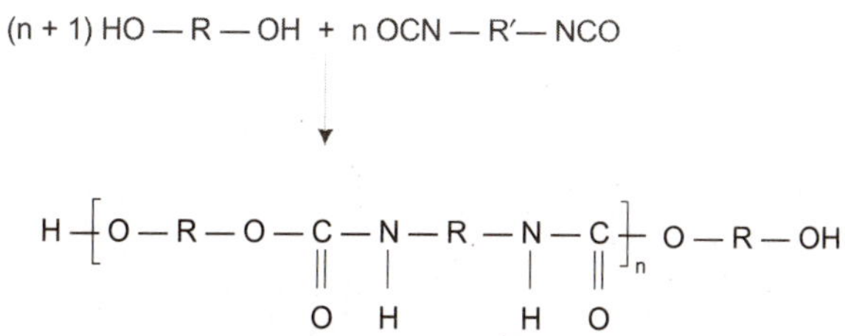

Polyurethanes are used to make foams, coatings, adhesives and elastomers.

9.14 POLYUREAS

Polyureas are polyamides of carbonic acid and are made of —NHCONH— repeat units. They result by diamines reacting with diisocynates as follows:

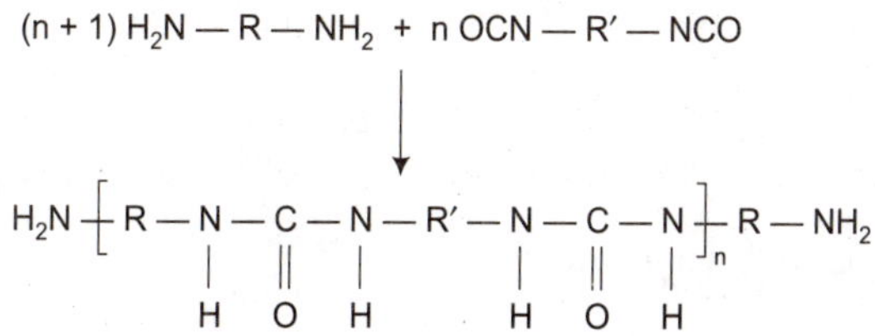

Polyureas have higher melting points, compared to the corresponding polyurethanes or polyamides, as shown:

$$\left[NHCONH - (CH_2)_5 \right]_n \qquad \text{Polyurea m.p. } 270°C$$

9.15 POLYETHYLENE GLYCOLS AND POLYPROPYLENE GLYCOLS

Ethylene oxide or propylene oxide can be polymerised with sodium methoxide or sodium hydroxide to give polymers called polyethylene glycols (PEG) and polypropylene glycols (PPG):

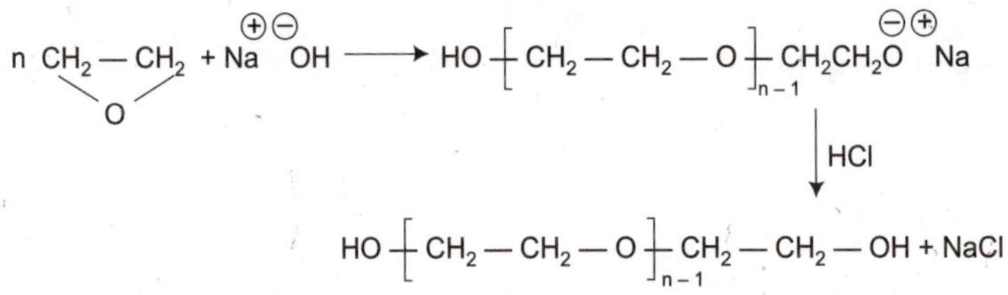

Polyethylene glycol

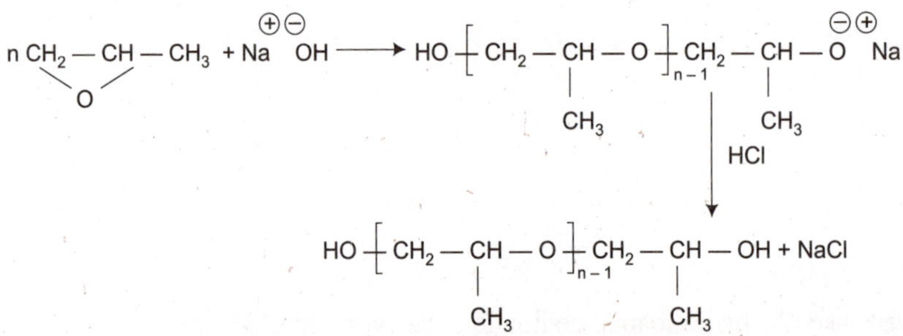

Polypropylene glycol

PEG are waxy materials available in different molecular weight grades and are commercially known as 'carbowax'. They are soluble in water and many organic solvents. They find application in pharmaceutical preparations and cosmetics.

PPG are liquid resins and are used as the diols in the preparation of unethane polymers.

9.16 POLYACRYLIC ACID

The polyacrylic acid has the following structure:

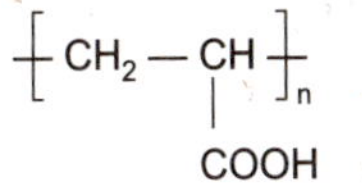

Acrylic acid monomer is obtained from ethylene cyanhydrin. A mixture of ethylene cyanhydrin and sulphuric acid is blown with steam, when 50% acrylic acid solution distills out. Usually, it is the sodium salt of acrylic acid which is polymerised and then subsequently acidified to generate polyacrylic acid. It is insoluble in its monomer but highly soluble in water. Further, polyacrylic acid exhibits the typical behaviour of a polyelectrolyte. Polyelectrolyte, by definition, means any polymer with an ionisable group present in its repeat units. Thus, polyacrylic acid when dissolved in water, attains the following structure:

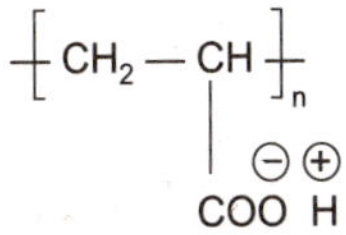

Due to ionisation, as shown above, polyacrylic acid shows very high viscosities in solutions, it is used as a thickening agent for adhesives. Since it is extremely soluble in water, its use as a plastic is very much limited.

9.17 POLYMETHACRYLIC ACID

Polymethacrylic acid has the following structure:

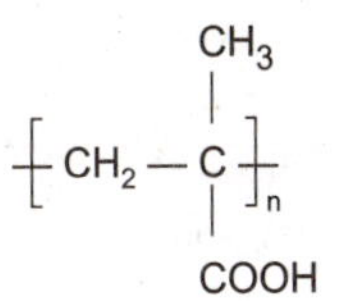

The starting material is methacrylic acid which is obtained from acetone cyanhydrin. The polymerisation is similar to that of acrylic acid. Polymethacrylic acid is also insoluble in its monomer but readily soluble in water, and too, has a limited scope as plastic. Both polyacrylic acid and polymethacrylic acid do not strictly behave as thermoplastic upon heating but cross-link, char and decompose. In the solution state, polymethacrylic acid also exhibits the behaviour of a polyelectrolyte.

9.18 POLYVINYL ACETATE

The structure of polyvinyl acetate is as follows:

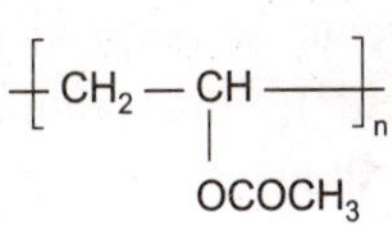

The monomer is prepared by reaction between acetaldehyde and acetic anhydride:

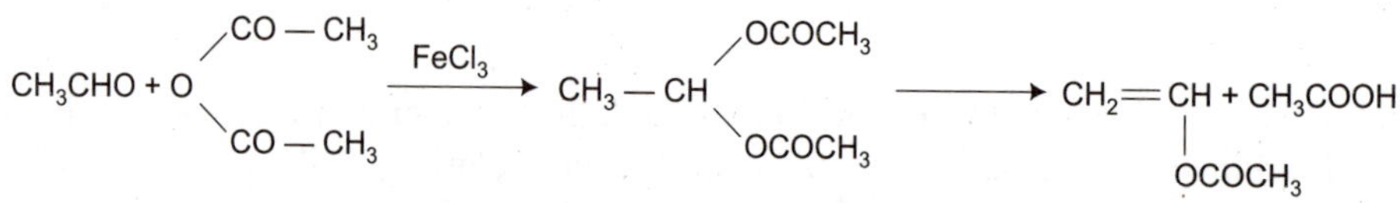

Bulk, emulsion or solution polymerisation methods can be used for preparing the polymer from the vinyl acetate monomer. The polymer is soluble in aromatic solvents and also in alcohols or esters. Polyvinyl acetate is mainly used for the manufacture of polyvinyl alcohol. Along with other polymers and compounds, it is used in adhesives, lacquers and for the large-scale manufacture of gramophone records.

9.19 POLYVINYL ALCOHOL

Polyvinyl alcohol has the following structure:

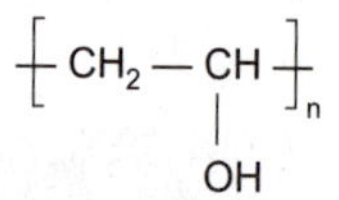

The polymer is manufactured by the hydrolysis of polyvinyl acetate, in the presence of acids or alkalies, as shown:

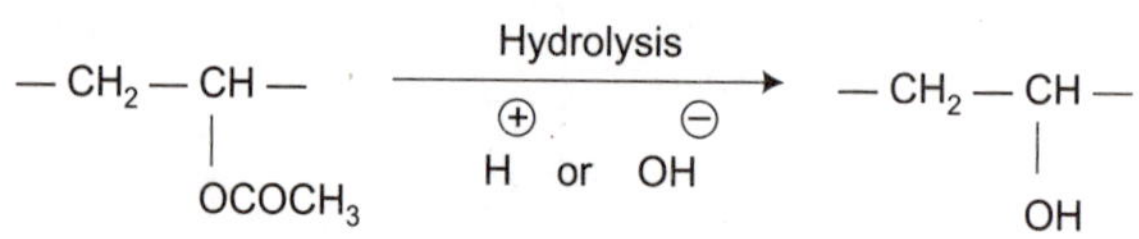

Even after a prolonged hydrolysis, polyvinyl alcohol(PVA) contains around 1 to 2 mole per cent of acetyl groups.

Polyvinyl alcohol, unlike many polymers, is soluble in water. It dissolves slowly in cold water, but, at higher temperatures, it goes fairly fast into solution, more so around 90°C.

Polyvinyl alcohol solution, like starch solution, gives an intense blue colour when reacted with iodine. Polyvinyl alcohol is a good protective colloid for aqueous emulsions and is employed for this purpose in a large variety of emulsion and suspension systems. It also finds use in wet strength adhesives. Further, it is used in the manufacture of textile fibres, where one major aspect to be taken care of is that the fibre as the final product should be made insoluble in water. This is done by treating the fibre with a concentrated aqueous solution of sodium sulphate containing calculated quantities of formaldehyde and sulphuric acid. By this treatment, the hydroxyl groups in the polymer chain are converted into formal groups, as shown:

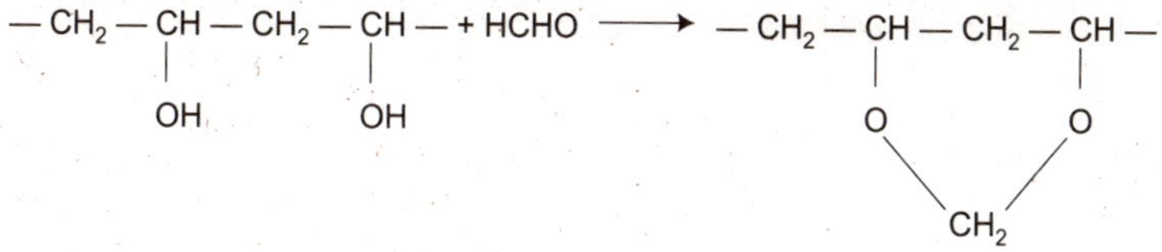

Formal group formation between —OH groups of two or more neighbouring chains are also possible. This renders the fibre insoluble in water. The fibres made from PVA are found to possess a higher water absorption property in comparison to other synthetic fibres. It is well known that cotton fibres and hence, clothes made of them can absorb large quantities of sweat. This property of PVA fibre makes it possible for it to substitute cotton fibres, wherever necessary. PVA fibres have excellent dimensional stability. Apart from this, they can be very easily washed and dried. The fibres are also found to possess excellent resistance to abrasion and remarkable tenacity.

9.20 POLYVINYL ACETALS

Polyvinyl acetals are produced by the reaction between the appropriate aldehydes and polyvinyl alcohol:

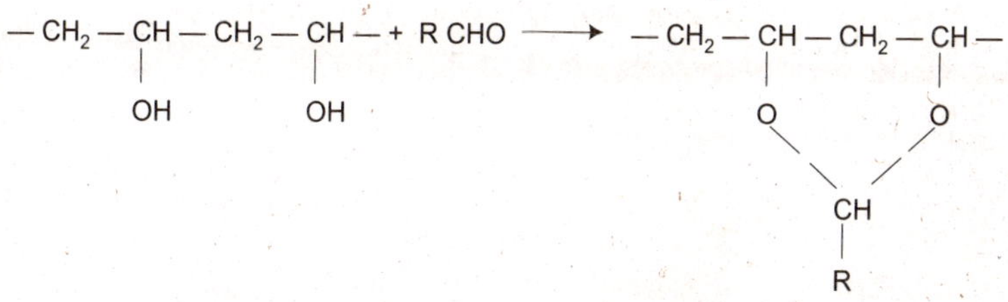

The reaction takes place in a suitable solvent medium with acid catalysts.

Many polyvinyl acetals are known, but two of them are of considerable industrial significance: polyvinyl formal and polyvinyl butyral. The first polyvinyl formal, is soluble in dioxane, acetic acid, phenols, chlorinated hydrocarbons, and the like. This polymer is well known for its excellent heat resistance and dielectric properties. Besides this, it also has excellent abrasion resistance. The main use of this polymer is in electrical insulation coatings. The second, polyvinyl butyral, is soluble in acetic acid, pyridine, cyclohexanone, ethyl alcohol, and so on, and is used in adhesive compositions. It is also used in the manufacture of shatter-proof glass. For this purpose, the polymer is made into a thin film of a thickness around 0.5 to 1.0 mm. This film is placed between two glass sheets and the whole assembly is subjected to moderate heat and pressure to make it a single cohesive system. The two glass sheets thus get firmly bonded through the butyral film. Even if the glass sheets break owing to some impact, the splinters remain stuck to the butyral film.

9.21 POLYVINYL CARBONATE

Vinyl carbonate can be polymerised with peroxides or azo compounds to yield polyvinyl carbonate:

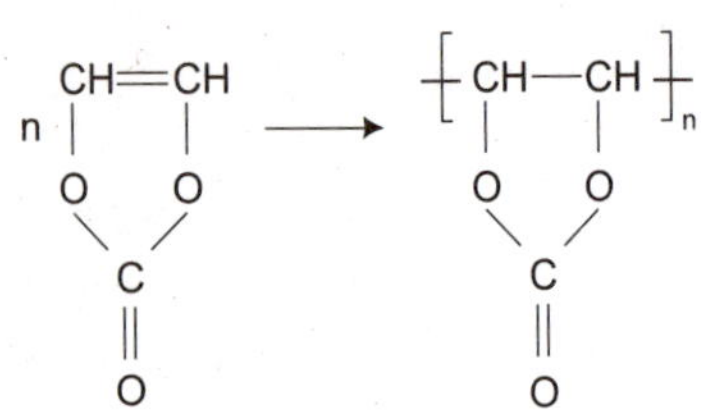

The polymerisation can be done in bulk, solution or emulsion systems.

Polyvinyl carbonate is a solid with a melting point around 200°C; melting is, however, accompanied by decomposition. The polymer is soluble in dimethyl sulphoxide, acetone, dimethyl formamide, and so on. Polyvinyl carbonate undergoes rapid saponification reaction with alkalies or acids to produce polyvinyl alcohol. Polyvinyl carbonate yields fibres with high tensile strength. Many copolymers of vinyl carbonate with other monomers like vinyl chloride, vinyl acetate or vinyl thiocyanate are known. A copolymer of vinyl carbonate with vinyl thiocyanate is extremely useful in protecting tissue cells from high energy ionising radiations.

9.22 POLYVINYL CHLORIDE

The polymer has the following structure:

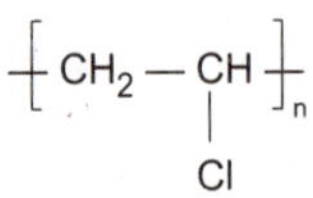

The monomer, vinyl chloride, boils at –14°C. There are two methods for the production of vinyl chloride: one is by cracking ethylene dichloride in vapour phase and the other is by reacting acetylene with hydrogen chloride, in the presence of a catalyst (the well known addition reaction of acetylene):

$$CH_2Cl-CH_2Cl \xrightarrow[\text{Pressure}]{500°C} CH_2{=}CHCl + HCl$$

$$CH{\equiv}CH + HCl \xrightarrow[\text{Catalyst}]{150°C} CH_2{=}CHCl$$

Industrial polymerisation of vinyl chloride is carried out either in suspension or emulsion. Limited quantities of PVC are also made by bulk polymerisation. Redox initiators are used for emulsion systems, while azo compounds are employed in bulk and suspension polymerisation. The emulsion system has the advantage that the polymerisation could be done at low temperatures (say, 20°C) whereas higher temperatures (say, 50°–80°C) are required for the suspension and bulk polymerisation. Since the monomer is a gas at these temperatures, polymerisation has to be conducted in a pressure reactor or an autoclave.

Structurally, the PVC molecule is partially syndiotactic and does not have a completely regular structure. This is why PVC has low crystallinity. The polymer molecules are either linear or only slightly branched. PVC is insoluble in vinyl chloride, alcohols, water and hydrocarbons. Acids and Alkalis have practically no effect on PVC, at least up to 20°C. PVC, however, dissolves in ketones, chlorinated hydrocarbons and the like. A mixture of acetone and carbon disulphide is found to be an excellent solvent for PVC.

PVC is one of the cheapest and most widely used plastics globally. It is used for the large-scale production of cable insulations, equipment parts, pipes, laminated materials and in fibre manufacture. PVC is thermally not very stable and, beyond 200°C, it degrades with the evolution of HCl. This is further attended by the formation of conjugated double bonds along the chain, which will result in the discolouration of the polymer. To avoid such discolouration a suitable stabiliser is added to the polymer. The usual stabilisers are alkali earth oxides, organo-metallic salts, epoxy compounds and amine type compounds (see Section 10.3.3).

PVC, as such, is a horn-like material and difficult to process. It is, therefore, compounded with plasticisers. Depending on the plasticiser percentage, fully rigid to highly flexible finished products can be obtained. In many cases, the plasticiser content may be up to 30% by weight of the polymer.

Polyvinyl chloride finds extensive use in certain formulations called plastisols and organosols. These are paste-like compositions and are produced from emulsion-polymerised PVC resin particles. The plastisols are produced by dispersing PVC particles into plasticisers. The organosols, on the other hand, are produced by dispersing PVC particles in a suitable mixture of volatile organic liquid and plasticisers. Plastisols and organosols also contain other compounds such as colourants, fillers, blowing agents and stabilisers. At ordinary temperatures, the PVC particles present in plastisols or organosols do not dissolve in the liquid medium, but only remain as a dispersed system. When the temperature is raised to 150–200°C; PVC fuses with the plasticiser and forms a physically crosslinked structure in the shape of the ultimate plastic article.

The total chlorine content in PVC is around 56.8% and can be further increased to 60–65%, by dissolving it in a suitable solvent such as chlorobenzene and chlorinating at 100°C. The polymer thus formed is called chlorinated PVC. It is assumed that, during chlorination, the chlorine replaces the hydrogen atoms in the $—CH_2—$ units rather than in the —CH—Cl— units of the molecule. The principal advantage of chlorination is that the polymer becomes more resistant to acids and bases. However, the stability towards heat and light is found to decrease. The main use of chlorinated PVC is in the manufacture of adhesives, coatings and fibres.

9.23 POLYVINYLIDENE CHLORIDE

The polyvinylidene chloride polymer has the following structure:

$$\left[CH_2 - CCl_2 \right]_n$$

The vinylidene chloride monomer can be obtained from trichloroethane by pyrolysis:

$$CHCl_2CH_2Cl \xrightarrow[\text{Pyrolysis}]{400°C} CH_2{=}CCl_2 + HCl$$

The polymer is prepared by radical polymerisation using the emulsion technique. This polymer is insoluble in many solvents. Its main use is in making fibres. The vinylidene chloride homopolymer and its copolymers with vinyl chloride, vinyl acetate or acrylonitrile, come under the special category of chlorine containing fibre-forming polymers and are extensively used as fibres. Copolymers with ethyl acrylate, however, form thermoplastic materials.

9.24 POLYVINYL FLUORIDE AND POLYVINYLIDENE FLUORIDE

Polyvinyl fluoride has the following structure:

$$\left[CH_2 - \underset{\displaystyle F}{\underset{|}{CH}} \right]_n$$

and vinyl fluoride is obtained from acetylene:

$$CH{\equiv}CH + HF \xrightarrow{\text{Catalyst}} CH_2{=}CHF$$

The monomer is a gas at ambient conditions (b.p. –72°C) and hence, free-radical polymerisation using benzoyl peroxide is carried out under pressure at 85°C.

Polyvinyl fluoride is a crystalline polymer with a melting point of around 200°C. It has excellent outdoor life. It is highly resistant to chemicals and in that respect, resembles other fluorocarbon polymers such as Teflon. It maintains good structural integrity even above 150°C and has good thermal stability. Gas permeability is also sufficiently low. It has already proved its merit as a protective coating as well as a glacing material over cement mortars in the building industry. For durability, even in extremely small thickness, it is far better than conventionally use materials such as paints and other surface coatings.

Polyvinylidene fluoride has the following structure:

$$\left[CH_2 - CF_2 \right]_n$$

It is a crystalline polymer with a melting point of around 170°C. It does not undergo creep or distortion over a wide temperature range. Like polyvinyl fluoride, it is resistant to many solvents and finds use in coating compositions.

9.25 POLYTETRAFLUOROETHYLENE

Polytetrafluoroethylene (PTFE), also known as 'Teflon' in trade, may be visualised as polyethylene with all its hydrogen atoms substituted by fluorine, as shown:

$$\left[CF_2 - CF_2 \right]_n$$

The monomer used is tetrafluoroethylene (TFE), which is a gas at room temperature, with a boiling point of –76°C. There are many methods for the preparation of this monomer. In one method, it is produced by the dechlorination of sym. dichloro tetrafluoroethane. In another method, it is obtained by the decarboxylation of sodium perfluoropropionate. Chloro difluoromethane on pyrolysis also yields this monomer. Yet another method is to permit tetrafluoromethane to decompose under the influence of an electric arc.

Polymerisation is carried out by the emulsion method using peroxide initiators. TFE can also be polymerised with redox initiators such as hydrogen peroxide and ferrous sulphate. In some cases, oxygen is used as the initiator. The polymerisation is associated with the release of large quantities of heat and if sufficient precautions are not taken, this can result in violent explosions.

The PTFE molecule has a highly regular structure with a configuration leading to a 13_6-helix. It is also a linear polymer with practically no branching. This is because branching requires the breakage of the strong C—F bonds, which is quite unlikely under the polymerisation conditions. It is a highly crystalline polymer with a melting point of around 330°C. The mechanical strength of this polymer remains practically unchanged over a wide range of temperature from –100°C to + 350°C. It does not dissolve in any of this strong acids, including hot fuming nitric acid. It is resistant to corrosive alkalis and known organic solvents. Out of thousands of chemicals tried, only molten alkali metals were found to react with the polymer to any significant extent. A solution of some alkali metals in liquid ammonia is also capable of reacting with this polymer. Probably, this is because fluorine atoms from the polymer chain get removed by the alkali metals. This polymer has a very low dielectric constant. It should be noted that the conventional techniques used for the processing of other polymers cannot be applied to PTFE because of its very low melt flow rates. The finished products are, in fact, produced by suitably pressing the polymer powder to achieve the required shape, followed by a process, of sintering at temperatures higher than 380°C. This yields highly homogeneous articles. PTFE is used in making articles such as pump valves and pipes where chemical resistance is required. It is used in non-lubricated bearings and its fibre is used to form belts, filter cloth, and similar products where resistance to acid and alkalies is needed.

9.26 POLYVINYL CARBAZOLE, POLYVINYL PYRROLIDONE AND POLYVINYL OXAZOLIDINONE

Polyvinyl carbazole is produced from N-vinyl carbazole. It has the following structure:

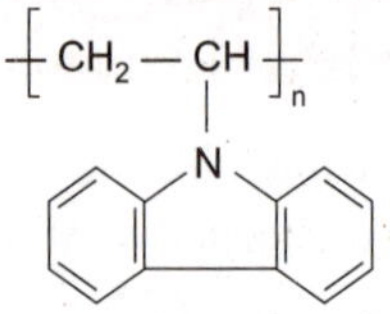

The initiator used for the polymerisation is benzoyl peroxide. The polymer dissolves in ketones, ethers, chlorinated hydrocarbons, and so on. It is generally, resistant to attack by many strong chemicals and is a good substitute for materials such as asbestos and mica.

Polyvinyl pyrrolidone is prepared from its monomer N-vinyl pyrrolidone:

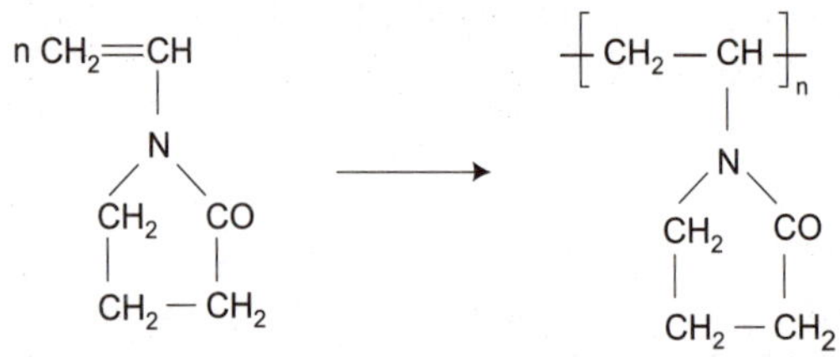

This polymer finds excellent use in life-saving substances such as blood plasma. An appropriate solution with a concentration of 3.5% of the polymer can be used as a substitute for blood plasma. This solution is very stable over a long period of time and can be given to patients, irrespective of their blood groups. Another useful application of this polymer is in the form of additives to many basic dye compositions, where the polymer deepens or intensifies the original colour of the dye.

The monomer for the synthesis of polyvinyl oxazolidinone is N-vinyl-5-methyl-2-oxazolidin-one:

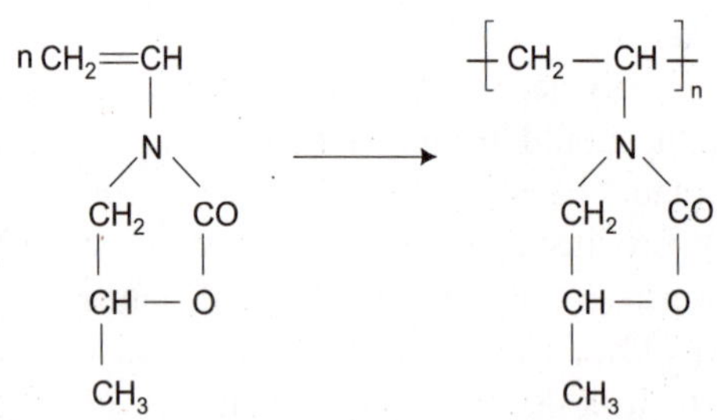

The polymer and the copolymers of this monomer are water soluble and are frequently used as complexing agents to a number of inorganic as well as organic compounds.

9.27 POLYBUTYLENE AND POLYISOBUTYLENE

Polybutylene (or polybutene) has the following structure:

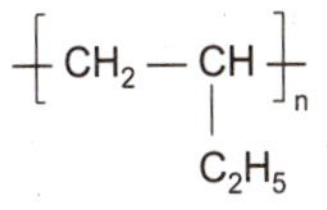

Butylene is the starting material. Polymerisation is done by the Ziegler–Natta catalyst. The commonly used catalyst systems are $Al(iso—C_4H_9)_3$ and $TiCl_3$ or $Al(C_2H_5)_3$ and $TiCl_3$.

The isotactic polybutylene is highly crystalline with a density of 0.91g/cc, and has a melting point of around 125°C.

Polybutylene finds extensive use in rubber-based industries as it has properties very similar to those of natural rubber. This polymer offers good resistance to acids, alkalis and oils.

Polyisobutylene (or polyisobutene) has the following structure:

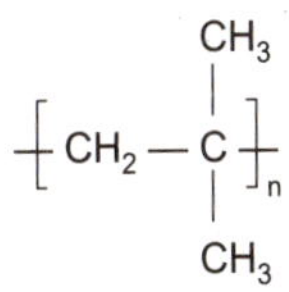

It is produced by the cationic polymerisation of isobutylene using BF_3 or $AlCl_3$ catalysts in a suitable solvent such as hexane or methylene dichloride at −80°C. This polymer dissolves in many hydrocarbons and halogenated hydrocarbons. Polyisobutylene is chemically one of the most inert polymers known so far. Moisture, acids and alkalis have practically no effect on this polymer.

Polyisobutylene is widely used as an insulation material in the electrical industry. It is also added to motor oils to suppress the sharp fall in viscosity with increasing temperature.

A copolymer of isobutylene with 2–5% of isoprene is made under the commercial name of 'butyl rubber'. Due to the presence of the double bonds in the molecule butyl rubber can be vulcanised to form cross-linked structures (see Section 11.9.2) with improved mechanical properties.

9.28 POLYISOPRENES

The structure of polyisoprene is as follows:

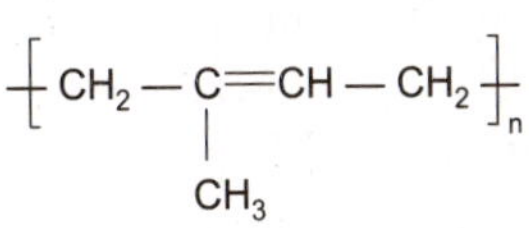

Both the cis- as well as trans-varieties occur in nature as natural rubber and gutta-percha, respectively (see Section 7.7 for their configurations).

Natural rubber is a highly soft and elastic material. It is soluble in carbon disulphide and petrol. Gutta-percha, on the other hand, is a hard thermoplastic solid and dissolves in petrol only on heating and is thrown out of solution when cooled.

Natural rubber is obtained in the form of latex from rubber trees (*Hevea brasiliensis*). The latex normally contains 30–60 % rubber. It can be used, as such, in the latex form or the solid rubber can be coagulated from the latex using 1% acetic acid solution. The coagulated crumbs of solid rubber are washed and sent through corrugated rollers, when 'pale crepe' rubber sheets are obtained. These sheets are then smoked in vapours of tar oil so as to obtain the commercial grade of 'smoked sheets'. (Smoking is done to improve the storage life of the rubber against attack by microorganisms). Smoked sheets are then compounded with chemicals such as sulphur, accelerators and anti-oxidants and then converted into utility articles through moulding, extrusion and other suitable processing techniques.

Polyisoprene are also synthesised commercially, starting from isoprene monomer. Isoprene is synthesised from propylene as follows:

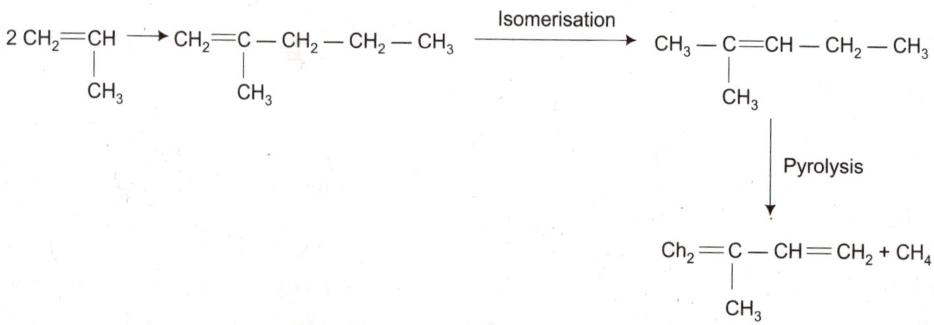

Polymerisation of isoprene yields polymers containing varying degrees of cis-1, 4 and trans-1, 4 as well as 1, 2 or 3, 4-vinyl units in the molecule, depending upon the initiator and the solvent systems employed. Under controlled conditions, polyisoprene containing up to 96% cis-configuration can be obtained by using a lithium alkyl initiator and anhydrous, oxygen-free aliphatic hydrocarbon solvents.

9.29 POLYBUTADIENES

Polybutadienes are produced by the polymerisation of butadiene monomer. Butadiene is obtained through several routes. For instance the C_4 fractions from petroleum cracking are a major source of butadiene. It can also be manufactured from ethanol:

$$C_2H_5OH \rightarrow CH_3CHO \rightarrow CH_3CH{=}CHCHO \rightarrow CH_3-CH{=}CH-CH_2OH \rightarrow CH_2{=}CH-CH{=}CH_2$$

Acetylene is yet another starting material from where butadiene can be prepared:

$$2CH{\equiv}CH \rightarrow CH_2{=}CH-C{\equiv}CH \rightarrow CH_2{=}CH-CH{=}CH_2$$

Industrially, different techniques are adopted for synthesising polybutadiene from butadiene, each giving different grades of the polymer. The oldest method of producing butadiene rubber uses sodium metal (as fine suspension in mineral oil) as the initiator. The name 'Buna rubber' for polybutadiene rubbers was based on this process (butadiene-natrium polymer).

In modern technology, anionic polymerisation of butadiene with lithium metal or Ziegler–Natta catalysts is employed to prepare syndiotactic 1, 2-polybutadiene (with 90% vinyl units). Ziegler polymerisation employing a $VOCl_2 + (C_2H_5)_2AlCl$ catalyst system gives a very high 'cis'-polybutadiene. 'Cis'-polybutadienes have very high abrasion resistance and hence, find applications in automobile tyres. Polybutadienes as such, however, are not popular synthetic rubbers, whereas copolymers of butadiene and styrene (GRS or SBR) or those of butadiene and acrylonitrile (nitrile rubber) are the two commercially important synthetic rubbers. Both these rubbers are made by the free-radical polymerisation in an emulsion system using redox initiators such as hydrogen peroxide and ferrous sulphate. Dodecyl mercaptan is often used as the chain modifier in order to control the molecular weight. The resultant copolymers usually contain 80% of 1, 3 (both cis- as well as tans-) and 20% of 1, 2 or 3, 4 repeat units. The polymerisation can be done either at low temperatures (5°C) or at ambient temperatures (say, 30°C). The rubbers obtained in the former case are known as 'cold rubbers' and those obtained in the latter case are known as 'warm rubbers'. The product is obtained as latex and can be used as such or can be coagulated to get solid rubbers. Both of them can be vulcanised by the conventional method using sulphur. GRS rubbers blend very well with natural rubber and hence, are a good substitute for it. Nitrile rubbers (with an acrylonitrile content of more than 25%) are resistant to hydrocarbons and oils and hence, are used as oil seals and for other oil-resistant applications.

Low molecular weight liquid polybutadienes with specific terminal functional groups find special applications such as adhesives or solid propellant fuel binders. Carboxyl-terminated and hydroxyl-terminated polybutadienes (CTPB and HTPB) are two examples in this context:

$$HOOC-R\left[CH_2-CH{=}CH-CH_2\right]_n R-COOH$$

$$HO-R\left[CH_2-CH{=}CH-CH_2\right]_n R-OH$$

where n is around 200.

CTPB can be synthesised either by free-radical polymerisation, using azo-initiators carrying carboxyl groups in its molecule, or by anionic polymerisation initiated by dianions and terminating the growth with CO_2, followed by hydrolysis (see Section 11.10.2 for reactions).

HTPB can be prepared either by free-radical polymerisation with hydrogen peroxide at elevated temperatures under pressure or by the anionic technique using dianion as in the case of CTPB, but terminating with ethylene oxide followed by hydrolysis (see Section 11.10.2 for reactions).

Similarly, a liquid terpolymer of butadiene, acrylic acid and acrylonitrile (known as PBAN resin) can be prepared by emulsion polymerisation of the three monomers using high percentages of azo-initiators and mercaptan chain modifiers:

$$-\!\!\left[\text{BD}\right]_{n_1}\!\!-CH_2-\underset{\substack{|\\ CN}}{CH}-\!\left[\text{BD}\right]_{n_2}\!\!-CH_2-\underset{\substack{|\\ COOH}}{CH}-\!\left[\text{BD}\right]_{n_3}$$

$$\text{where } -\!\left[\text{BD}\right]\!- = -\!\left[CH_2-CH{=}CH-CH_2\right]\!-$$

The carboxyl and hydroxyl groups in CTPB, PBAN or HTPB are utilised for cure reactions with suitable compounds to get cross-linked elastomeric structures (see Section 11.9.3 for cure reactions).

9.30 POLYCHLOROPRENE

Polychloroprene, also known as neoprene, has the following structure:

$$-\!\left[CH_2-\underset{\substack{|\\ Cl}}{C}{=}CH-CH_2\right]_n\!-$$

Polychloroprene is produced from chloroprene. Industrially, chloroprene is manufactured by the addition of hydrogen chloride to vinyl acetylene. Vinyl acetylene, in turn, can be produced by dimerisation of acetylene:

$$2CH{\equiv}CH \rightarrow CH{\equiv}C-CH{=}CH_2 \xrightarrow[\text{CuCl}]{+\text{ HCl}} CH_2{=}\underset{\substack{|\\ Cl}}{C}-CH{=}CH_2$$

The polychloroprene is synthesised by the emulsion polymerisation technique. Trans-1, 4 is the preferred configuration produced during the polymerisation. It is, therefore, an easily crystallisable elastomer. It can be readily vulcanised with the help of zinc oxide or magnesium oxide (see Section 11.9.2). The vulcanised products are found to have excellent tensile strength, even when carbon black is not added. Fillers, in fact, have very little reinforcing effect on chloroprene rubbers. They have a higher oil resistance than many rubbers, including nitrile rubber.

This elastomer is principally used for providing oil-resistant insulation coatings to wires and cables and for producing shoe soles, solid tyres, gloves and industrial hoses.

9.31 POLYSULPHIDES

Alkyl dihalides can react with alkali polysulphides to give polysulphide elastomers. Polysulphides have excellent oil-resistant properties and are linear condensation polymers. During the reaction, the sodium salt gets eliminated as shown:

$$n\,RCl_2 + n\,Na_2S_x \longrightarrow -\!\left[R-S_x\right]_n\!- + 2n\cdot NaCl$$

In the actual synthesis, the dichloroalkyl compound is reacted with an excess quantity of sodium tetrasulphide in the presence of magnesium chloride and sodium hydroxide, so that the condensation takes place in an emulsion system, the emulsion being stabilised by the magnesium hydroxide formed. Owing to the excess quantity of sodium polysulphide, the product will have —SNa end groups as shown:

$$NaS_4R \sim\sim\sim\sim\sim\sim\sim\sim RS_4Na$$

This type of molecules can undergo self-condensation, leading to a further increase in molecular weight:

$$2NaS_4R \sim\sim\sim \rightleftharpoons \sim\sim\sim RS_4R \sim\sim\sim + Na_2S_4$$

The high molecular weight tetrasulphide resin formed is then stripped down to the disulphide resin by treatment with excess sodium sulphite:

$$\sim\sim\sim RSSSSR \sim\sim\sim + 2Na_2SO_3 \longrightarrow \sim\sim\sim RSSR \sim\sim\sim + 2Na_2S_2O_3$$

The high molecular weight disulphide polymer is then split down to low molecular weight liquid resins with thiol end groups by treatment with sodium hydrosulphide:

$$\sim\sim\sim RSSR \sim\sim\sim + NaSH + Na_2SO_3 \longrightarrow \sim\sim\sim RSNa + \sim\sim\sim R\text{:}$$

The extent of 'splitting' and hence, the molecular weight of the final product can be controlled by the quantity of sodium hydrosulphide taken. After the splitting step, the latex is acidified, when the polymer is thrown out and the —SNa groups get converted into —SH. The final product has the following structure:

$$HS\!\left[R-SS\right]_n\!RSH$$

If a small percentage of trichloroalkyl compound is mixed with the dichloro compound, we get a resin with a small percentage of pendant thiol groups:

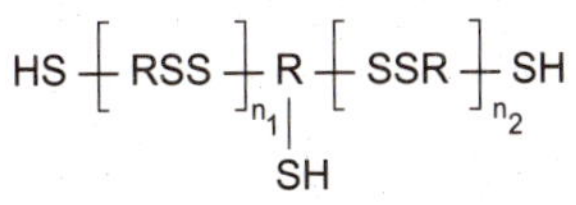

The thiol groups are utilised in the cure reaction (see Section 11.9.3) to obtain a three-dimensional network structure. The physical properties of the polysulphide will vary with the length of the aliphatic group and the number of sulphur atoms present in the chain.

The polysulphides are usually cured with the help of lead dioxide and unlike rubbers, sulphur does not vulcanise them. Sulphur, however, can increase the rate of the cure reaction. Carbon black is frequently employed as a reinforcing material for polysulphide.

Polysulphides are mainly used to make sealants, gaskets, balloons, fabric coatings and gasoline hoses. They have excellent oil resistance and very low gas permeability and are practically unaffected by ozone. Also, aging is very slow in the case of this polymer. Polysulphides form excellent fuel material and, upon mixing with inorganic oxidisers such as ammonium perchlorate, they become solid propellants for rockets.

9.32 PHENOL-FORMALDEHYDE RESIN

Phenol-formaldehyde (PF) resins are formed by the polycondensation between phenol and formaldehyde. The polycondensation reaction can be catalysed either by acids or bases. The acid-catalysed reaction proceeds as follows:

(*i*) $HCHO + \overset{\oplus}{H} \longrightarrow \overset{\oplus}{C}H_2OH$

(*ii*)

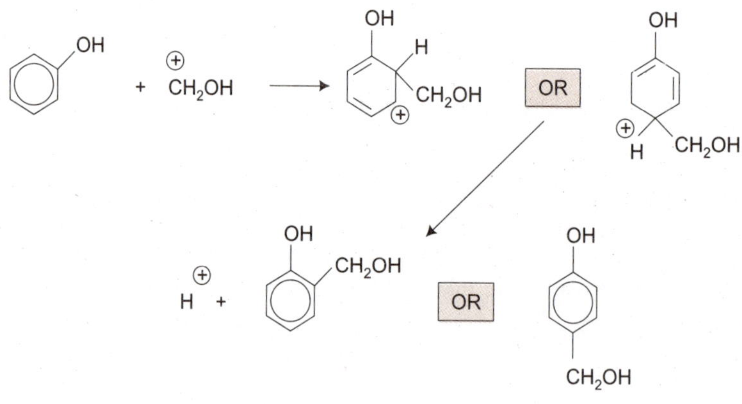

(Monomethylol phenols)

(*iii*)

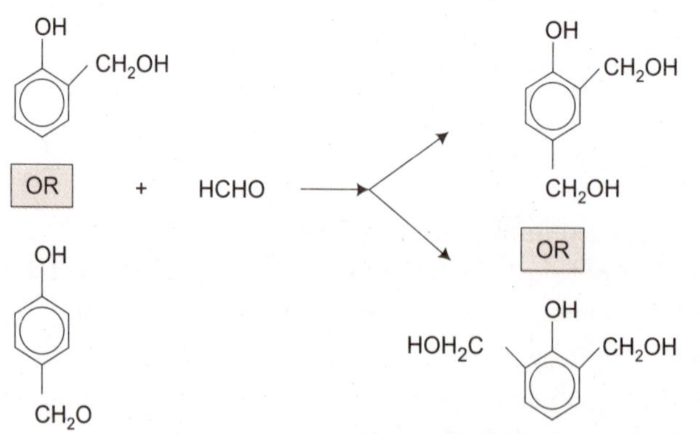

(Dimethylol phenols)

Similarly, the alkali-catalysed reaction can be represented as follows:

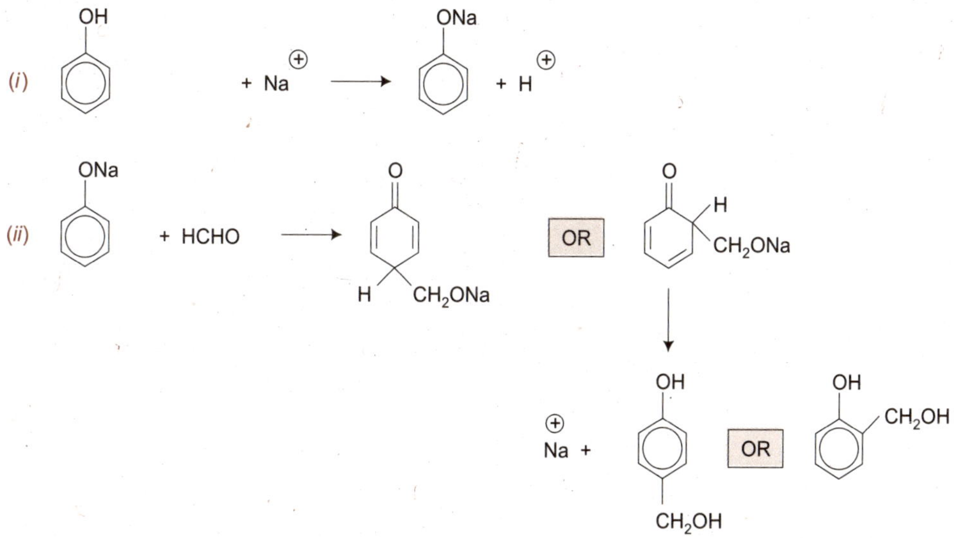

In both the cases, further condensation between methylol phenols and phenol occurs to give the methylene compound:

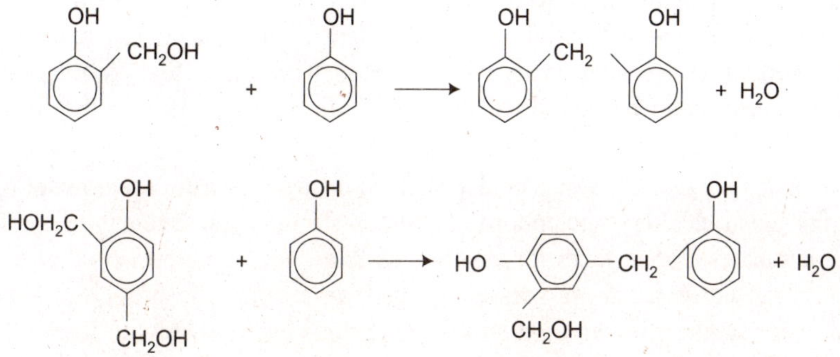

If excess of formaldehyde is taken and the reaction allowed to continue to almost all the available ortho- and para-positions of the phenol molecules, a cross-linked three-dimensional polymer will ultimately result:

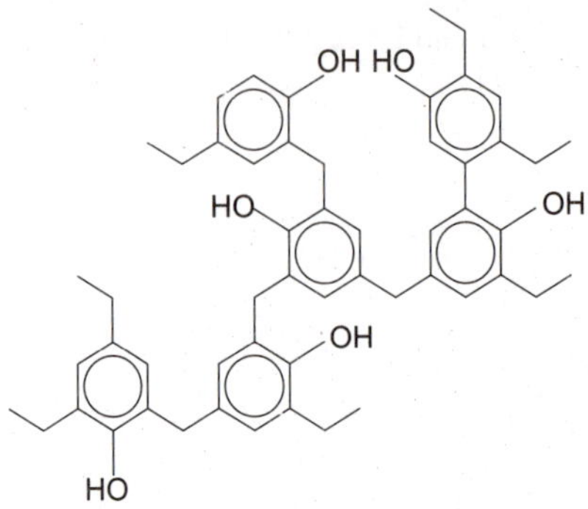

In actual practice, however, the reaction is carried out to a relatively low stage or polymerisation, leading only to linear molecules. This is done in two ways. In the first method, phenol is reacted with 75% of the stoichiometric quantity of formaldehyde using an acid catalyst. In this case, owing to an insufficient quantity of the formaldehyde, only linear molecules are formed:

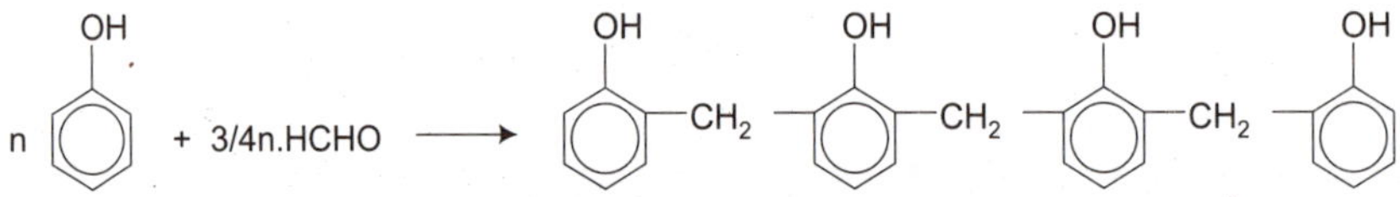

The resin formed can be stored for any length of time without hardening, and further cross-linking, whenever necessary, can be effected by adding excess of formaldehyde and then heating. These types of PF resins are called 'novolak' resins or 'Bakelite' resins.

In the second method, equimolar quantities of phenol and formaldehyde are reacted under alkaline conditions and the reaction rate and the extent of conversion are carefully controlled so that the reaction is carried out only up to linear stage. The resin formed, however, contains the entire quantity of methylol groups at the ortho- and para-positions and hence, on storage, slow reaction will continue leading to the hardening of the resin. This type of resin, which is called 'resol', has, therefore, a limited 'shelf life'. Resols can be converted into the final cured product merely by heating. Cure reactions of 'resols' and 'novolak' are given in Section 11.9.3.

The Bakelite resins are usually compounded with fillers such as asbestos powder or sawdust and used for moulding electrical items, telephone instruments and so on (see Fig. 1.1). The resols are used for making laminates. Paper, fabric or asbestos cloth are impregnated with the resols, and many layers of the impregnated material are pressed together in a hydraulic press and heated to around 150°C. The resin cures under pressure and bonds the various layers firmly, to form a single sheet of a greater thickness. Phenolic laminates are used in electrical circuits. Resols as well as novolak resins are used as coating compositions also. Mixed with sand, they find use as core binders in foundries.

9.33 UREA-FORMALDEHYDE AND MELAMINE-FORMALDEHYDE RESINS

Urea, too, can condense with formaldehyde to form what are known as urea-formaldehyde (UF) resins.

Actually, the reaction proceeds in a manner similar to the formation of the phenol-formaldehyde (PF) resin. Here also, mono-and dimethylol derivatives are formed, which further condense with urea to give the final resin structure.

(*i*) 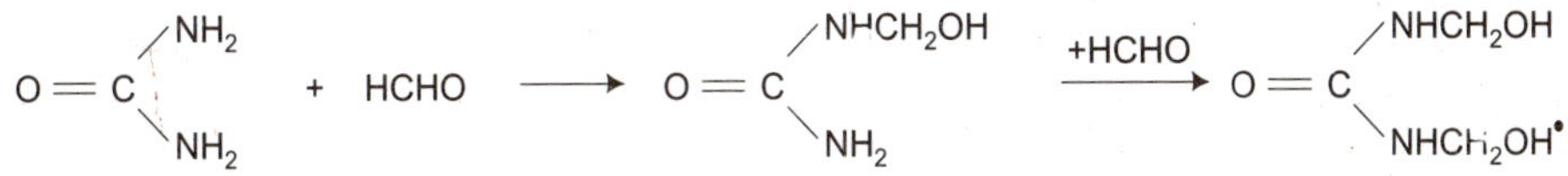

(*ii*) 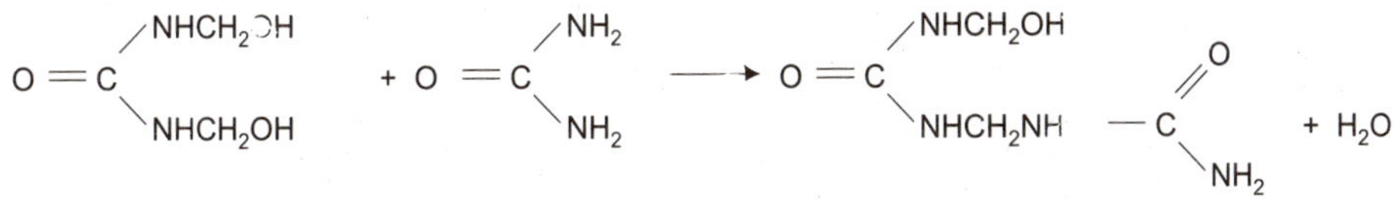

and so on

(*iii*) $$n.\ HCHO + n.\ NH_2CONH_2 \xrightarrow[-\ H_2O]{} H\left[NHCO-NH-CH_2\right]_n OH$$

The hydrogen atoms of the imide group in the linear polymer can be further replaced with methylol groups if excess of formaldehyde is used:

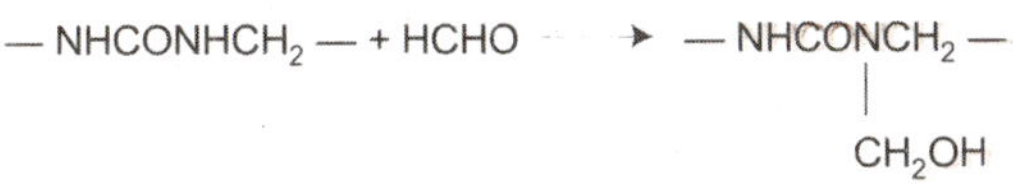

As in the case of PF resins, the final composition of the polymer depends on the ratio of the urea to formaldehyde. Thus, if the linear polymer is heated subsequently with more formaldehyde, a three-dimensional network of the polymer will be produced with the following structure:

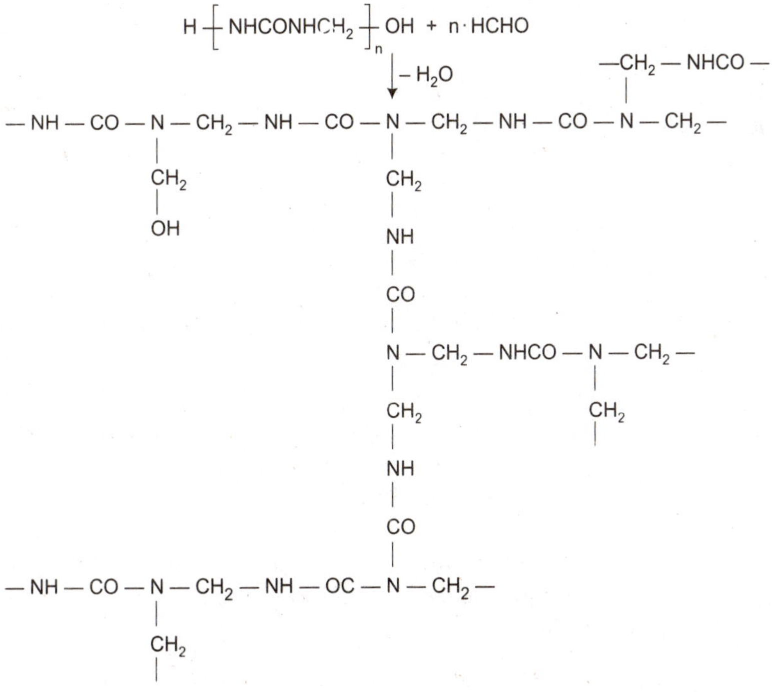

Melamine and formaldehyde can also react to give methylol derivatives of melamine, as shown:

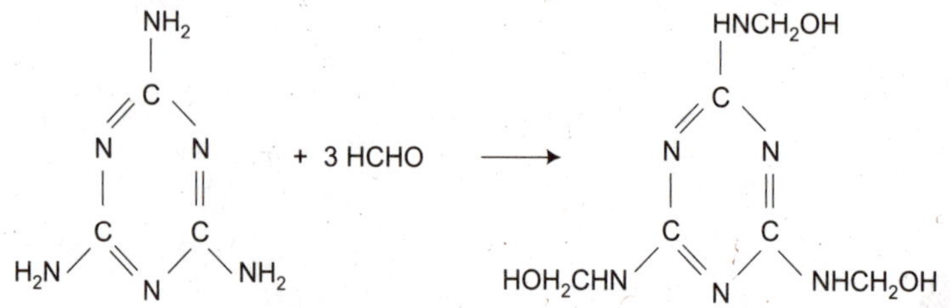

The polycondensation of methylol melamine derivatives with larger quantities of melamine, will give a linear polymer. This polymer, when reacted further with additional quantities of formaldehyde, will give a three-dimensional network as shown, which is found to be insoluble in many solvents:

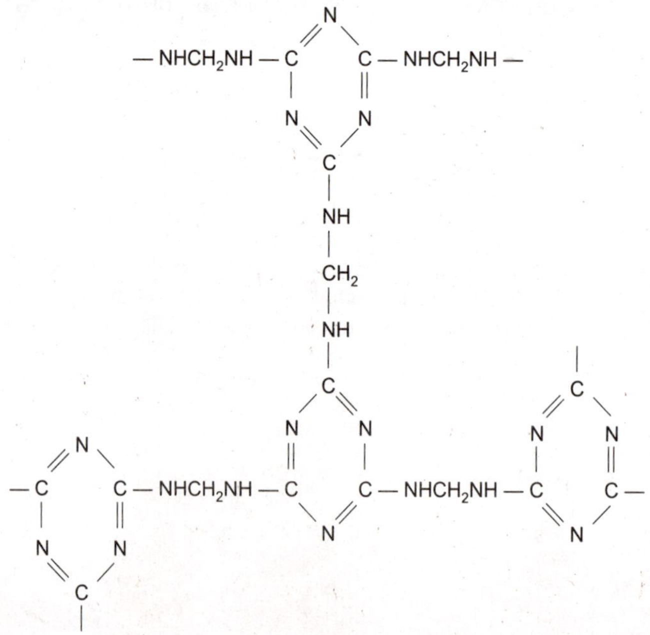

Both uncross-linked UF and MF resins (also known as amino resins) are water soluble and hence, find use as sizing agents and textile-finishing resins. They are also used in the paper industry to improve the wet strength of paper. In the plywood industry, they are used as adhesives.

Melamine resin also finds use in the manufacture of decorative laminates. They are also used for the preparation of lacquers. Compared to the phenolic resins, the amino resins, in general, are colourless, and clearer and hence, objects of lighter shades can be made out of them. Similarly, the tensile strength and hardness are also much better compared to the phenolics. However, their heat and moisture resistances are definitely lower. Between urea-formaldehyde and melamine-formaldehyde polymers, the latter have better heat and moisture resistance. The hardness is also higher in the case of melamine resins.

9.34 EPOXY POLYMERS

The epoxy polymers are basically polyethers. One type of epoxy polymer (or epoxy resins as they are generally called) is prepared from epichlorohydrin and bisphenol-A. The reaction is carried out with excess of epichlorohydrin. The scheme is as follows:

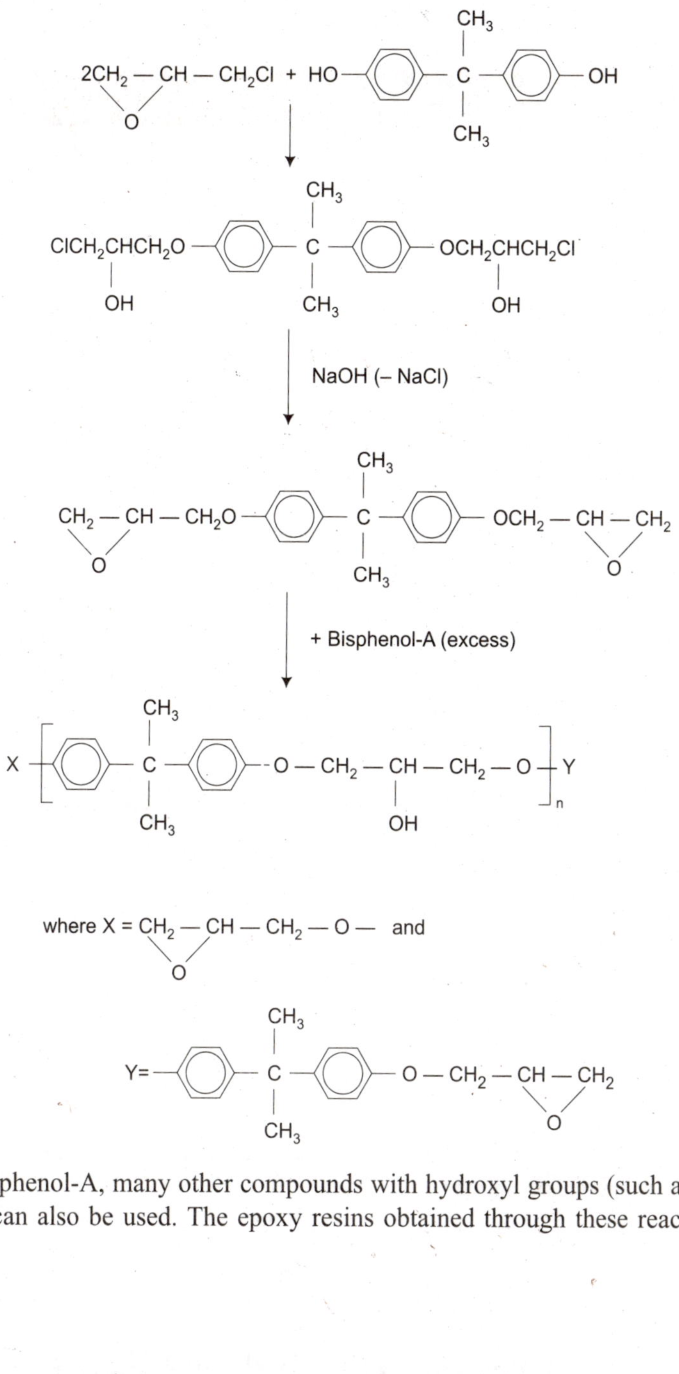

Instead of bisphenol-A, many other compounds with hydroxyl groups (such as glycols, glycerols and resorcinols) can also be used. The epoxy resins obtained through these reactions will be either highly viscous liquids or solids with high melting points. The epoxy resins can be further cured with substances such as amines, polysulphides and polyamides (see Section 11.9.3). Epoxy resins find a large number of uses because of their remarkable chemical resistance and good adhesion. Epoxy resins are excellent structural adhesives. When properly cured, epoxy resins can yield very tough materials. They are used in industrial floorings, foams, potting materials for electrical insulations, etc. One of the principal constituents in many of the fibre-reinforced plastics (FRP) is an epoxy polymers.

9.35 SILICONE POLYMERS

Like carbon, silicone also comes in the fourth group of the periodic table. It has a valency of 4 and there are many compounds of silicone which apparently resemble carbon compounds. For example, carbon combines with hydrogen to form methane. Likewise, silicone also combines with hydrogen to form SiH_4, which is called silico-methane. Unlike carbon, however, silicone cannot form Si = Si double bonds and hence, the polymers of silicone are produced through condensation. The Si—Si linkage is thermally unstable in the case of large molecules and hence, silicone forms polymers mainly through silicone bonds represented usually as —Si—O—Si—.

Quartz, strictly speaking, is a polymer of silicone dioxide. It may be thought of as being produced from orthosilicic acid, H_4SiO_4, through a condensation reaction. It has the following structure:

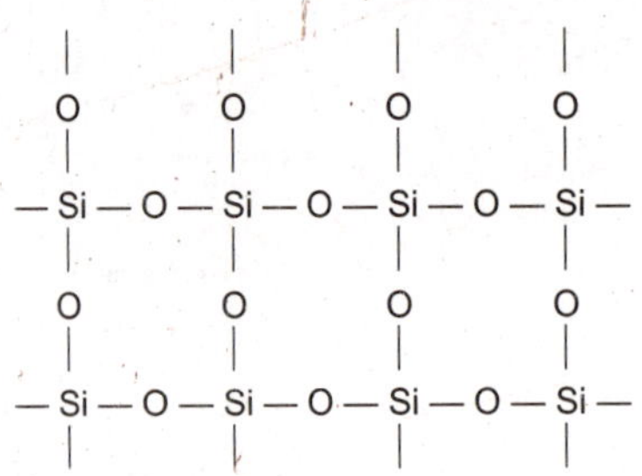

The starting materials for producing silicone polymers are alkyl chlorosilanes, aryl chlorosilanes or substituted esters of orthosilicic acid. In such compounds, there are groups which are easily hydrolisable to form hydroxyl groups. These compounds when treated with water, form hydroxyl compounds:

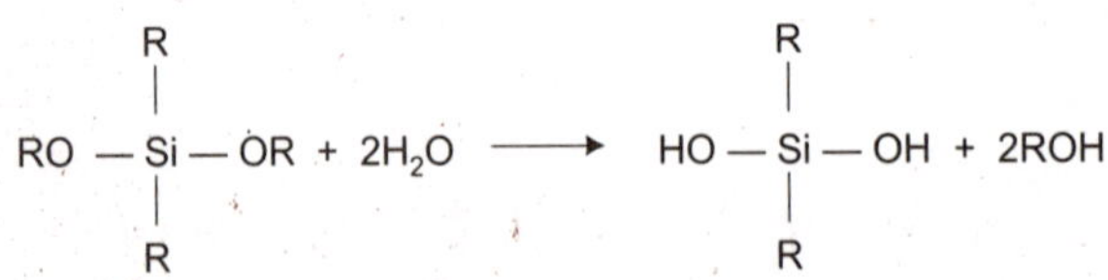

These hydroxyl compounds are called silanols. The silicone polymers are prepared from silanols through condensation reactions as shown:

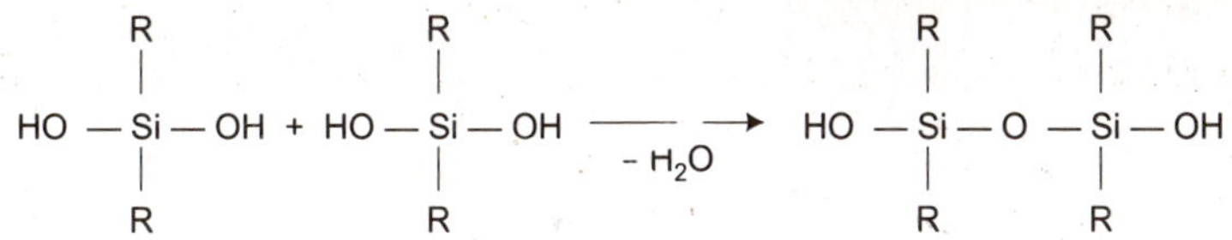

Polycondensation can proceed in the same way, resulting in the formation of the polymer with a final structure as follows:

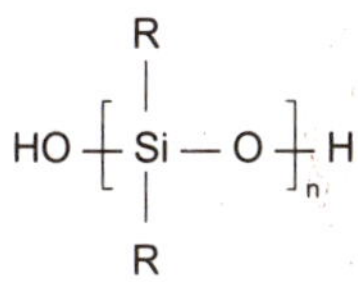

Silicone polymers are thermally very stable. They are available in liquid, waxy and rubbery forms. These polymers find use in surface coatings, where water repellency and chemical resistance are of prime concern. Silicone polymers are also used to manufacture laminates which can stand fairly high temperatures without degradation.

Light weight foams are made from silicone resins by using foaming agents. Such foams are used in aeroplanes and missiles. Further silicon polymers are used as elastomers and can be vulcanised with the help of peroxide initiators (see Section 11.9.2). Silica, in the form of fine powder, is used as a filler to reinforce silicone rubbers. Such rubbers remain flexible even at temperature as low as –90°C and as high as +150°C.

9.36 ORGANOTIN POLYMERS

Organotin polymers are also known as polystanaoxanes and have the following structure:

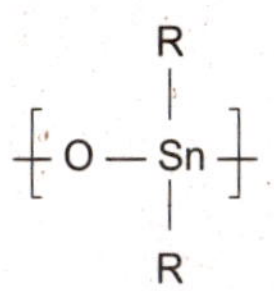

They are produced by the hydrolysis of substituted chlorostannanes, followed by polycondensation:

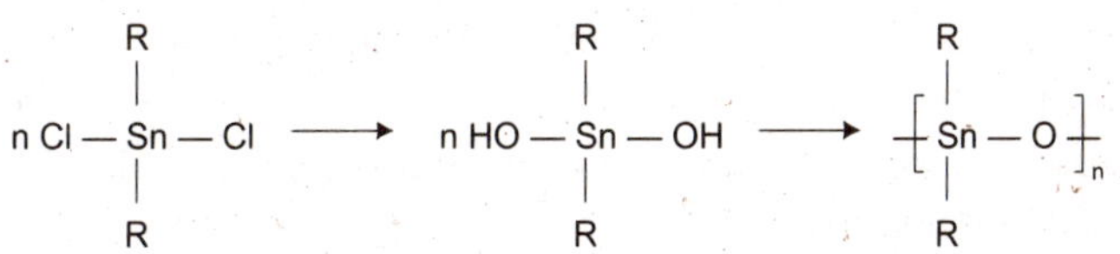

The organotin polymers are found to be brittle. They generally melt around 350°C.

9.37 ORGANOTITANIUM POLYMERS

Organotitanium polymers are known as polytitanoxanes and are prepared from organic titanates through a partial hydrolysis followed by polycondensation:

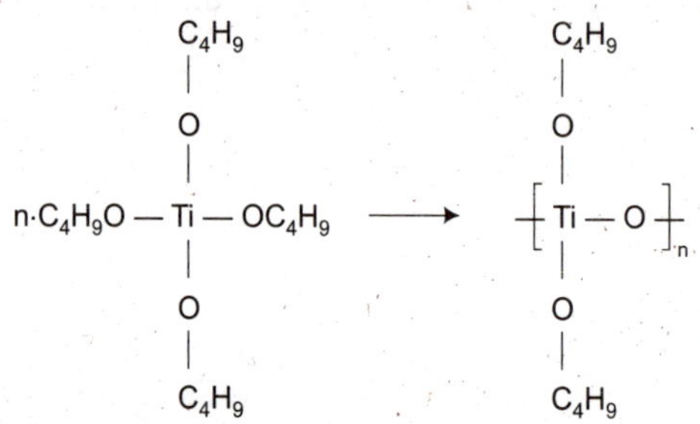

These polymers are transparent heat-resistant materials.

9.38 CELLULOSE AND ITS DERIVATIVES

9.38.1 Cellulose

Cellulose is a naturally occurring, linear, stereo-regular polysaccharide, made-up of β*D*(+) glucose residues. Its structure can be written as follows:

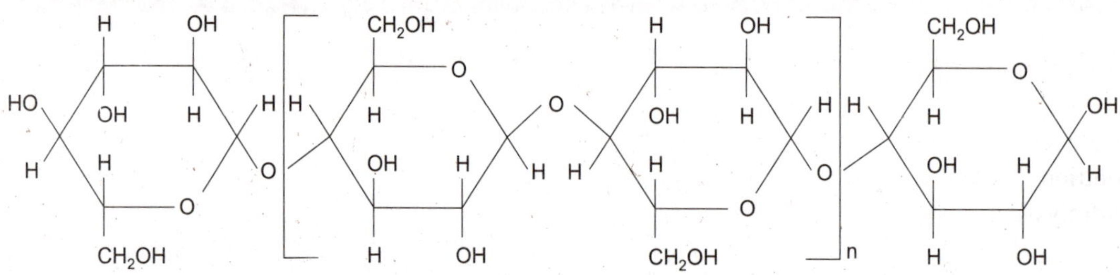

Cellulose is a versatile polymer which is found in plenty in nature in the form of cotton, hemp, jute, flax, etc. Also, a good percentage of wood consists of cellulose. Cellulose has an extremely high degree of crystallinity and this is due to the formation of several hydrogen bonds originating from its hydroxyl groups. Its high stereo-regularity is also a contributory factor for its high crystallinity.

Cellulose has a very high melting point and in fact, it decomposes before beginning to melt. It is generally resistant to dissolution in several solvents, though it can swell in some, such as water, owing to the presence of hydrogen bonds. A closer analysis of the swelling process shows, that it takes place only in the amorphous regions, whereas the crystalline regions are practically unaffected. Mostly, cellulose is dissolved through chemical reactions. In industry, cellulose is usually dissolved and subsequently reprecipitated in a pure form. This form of cellulose is known as the 'regenerated cellulose'. Cotton contains the highest percentage of cellulose with only small amount of other substances such as protein and pectin. Cotton fabrics launder well compared to many of the synthetic fibres, which tend to lose strength in the wet condition. In fact, it is found that the strength of cotton fibre is around 25% higher in the wet condition than when it is dry. Another interesting point in favour of cotton fabrics is that it can be folded and unfolded several times, without loss of strength. In comparison to silk or wool, cotton is a better conductor of heat and has also excellent moisture absorption capability. These two properties make it an ideal clothing material for wear in humid summer weather. Cotton fabrics can be easily dyed and hence, are easily amenable for appealing and elegant designs.

9.38.2 Rayon

Rayon (also known as 'viscose') is, in fact, regenerated cellulose. In an actual manufacturing process, the initial step is the production of alkali cellulose. This is effected by treating cellulose obtained from wood pulp with strong alkalis at relatively low temperatures. The alkali cellulose thus obtained is further treated with carbon disulphide to produce what is known as cellulose xanthate. Representing cellulose as Cell—OH, the reactions involved can be written as follows:

$$\text{Cell—OH} + \text{NaOH} \longrightarrow \underset{\text{Alkali cellulose}}{\text{Cell—ONa}} + H_2O$$

$$\text{Cell—ONa} + CS_2 \longrightarrow \text{Cell—O—C}\begin{matrix}\text{//S} \\ \text{\\SNa}\end{matrix}$$

Cellulose xanthate

The alkali cellulose is held in excess carbon disulphide for 3 hours in a tub and, at the end of the process, the unreacted carbon disulphide is removed. The cellulose xanthate formed is dissolved in a dilute solution of sodium hydroxide. The solution thus formed is industrially known as 'viscose solution'. Cellulose xanthate is an unstable product and will generate pure cellulose through hydrolysis as shown:

$$\text{Cell—O—C}\begin{matrix}\text{//S} \\ \text{\\SNa}\end{matrix} + H_2O \longrightarrow \text{Cell—O}\underset{\underset{\text{S}}{\|}}{\text{C}}\text{SH} + \text{NaOH} \xrightarrow{+H_2O} \text{Cell—OH} + CS_2$$

The process of hydrolysis and the simultaneous regeneration of cellulose are achieved by treating the viscose solution with a dilute solution of sulphuric acid containing sodium or ammonium sulphate. If the viscose solution is injected into a bath of sulphuric acid and sulphate in the form of fine jets, the cellulose is regenerated in the form of glossy filaments.

9.38.3 Cellophane

Cellophane, a transparent film commonly used in our day-to-day life, is also regenerated cellulose made in the form of a thin film. In the manufacturing process, the viscose solution is extruded into a thin film and made to pass through a trough containing sulphuric acid, ammonium sulphate and sodium sulphate. During this process, the xanthate groups are eliminated and the cellulose is precipitated, as explained earlier. Subsequently, the film streams through another trough containing glycerol. Here, an appropriate quantity of glycerol will be absorbed by the film, which in turn, serves as a good plasticiser for cellophane. On drying this, a thin film of cellophane is produced. Its well-known uses are in the package and wrapping industries.

9.38.4 Cellulose Nitrate

Cellulose nitrate polymer, also known as nitrocellulose (NC), is a nitric acid ester of cellulose. Cellulose is nitrated by a mixture of strong nitric acid, sulphuric acid and a small quantity of water. At elevated temperatures, the reaction rate increases but so does the chain degradation. To overcome this problem, nitration is done at around 20°C. Usually, the reaction will be complete within 45 minutes, after which, the reaction mixture is transferred and centrifuged to separate out the product. Next, the product is purified by washing and is stabilised with dilute alkali, and subsequently bleached. The last traces of the water present are removed by washing with alcohol.

Depending on the degree of nitration, cellulose nitrate has different applications. For example, NC with a nitrogen content of more than 12.3% is used as explosives (*e.g.*, gun-cotton), whereas NC with 10.5% to 11% nitrogen content is used as celluloid plastics, 11% to 12% as nitrolacquers and 12% to 12.3% as photographic films. Celluloid has excellent impact resistance and is one of the most inexpensive plastics available. It is used to make moulded decorative articles, fountain pens, frames for spectacles, etc. Nitrolacquers find application in paint and surface coating industries. The greatest disadvantage of cellulose nitrate is its very high inflammability, and for this reason, the present-day photographic films are made of cellulose acetate instead of cellulose nitrate.

9.38.5 Cellulose Acetate and Other Related Polymers

Another important cellulosic plastic is cellulose acetate. It is produced by the action of acetic anhydride on cellulose. The catalyst used for the reaction is either sulphuric acid or perchloric acid:

$$\text{Cell—OH} + (CH_3CO)_2O \rightarrow \text{Cell—O—COCH}_3 + CH_3COOH$$

The product obtained by complete acetylation (where all the three hydroxyl groups in each glucoside ring of cellulose are acetylated) is called primary cellulose acetate or cellulose triacetate. This plastic is used for the manufacture of fibres. Partial saponification of primary cellulose acetate leads to the formation of secondary cellulose acetate containing 2.2 to 2.5 acetate groups per glucoside ring. It is a very important plastic and is used for the manufacture of non-flammable safety cinematographic films, moulded articles and fibres. Cellulose acetate-butyrate is another important cellulosic plastic. It is made by the esterification of cellulose with a mixture of acetic anhydride and butyric anhydride. The esterification is controlled in such a way that the final polymer contains 26% to 29% of acetyl and 17% to 48% of butyl groups. Cellulose acetate-butyrate has several better properties, such as a high impact strength, good dimensional stability, excellent compatability with plasticisers and much lower moisture absorption, as compared to cellulose acetate. This plastic is extensively used in the manufacture of automobile components. Two other cellulose-based plastics are cellulose propionate and cellulose acetate-propionate.

9.38.6 Ethyl and Methyl Cellulose

Ethyl cellulose and methyl cellulose are the ether derivatives of cellulose and are prepared from alkali cellulose by treatment with methyl chloride or ethyl chloride:

$$\text{Cell} — \text{ONa} + CH_3Cl \longrightarrow \text{Cell} — O — CH_3 + NaCl$$

$$\text{Cell} — \text{ONa} + C_2H_5Cl \longrightarrow \text{Cell} — O — C_2H_5 + NaCl.$$

Ethyl cellulose is flexible and tough even at very low temperatures. It has better mechanical and electrical properties compared to other cellulosic plastics. Both methyl cellulose and ethyl cellulose find application as finishing materials (in the textile industry) and as paint bases.

9.38.7 Carboxymethyl Cellulose

Carboxymethyl cellulose, usually known as CMC, is produced as its sodium salt, by reacting cellulose with NaOH and chloroacetic acid:

$$\text{Cell} — OH + 2NaOH + ClCH_2COOH \rightarrow \text{Cell} — OCH_2COONa + NaCl + 2H_2O$$

CMC is used as a thickening agent in textile additives, cosmetics, cakes, ice-creams, etc., and as a protective colloid in suspension and emulsion polymerisation systems.

CHAPTER 9

Books for Further Reading

Ott, E., Spurlin, H.M. and Craffin, M.W., Eds., *Cellulose and Cellulose Derivatives*, Vols. 1, 2 and 3, Interscience, New York, 1954; Vol. 4 and 5 Bikales, N.M. and Segal, L. Eds., 1971.

Whitby, G.S., *Synthetic Rubber*, John Wiley and Sons, New York, 1954.

Raff, R.A.V. and Allison, J.B., *Polyethylene*, Interscience, New York, 1956.

Schildknecht, C.E. Ed., *Polymer Processes*, Interscience, New York, 1956.

Fisher, H.L., *Chemistry of Natural and Synthetic Rubbers*, Reinhold, New York, 1957.

Teach, W.C. and Kiessling G.C., *Polystyrene*, Reinhold, New York, 1960.

Horn, M.B., *Acrylic Resins*, Reinhold, New York, 1960.

Christopher, W.F., *Polycarbonates*, Reinhold, New York, 1962.

Fettes, E.M.Ed., *Macromolecular Syntheses*, Vols. 1–7 (continuing series) John Wiley and Sons, New York, 1963–1979.

Smith, W.M., Ed., *Manufacture of Plastics*, Vol. 1, Reinhold, London, 1964.

Yarsley, V.E., Flavell, W., Adamson, P.S. and Perkins, N.G., *Cellulose Plastics*, Illiffe, London, 1964.

Dombrow, B.A., *Polyurethanes*, Reinhold, New York, 1965.

Korshak, V.V. and Vinogradova, S.V., *Polyesters*, Pergamon, Oxford, 1965.

Boenig, H.V., *Polyolefins*, Elsevier, Amsterdam, 1966.

Floyd, D.E., *Polyamide Resins*, Reinhold, New York, 1966.

Whitehouse, A.A.K., Pritchett, E.G.K. and Barnett, G., *Phenolic Resins*, Illiffe, London, 1967.
Frazer, A.H., *High Temperature Resistant Polymers*, John Wiley, London, 1968.
Ritchie, P.D., *Vinyl and Allied Polymers*, Vol. 1, Illiffe, London, 1968.
Frank, H.P., *Polypropylene*, Gordon and Breach, New York, 1968.
Sarvetnik, H.A., *Polyvinyl Chloride*, Reinhold, New York, 1969.
Blow, C.M., *Rubber—Technology and Manufacture*, Butterworths, London, 1971.
Barker, S.J. and Price, M.B., *Polyacetals*, American Elsevier, New York, 1971.
Ranney, M.V., *Polyimide Manufacture*, Noyes Data Corporation, Park Bridge, 1971.
Penn, W.S., *PVC Technology*, MacLaren, London, 1972
May, C.A. and Tanaka, Y. Eds., *Epoxy Resins—Chemistry and Technology*, Dekker, New York, 1973.
Kohan, M., *Nylon Plastics*, Wiley-Interscience, New York, 1973.
Bruins, P.F., *Unsaturated Polyester Technology*, Gordon and Breach, New York, 1975.

10

POLYMER DEGRADATION

10.1 WHAT IS POLYMER DEGRADATION?

You may have noticed that a plastic bucket left for long in the sun and rain loses its lustre and strength. This deterioration in properties is due to a phenomenon called *'polymer degradation'*, which is characterised by an uncontrolled change in the molecular weight or constitution of the polymer. Conventionally, the term *'degradation'* is taken to mean a reduction in the molecular weight of the polymer.

A polymer can suffer degradation mainly at two stages of its life. First, during the fabrication process and secondly, during its daily usage. Many of the fabrication processes such as extrusion or moulding involve heating, which the polymer may not be able to endure and as a consequence, may just degrade. Similarly, in the hustle-bustle of modern life, plastic articles have to face mechanical stresses, solar radiations, atmospheric oxygen, moisture, etc. and this can degrade the polymer.

While polymer degradation is generally undesirable, it can sometimes be put to advantage also. For example, when a spacecraft re-enters the earth's atmosphere, intense heat is generated, which is sufficient to melt the metal and even make the craft explode. To avoid such a situation, the spacecraft is covered by a *'heat shield'*, which is a polymer-based material. As the heat shield encounters a high temperature on re-entry, the material chars and degrades and in the process, takes away the severe heat and thus protects the spacecraft.

10.2 TYPES OF DEGRADATION

Polymer degradation is broadly of two types: (*i*) chain-end degradation and (*ii*) random degradation. In the first type, the degradation starts from the chain ends, resulting in successive release of the monomeric units (Fig. 10.1). This phenomenon is actually the reverse of the propagation step in chain polymerisation. For this reason, this type of degradation is also called *'depolymerisation'*. We may imagine the propagation step in polymerisation to represent the *'closing of a zip'* and depolymerisation the '*opening of the zip*'. Indeed, this type of degradation is also termed as '*unzipping*'. The scission is almost like cutting off successively a long strip of paper into small bits from its end. The result is that the molecular weight of the polymer decreases slowly and a large quantity of the monomer is liberated simultaneously.

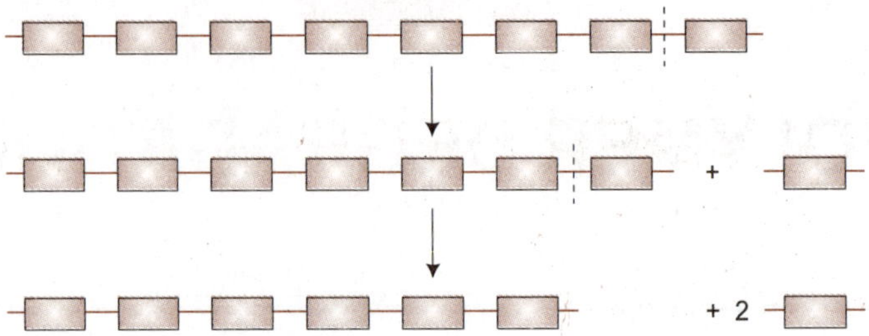

Fig. 10.1 Schematic representation of a polymer molecule undergoing chain-end degradation. (Each block in the diagram represents one monomeric unit).

A good example of unzipping or chain-end degradation is provided by the thermal depolymerisation of non-terminated (*i.e.*, living) poly α-methylstyrene synthesised through the anionic mechanism. This polymer in a solvent such as anhydrous tetrahydrofuran, upon heating from −70°C to +60°C, undergoes degradation from the chain end carrying the active carbanion in such a manner that, at 60°C, the entire polymer is converted into the monomer (cf. depropagation at the ceiling temperature in Section 4.2.5). The important precaution to be taken is that the system should be kept free from moisture and other impurities such as oxygen and carbon dioxide. The process is, interestingly enough, reversible. Hence, by simply cooling the system back to −70°C, the polymer can be obtained qualitatively. Polymers formed by chain reaction polymerisation are more susceptible to unzipping (or chain-end degradation). But here again, all such polymers do not undergo unzipping straightaway. For example, it is well known that polymethyl methacrylate, when heated to 300°C under vacuum, depolymerises to give back the monomer in an almost quantitative yield. The unzipping, however, sets in only after a weak link somewhere in the polymer backbone is broken under the influence of heat to generate a chain end carrying a free-radical:

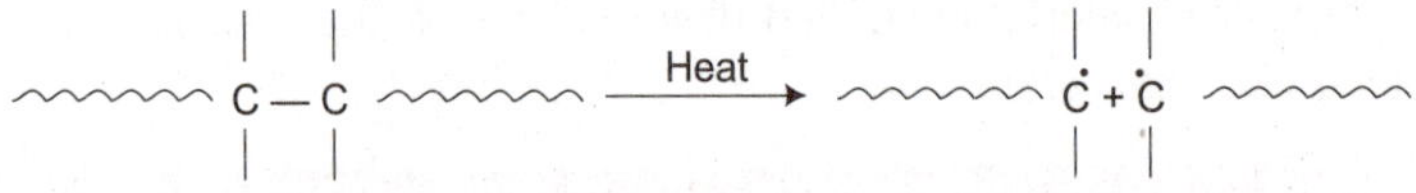

The unzipping now starts from the active chain ends formed. In fact, depolymerisation is a useful method for recovering the monomer from waste polymethyl methacrylate with a recovery of almost 100%. Polystyrene, polybutadiene and polyisoprene also undergo depolymerisation to give back the monomer, although to a limited extent. Polycaprolactam (nylon 6), which is not a '*chain reaction*' polymer, can also undergo depolymerisation to a limited extent, giving back the caprolactam monomer up to 6%. In general, it can be said that chain-end degradation occurs when the backbone bonds are weaker than the bonds of the side groups and only with polymer molecules carrying active chain ends with a free-radical, cation, anion, etc. Schematically, we can write this as follows:

$$M_n^* \rightarrow M_{n-1}^* + M$$
$$M_{n-1}^* \rightarrow M_{n-2}^* + M \text{ and so on}$$

where M is the monomer and M_n^*, M_{n-1}^* and so on represent the polymer chain made up of so many monomeric units and carrying an active end group.

The enzymatic degradation of certain natural products also follows the *'unzipping'* mechanism: For example, the enzyme called β-amylase degrades starch to maltose, beginning with the chain ends.

The second type of degradation occurs at any random point along the polymer chain instead of at the chain ends (Fig. 10.2) and hence, the name *'random degradation'*. This type of degradation could be thought of as the reverse of the polycondensation process as illustrated by the following equation:

$$M_n \rightarrow M_x + M_y$$

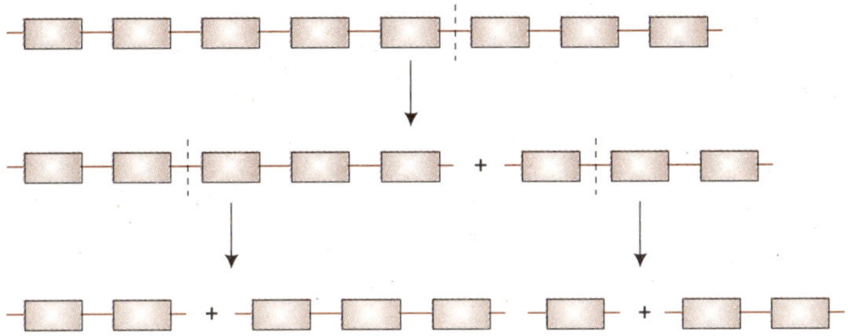

Fig. 10.2 Schematic representation of a polymer molecule undergoing random degradation. (Each block in the diagram represents one monomeric unit).

Here, the polymer degrades to lower molecular weight fragments but (unlike in chain-end degradation) practically no monomer is liberated.

Almost all polymers can undergo random degradation. Polyesters, for example, undergo hydrolytic degradation, resulting in chain scission:

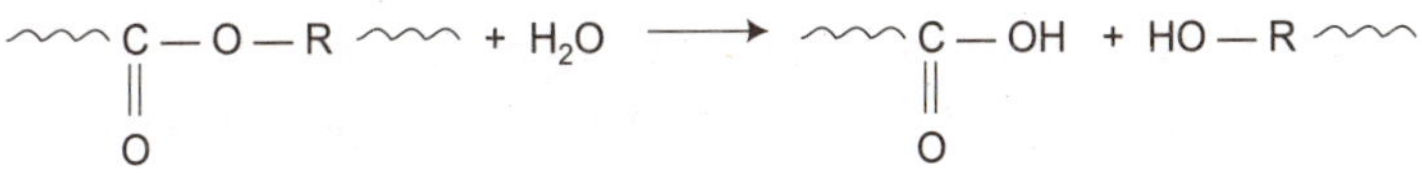

Further, in random degradation, of polyethylene for instance, a hydrogen atom may migrate from one carbon to another and yield two fragments, as shown:

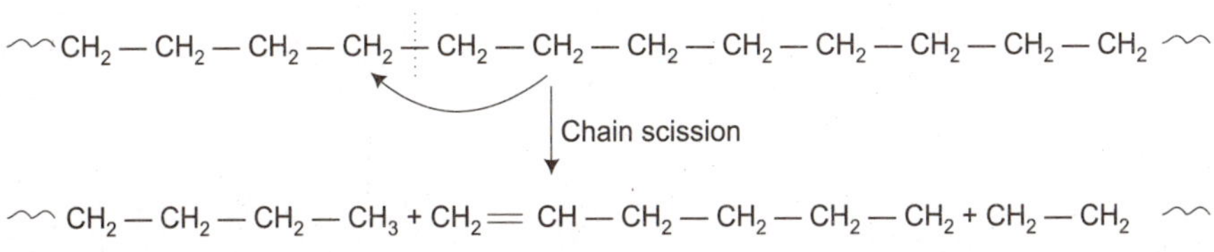

For random degradation to occur, the polymer chain need not necessarily carry any active site.

It would thus be noted, that the polymer molecule breaks up while degrading. The process of breaking up is decided by the type of the degradation. In chain-end degradation, the molecular weight drops very slowly and a large quantity of the monomer is released; in random degradation, there is a sudden drop in the molecular weight with little or no release of the monomer.

Degradation of polymers may be brought about either by physical factors (such as heat, light or mechanical stress) or by chemical agents (such as oxygen, ozone, acids or alkalies). In the following sections of this chapter, we will study how these factors or agents bring about degradation.

10.3 THERMAL DEGRADATION

Thermal degradation of polymers may follow either *'unzipping'* or the *'random'* route. The *'unzipping'* mechanism gives the pure monomer, while random degradation leads to the formation of a host of products, depending on the structure of the polymer. Since many polymers have a carbon-carbon (C—C) chain as the backbone, their thermal stability is dependent on the stability of the C—C bond. So, let us examine the factors which affect the stability of such a bond.

10.3.1 Factors Affecting the C—C Bond Stability

Several factors can affect the C—C bond stability in a polymer. Consider the following example:

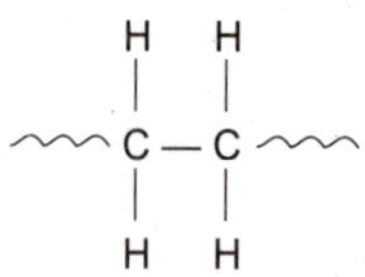

Polyethylene

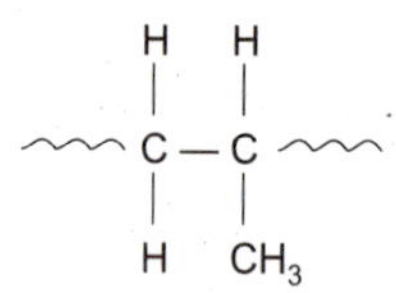

Polypropylene

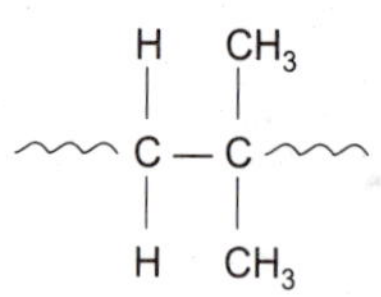

Polyisobutylene

Here, as we proceed from left to right, the number of substituents increases and the stability of the polymer backbone bond decreases. The result is that while polyethylene is thermally stable, polypropylene is less stable and polyisobutylene the least stable. In the above example, one H atom in ethylene is substituted by a methyl group to form propylene. The same H atom in ethylene can also be substituted by a phenyl group to give styrene, which will obviously be thermally less stable than ethylene. Now, if one hydrogen atom in styrene is, in turn, substituted by a methyl group, we get methyl styrene which, with one more substituent than in styrene, is thermally the least stable among the three compounds:

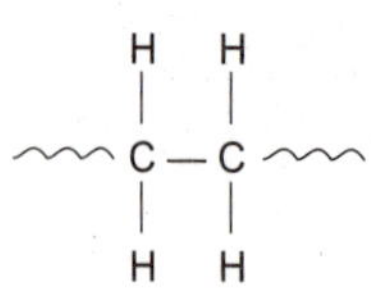

Polyethylene

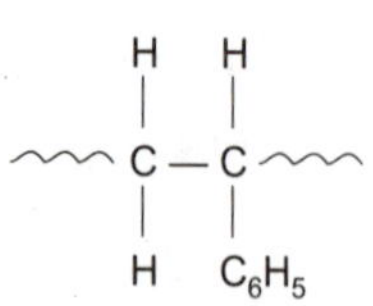

Polystyrene

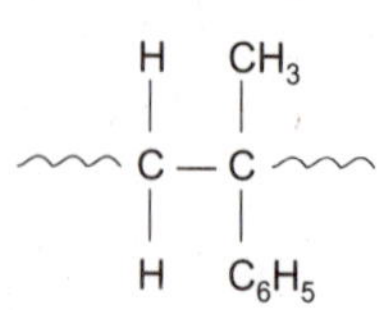

Poly α-methyl styrene

The effect of substituent groups on the stability of the backbone C—C bond is apparent from a comparison of the bond dissociation energies (in kcal/mol) for C—C in CH_3—CH_3 = 88, CH_3CH_2— CH_3 = 85, $(CH_3)_3$ C—CH_3 = 80 and $C_6H_5CH_2$—CH_3 = 70.

A word of caution would be in order. All substituents do not always reduce the thermal stability of the polymeric system. For instance, polytetrafluoro ethylene (Teflon) has all the hydrogen atoms of ethylene substituted by fluorine, but is one of the most stable polymers:

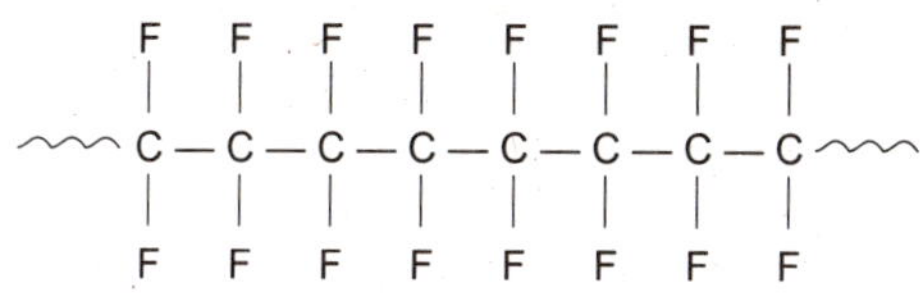

(Teflon)

Teflon can withstand temperatures as high as 400°C, without undergoing degradation; this property incidentally makes Teflon highly useful for domestic and industrial applications. Its high thermal stability is attributed to the very high dissociation energy (108 kcal/mol) of the C—F bonds. Besides, the highly electronegative nature of fluorine protects the C—C bond in Teflon from an external attack.

10.3.2 Other Factors Affecting Thermal Stability

Generally, the aromatic groups in a polymer backbone increase the thermal stability. Polycarbonate is a case in point:

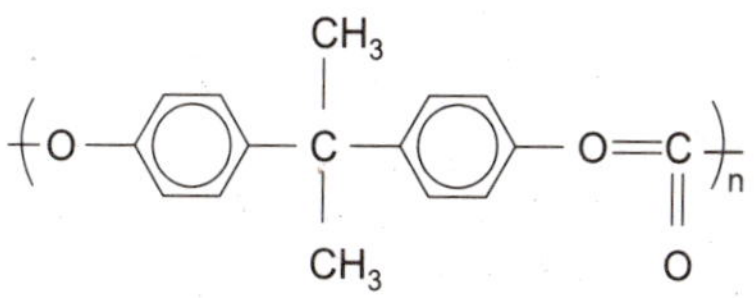

(Polycarbonate)

A substance like polyphenylene, where the chain backbone is formed entirely of aromatic groups, is, by the same token, thermally more stable than polycarbonate:

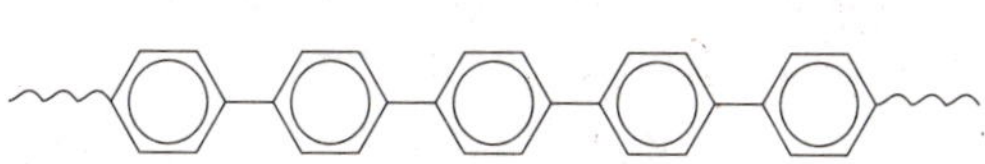

(Polyphenylene)

The high thermal stability of polytetrafluoro ethylene and that of polyphenylene are combined in polytetrafluoro phenylene, which is a relatively new polymer, capable of withstanding temperatures up to 500°C.

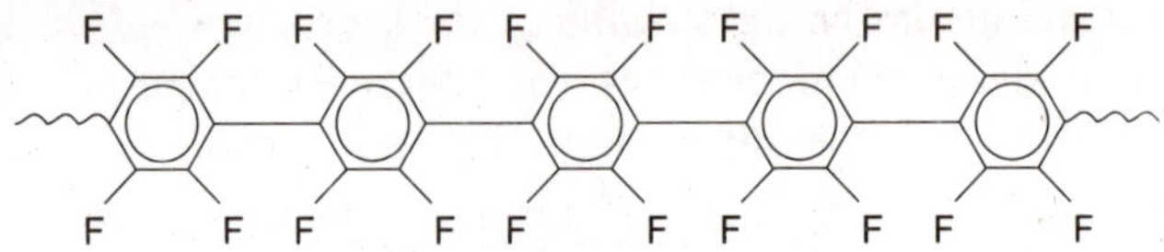

(Polytetrafluoro phenylene)

The branching and the presence of oxygen atom in the polymer chain are other important factors which make polymers susceptible to thermal degradation. The easy degradability of the polyethylene oxide molecule, for instance, is attributable to the presence of the oxygen atom in the polymer chain backbone:

$$\sim\sim O-CH_2-CH_2-O-CH_2-CH_2-O-CH_2-CH_2-O-CH_2-CH_2\sim\sim$$

(Polyethylene oxide)

10.3.3 Polymer Degradation Involving Substituent Groups

There are some reactions involving polymer degradation by breaking the substituent groups but not the main chain. A typical example is PVC which degrades around 200°C, giving HCl. Although the mechanism of PVC degradation is highly complex in nature, the free-radical path seems to play a major role. What apparently happens is that some free-radical ($\dot{R}$), produced either from impurities or because of the reaction temperature being reached, attacks the methylene group and abstracts a H with a simultaneous transfer of the free-radical site on to the chain:

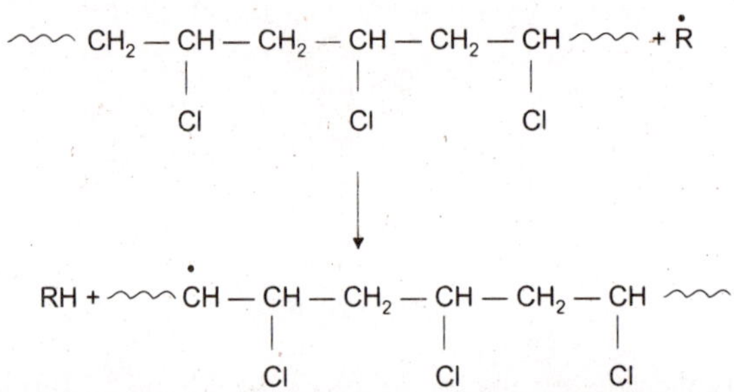

The labile chlorine atom, which is at a β-position with respect to the free-radical carbon atom, is now released as a free-radical so as to stabilise the structure:

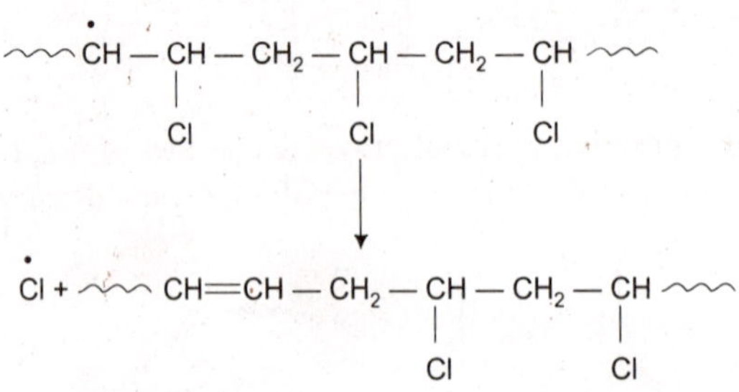

The chlorine free-radical can now attack a methylene group, abstract a H and produce yet another labile chlorine atom:

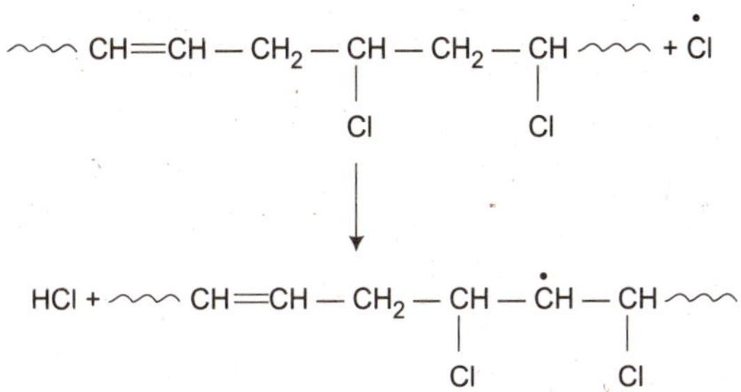

The chain reaction proceeds in a similar way, leading to the production of a polymeric chain having conjugated double bonds:

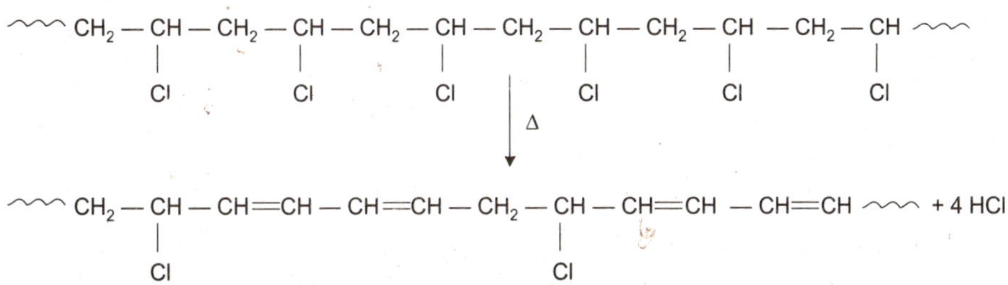

PVC, which is basically colourless, thus turns deep yellow, orange, brown or black, depending on the concentration of the conjugated double bonds which in turn, depends on the degree of degradation. The role of HCl in this reaction is catalytic since the degradation becomes faster as soon as HCl is evolved. Any material which can take away HCl from the reaction sphere can, therefore, arrest the degradation and stabilise the polymer. It is for this reason that PVC available in market contains certain additives called *'stabilisers'* which react with HCl and remove it from the system. Zinc octoate, sodium carbonate, certain salts of calcium or barium or lead, many amines, certain organo-tin compounds and also some epoxidised vegetable oils are used as stabilisers for PVC. The mechanism of their action can be depicted as follows:

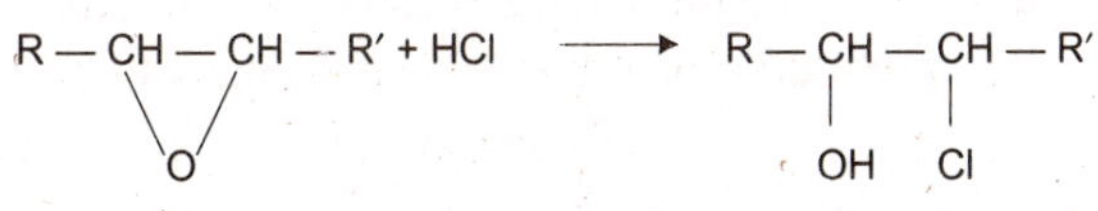

(Epoxidised oil)

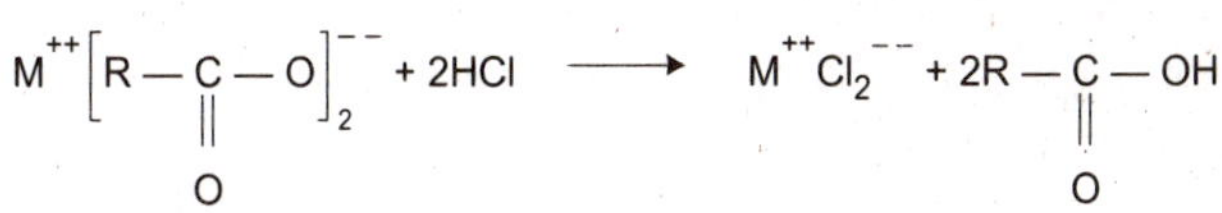

(Metal salts)

Although PVC has a tendency to degrade around 200°C, it is still the world's most widely used plastic.

10.4 MECHANICAL DEGRADATION

A simple compound such as water or benzene cannot be broken up by subjecting it to mechanical stresses arising from, say, high-speed stirring or milling. But a polymer such as polystyrene dissolved in a solvent, when subjected to vigorous stirring or beating, undergoes considerable molecular degradation or fragmentation. In fact, in the rubber industry, rubber is masticated by passing it through two rotating rollers to reduce its molecular weight and make it more processable. This mastication converts hard and tough rubber into a soft, supple and even semisolid mass. Now, what happens to the rubber at a molecular level when it is so masticated?

Let us first remember that rubber is polyisoprene with the following structure:

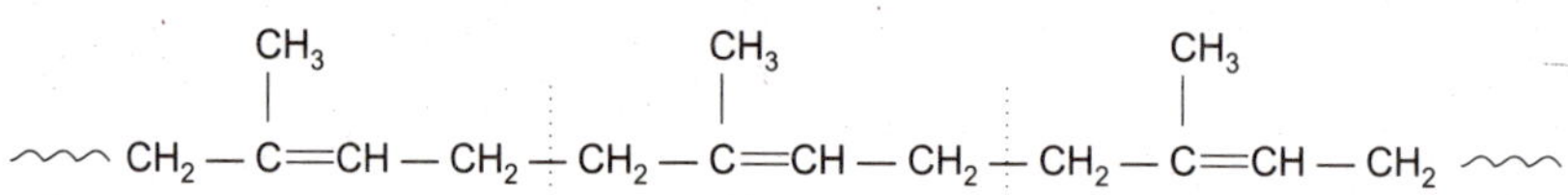

(Isoprene unit) (Isoprene unit) (Isoprene unit)

(Polyisoprene)

In polyisoprene, the bonds which are most vulnerable for scission are the CH_2—CH_2 links between the isoprene units. These bonds in the susceptible regions cleave when the polymer is subjected to mechanical stresses in the form of milling or mastication. The net effect is that the big molecule is broken or scissioned into small parts. This scission at the molecular level reflects on the gross properties of the polymer.

Besides milling and mastication, other mechanical methods also make the polymer molecules degrade. For example, when a polymer solution or melt throttles through a small orifice under a fairly high pressure, fragmentation of the polymer molecule can occur. Agitation, grinding or extrusion are other well-known methods for effecting the mechanical degradation of the polymer. The basic phenomenon in all these is the same; namely,. to subject the polymer to a very powerful shearing force that breaks the molecule. Many interesting observations have been made by researchers on mechanical degradation. For instance, the bigger molecules are found to be affected much more than the smaller ones during mechanical degradation. This means that the larger the initial molecular weight, the greater the molecular weight drop due to mechanical degradation.

The way in which a polymer sample attains a constant limiting molecular weight value, when subjected to a mechanical stress for a sufficiently long time, can be seen from Fig. 10.3. Mechanical degradation can reduce the average molecular weight, give a narrower molecular weight distribution and make the polymer easily processable.

Studies have provided some interesting clues on the degradation mechanism. For instance, degradation is found to be almost totally absent when rubber is masticated in an atmosphere of

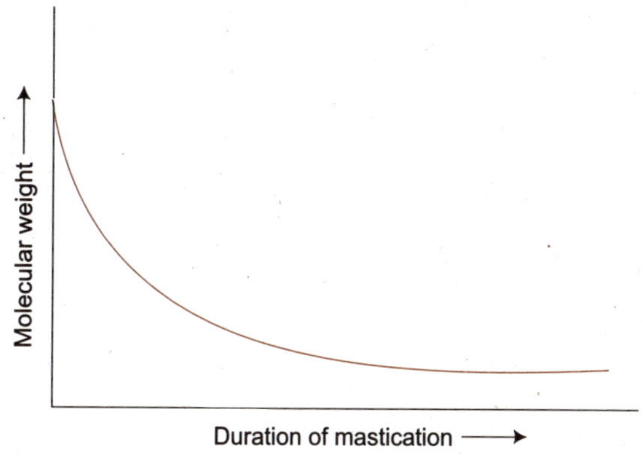

Fig. 10.3 Effect of mastication time on molecular weight of a polymer undergoing mechanical degradation.

nitrogen at an ambient temperature. However, when the process is repeated in the presence of a small quantity of oxygen or air, degradation is very quick and significant. The reason for this is simple. During degradation free-radicals are produced at the end of the scissioned fragments of the chain. In the inert atmosphere of nitrogen, these radicals recombine to form bigger molecule, thereby almost totally nullifying the effect of the degradation. When, however, oxygen is present, due to its radical nature, it readily reacts with the free-radicals formed at the fragmented ends of the polymer molecule and renders the chain break permanent.

10.5 DEGRADATION BY ULTRASONIC WAVES

Have you heard of a *'silent'* sound? Sounds paradoxical, doesn't it? But this is how ultrasonic sound waves are sometimes referred to because of their very high frequencies (say, above 20,000 hertz) which is beyond the audible range of the human ear.

When a dilute solution of a high molecular weight polymer is subjected to ultrasonic waves, the polymer begins to degrade, thereby reducing its molecular weight. A close look at the process taking place at a molecular level shows that ultrasonic degradation is a special case of mechanical degradation as the polymer here is subjected to very high vibrations, which are only mechanical forces. As ultrasonic waves pass through the solution, the localised shear gradients produced tear off the polymer molecules leading to chain scission. As a result, the average molecular weight of the sample goes down. Another point of resemblance between ultrasonic and mechanical degradations is that the decrease in the molecular weight of the polymer does not go below a certain value even after a long exposure to ultrasonic or mechanical stresses.

Similarly, both in mechanical and ultrasonic degradations, bigger molecules are affected more than the smaller ones. This can be seen from Fig. 10.4, which shows the pattern of ultrasonic degradation with time in the case of two polymeric samples with different initial molecular weights.

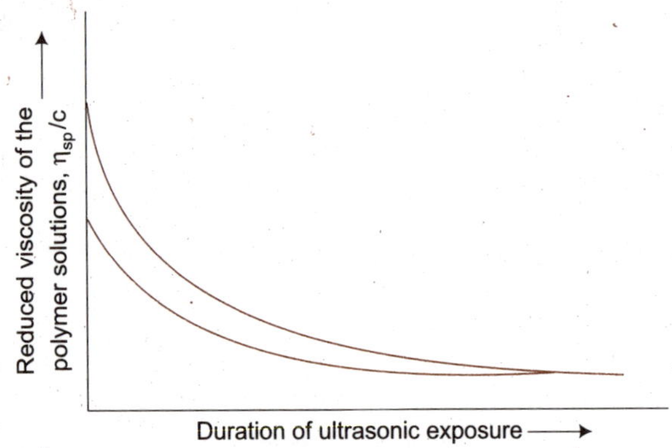

Fgi. 10.4 Effect of ultrasonic degradation of two samples of the same polymer with different initial molecule weights. Here, molecule weight is represented in terms of reduced viscosity (see Section 12.6.1).

10.6 PHOTODEGRADATION

Do you recall an old, domestic, laundry problem—a dazzling white synthetic garment yellowing after long use? The yellowing was not always due to a 'lazy' laundryman, but due to a phenomenon called *'photodegradation'*—a molecular degradation process brought about by ultraviolet light. The yellowing, and the embrittlement on storage, of some of the transparent plastics or coloured rubber articles are also due to their interaction with ultraviolet light.

An interesting example of photodegradation by ultraviolet (UV) radiation is of polymethyl methacrylate (PMMA). This polymer in a molten state gives on irradiation with UV rays an almost quantitative yield of the monomer. It is generally believed that free-radicals are initially formed during photodegradation, and in many cases, the subsequent course of the reaction of the free-radicals depends on the physical state of the material. Thus, while the molten polymer gives a quantitative yield of the monomer, the solid polymer does not give any significant quantity of the monomer. Small quantities of methyl formate are, however, formed during the photodegradation of the solid PMMA, as shown here:

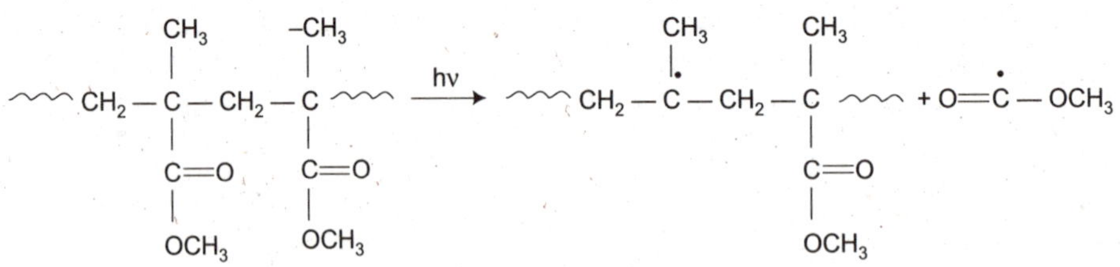

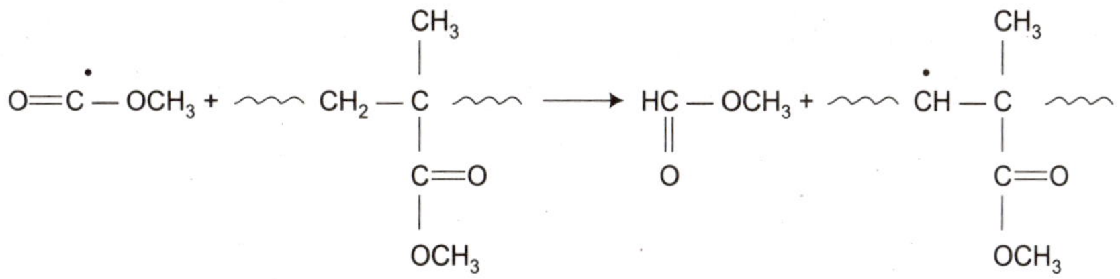

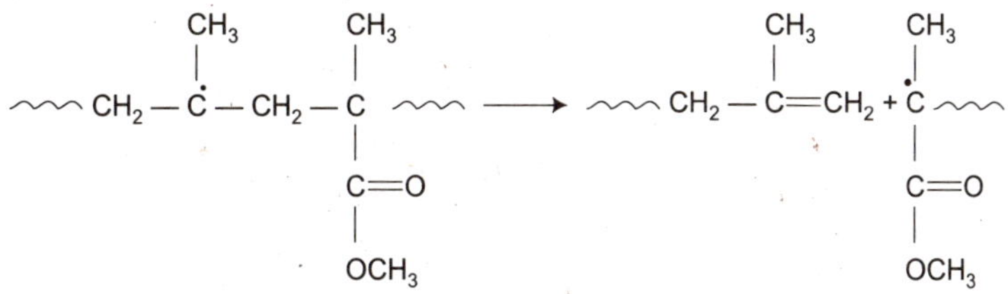

A similar type of photodegradation is observed in the case of poly-α-methyl styrene also.

It has been shown that polyethylene molecules, due to structural defects or impurities, quite often contain a C=C or C=O linkage, which makes the nearby bonds susceptible to photodegradation as follows:

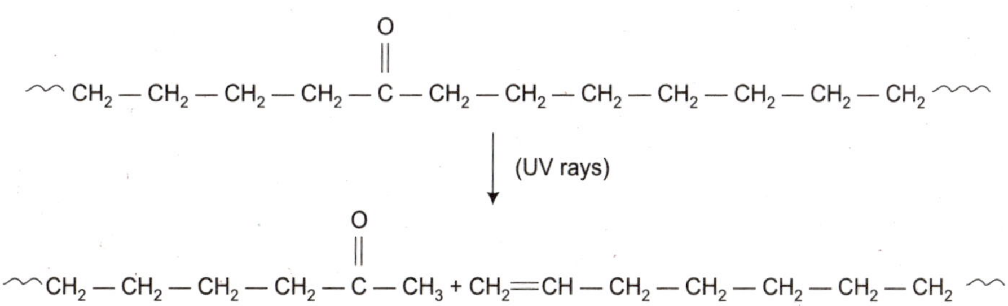

10.6.1 Photostabilisers

Photostabilisers protect polymers from the deteriorating effects of light to a considerable extent. Phenyl salicylate, known as *'Salol'*, is one of the early stabilisers. Stabilisers such as 2, 4-hydroxy benzophenone (Uvinul 400), 2-hydroxy, 4-methoxy benzophenone (Virnul M 40) and 2-hydroxy, 4-dodecyloxy benzophenone (DOBP) are used in the plastic industry.

The function of a photostabiliser (or photoabsorber) is to absorb the UV radiation and dissipate the energy thus absorbed to the environment in some harmless form. The absorbed energy is transmitted back as heat or radiations of a longer wavelength (consisting of quanta of lesser energy). In doing so, the stabiliser acts as a filter and does not allow the radiation energy to attack the polymer molecules.

Stabilisers usually are molecules with aromatic rings having hydroxy groups and ketonic groups. The following is an example:

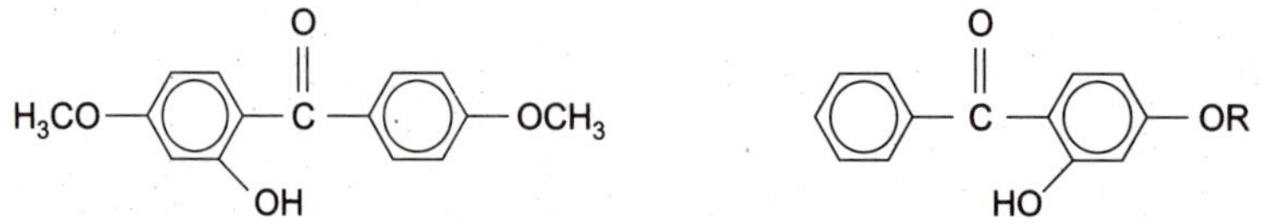

(substituted 2-hydroxy benzophenones)

Most stabilisers have the hydroxyl and ketonic groups in the ortho-position, which encourages the formation of the hydrogen bond. In addition, stabilisers have an absorption coefficient higher than that of the polymer in the near ultraviolet range of the electromagnetic radiation. Hence, they absorb most of the light and subsequently convert it into heat. This is effected through the following simple reversible reaction:

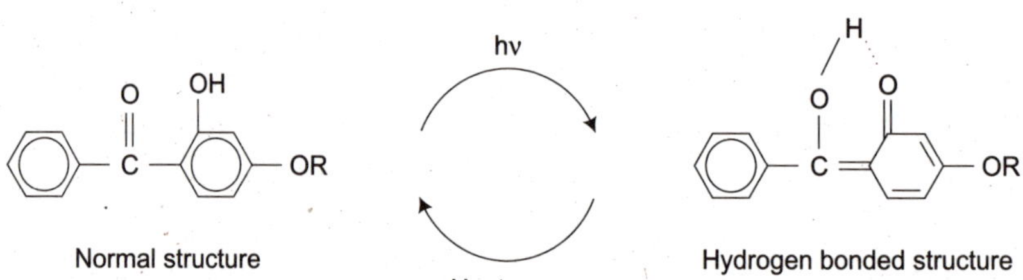

If we represent the stabiliser by A and its hydrogen bonded structure by B, the overall reaction may be represented as follows:

$$A \xrightarrow{h\nu} B$$

$$B \longrightarrow A + \text{Heat}$$

This absorption of energy and the subsequent dissipation as heat will take place even at extremely low concentrations of the stabiliser and this is precisely how the polymer is protected from the harmful effects of the UV rays.

There are some stabilisers, such as 2-(hydroxy phenol) benzotriazole, which do not contain a ketonic group but are still capable of forming a hydrogen bonded structure as shown:

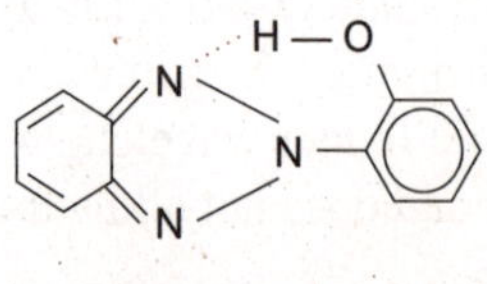

Photostabilisers are normally used to protect polymeric products of bright colours. They should, therefore, be of *'non-staining'* character so that they do not introduce any unwanted colouration or discolouration of the product. This is an important consideration in selecting a photostabiliser.

Carbon black, for instance, is a very good photostabiliser but obviously, cannot be used for protecting objects with a bright colour. Non-toxic nature and chemical inertness are other important considerations for the stabiliser if the final product is to come in contact with skin or food.

10.7 DEGRADATION BY HIGH-ENERGY RADIATION

X-rays, gamma rays, alpha rays and beta rays are among the well-known high-energy radiations. Like UV rays, X-rays and gamma rays are electromagnetic radiations, but their energy level is much higher than that of UV rays. Beta rays are fast moving electrons, and alpha rays are fast moving helium nuclei.

Degradation by the high-energy radiations is more massive than that by the lower energy (UV) radiations. We can say that high-energy radiations smash a polymer molecule to tiny fragments like what a fast moving cricket ball can do to a sheet of glass. When high-energy radiations are so *'brutal'*, the polymer responds in a complex manner. It lets all sorts of things to happen: from breaking of its bonds or scission of its chain to cross-linking. If the molecule is scissioned, the polymer degradation is obviously associated with a reduction in the molecular weight. On the other hand, cross-linking can take place between polymer molecules, building up the polymer network with a resultant increase in the molecular weight. Polymers which degrade on irradiation are polyisobutylene, polytetrafluoro ethylene, cellulose, polymethacrylates, poly α-methyl styrene, etc. On the other hand, polyethylene, polypropylene, polyamides, polyacrylates, polyisoprene, polybutadiene, etc. get cross-linked when subjected to high-energy radiations. Many polymers, on being exposed to radiation, give out gases such as H_2, CO_2,CO, CH_4, NH_3 and $(CN)_2$. For example, polyethylene and polystyrene, when subjected to high-energy radiations, give out hydrogen which comes from the nearby macromolecules as shown by the following structure (the removal of hydrogen leads to cross-linking of the polymeric chains):

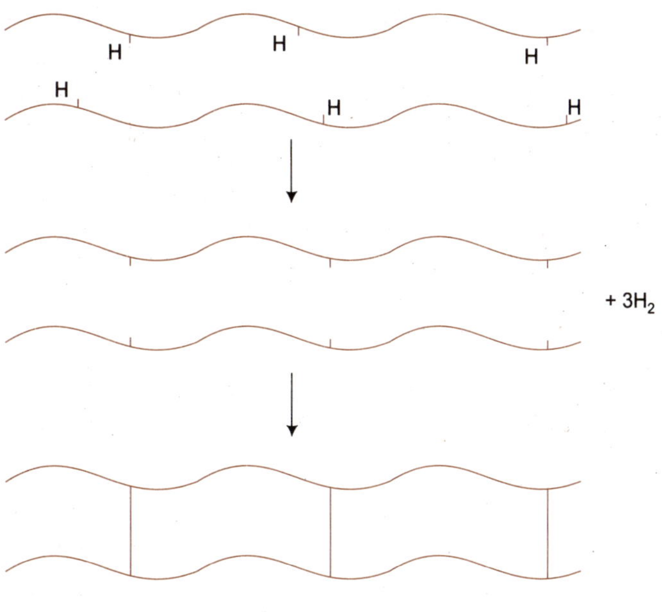

CHAPTER 10

When, however, polyisobutylene is subjected to high-energy radiations, methane and hydrogen are evolved. The IR spectrum of the irradiated sample shows the presence of double bonds and a simultaneous reduction in the number of methyl groups. The reaction could be given as:

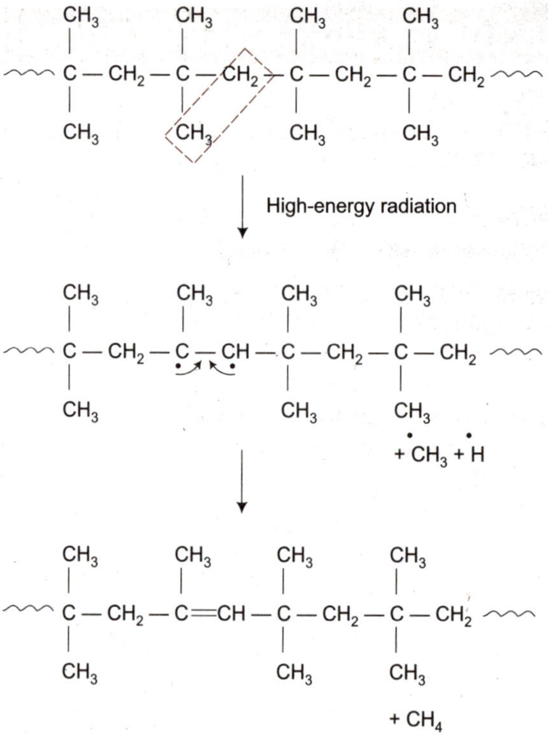

As against this, the behaviour of polymethyl methacrylate towards high-energy radiation is interesting in some other ways. First, its molecular weight decreases considerably (with the evolution of gases such as (CO_2, CO, H_2 and CH_4), which means that the polymer undergoes degradation during irradiation. Secondly, the polymer develops a yellow colour, which turns some what reddish brown with a higher radiation dosage. The colour is retained in vacuum but vanishes when exposed to air. Disappearance of the colour starts from the periphery of the polymer block and moves to its interior at room temperature. If the temperature is raised to 80°C; the discolouration phenomenon becomes faster. This discolouration is attributed essentially to the formation of free-radicals (during the polymer irradiation) which, when trapped in the polymer bulk, react with oxygen and get destroyed. Enhanced mobility of these radicals at elevated temperatures is said to be the reason for the disappearance of the colour when the polymer is in a molten stage. Although the presence of the free-radicals in the solid polymer (after irradiation) has been established by spectroscopic methods, the way in which the radicals are created and ultimately destroyed has not been completely understood.

10.8 OXIDATIVE DEGRADATION

Oxidative degradation usually leads to hardening, discolouration as well as surface changes. The ease of oxidative degradation of the polymer depends primarily on its structure. Thus, unsaturated polymers such as polyisoprene or polybutadiene containing double bonds are easily attacked by oxygen.

The early stage of oxidative degradation is the formation of free-radical sites on the chain backbone by the attack of molecular oxygen, ozone or some free-radical produced by the thermal decomposition, of some additives (say, initiator molecules trapped inside the polymer matrix):

$$\mathrm{P} \rightarrow \dot{\mathrm{P}}$$

Where P represents the polymer chain.

Once the free-radical sites are formed, attack by oxygen becomes easier. Oxygen, because of its biradical nature, adds on at the free-radical site, forming a peroxide radical:

$$\dot{\mathrm{P}} + \mathrm{O_2} \rightarrow \mathrm{PO\dot{O}}$$

The peroxy radical can now attack the neighbouring segment, abstract hydrogen and form a hydroperoxide and a new free-radical site:

$$\mathrm{PO\dot{O}} + \mathrm{P} \rightarrow \mathrm{POOH} + \dot{\mathrm{P}}$$

It can also attack a neighbouring double bond:

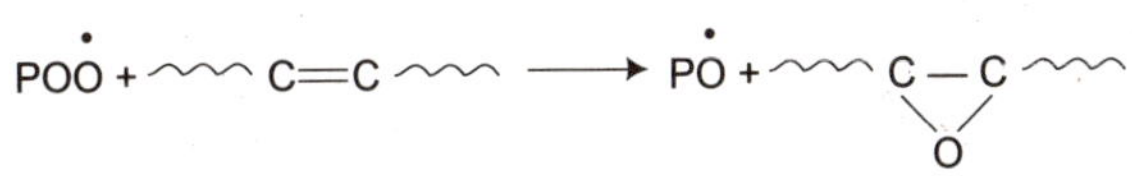

The hydroperoxide formed can lead to the formation of several new free-radical sites:

$$\mathrm{POOH} \rightarrow \mathrm{P\dot{O}} + \mathrm{\dot{O}H}$$

$$\dot{\mathrm{P}} + \mathrm{POOH} \rightarrow \mathrm{P\dot{O}} + \mathrm{POH}$$

$$\mathrm{P\dot{O}} + \mathrm{POOH} \rightarrow \mathrm{POH} + \mathrm{PO\dot{O}}$$

With so many free-radicals formed, their recombination can occur and result in the termination of the chain reactions:

$$2\mathrm{PO\dot{O}} \rightarrow \mathrm{POOP} + \mathrm{O_2}$$

$$2\mathrm{P\dot{O}} \rightarrow \mathrm{POOP}$$

$$\dot{\mathrm{P}} + \mathrm{\dot{O}H} \rightarrow \mathrm{POH}$$

$$2\dot{\mathrm{P}} \rightarrow \mathrm{P{-}P}, \text{ etc.}$$

10.8.1 Mechanism of Rubber Oxidation

Rubber, on oxidation, yields a variety of products. One common product is levulinaldehyde; other products being formic acid, acetic acid and carbon dioxide. The following scheme will explain the oxidative degradation of rubber:

Step 1

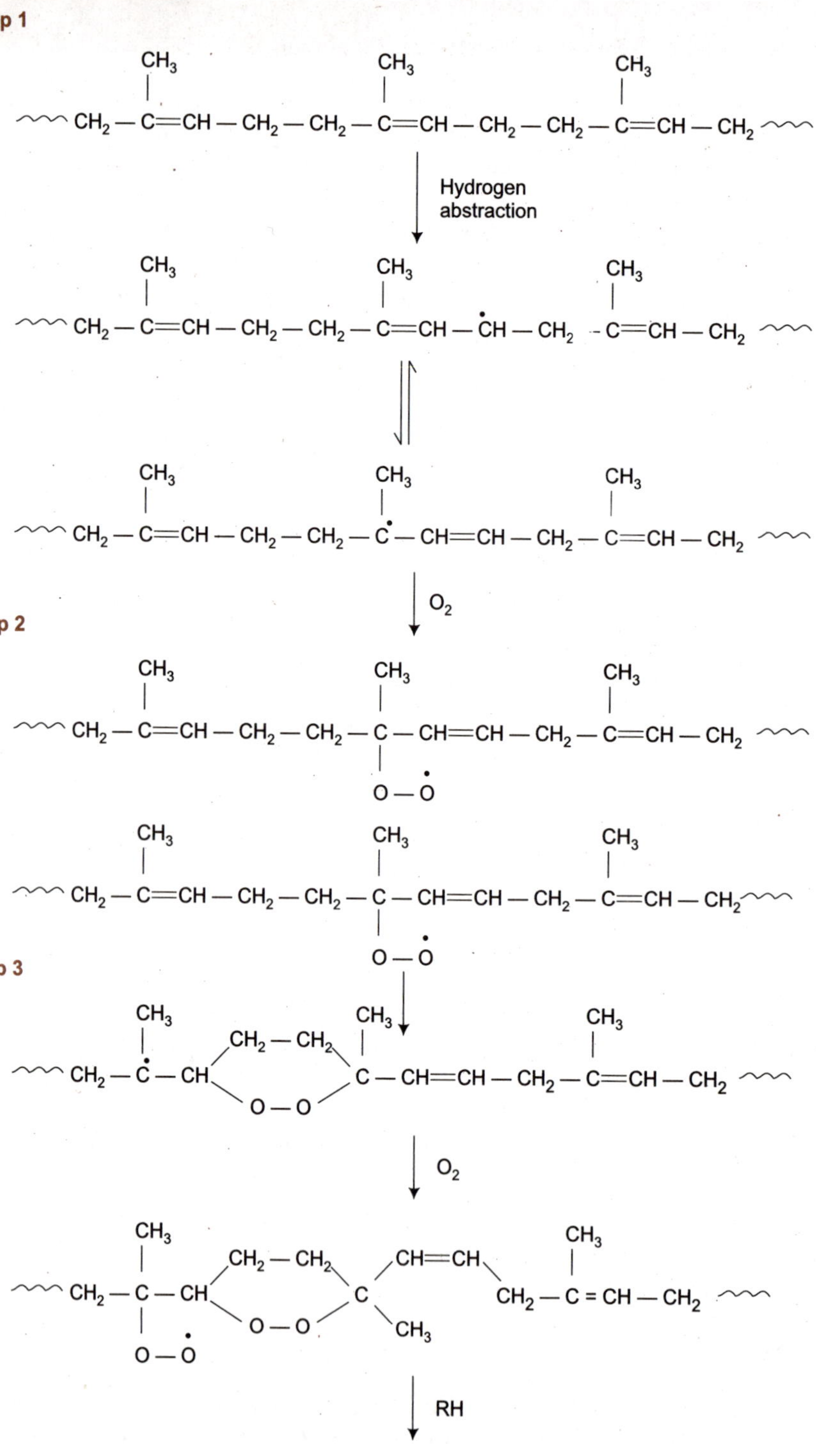

Step 4

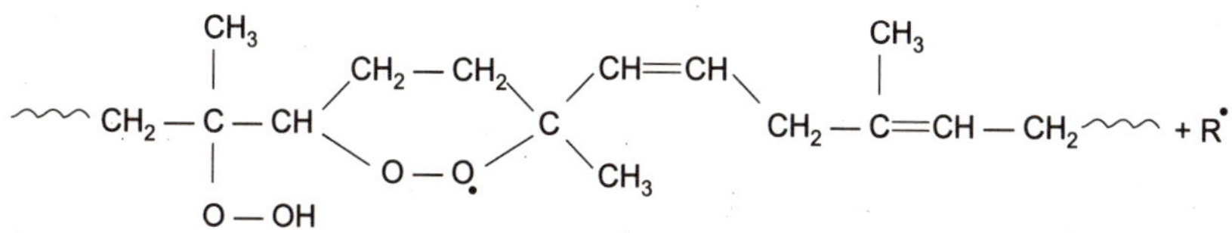

Step 5

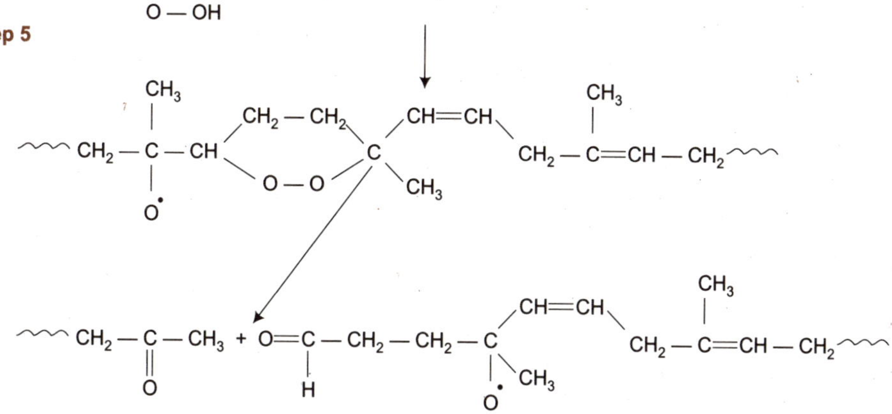

Step 6

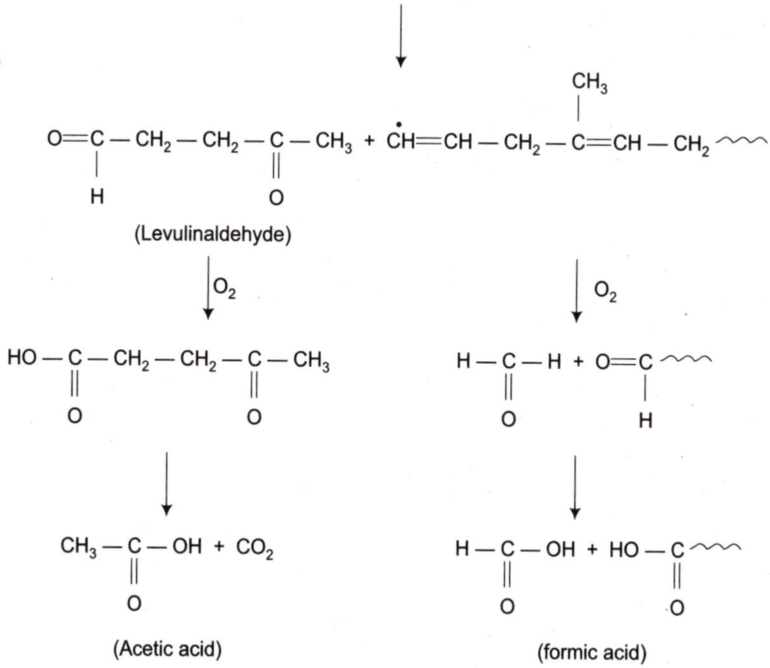

The formation of peroxy cyclic intermediate explains how simple compounds are produced during the oxidation of rubber.

10.8.2 Ozone Oxidation

Oxidative degradation caused by ozone has also received fairly good attention. Many elastomers when exposed to ozone were found to crack on stretching. Cracks developing in rubber articles during storage are primarily due to the atmospheric ozone. Ozone adds itself to the double bond to form an ozonide having the following structure:

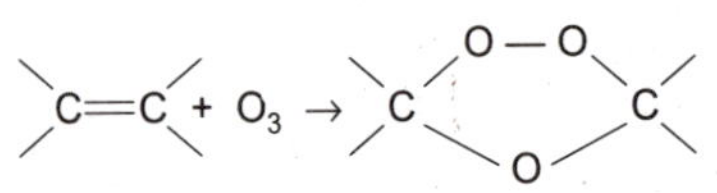

When this ozonide is hydrolysed, the original double bond (from which the ozonide is formed) will be cleaved to produce ketonic and aldehydic groups as shown:

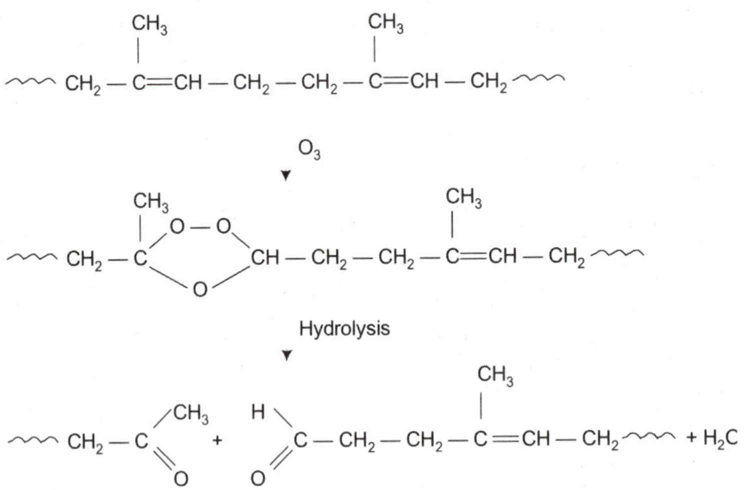

The process in which ozone cleaves the molecules into smaller ones is called *ozonolysis*.

Ozonolysis can be used to obtain low molecular weight liquid polymers, which are suitable for specific application, from high molecular weight solid polymers. Another method of getting such liquid polymers would be the oxidative degradation, the scheme for which is as follows:

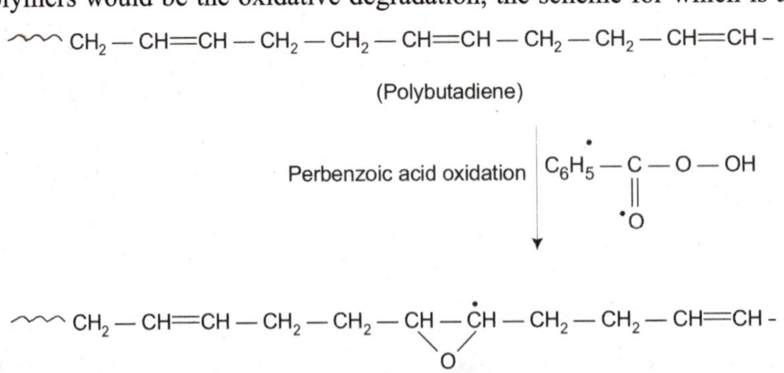

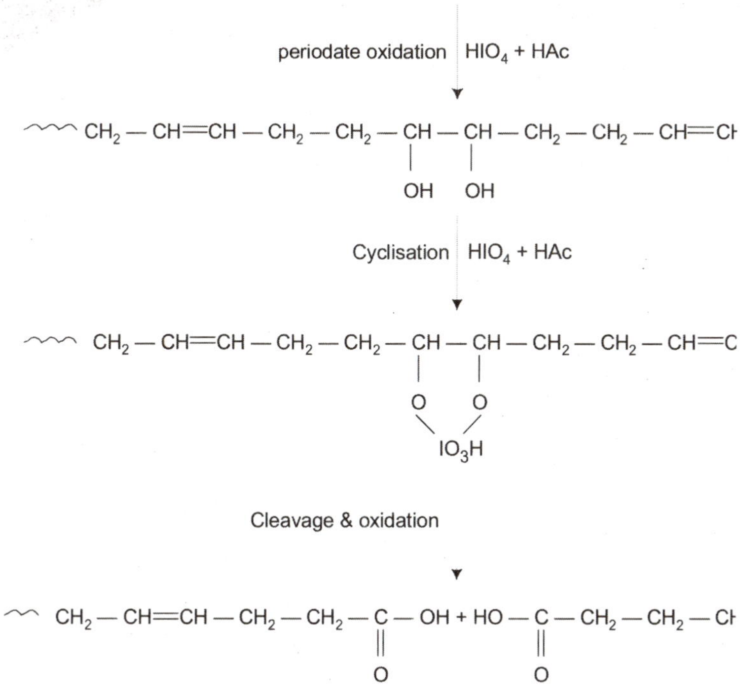

The product of oxidation in this case is a polybutadiene liquid polymer with carboxyl groups at its two ends as follows:

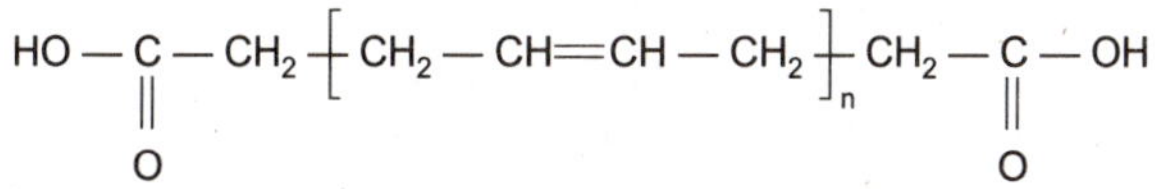

(CTPB)

It is carboxyl-terminated polybutadiene (CTPB) which, incidentally, is one of the most important prepolymers used in high-energy solid propellants.

10.8.3 Oxidative Degradation of Saturated Polymers

Unsaturated polymers are very much oxidation-prone. This, however, is not an exclusive property of unsaturated polymers. Saturated ones, such as polystyrene which normally resist oxidation, also get oxidised in the presence of sunlight or at elevated temperatures (around 200°C):

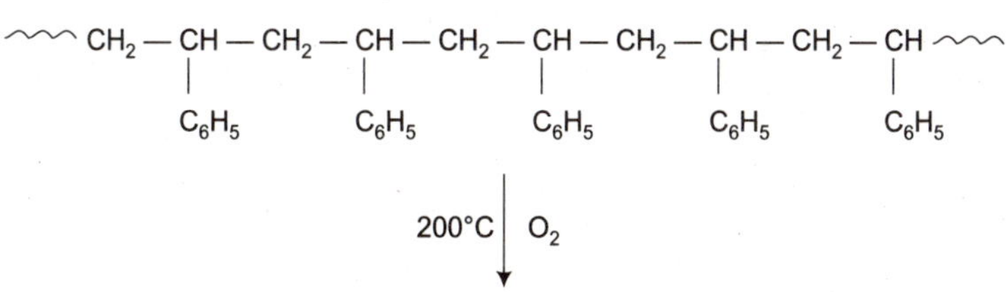

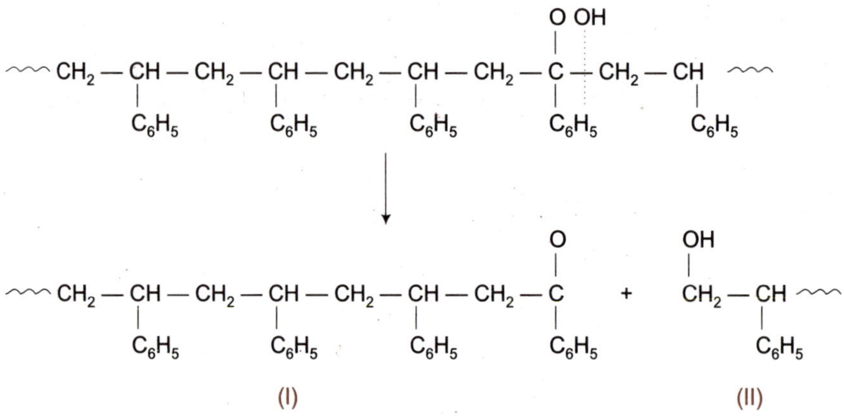

The evolution of benzaldehyde during (thermal) oxidative degradation of polystyrene (now shown) strongly suggests a hydroperoxide mechanism. Benzaldehyde is produced by the cleavage, occurring at the ketonic end of structure I:

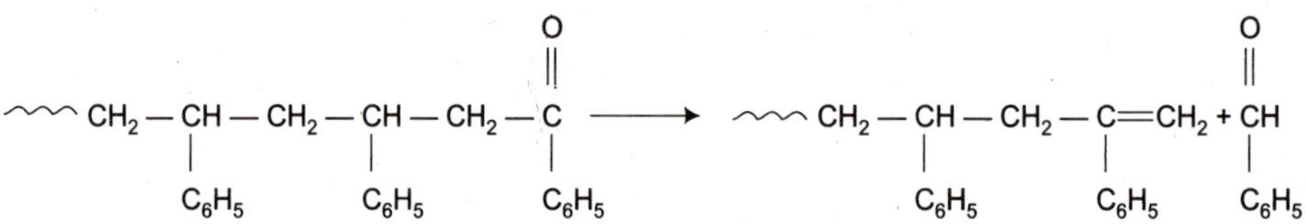

In fact, the formation of hydroperoxide as the first step in the oxidative degradation of saturated polymers has been confirmed in many cases. The degradation of polyethylene depends on its phase state. The amorphous polyethylene is more susceptible to oxidation than its crystalline variety. The tertiary hydrogen atoms at the site of branching are the ones which form the hydroperoxide during oxidation. This hydroperoxide subsequently cleaves the chain as follows:

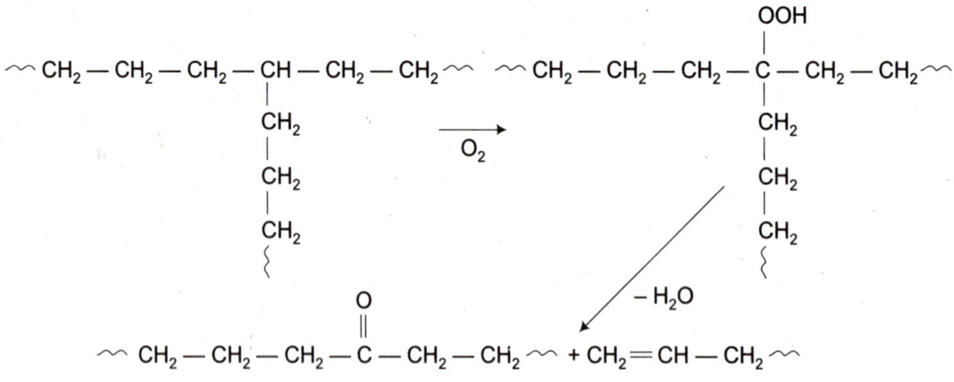

10.8.4 Oxidation of Phenol Formaldehyde

So far, we have seen the effect of oxygen on certain saturated and unsaturated polymers. Now, let us see how oxygen attacks a common resin such as phenol formaldehyde. The structure of phenol formaldehyde is:

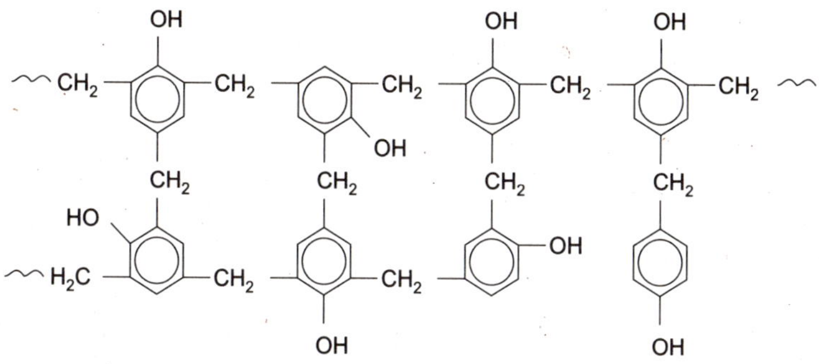

(1)

It is observed that attack by oxygen commences at 200°C. Methylenic linkages are the ones susceptible to oxidation:

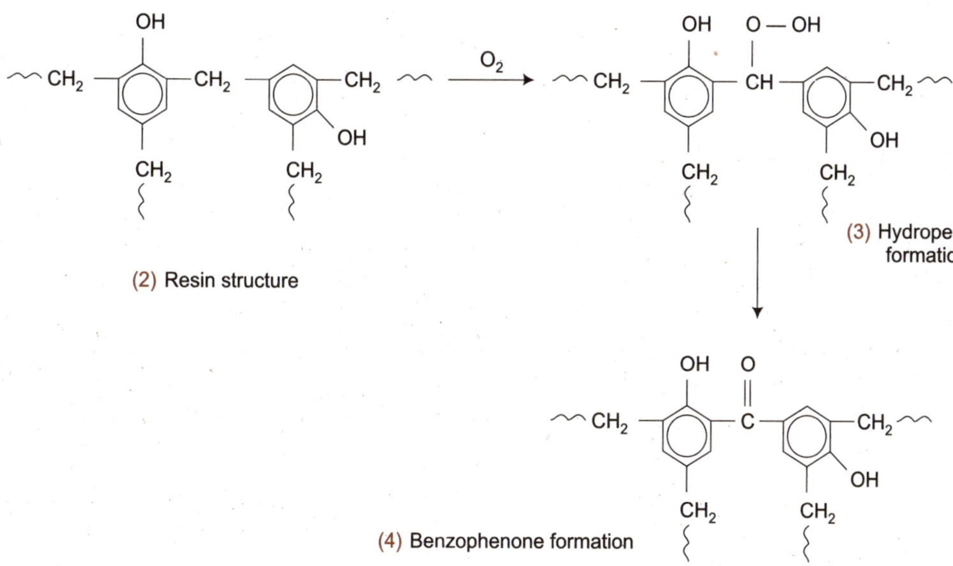

(2) Resin structure

(3) Hydroperoxide formation

(4) Benzophenone formation

The formation of benzophenone linkage as in Structure (4) has been confirmed by the analysis of the product formed. On continuing the oxidation, the following changes are indicated by IR and other analytical methods:

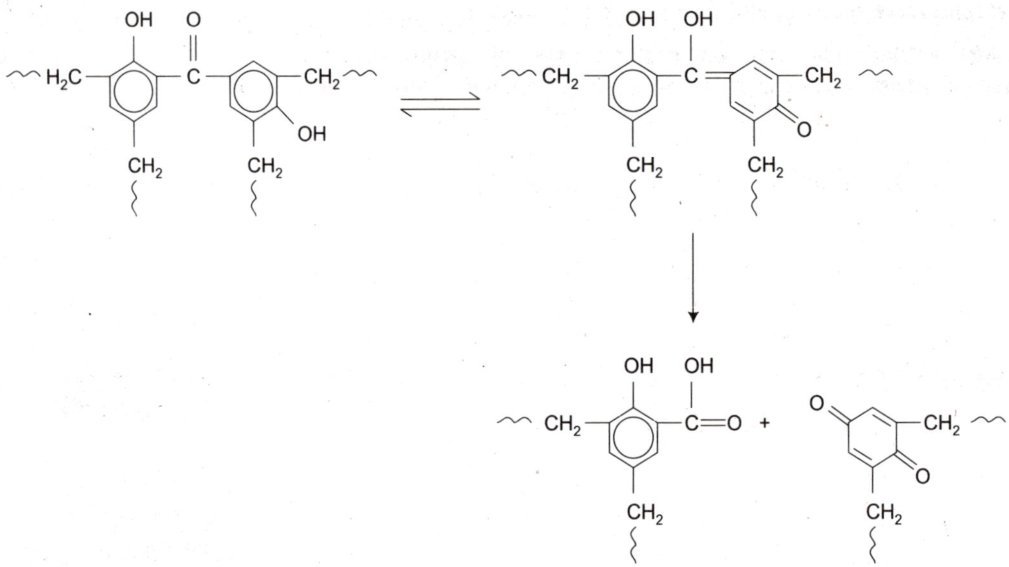

The appearance of a carboxyl group has been well identified by the IR analysis. Thus, how a resin such as phenol formaldehyde is chemically degraded by oxygen at elevated temperatures is seen above.

10.8.5 Antioxidants

Many polymers are protected against oxidative degradation by incorporating certain chemical compounds called *'antioxidants'*. Even in small amounts, antioxidants are significantly effective in preserving the original properties of the polymer. Antioxidants oppose oxidation and in a good number of cases suppress many undesirable reactions, promoted either by oxygen or peroxides. The formation of peroxides and hydroperoxides by the attack of oxygen has already been shown in Section 10.8. The various reactions, now shown, give an idea of the different routes by which antioxidants can arrest a reactive polymer radical and prevent the polymer P from degradation through these active radicals.

$$\dot{P} + XH \qquad PH + X^{\bullet}$$

$$P\dot{O} + XH \qquad POH + X^{\bullet}$$

$$PO\dot{O} + XH \qquad POOH + X$$

$$H\dot{O} + XH \qquad HOH + X^{\bullet}$$

Reactive radicals + Anti-oxidant   Stable products + Stabilised radicals

The antioxidant (XH) interacts with reactive radicals like $\dot{P}$, $P\dot{O}$, $PO\dot{O}$, $H\dot{O}$ (formed along the polymer chain due to the attack by oxygen and subsequent chain reactions) and produces stable products. The radicals ($\dot{X}$) formed from the antioxidant are incapable of attacking a new polymer chain, since they are resonance stabilised. However, they can recombine:

$$\dot{X} + \dot{X} \rightarrow X—X$$

The structures of a few commonly used antioxidants are listed as follows:

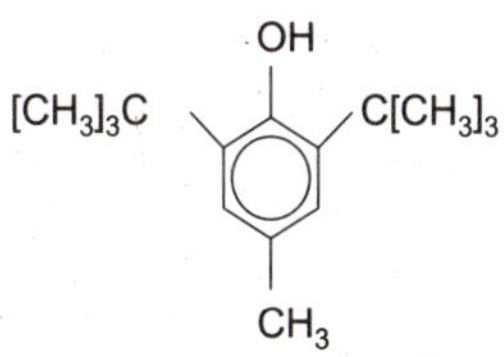

(Di-t-butyl p-cresol)
(TBC)

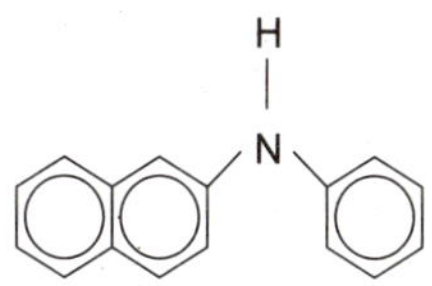

(Phenyl β-naphthyl amine)
(PBNA)

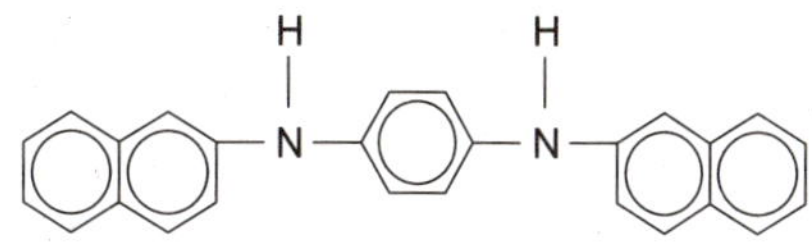

(Sym di-β-Naphthyl p-phenylene diamine)
(BNPD)

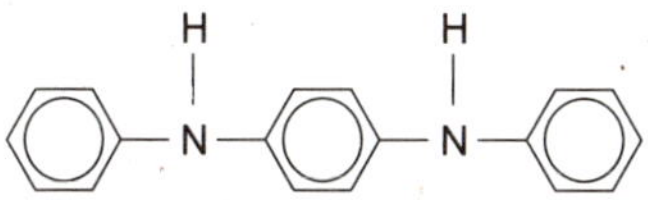

(Diphenyl p-phenylene diamine)
(DPPD)

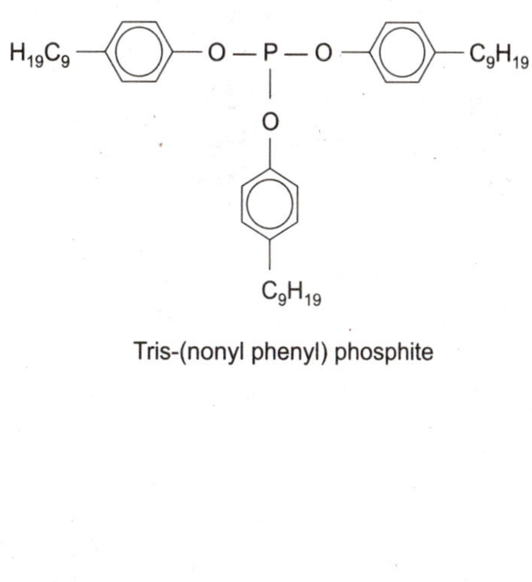

Tris-(nonyl phenyl) phosphite

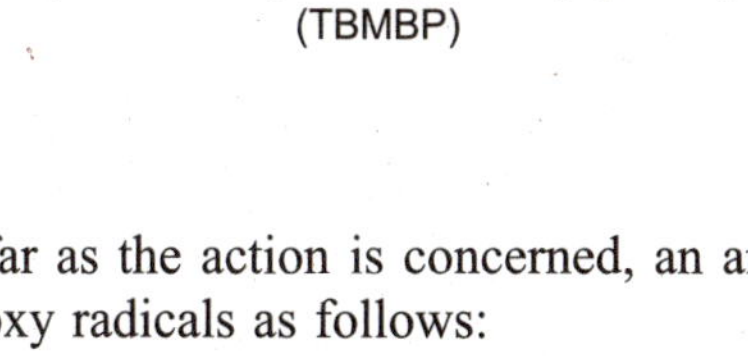

2, 2′-Thio bis (4-Methyl 6-t-butyl phenol)
(TBMBP)

As far as the action is concerned, an antioxidant such as di-t-butyl-p-cresol (TBC) can destroy two peroxy radicals as follows:

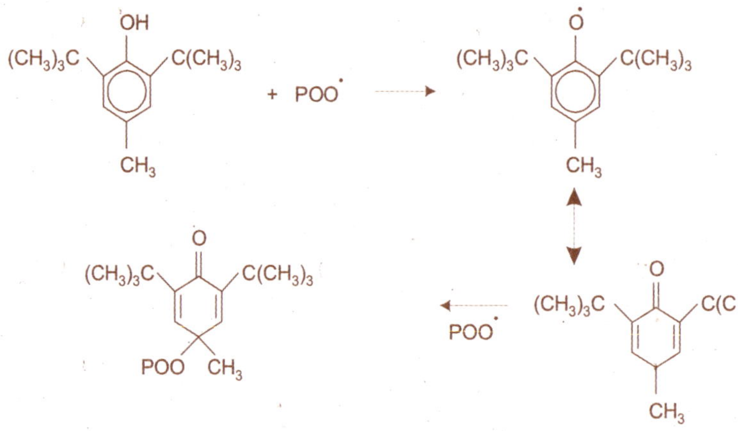

Phenol derivatives are very widely used as antioxidants in polymer industry, but they are not effective at high temperatures because of their susceptibility to thermal decomposition and because they generate free-radicals. The secondary aryl amines and hindered phenols shown above, which are usually used as common antioxidants, have reactive N—H and O—H functional groups. They are capable of transferring hydrogen to peroxy radicals as follows:

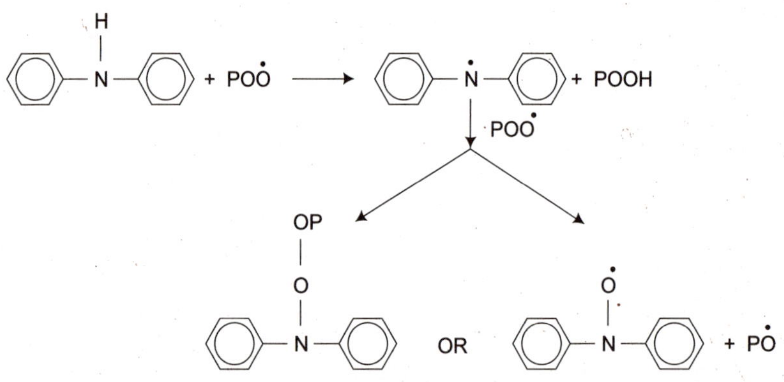

An overall picture of the reduction in oxygen absorption by a polymer, containing small amounts of antioxidants, is shown in Fig. 10.5.

Many antioxidants such as PBNA are capable of staining and for this reason, compounds such as 2, 2′-thiobis (4-methyl-6-tert-butyl phenol) are used, which come under the non-staining class. It should be noted that TBC is also a non-staining antioxidant.

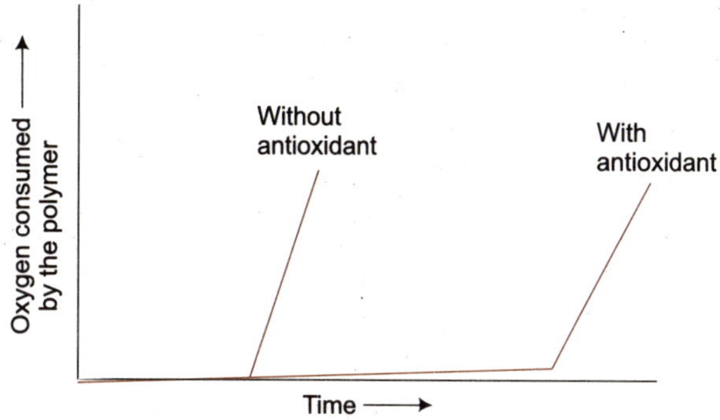

Fig. 10.5 Effect of antioxidant on oxidative degradation of a polymer. (The amount of oxygen consumed is a direct measure of degradation undergone by the polymer).

10.9 HYDROLYTIC DEGRADATION

As explained while discussing random degradation, polymer chains containing ester functional groups in their backbone undergo degradation by hydrolysis. Also, polymer chains containing amide and acetal groups can undergo hydrolytic degradation. Acidolysis and aminolysis of these polymers also lead to random degradation of the polymer chain (for reactions see Sections 11.2 to 11.4).

Books for Further Reading

Grassie, N., *Chemistry of High Polymer Degradation Processes,* Butterworths, London, 1956.

Madorsky, S.L., *Thermal Degradation of Organic Polymers,* Interscience, New York, 1964.

Scott, G., *Atmospheric Oxidation and Antioxidants*, Elsevier, Amsterdam, 1965.

Hawkins, W.L., Ed., *Polymer Stabilisation,* Wiley-Interscience, New York, 1972.

Geuskens, G. Ed., *Degradation and Stabilisation of Polymers*, Halsted, New York, 1975.

Ranby, B. and Rabek, J.F., *Photodegradation, Photo-oxidation and Photostabilisation of Polymers*, John Wiley and Sons, New York, 1975.

Jellinek, H.H.G., Ed., *Aspects of Degradation and Stabilisation of Polymers*, Elsevier, Amsterdam, 1978.

11

POLYMER REACTIONS

11.1 INTRODUCTION

The functional groups present in a polymer molecule can, depending on its chemical nature, undergo a variety of reactions leading to a polymer structure with entirely new properties and applications. An interesting example of how we can modify a polymer for new applications can be as follows: We are familiar with polyethylene. If we can make a copolymer of ethylene and a monomer containing an —SH group, we get a polymer with a lot of mercapto side groups:

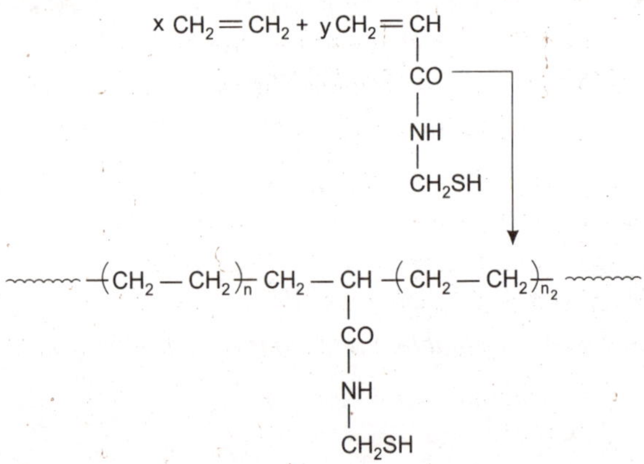

This is a linear polymer and, like polyethylene, can form a film. The reactivity of the mercapto side groups in this polymer can be used for an exciting application. The mercapto groups on mild oxidation form a disulphide, bridging the two linear molecules:

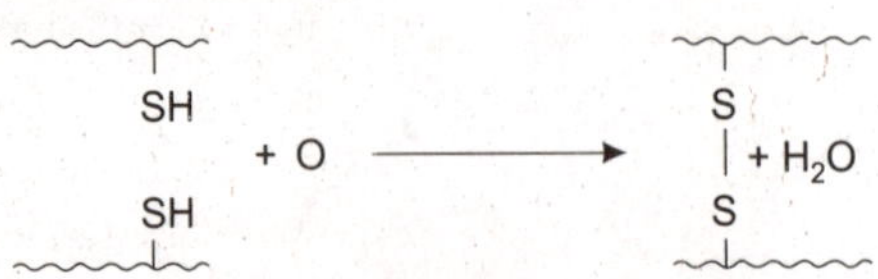

Thus, on mere exposure to air, polymer oxidation takes place and we get a cross-linked polymer which becomes insoluble. This polymer reaction is used in hair sprays in order to set the hair curls. A solution of the mercapto polymer is sprayed on the hair, which is then set to the required curl pattern and allowed to stay in that set condition for sometime. The atmospheric oxygen oxidises the polymer and, because of the cross-links, the film formed on the hair becomes resilient and insoluble and keeps the curl intact as long as desired. For setting a new curl pattern, all we have to do is to wash it off with a solution of a mild reducing agent such as thioglycolic acid when the S—S cross-linkings are broken. The polymer is then washed off with water and we are ready to set a new curl pattern.

The behaviour of functional groups in a polymer molecule normally is not very much different from those in a simple organic molecule. For this reason, many interesting reactions are possible with polymer molecules too. Now, let us examine some such reactions.

11.2 HYDROLYSIS

Polymers with amide, ester and acetal linkages can be easily hydrolysed in the presence of acids and alkalis. When such groups are present in the chain backbone, the hydrolysis leads to chain scission or degradation:

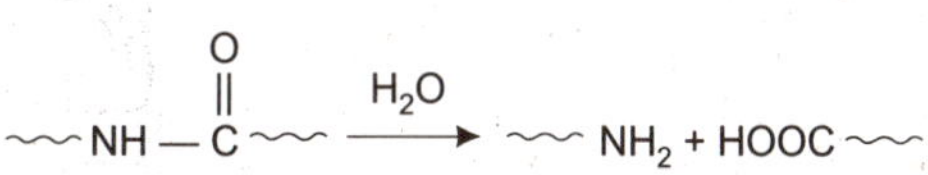

On the other hand, when these hydrolysable groups form the side groups of the polymer chain, hydrolysis leads only to a change of functional group and not to any chain scission. An example for this type of hydrolysis is provided by polyvinyl pyrrolidone:

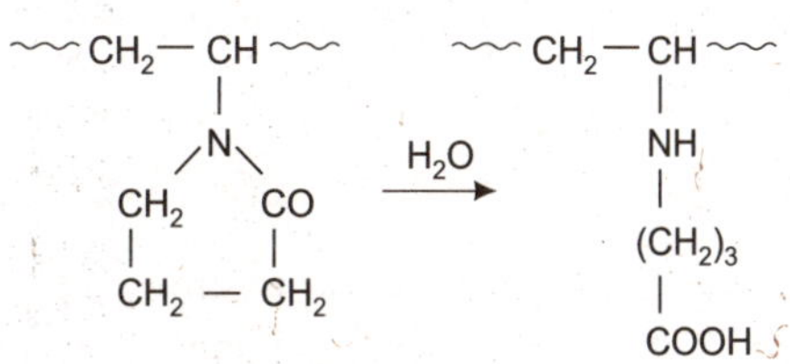

Polyvinyl acetate is another well-known polymer which can be hydrolysed with an alkali to give polyvinyl alcohol:

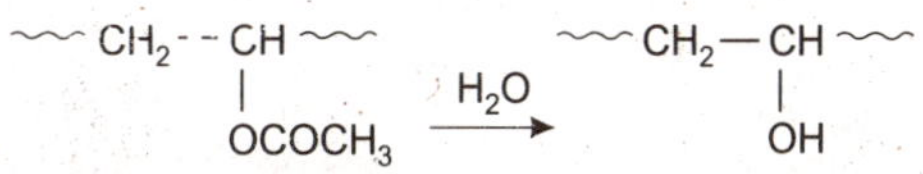

Reactions of this type, which lead only to changes in functional groups and not to any chain degradation, are called the *polymer analogue reactions*. In such reactions, the chemical composition of the polymer repeat unit is changed, but the degree of polymerisation remains unchanged.

11.3 ACIDOLYSIS

Acidolysis involves the splitting of the polymer molecule using carboxylic acids:

$$\sim NH-(CH_2)_x-NH-CO-(CH_2)_y-CO\sim$$

$$\big\downarrow \; HOOC-R-COOH$$

$$\sim NH-(CH_2)_x-NH-CO-R-COOH + HOOC-(CH_2)_y-CO\sim$$

In the above example, we see that a high molecular weight polyamide is degraded into a low molecular weight polymer with terminal carboxyl groups.

11.4 AMINOLYSIS

Aminolysis involves the splitting of the polymer molecule with an amine and this also proceeds in a manner similar to acidolysis:

$$\sim NH-(CH_2)_x-NH-CO-(CH_2)_y-CO\sim$$

$$\big\downarrow \; H_2N-R-NH_2$$

$$\sim NH-(CH_2)_x-NH_2 + H_2N-R-NH-CO-(CH_2)_y-CO\sim$$

In this case, we get a low molecular weight polymer with amine terminal groups.

11.5 HYDROGENATION

The reader may be aware of the hydrogenation of unsaturated organic molecules. The same hydrogenation technique is used to obtain many new polymers from the existing ones:

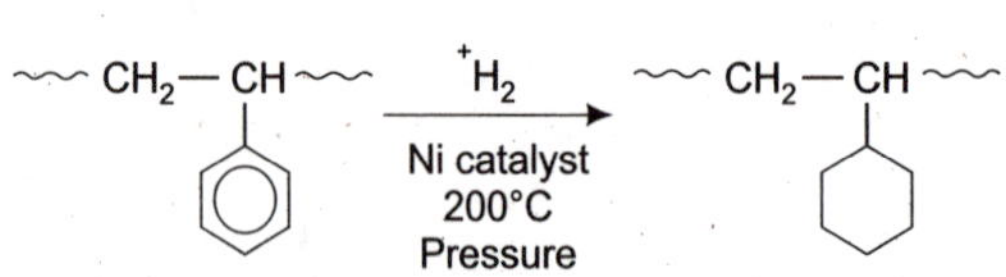

Similarly, polymers or copolymers with double bonds on the backbone can also be hydrogenated, as now discussed.

(*i*) Polyisoprene can be hydrogenated to get a polymer resembling an alternating copolymer of ethylene and propylene:

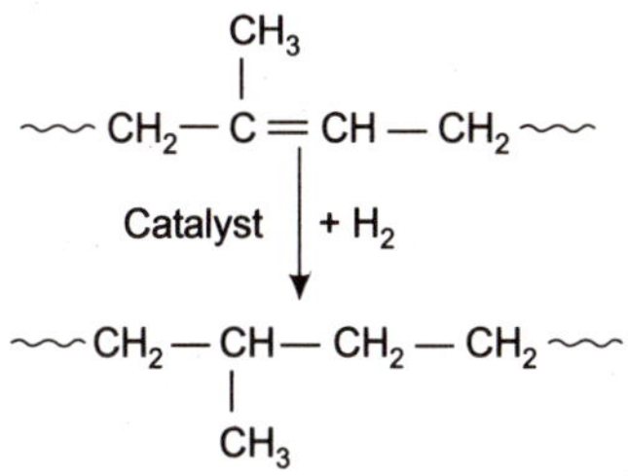

(*ii*) Styrene-butadiene rubber can be hydrogenated to a copolymer of styrene and ethylene:

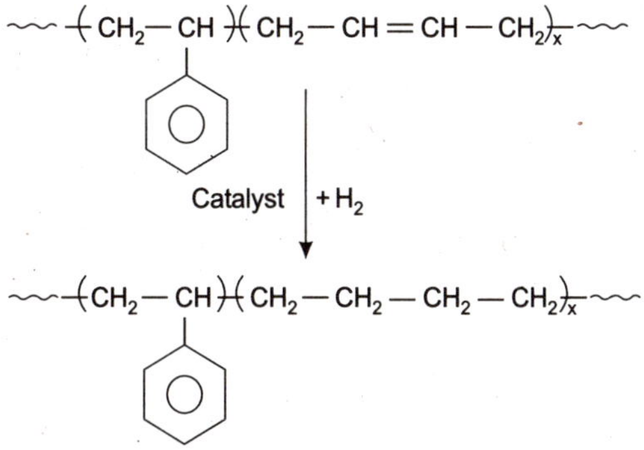

11.6 ADDITION AND SUBSTITUTION REACTIONS

Here, we consider various addition and substitution reactions, which are as follows:

(*i*) Addition of oxygen, chlorine, bromine, ozone or hydrogen chloride to unsaturated polymers is also common. Thus, polybutadiene can be epoxidised with perbenzoic acid, peracetic acid, etc. as given:

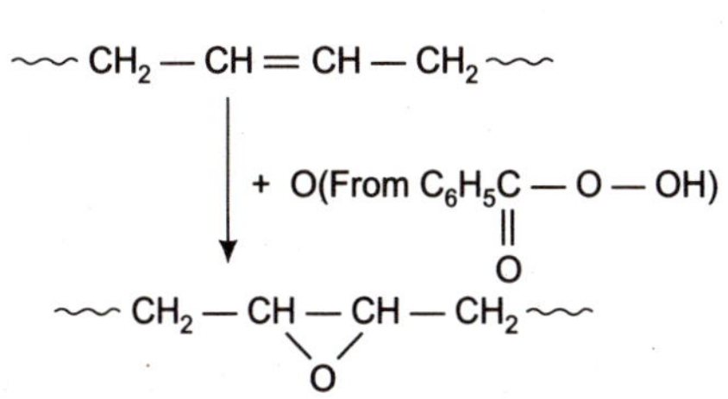

In this reaction, oxygen from the perbenzoic acid adds to the double bond in the polymer molecule.

(*ii*) Ozone adds to unsaturated polymers to form an ozonide, which, in turn, can be hydrolysed to cleave the molecule:

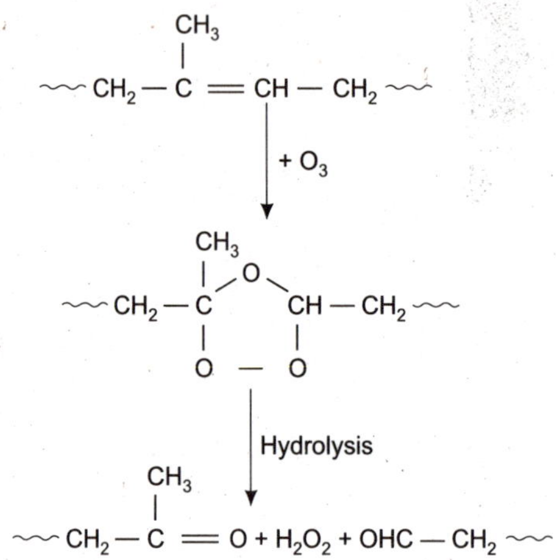

(*iii*) Chlorine adds to natural rubber to form chlorinated rubber as follows:

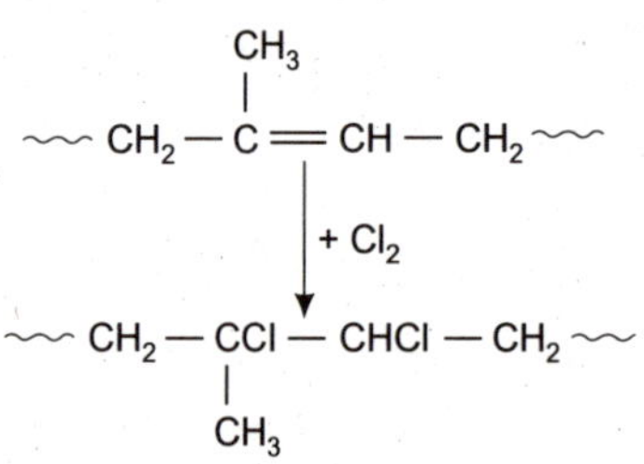

(*iv*) Likewise, hydrogen chloride can also add across the double bond, as in the following reaction:

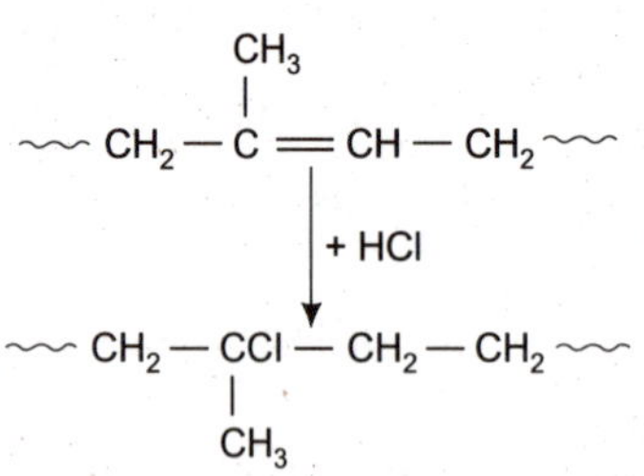

(*v*) With saturated polymers such as polyethylene, chlorination can be carried out by a substitution reaction through a radical chain process:

$$\sim CH_2-CH_2\sim + \dot{R} \longrightarrow \sim \dot{C}H-CH_2 \sim + RH$$

$$\sim CH-CH_2\sim + Cl_2 \longrightarrow \sim CHCl-CH_2 \sim + \dot{C}l$$

Chlorinated polyethylene containing around 30% of chlorine, is a rubbery material in contrast to polyethylene which is a tough thermoplastic. The irregular substitutions of chlorine in the backbone reduces the crystallinity of polyethylene and makes it rubbery. Chlorination beyond 50%, however, makes the polymer more and more rigid and at around 70% chlorination, the polymer becomes quite brittle.

(*vi*) Polyethylene can also be chlorosulfonated through a substitution process:

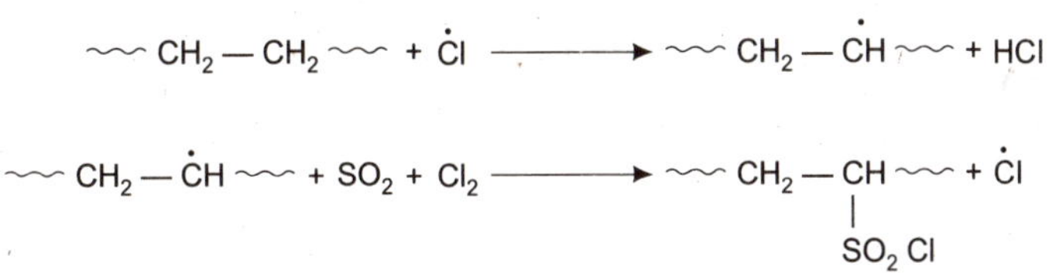

Chlorosulfonated polyethylene is known for its solvent, chemical and flame-resistant characteristics.

11.7 REACTION OF VARIOUS SPECIFIC GROUPS

11.7.1 Hydroxyl Groups

(*i*) Polymer molecules with hydroxyl groups, as in polyvinyl alcohol,can react with metallic sodium. The product thus obtained can be further reacted with alkyl halides to produce polyvinyl ethers:

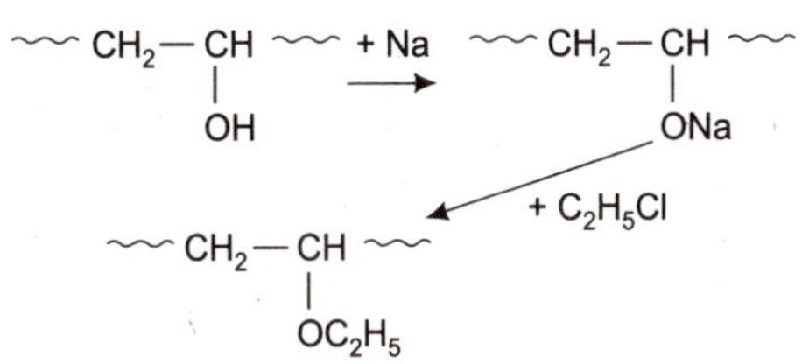

(*ii*) Polyvinyl alcohol can also react with ethylene oxide under alkaline conditions to yield ethoxy derivatives of the polymer:

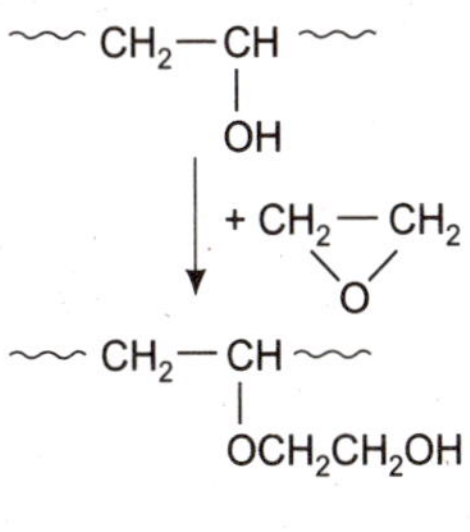

(*iii*) The hydroxyl groups in polyvinyl alcohol can undergo the usual esterification reaction too. Thus, starting from polyvinyl alcohol, we can get back polyvinyl acetate by reacting it with acetic anhydride and anhydrous sodium acetate, as shown:

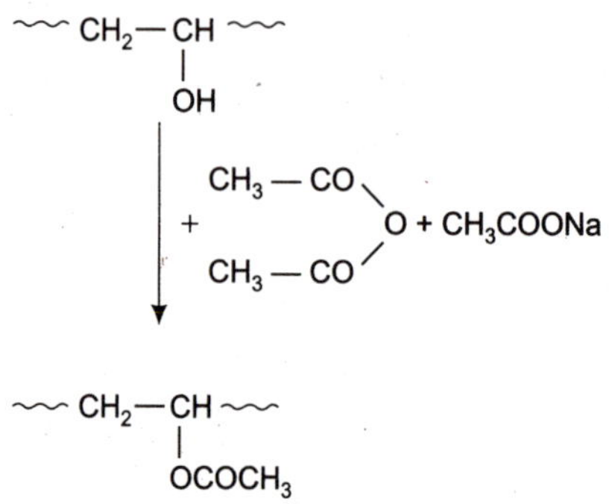

(*iv*) Polyvinyl alcohol can also react with isocyanates to form derivatives of urethanes:

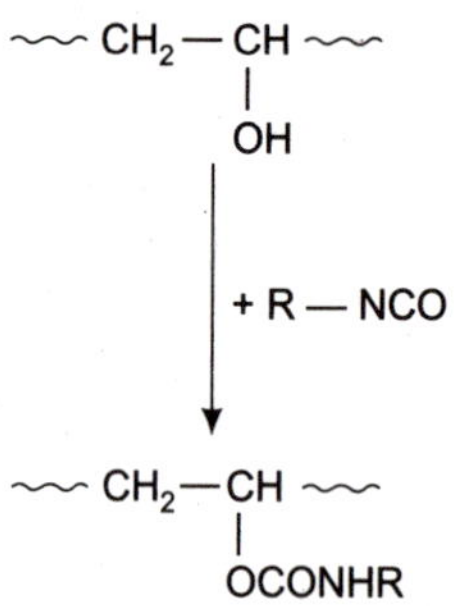

Similarly, the hydroxyl groups in polypropylene glycol can react with isocyanates to form polyurethanes.

11.7.2 Aldehyde Groups

(*i*) Aldehyde groups are present along the chain of polyacrolein and hence, they can react in a manner similar to that of a simple aldehyde. Sodium bisulphite adds to polyacrolein as follows:

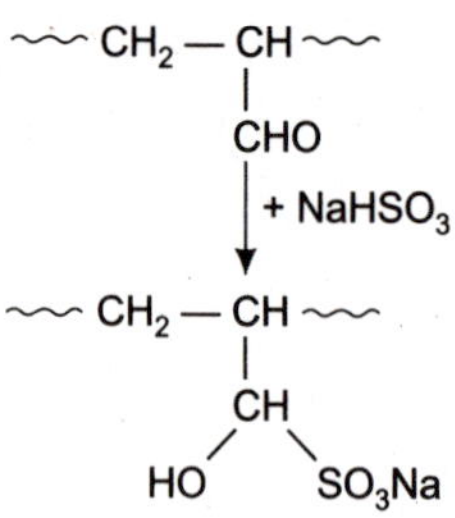

This product is soluble in water as against polyacrolein. We can thus convert a water-insoluble polymer into a water-soluble one.

(*ii*) Similarly, polyacrolein reacts with hydroxylamine as well as phenyl hydrazine:

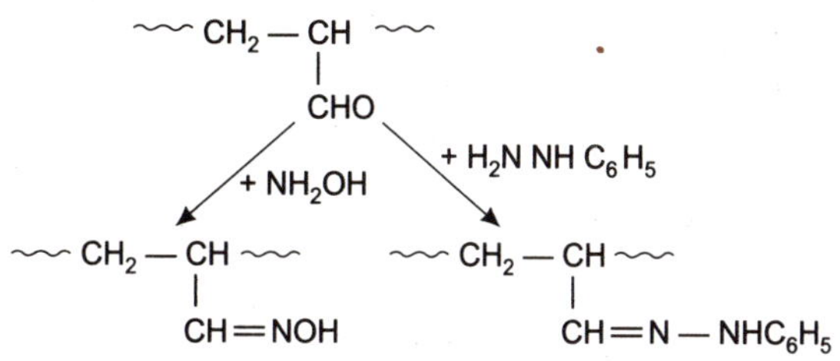

(*iii*) Acetals are produced by treating polyacrolein with alcohols;

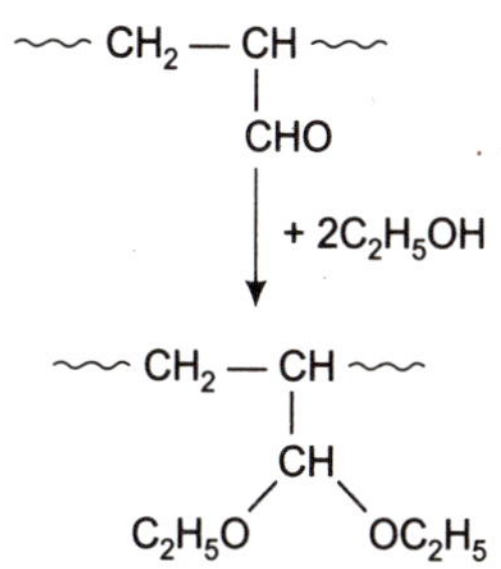

11.7.3 Ketonic Groups

Polyvinyl methyl ketone undergoes reactions characteristic of a ketone. It reacts with hydroxylamine as follows:

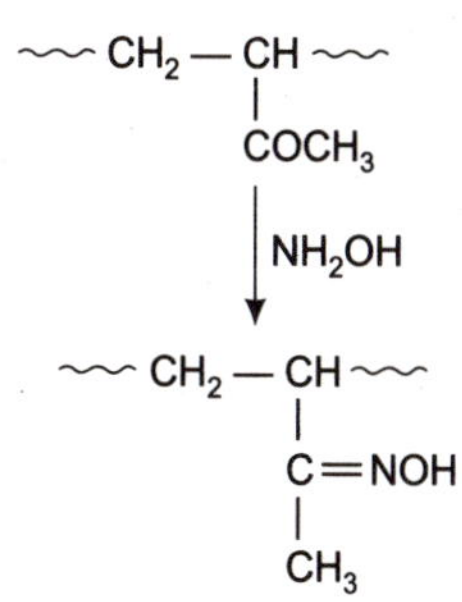

11.7.4 Carboxylic Groups

(*i*) Polyacrylic acid is a polymer with a large number of —COOH groups. The properties of this polymer are similar to those of simple organic acids. For example, polyacrylic acid can react with alkalis to form salts:

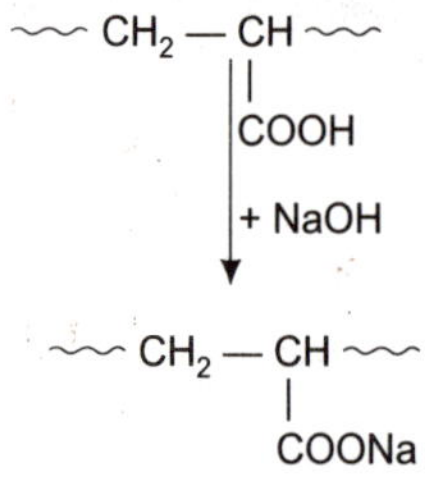

This sodium salt and polyacrylic acid are both soluble in water and can undergo ionisation as shown:

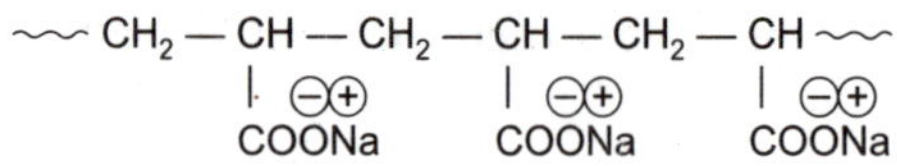

These kind of polymers, having ionisable groups are known as *'polyelectrolytes'*.

(*ii*) Carboxylic groups can also be reacted with epoxy groups to form esters:

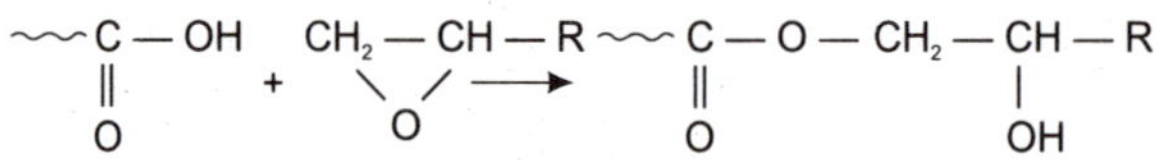

This reaction is utilised to cure liquid butadiene resins containing carboxylic groups, the details of which can be found in Section 11.9.3.

11.7.5 Other Groups

(*i*) The amino groups present in polyvinyl amine can be reacted with reagents such as acetic anhydride or benzene sulphonyl chloride to form the modified polymers:

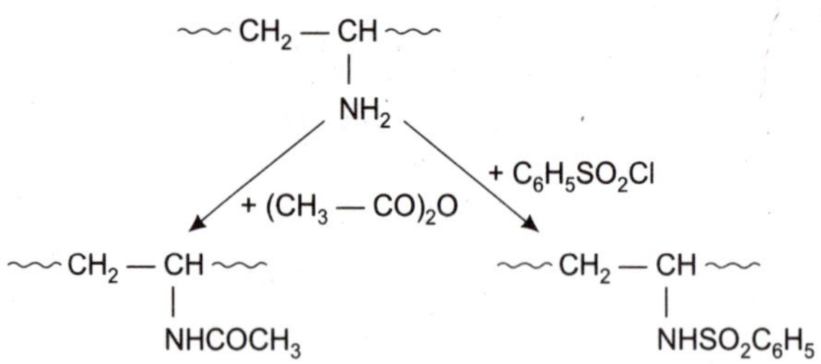

(*ii*) The nitro groups present in a polymer may be converted to amino groups by reduction. Hence, polynitropropylene could be converted to polypropylenamine as follows:

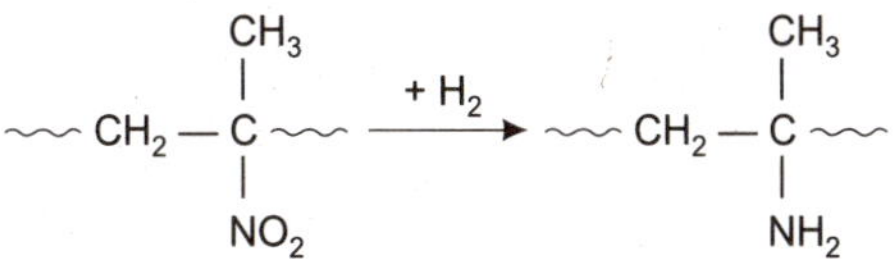

(*iii*) The amido groups present along a polymer chain, like in simple acid amides, can undergo the Hofmann degradation reaction to yield amino groups. Thus, polyacrylamide can be converted into, polyvinyl amine:

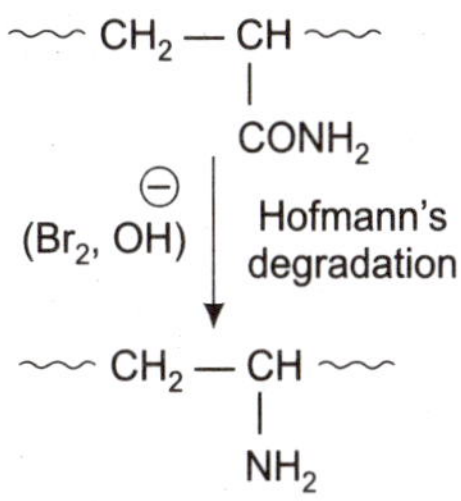

(*iv*) Many polymers contain aromatic rings and can undergo reactions characteristic to aromatic nucleus. The benzene nucleus of polystyrene undergoes many of the reactions we come across in aromatic chemistry. Thus, polystyrene can be sulphonated, as shown:

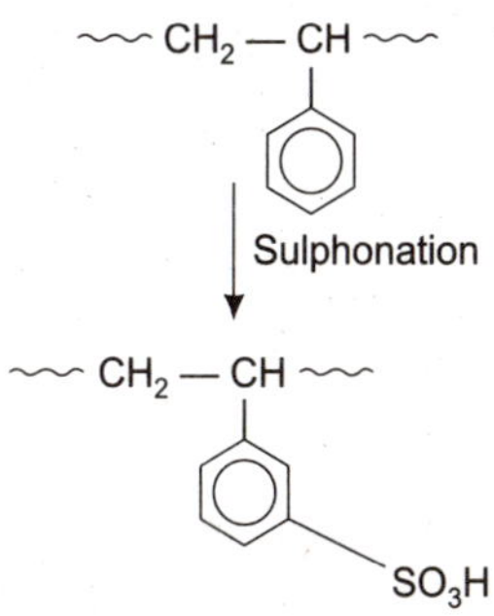

Similarly, polystyrene can be nitrated to polynitrostyrene and chlorination of polystyrene can be done in the presence of $AlCl_3$.

By employing Friedel-Crafts' reaction, polyvinyl cyclohexyl benzene is synthesised from polystyrene and cyclohexene as follows:

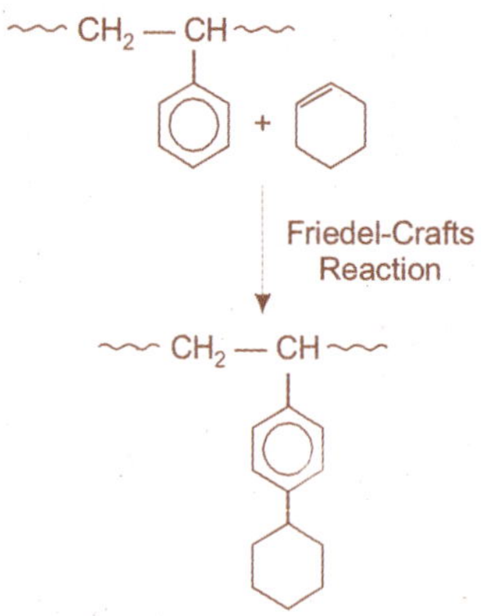

A unique polymer, called perfluoropolystyrene, is synthesised by fluorination of polystyrene. During this reaction, hydrogen fluoride is eliminated:

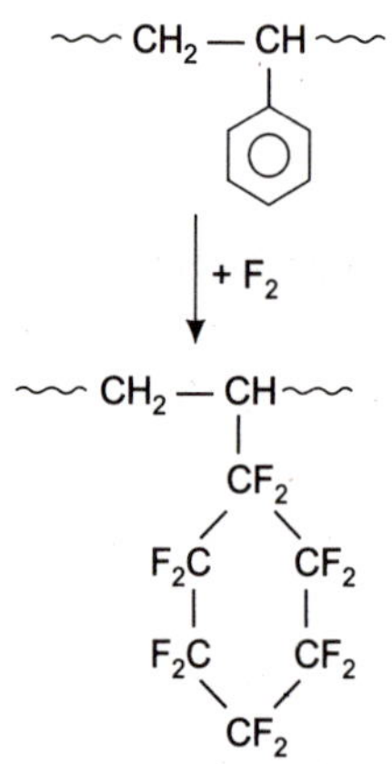

11.8 CYCLISATION REACTIONS

Cyclisation is another important reaction that polymers undergo. Some of the cyclisation reactions are as follows:

(*i*) Polyvinyl alcohol when reacted with formaldehyde gives polyvinyl formal:

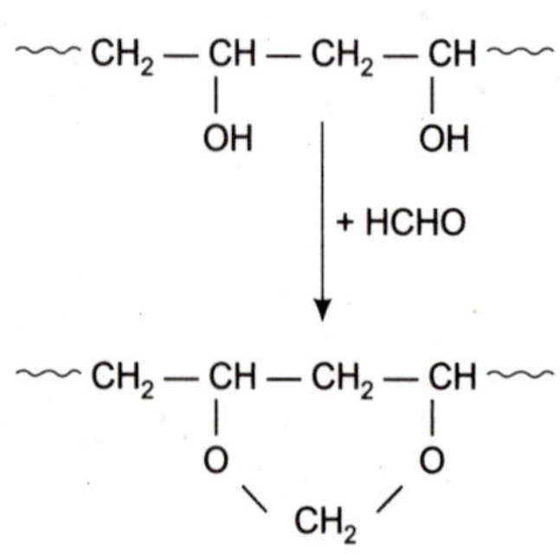

When butyraldehyde is used in place of formaldehyde, polyvinyl butyral results:

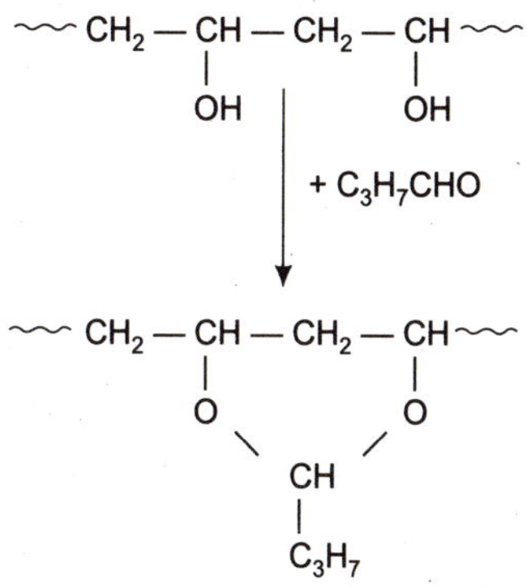

(*ii*) Polyacrylonitrile, upon heating, cyclises to form a ladder polymer:

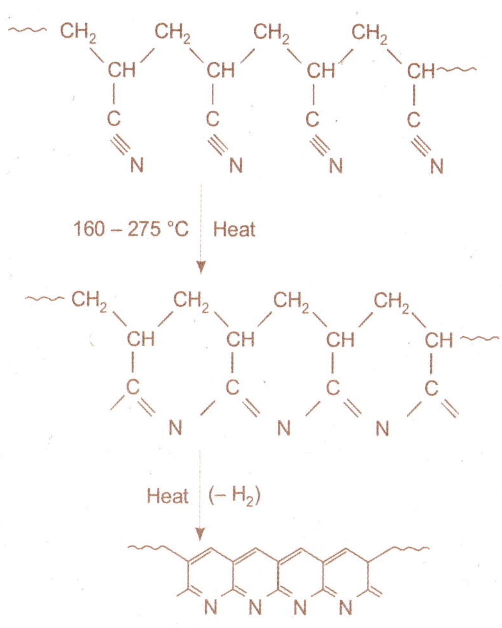

This polymer is highly flame-resistant.

(*iii*) 1, 2-polybutadiene cyclises in acid medium as follows:

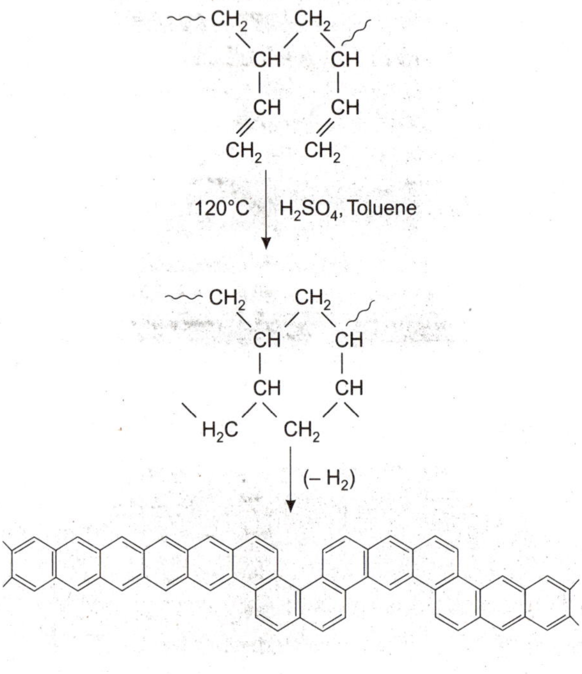

Intramolecular rearrangement occurs, forming a kinked structure since acenes with more than six linearly condensed rings become unstable.

(*iv*) Natural rubber also cyclises in an acid medium to form what is called cyclised rubber, which is industrially useful in the preparation of adhesives and coatings. Cyclisation converts natural rubber to a soluble plastic. The cyclisation process involves a carbonium ion mechanism and will lead to the two possible structures (*a*) and (*b*) as shown in Fig. 11.1.

If product (*a*) is formed, a polymer chain with a polycyclic ring structure will result and if product (*b*) is formed, a polymer chain with a monocyclic structure will result.

11.9 CROSS-LINKING REACTIONS

Of all reactions that we come across in polymer chemistry, cross-linking reactions are by far the most useful industrially. Bridgings formed between linear polymer chains, leading to a three-dimensional network structure, are known as *'cross-links'*. The cross-links between polymer molecules can be either through regular covalent bonds or through secondary valence type linkings such as hydrogen bonds. The former type is called chemical cross-linking or irreversible cross-linking and the latter, the physical or reversible cross-linking. The physical or reversible cross-linking, in fact, plays a vital role in polymer crystallisation. The intermolecular connections effected through secondary valence bonds, such as hydrogen bonds, are undoubtedly weaker compared to the regular

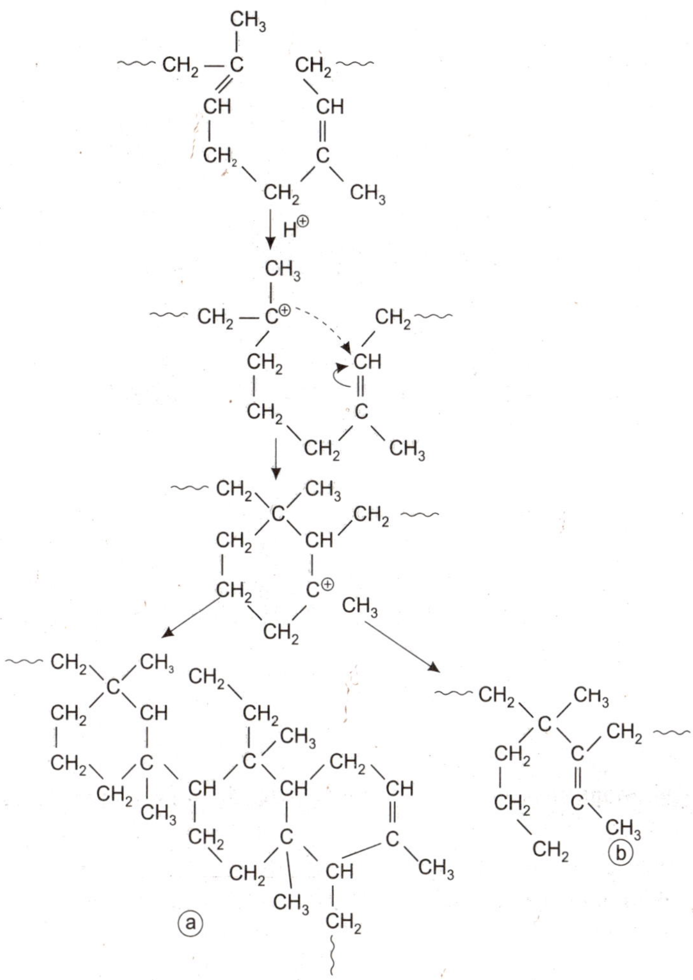

Fig. 11.1 Mechanism showing cyclisation in natural rubber.

covalent bonds, and hence, can be removed with much ease, such as by dissolving the polymer in appropriate solvents or by simply heating the polymer above its T_m. A physically cross-linked (and also an uncross-linked) polymer could be melted to a liquid and again cooled back to a solid reversibly. On the other hand, chemical cross-linking through covalent bonds is irreversible. A chemically cross-linked polymer can neither be melted by heating nor dissolved in a solvent. Prolonged or drastic heating will, however, lead to the degradation of the polymer. In this section, we shall discuss only the various reactions leading to irreversible cross-linking.

11.9.1 Cross-linking during Polymerisation

Cross-linking could be made to occur during polymerisation.

Thus, styrene when polymerised with small amounts of divinyl benzene results in a cross-linked polymer:

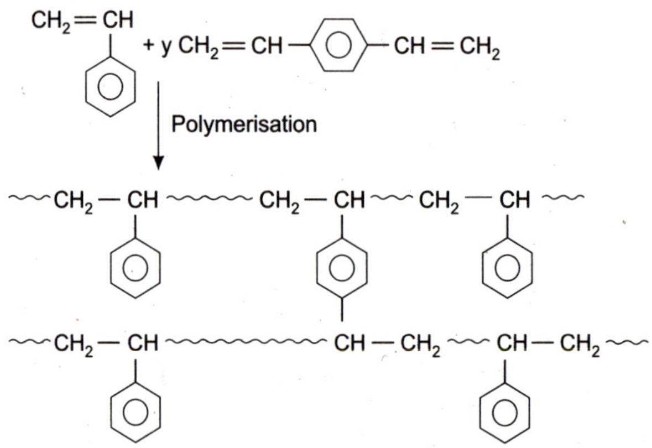

Similarly, incorporation of a trifunctional monomer along with the difunctional monomers could lead to cross-linking during a polycondensation reaction as shown:

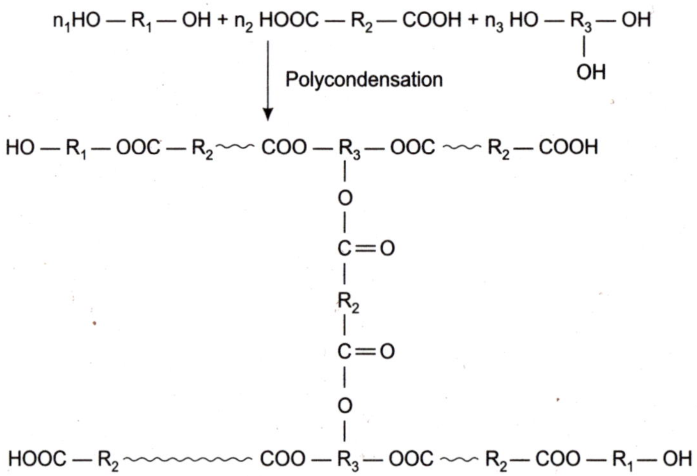

11.9.2 Vulcanisation

Vulcanisation is very widely used in rubber technology to effect cross-linking.

The vulcanisation process was discovered by Goodyear in 1839. He used sulphur to cross-link polyisoprene molecules in natural rubber. Ever since, many studies have been carried out on vulcanisation of natural as well as synthetic rubbers. Many reactions are probable. For instance, sulphur, or sulphur-related compounds react with rubber molecules to effect cross-linking as follows:

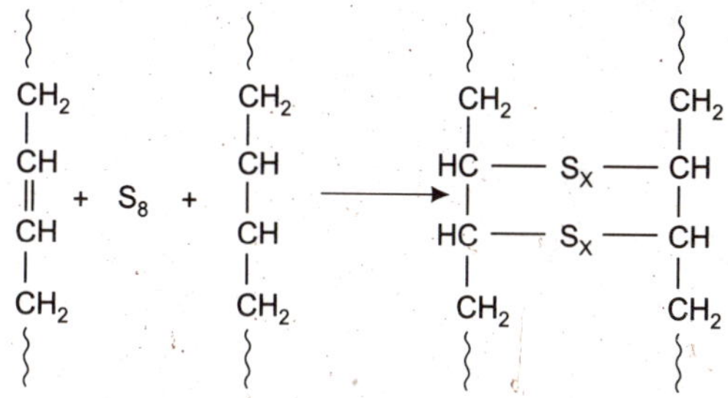

Sulphur can also bring about vulcanisation without affecting the double bonds in the polymer molecule, by eliminating H_2S:

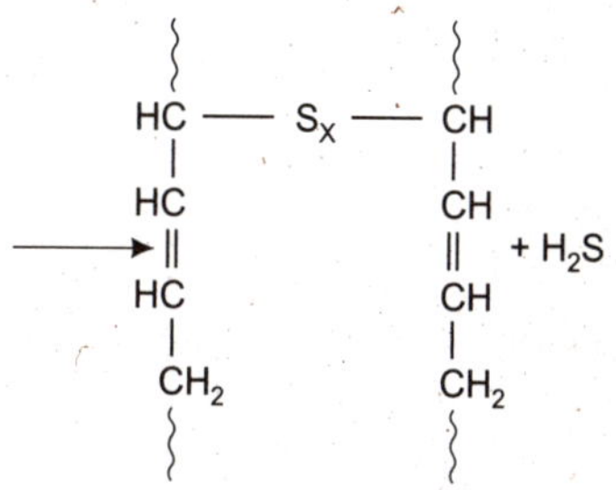

S_2Cl_2 can also be used to give the following cross-linking reaction:

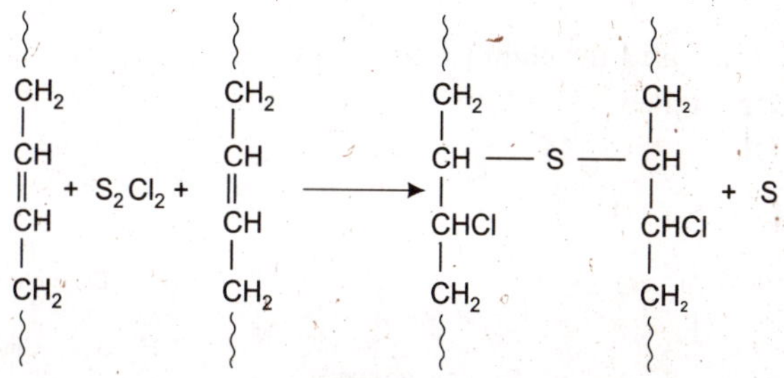

The liberated S can further affect cross-linking.

There is yet another kind of vulcanisation where certain sulphur compounds such as thioacids, mercaptans or mercaptoacids add to the double bonds of the elastomer as follows:

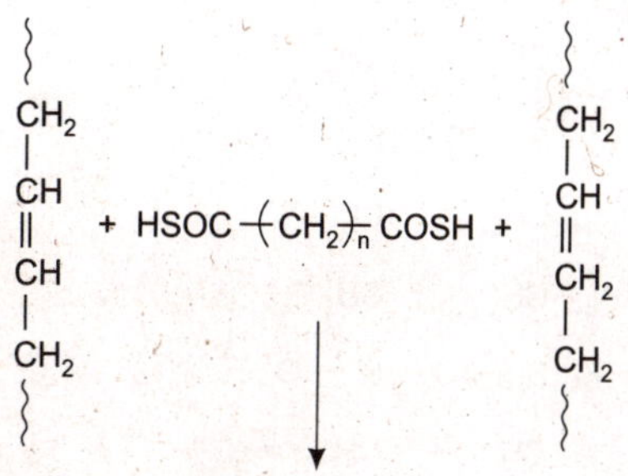

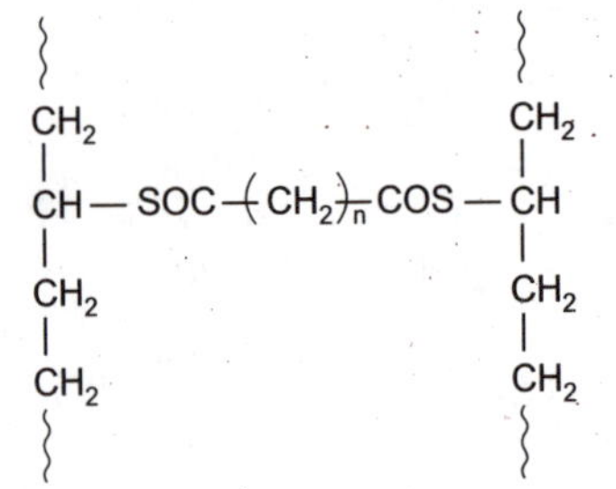

In the above case, the cross-link is effected between two linear chains through a single bisthiol acid.

Elastomeric cross-linking can also be brought about without employing sulphur. Thus, for instance, polychloroprene can be cross-linked with zinc oxide, as follows:

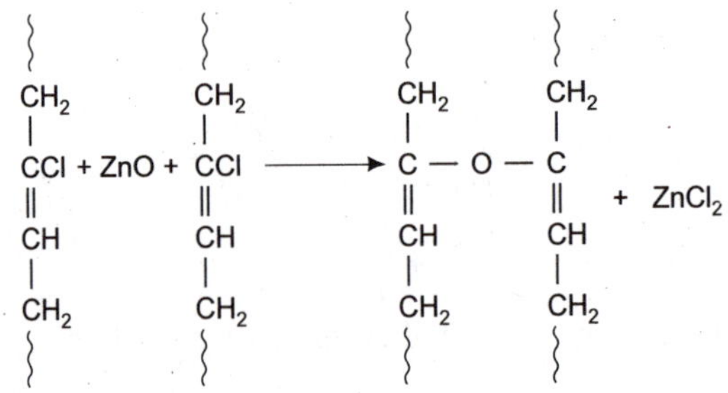

Silicone rubbers are vulcanised through a free-radical mechanism, either by employing benzoyl peroxide or by radiation techniques. The ultimate cross-linking may be as follows:

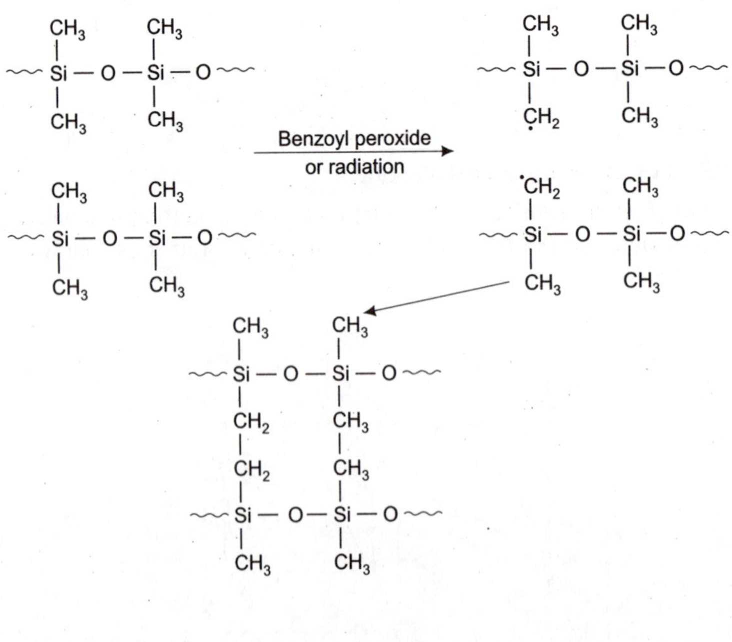

11.9.3 Cure Reactions

Cross-linking in certain polymers can be also effected through what are known as *'cure reactions'*. These uncross-linked linear polymers contain either reactive functional groups or double bonds in their molecules. They are usually not of a very high molecular weight and hence, are in the form of liquid resins. Such liquid polymers possessing reactive functional groups or double bonds are referred to as 'prepolymers'. When these prepolymers are reacted with low molecular weight or polymeric substances containing appropriate functional groups capable of reacting with active groups of the prepolymer, *'curing'* takes place, resulting in a cross-linked solid polymeric mass. The low molecular weight or the polymeric material used to bring about the cure reaction is called the *'curative'* or *'curing agent'*. The functionality of both the prepolymer and the curative plays an important role in determining whether the reaction between them will result in cross-linking or not. For cross-linking to take place, both the prepolymer and the curative should not only have a minimum functionality of two, but one or both of them should have some trifunctional molecules also. Unless such molecules are present in the system, no cross-linking takes place but only the linear coupling of the prepolymer molecules through the curative molecules results. Linear coupling will result only in another linear molecule of a longer chain length. The trifunctional molecules provide for sites where linear molecules can be connected through interchain links leading to cross-linking. The amount or degree of cross-linking depends upon the ratio of the bifunctional to the trifunctional molecules present in the system.

Some important examples of cure reactions are enumerated in the forthcoming sections. To avoid complicated chemical formulae, the repeat units of the polymer backbone are shown by a wavy line and the functional groups alone are indicated by chemical formulae. (We have already come across the complete chemical formulae of the prepolymers in Chapter 9.)

(*i*) Curing of an unsaturated polyester resin using styrene and a free-radical initiator.

(*a*) Prepolymer structure:

(*b*) Curative structure:

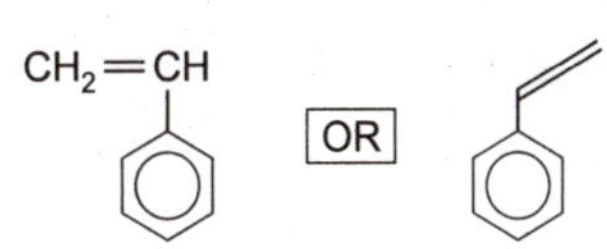

(*c*) Cure reaction:

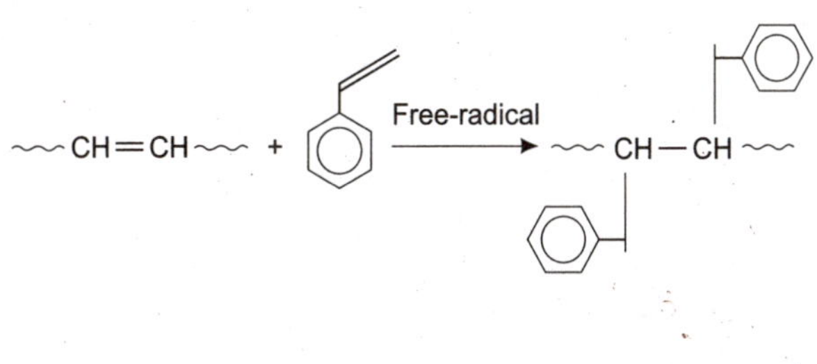

(*d*) Cured network structure:

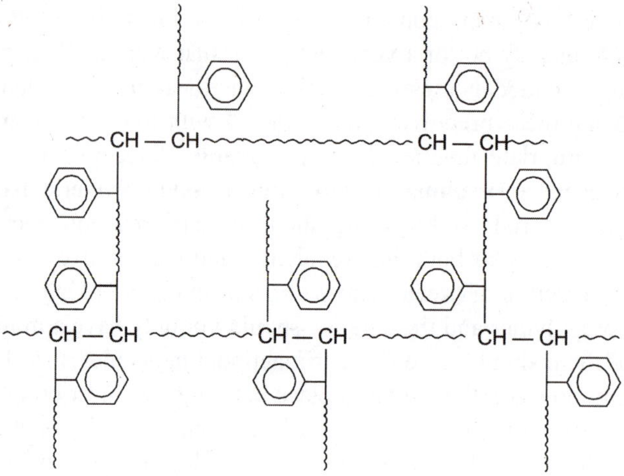

〰 = Polyester backbone

〰〰 = Polystyrene backbone

(*ii*) Curing of the thiol-terminated polysulphide resin with lead dioxide.

(*a*) Prepolymer structure:

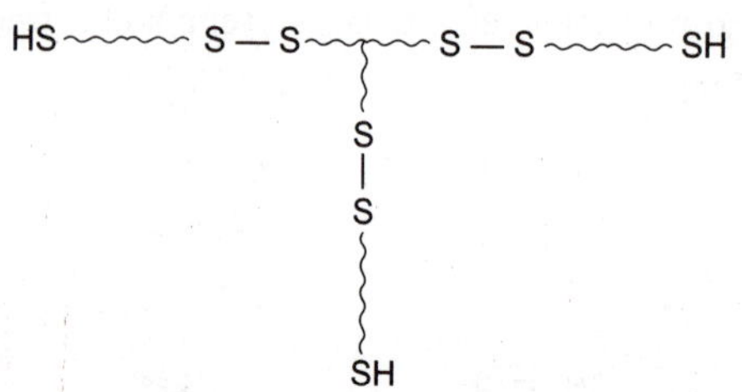

(*b*) Curative structure:

PbO_2

(*c*) Cure reaction:

$$\sim\sim SH + PbO_2 + HS \sim\sim$$
$$\downarrow$$
$$\sim\sim S-S \sim\sim + PbO + H_2O$$

(*d*) Cured network structure:

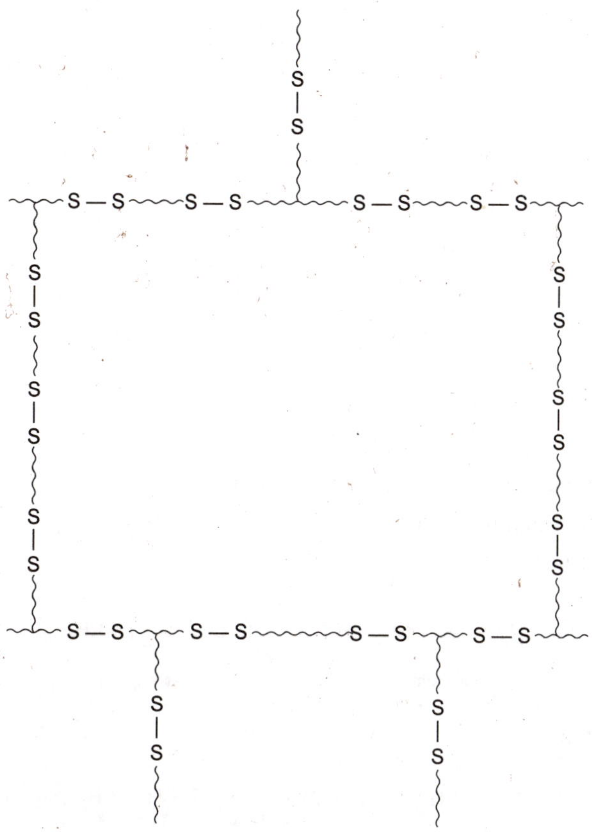

(*iii*) Curing of a thiol-terminated polysulphide resin using a diepoxide

(*a*) Prepolymer structure: Same as in *ii* (*a*).

(*b*) Curative structure:

(*c*) Cure reaction:

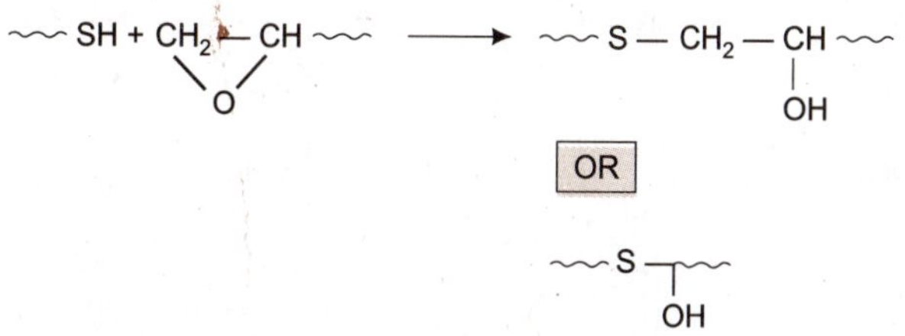

(*d*) Cured network structure:

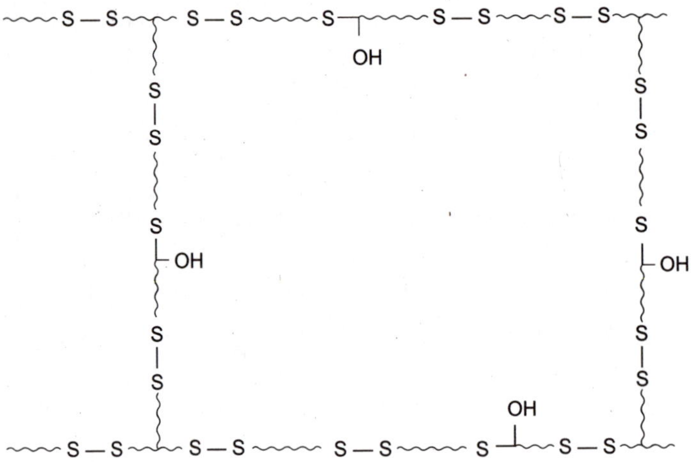

(*iv*) Curing of a diepoxide resin using a primary diamine.

(*a*) Prepolymer structure: Same as *iii* (*b*).

(*b*) Curative structure:

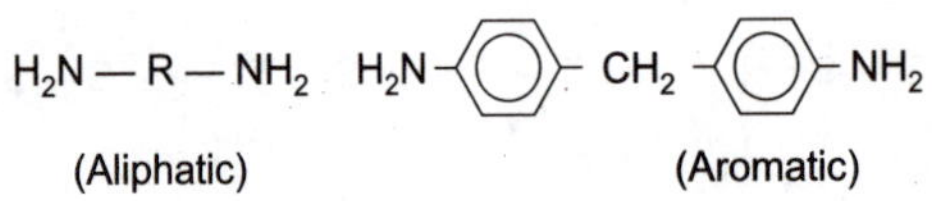

(*c*) Cure reaction:

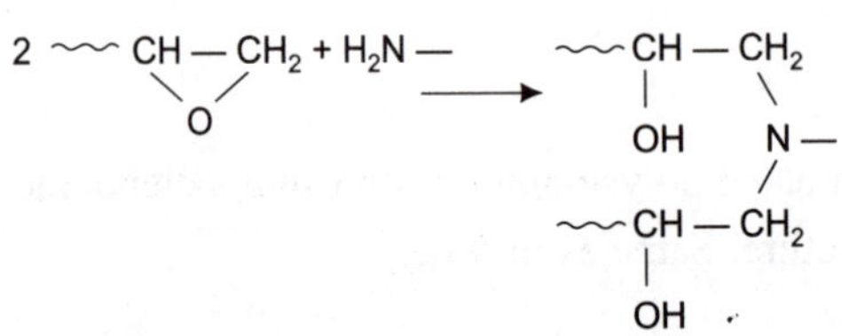

(*d*) Cured network structure:

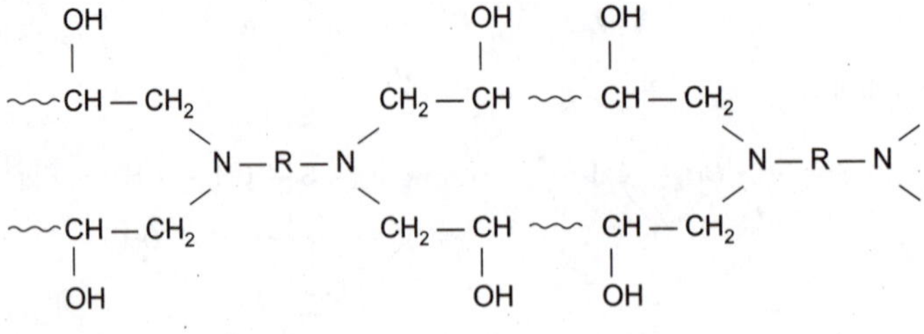

OR

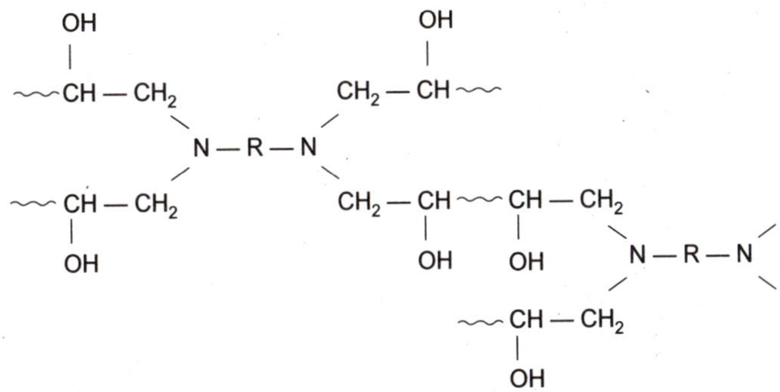

Under favourable conditions, the hydroxyl groups formed during the cure reaction can also react further with more molecules of the epoxy resin and introduce further cross-linking:

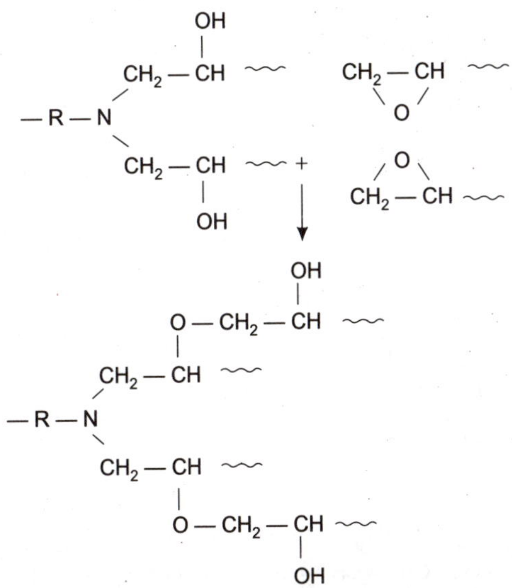

(*v*) Curing of a carboxyl-terminated prepolymer with a triepoxide or a triaziridine.

(*a*) Prepolymer structure:

HOOC ~~~~~~ COOH

(*b*) Curative structure:

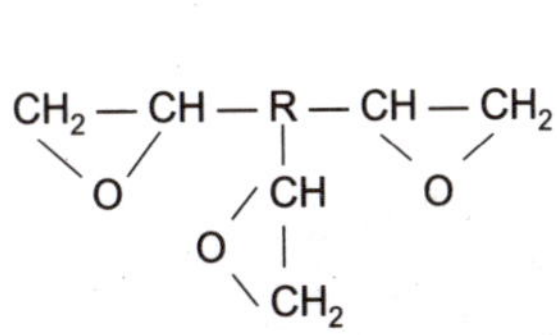

(Triepoxide)

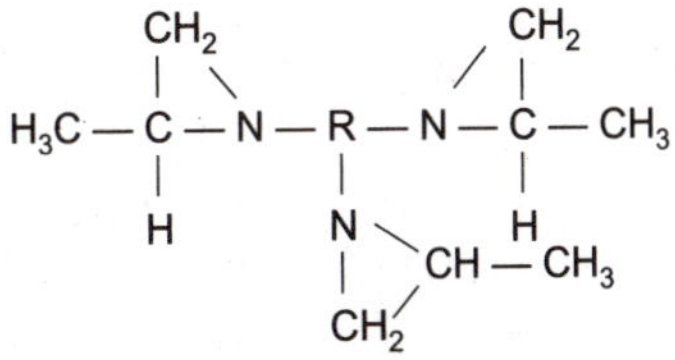

(Triaziridine)

(*c*) Cure reaction:

(*i*) with epoxy

(*ii*) with aziridine

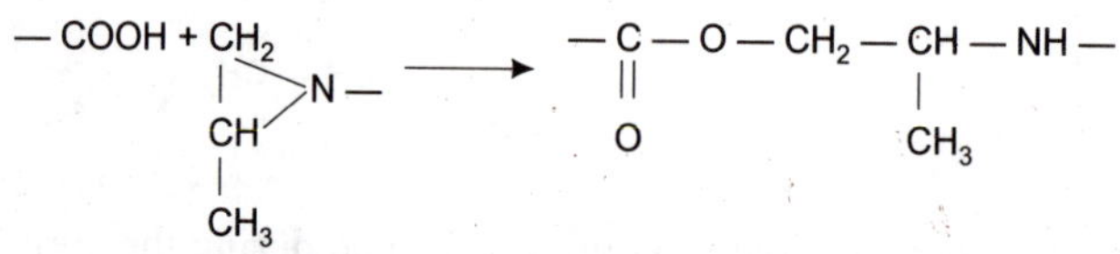

(*d*) Cured network structure:

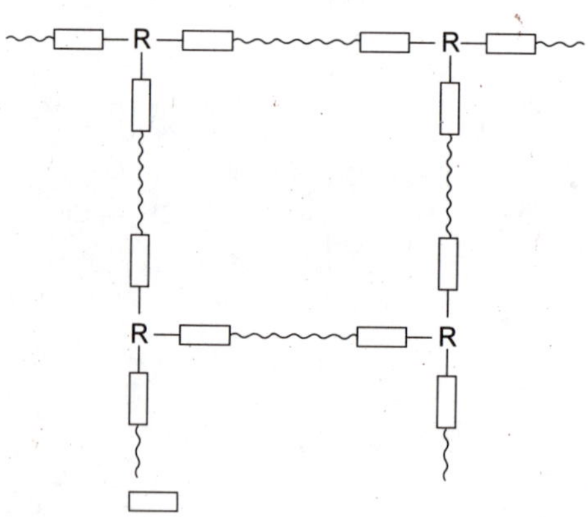

where ▭ represents either

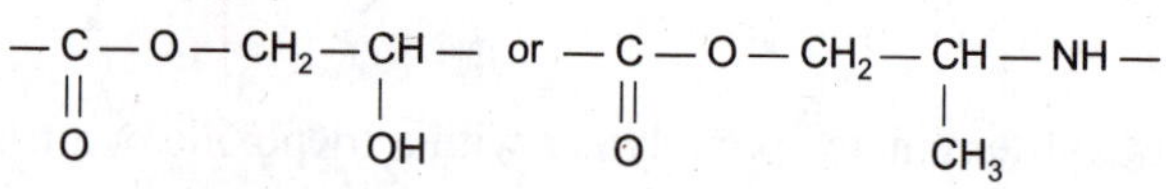

(*vi*) Curing of a prepolymer containing carboxylic side groups using a diepoxide or a diaziridine.

(*a*) Prepolymer structure:

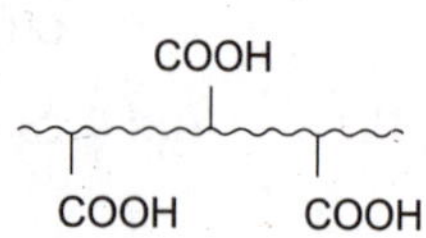

(*b*) Curative structure:

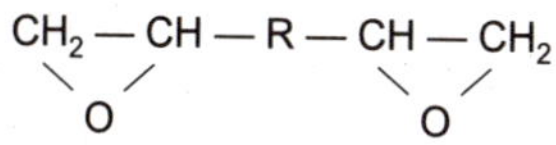

(Diepoxide)

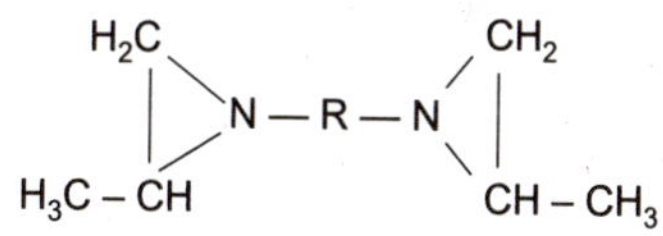

(Diaziridine)

(*c*) Cure reactions:

(*i*) with epoxy [same as *v* (*c*)].

(*ii*) with aziridine [same as *v* (*c*)].

(*d*) Cured network structure:

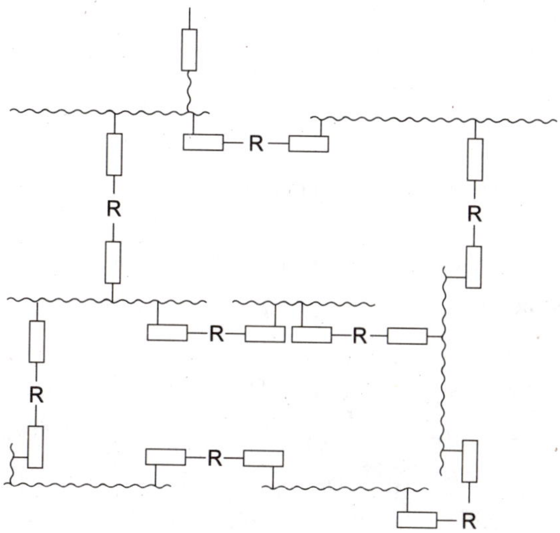

(*vii*) Curing a mixture of dihydroxy and trihydroxy polymers (*i.e.*, diol and triol polymers) using diisocyanates.

(*a*) Prepolymer structure:

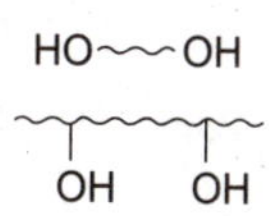

(Diol polymers)

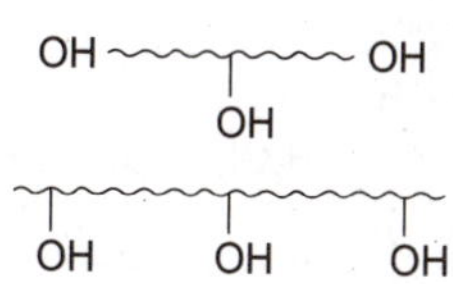

(Triol polymers)

(*b*) Curative structure:

$$O{=}C{=}N{-}R{-}N{=}C{=}O$$

(*c*) Cure reaction:

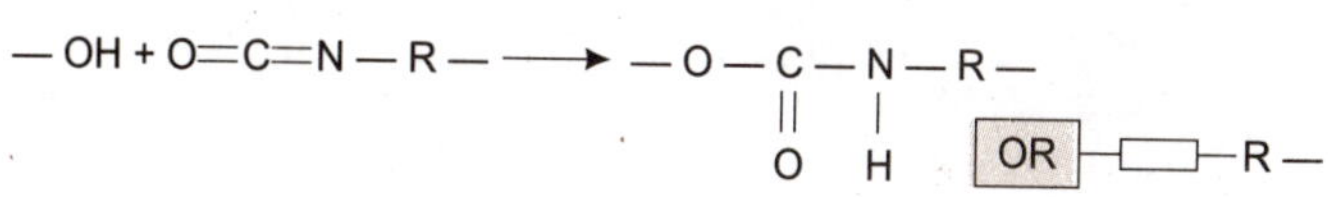

(*d*) Cured network structure:

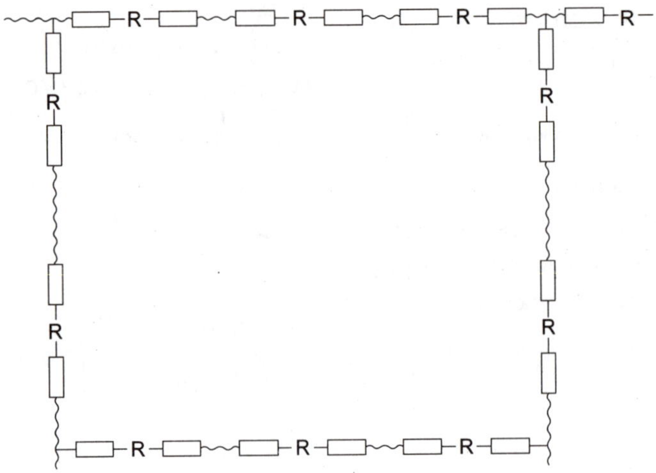

Similarly, a diisocyanate-terminated polymer can be cured with low molecular weight trihydroxyl compounds.

(*viii*) Curing of phenol-formaldehyde resins (resol type) by heating

(*a*) Prepolymer structure:

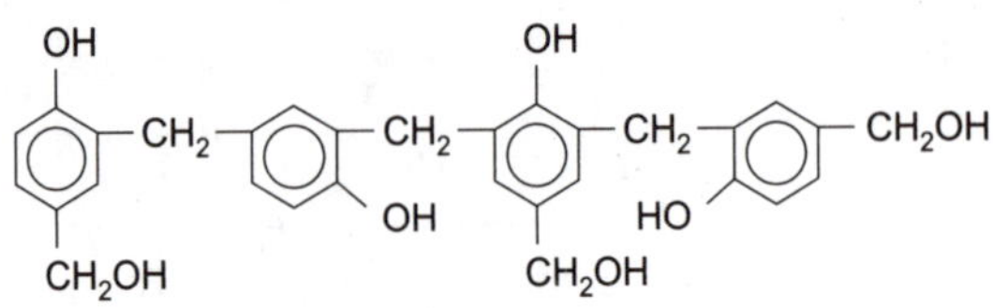

(*b*) Curative structure: no extraneous curative need be added.

(*c*) Cured reaction:

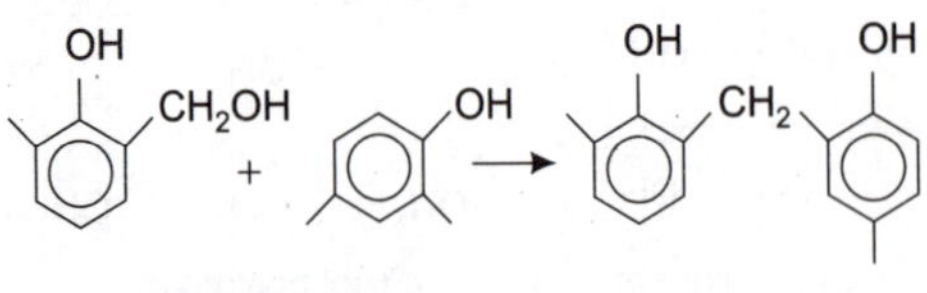

(*d*) Cured network structure:

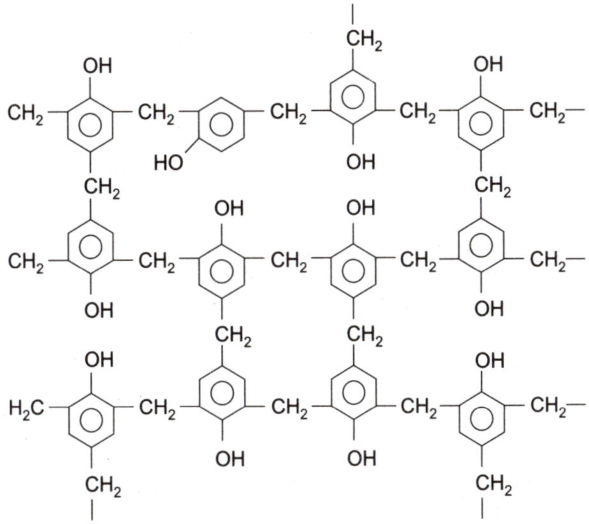

(*ix*) Curing of phenol-formaldehyde resins (novolak type) with formaldehyde.

(*a*) Prepolymer structure:

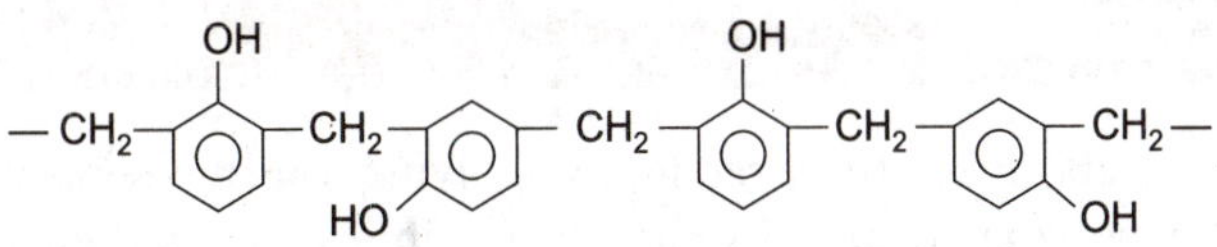

(*b*) Cure reaction:

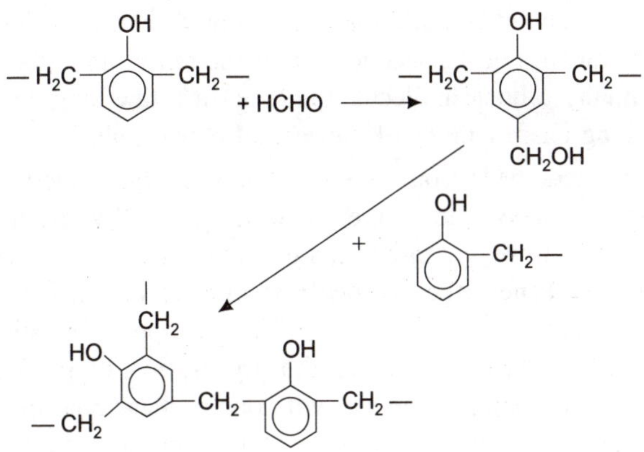

(*c*) Cured network structure: Same as *viii* (*d*).

(*x*) Curing of phenol-formaldehyde resins (novolak type) with hexamethylene tetramine (also known as *urotropin* or *hexa*).

(*a*) Prepolymer structure: Same as in ix (*a*).

(*b*) Curative structure:

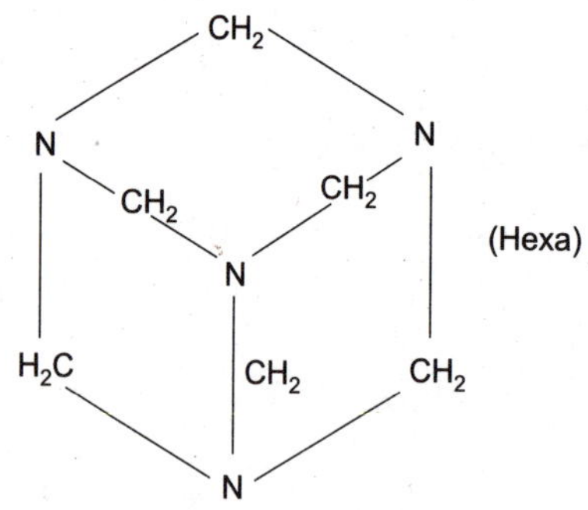

(*c*) Cure reaction:

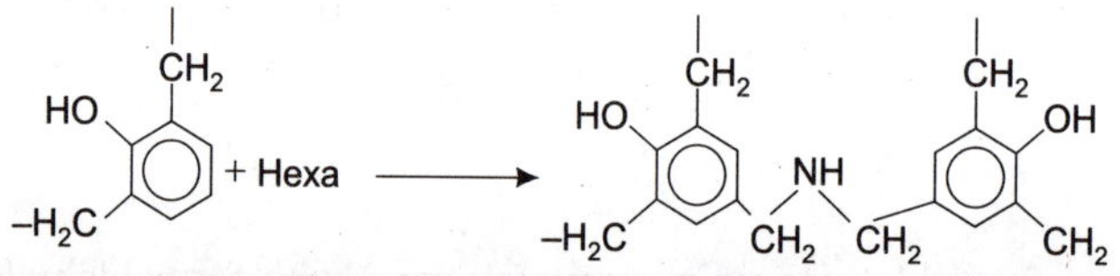

11.10 REACTIONS LEADING TO GRAFT AND BLOCK COPOLYMERS

Graft and block copolymers differ from random copolymers in the sequence of arrangement of the repeat units in the polymer chain. As the term implies, in random copolymers (linear or branched), the repeat units are distributed throughout the chain in a random fashion (except in alternating copolymers where the two units alternate strictly). A graft copolymer is a branched molecule wherein the main chain or the backbone chain is made entirely of one type of the repeat unit with short side chains or branches of the second type of the repeat unit attached at random points. A block copolymer is essentially a linear molecule made up of a few long chain lengths of one repeat unit separated by a few long chain lengths of the second repeat unit.

Graft and block copolymers made from two monomers exhibit properties that are quite different from the random copolymers made of the same monomers as well as from the two homopolymers. Methods of preparation are, however, rather tedious and hence, they have not assumed much industrial importance. We shall, nevertheless, describe in brief the various reactions used for their preparation in the following sections. The methods of direct copolymerisation of the two monomers to get random copolymers are not suitable for the preparation of graft and block copolymers. One usually starts with the two homopolymers or with one primary homopolymer and a comonomer and arrives at the block and graft polymers by one of the following techniques.

11.10.1 Grafting by Means of Free-Radicals

Primary homopolymer (PHP) is dissolved in the comonomer along with a free-radical initiator and polymerised thermally. The comonomer first starts polymerising and some of the growing polymer chains abstract a labile atom from the PHP molecule, creating a free-radical site on the same.

Comonomer units can now add on at this free-radical site, forming a graft growth. Similarly, some of the growing chains of the comonomer can also get terminated by coupling at this free-radical site. In the process, however, some homopolymer of the comonomer will also be formed at the time of creating a free-radical site on the PHP backbone. Since the process involves a chain-transfer mechanism, homopolymers made of monomeric units with 'labile' atoms are useful as the primary homopolymer on which a comonomer can be grafted. Grafting of styrene on polybutadiene is done by the following method:

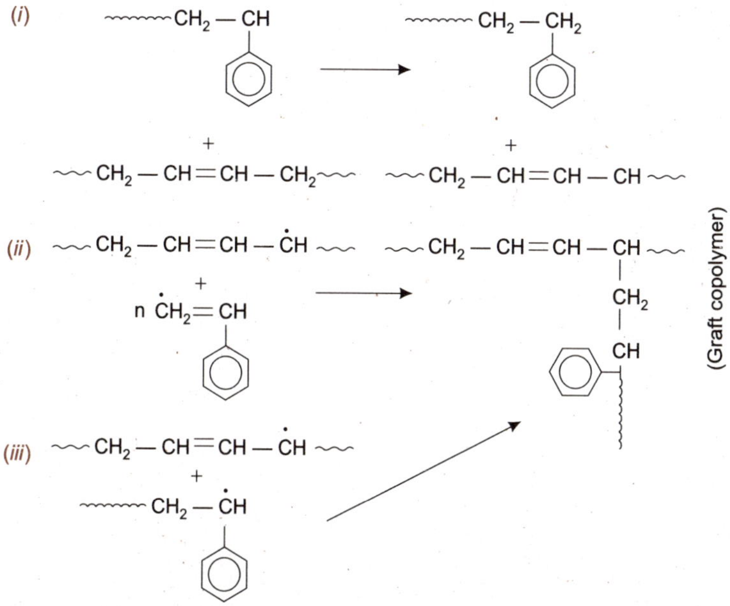

Free-radical sites can also be created on the primary homopolymer chain by irradiating a solution of the same in a comonomer:

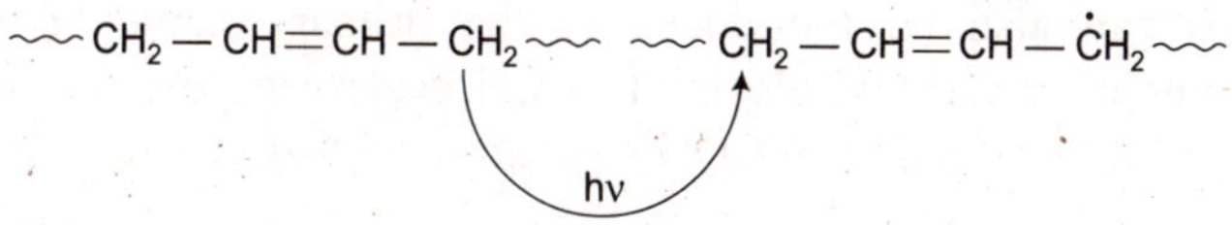

Yet another method of creating free-radical sites on the primary homopolymer is to incorporate reactive groups such as hydroperoxide groups and then to decompose the same:

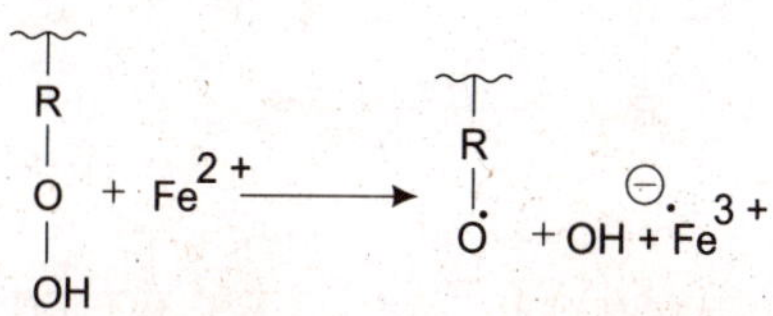

11.10.2 Grafting through Functional Groups

Graft polymer can be formed by a reaction through the functional groups in homopolymers. The primary homopolymer is made with one type of the functional group distributed randomly in the chain and the secondary homopolymer is made with another type of the functional group (that can react with the one present in the primary homopolymer) located at one end of the chain. By a suitable reaction between the two types of functional groups, the secondary homopolymer is grafted to the primary, resulting in a graft copolymer:

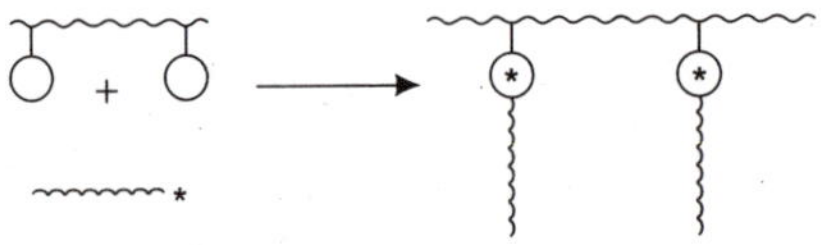

(O and * are functional groups.)

Functional groups can be introduced at random in the primary homopolymer by several methods. One method is to copolymerise the monomers with a small percentage of another monomer carrying the required functional group. For example, methyl methacrylate can be copolymerised with acrylic acid to get a primary homopolymer with a few —COOH groups placed randomly on the chain:

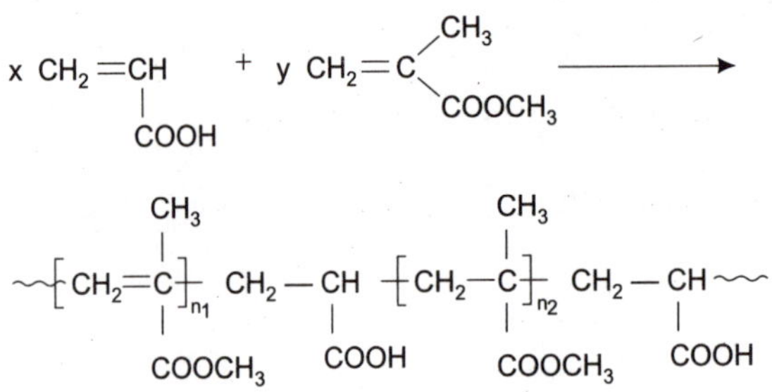

Similarly, hydroxyl, epoxy, isocyanate or acid chloride groups can be introduced by copolymerising monomers containing these functional groups.

Functional groups can also be introduced in the primary homopolymer by a chemical modification. For example, partial hydrolysis of —CN groups in polyacrylonitrile, or —COOR groups in polyacrylates, can give rise to —COOH groups, and that of vinyl esters to —OH groups:

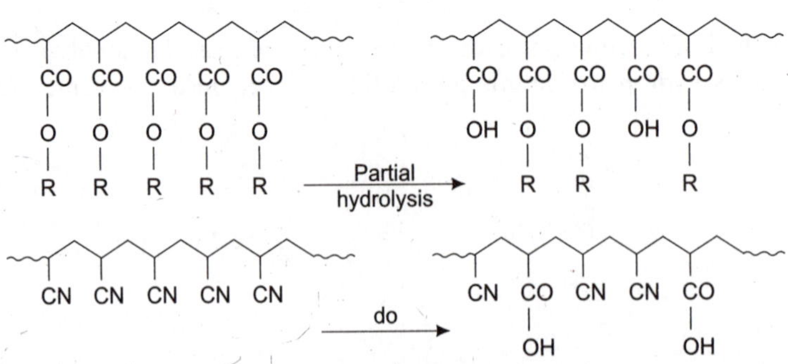

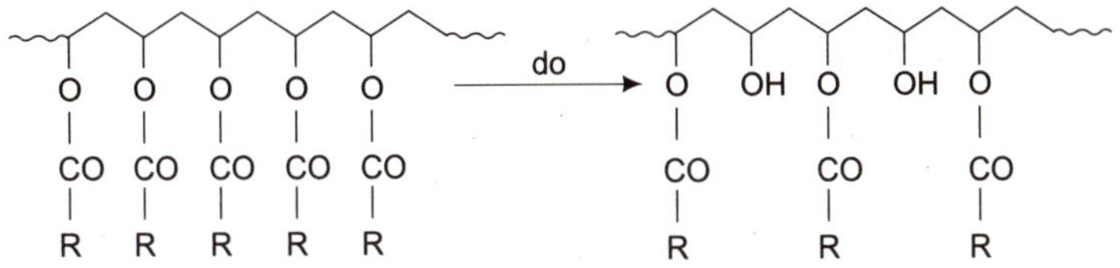

Polymers with functional groups at the chain terminal can be obtained by using an initiator containing functional groups. When the initiator decomposes, free-radicals having functional groups are formed:

(*i*) $HOOC-R-N{=}N-R-COOH \xrightarrow{\text{Heat}} 2HOOC-\dot{R} + N_2$

(*ii*) $HOOC-R-\underset{\underset{O}{\|}}{C}-O-O-\underset{\underset{O}{\|}}{C}-R-COOH \xrightarrow{\text{Heat}} 2HOOC-\dot{R} + 2CO_2$

(*iii*) $H_2O_2 \xrightarrow[\text{pressure}]{\text{Heat}} 2\dot{O}H$

Polymer chains grown with these free-radicals carry a functional group at one end. When chain termination occurs by coupling, a polymer with functional groups at both the ends is obtained. If the termination occurs by disproportionation, even then, one end of the polymer chain contains the functional group. For example, polymerisation of monomers with azobis cyanovaleric acid (ACVA) or succinic acid peroxide (SAP) results in polymers carrying carboxylic groups at the chain end, while polymerisation with hydrogen peroxide under suitable conditions can yield polymers with hydroxyl end groups. Functional groups at the chain end can also be introduced by anionic polymerisation. For example, a polymer chain grown with the anionic initiator butyl lithium can be terminated with carbon dioxide and hydrolysed with dilute hydrochloric acid to get a polymer with carboxyl group at the chain end:

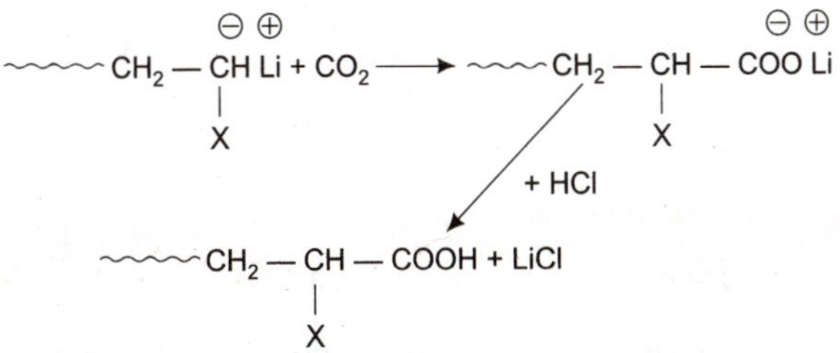

If the same growing anionic chain can be terminated with ethylene oxide and followed by hydrolysis a polymer with hydroxyl terminals can be obtained:

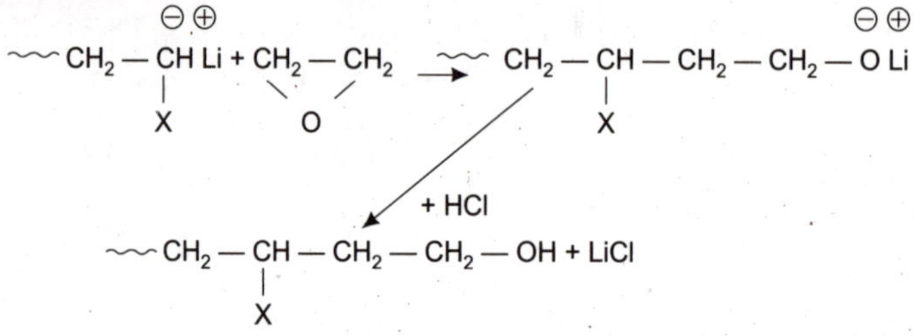

11.10.3 Block Copolymers through Anionic Polymerisation

While discussing anionic polymerisation, we have mentioned that if all impurities are carefully avoided the polymer chain growth continues till the entire monomer is consumed and then stops without any termination occurring, but still retaining the active carbanion site at the chain end. This phenomenon can be advantageously used for preparing block copolymers. The polymer chain is grown first using one monomer when a homopolymer chain still retaining the active end is formed. To this, the second monomer is added when the chain continues to grow, forming a polymer block of the second monomer. Again, when another lot of the first monomer is added, the block of the first monomeric units is added on to the chain. In this way, by adding alternately the two monomers in small lots, alternating blocks of the two monomeric chains can be introduced. The chain length of the alternating blocks can be controlled by adjusting the concentration of the initiator and the quantity of the monomer lots added. The thermoplastic styrene-butadiene block copolymer is made by this process on an industrial scale.

Block copolymers can also be prepared through functional groups. In this case, however, both primary as well as secondary homopolymers should have reactive functional groups attached at both the ends of the polymer chains:

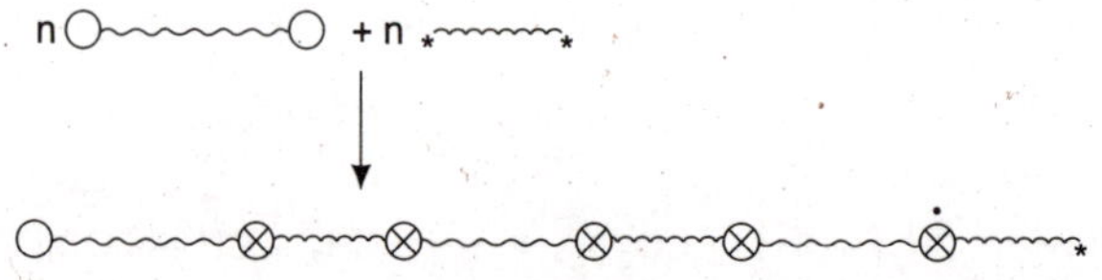

(Block copolymer)

11.11 MISCELLANEOUS REACTIONS

Makoto Okawara and others have studied several interesting reactions with polymers, some of which are now discussed.

11.11.1 Modification of Commercial Polymers

A reactive acid anhydride group can be introduced on to polystyrene by reacting the same with maleic anhydride. The reaction is brought either by a peroxide or by UV irradiation. The addition of the anhydride can either be on the chain backbone or on the aromatic side group:

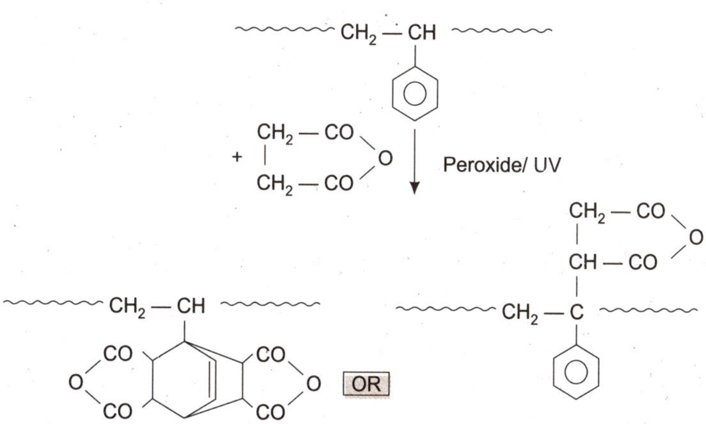

This is an example where an unreactive commercial polymer can be converted into a reactive polymer.

Similarly, a reactive aromatic group can be introduced into the commercial nylon 6:

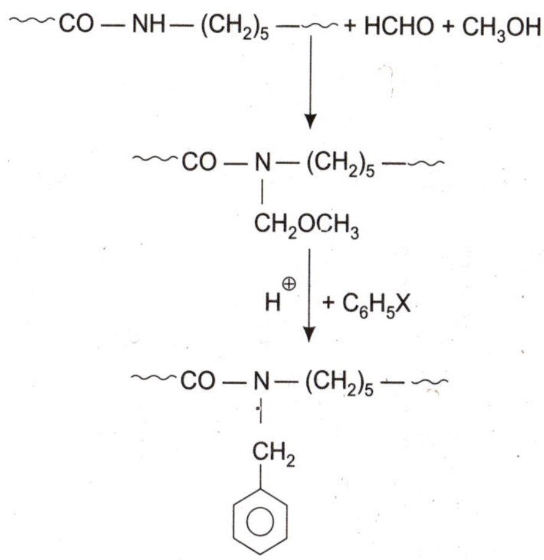

11.11.2 Introduction of Metal-Carbon Bond in Polymer Structures

The Ni—C sigma bond can be introduced in a polymer structure as shown:

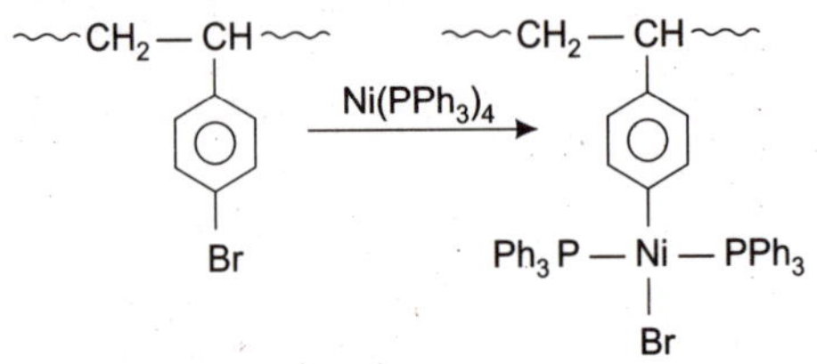

The above polymer is useful as a catalyst for the dimerisation of ethylene.

11.11.3 Substitution Reactions of Commercial PVC

We know that the chlorine atom in PVC can be easily eliminated as HCl by just heating. But the nucleophilic substitution of chlorine is rather difficult. Nevertheless, the same has been achieved and a number of substituted PVC derivatives have been made. We shall now take up a few of them.

(*i*) Substitution with, N, N-dialkyl dithiocarbamate has been done at 50–60°C in a dimethyl formamide medium (without any accompanied degradation of the polymer by dehydrochlorination):

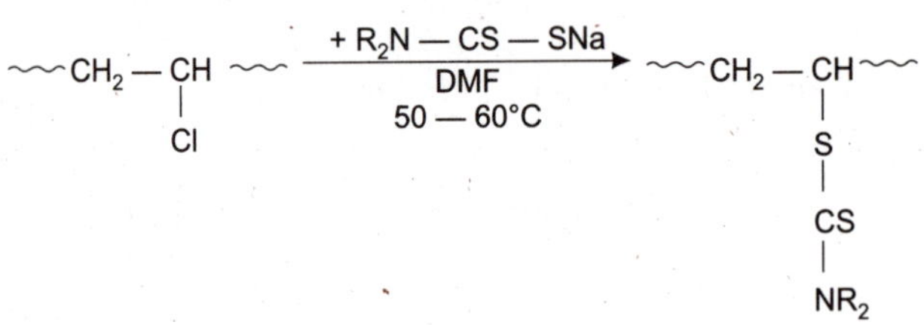

The above polymer can very easily be cross-linked by UV irradiation:

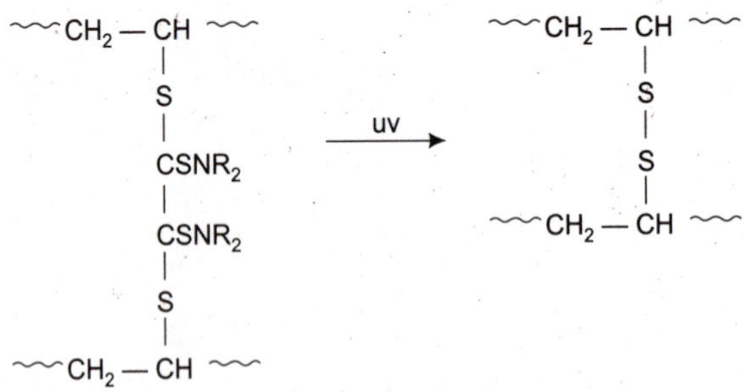

(*ii*) Similarly, a weak nucleophile like an azide anion ($\overset{\ominus}{N_3}$) also can substitute the Cl in PVC to give the azide derivative as shown. This polymer is capable of reacting with various reagents to give different polymers (as shown in structures a, b, c and d):

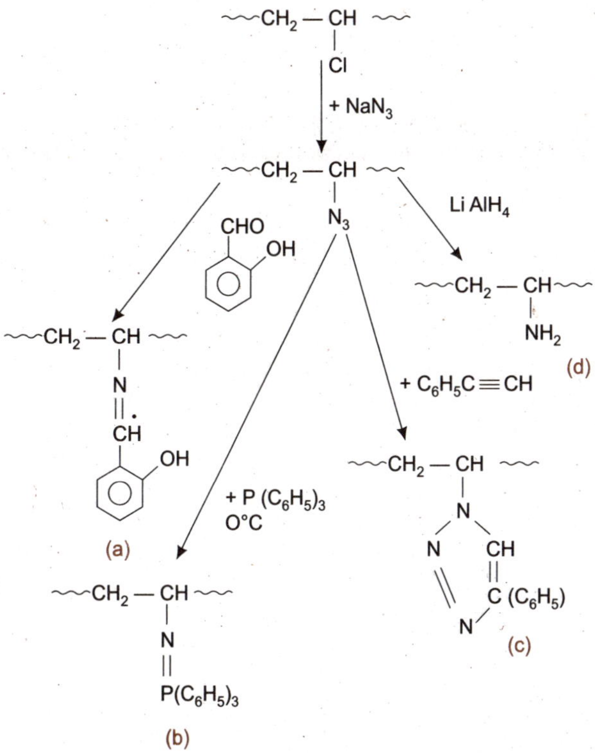

11.11.4 Photoreactive Polymer Containing the Benzoin Structure

The following scheme has been worked out by Makoto Okawara to prepare a photoreactive polymer:

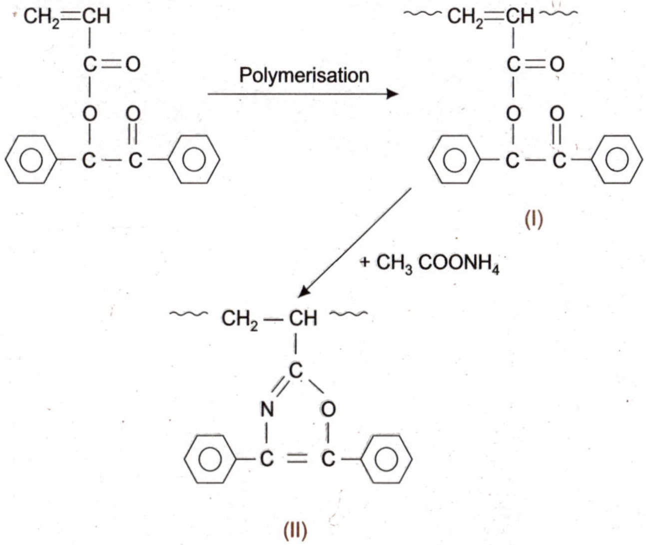

CHAPTER 11

These two polymers (structures I and II) exhibit certain interesting properties. The first polymer can be employed as a polymeric photosensitizer for photografting and photocross-linking. The second polymer exhibits intense fluorescence.

Books for Further Reading

Burlant, N.J. and Hoffman, A.S., *Block and Graft Copolymers*, Reinhold, New York, 1960.

Ceresa, R.J., *Block and Graft Copolymers*, Butterworth, London, 1962.

Fettes, E.M. (Ed.), *Chemical Reactions of Polymers*, Wiley-Interscience, New York, 1964.

Seligh, G. and Sjothun, I.J., *Vulcanisation of Elastomers*, Reinhold, New York, 1964.

Battaerd, H.A.J. and Treglar, G.W., *Graft Copolymers*, Wiley-Interscience, New York, 1967.

Moore, J.A. (Ed.), *Reactions on Polymers*, Reidel, Dordrecht, 1973.

12

POLYMER SOLUTIONS

12.1 THE PROCESS OF POLYMER DISSOLUTION

In Chapter 1, we have highlighted the striking difference between a polymer and a low molecular weight substance as far as their solubility characteristics are concerned. While low molecular weight substances can readily undergo dissolution, polymers take considerable time to do so. Even at fairly low concentrations, polymer solutions are highly viscous, compared to solutions of low molecular weight substances, which are far less viscous even at high concentrations.

In this section, too, pursue the examples given in Chapter 1. Add sodium chloride (a low molecular weight substance) to water in a beaker and stir. The salt crystals reduce in size and finally disappear. Obviously, sodium chloride molecules leave the crystal, lattice and disperse in water, forming a sodium chloride solution. (Actually, sodium chloride is ionised into Na^+ and Cl^- ions which are uniformly dispersed in water but, for simplicity, sodium chloride is described as being dispersed in water as molecules). As soon as the solute comes in contact with the solvent, the solute molecules start diffusing into the solvent phase; there is thus a progressive reduction in the volume of the solute phase, which ultimately merges with the solvent phase and forms the solution phase. Low molecular weight solids in solvents dissolve faster when stirred or agitated. Now, add solid polyvinyl alcohol to water and stir. For quite some time, nothing seems to happen to the solid fragments despite vigorous stirring. Slowly however, the fragments swell and increase in size. As against the fast dissolution undergone by the low molecular weight solutes, polymer fragments take time before undergoing any visible change. In the former case, the individual molecules (and in many cases, the ions) are discrete and separate from one another and are held together within the solid matrix only by van der Waal's forces or electrostatic forces of attraction. As soon as the solvent and the solute are brought together, the solvent molecules surround the solute molecules at the surfaces, establish solvent solute interaction and break the solute-solute attraction. As this happens the solute molecules, which are discrete, are isolated from the solid phase and as their size is comparable to that of the solvent molecules, they diffuse fast into the solvent phase. Dissolution here is almost instantaneous.

Polymer molecules, on the other hand, are giant molecules as compared to solvent molecules and are made up of hundreds of chain segments. These long molecules, moreover, are not in the form of extended straight chains, but are in the form of tightly folded random coils. Individual molecular coils are also not discrete and separate but are interpenetrating and entangled with one another. There are also varying degrees of cohesive and attractive forces between different segments of the same molecular coil as well as neighbouring coils. Forces such as dispersion, induction, dipole-dipole interaction and hydrogen bonding (both intramolecular and intermolecular) hold the molecular coils and their segments together tightly. Because of the large size and coiled nature of the polymer molecules and also because of strong forces of attraction between them, solvent molecules take time to establish interactions with polymer molecules, to overcome the forces of attraction, to release individual molecules out of the chain entanglement and get them out of the polymer phase. This explains the difference in the dissolution behaviour of low and high molecular weight substances.

Let us now take up, the process of polymer dissolution in detail and for the sake of simplicity, let us consider only polymers comprising linear flexible chains. When a polymer fragment is added to a solvent, forces of attraction or dispersion start acting between them, depending on their chemical nature, polarity and solubility parameter.* When this solvent-solute interaction becomes more powerful than the solute-solute attraction, the forces holding the polymer segments together become weak and the solvent molecules force their way between the segments, break the segment-segment contacts, surround individual segments and establish contact with them. In this process, the segments unfold or loosen out from their tightly coiled conformation. In other words, the segments which were *'aggregated'* in the absence of a solvent get now *'solvated'*.

The onset of the solvent-solute interaction and the resultant *'solvation'* and *'unfolding'* of the segments constitute a slow process. To begin with, a few segments of the coil here and there unfold and get solvated and the intersegmental space thus vacated is *'raided'* by more solvent molecules. As more solvent penetrates the intersegmental space, the solvent-solute interaction increases and more of the coil segments unfold and get solvated, slowly swelling up the polymer matrix.

The first stage in polymer dissolution is, therefore, characterised by a slow penetration of the solvent molecules into the interstices of the polymer coil and forcing them to *'swell'*. During this swelling stage—one might call it a pseudo-induction period—the volume of the polymer matrix increases and the solvent molecules leave the solvent phase and diffuse into the polymer matrix. The polymer molecules, on the other hand, remain within the matrix itself and do not diffuse out. The phenomenon of swelling depends purely on the forces of interaction between solvent molecules and polymer segments and is not influenced by stirring or agitation. The process can, however, be hastened by exposing more polymer area for solvent interaction. It is for this reason that in order to dissolve a polymer, it is put in the solvent in as many small pieces as possible and left overnight for swelling to occur. As swelling continues, more and more segments of the polymer molecule are solvated and loosened out. When all segments are solvated, the molecule as a whole, in the form of a loose coil, separates out from the swollen polymer. The loosened polymer molecule then diffuses slowly out of the polymer phase and disperses in the solvent phase, forming the solution.

Since molecules in a solid polymer are entangled with the neighbouring ones, polymer molecules during dissolution do not diffuse out of the swollen polymer one after another, but a stream of

*For 'solubility parameter', see Section 12.2.4.

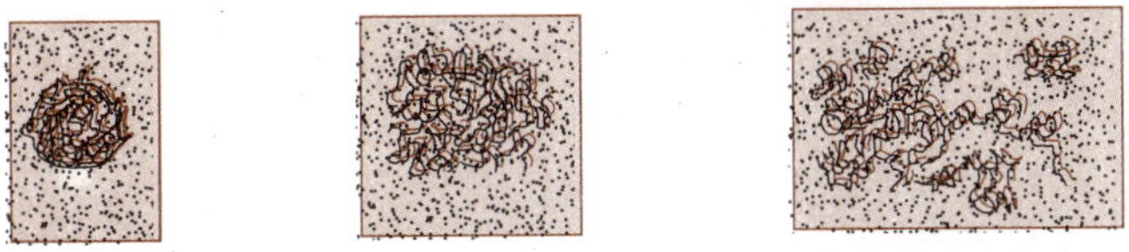

Fig. 12.1 Schematic illustration of different stages of polymer dissolution.

entangled molecules simultaneously diffuse and disperse in the solvent (Fig. 12.1). The rate of diffusion of polymer molecules into the solvent phase, and therefore the rate of dissolution, can be increased by agitation when the polymer is fully swollen. This stage of dissolution can be compared with the dissolution of low molecular weight substances.

Even though all the chain segments of a polymer molecule in solution are unfolded and fully solvated, the molecule does not assume the shape of an extended straight chain. The coiled nature of the molecule is still retained but with a very much expanded coil volume and with the empty space between the unfolded segments of the coil being occupied by solvent molecules. We can now visualise that the polymer coil, along with solvent molecules filled within, moves as a whole in the solvent phase, forming what is known as the *'hydrodynamic'* sphere or ellipsoid. The solvent held within the polymer coil can be called the *'bound'* solvent and that surrounding the coil as the *'free'* solvent. The *'bound'* solvent, however, is in equilibrium with the *'free'* solvent because solvent molecules diffuse in and out of the expanded polymer coil at a constant rate. The apparent volume occupied by the expanded coil (along with the *'bound'* solvent) is referred to as the *'hydrodynamic volume'* of the polymer molecule in the solution under flow.

The striking difference in the viscosities of a low molecular weight solution and a polymer solution can now be explained as follows: In the former case, the molecular size of the solute and that of the solvent are comparable and the solute does not swell in volume in the process of dissolution; there is a free molecular mobility with respect to the solute without much increase in intermolecular friction. Hence, the viscosity of the solution is not very much different from that of the solvent. In the case of polymer solutions, however, the solute molecules, which are gigantic compared to the solvent molecules, swell considerably in volume during the process of dissolution. The giant size and the increased volume of the polymer molecule restrict its molecular mobility in the solution and increase the intermolecular friction. The polymer solution, therefore, is highly viscous.

The states of solute molecules in low molecular weight and polymeric solutions are shown in Figs. 12.2 and 12.3.

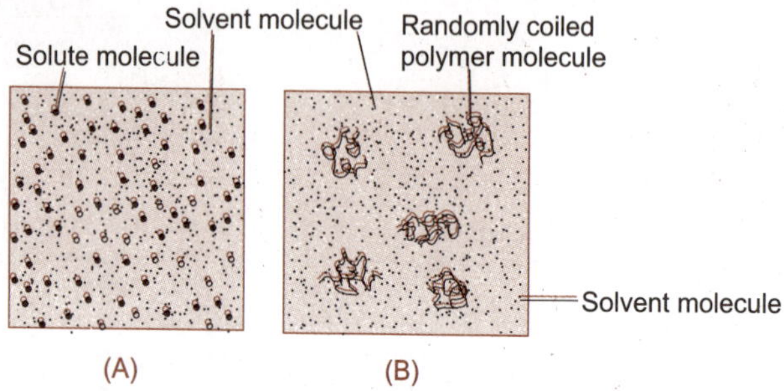

Fig. 12.2 Schematic representation of solute and solvent molecules present in dilute solutions of: (A) low molecular weight compounds and (B) polymers.

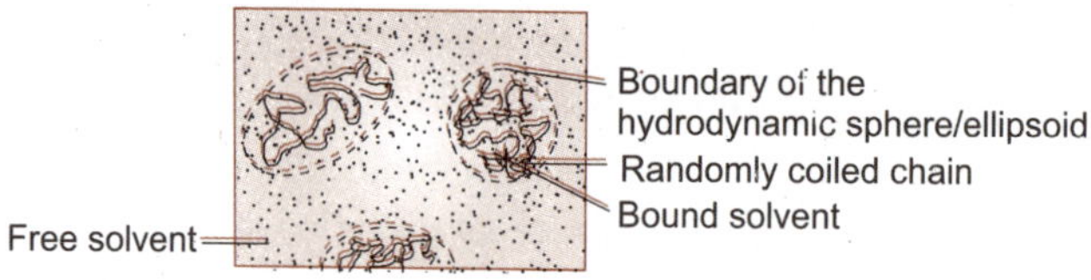

Fig. 12.3 Schematic details of dilute polymer solutions.

12.2 THERMODYNAMICS OF POLYMER DISSOLUTION

12.2.1 General Principles

Thermodynamic principles governing the dissolution of low and high molecular weight substances are essentially the same. Dissolution of a solute in a solvent involves a change of state and hence, is dependent on the changes in the entropy and the enthalpy of the system. We know that any given system tries to shift in the direction of a state in which the enthalpy or the heat content (H) is minimum, and/or the entropy or degree of disorderliness or randomness (S) is maximum. Any system can spontaneously shift from a state A to a state B if, in the process, there is a loss in the enthalpy ($-\Delta H$) and/or a gain in the entropy ($+\Delta S$). While the system changes from state A to state B, the associated entropy and enthalpy changes are independent of each other. During the process, the enthalpy can increase ($+\Delta H$), decrease ($-\Delta H$) or remain unchanged ($\Delta H = 0$). Similarly, during the same process, there can be either an increase or a decrease in the entropy factor. In the process, if there is a decrease in the enthalpy with a simultaneous increase in the entropy ($-\Delta H$ and $+\Delta S$), both of them support the change of state and the process occurs spontaneously. On the other hand, if there is an increase in the enthalpy along with a decrease in the entropy ($+\Delta H$ and $-\Delta S$), both the parameters oppose the change of state and the process does not occur. But what happens when both the entropy as well as enthalpy either increase or decrease? When such a situation arises, one of them supports the process, whereas the other opposes it, and the feasibility of the process depends on the magnitude of these counter factors. The resultant free energy change (ΔF) as per the equation will

$$\Delta F = \Delta H - T \cdot \Delta S \tag{12.1}$$

then decide the feasibility or otherwise of the process. When there is a resultant decrease in the free energy ($-\Delta F$), the process will occur.

After this recapitulation, let us first consider the process of dissolution of a low molecular weight solute. Dissolution always results in a positive entropy change or an entropy gain due to the freedom of mobility acquired by the solute molecules in the dissolved state as against the bound nature in the undissolved state. Hence, the only other parameter which will decide if the process can occur or not is the change in enthalpy. If in the process of dissolution, there is a decrease in the enthalpy (*i.e.*, when the heat of dissolution is negative or when it is an exothermic dissolution, $-\Delta H$), the ΔF values become more negative and the process becomes spontaneous. Similarly, the process of dissolution becomes spontaneous even when there is no heat change (*i.e.*, the process is neither exothermic nor endothermic) since the entropy gain alone results in a decrease in the free energy of the system. In a process where there is an increase in the enthalpy (*i.e.*, endothermic dissolution, $+\Delta H$), the process of dissolution will be opposed by the $+\Delta H$ factor, while the same is favoured by the $+\Delta S$, and the process can occur as long as $+\Delta H$ is less than $T{\cdot}\Delta S$. When $+\Delta H$ becomes equal to or greater than $T{\cdot}\Delta S$, the dissolution process cannot occur.

It can be seen from the above discussion, that whenever the dissolution process is exothermic, the solute is freely soluble in the solvent in all proportions. In endothermic dissolution, however, at the beginning of the dissolution process, the $+\Delta H$ value is less than the $T{\cdot}\Delta S$ value and that is why dissolution occurs in the first place. As the dissolution process proceeds, the entropy gain or $T{\cdot}\Delta S$ progressively decreases since the concentration of the solute phase is decreasing and that of the solution phase increasing. When the $T{\cdot}\Delta S$ value decreases and becomes equal to $+\Delta H$, ΔF becomes zero and hence, no further dissolution is possible. At this stage, an equilibrium is established between the entropy gain and enthalpy gain and the number of solute molecules going into solution equals those precipitating out. The solution at this point of equilibrium is said to be *'saturated'*. This is as far as dissolution of the low molecular weight solute is concerned. But what about the dissolution of macromolecular or polymeric substances?

The same basic principle applies to polymer dissolution also. Whenever strong attraction forces act between polymer segments and solvent molecules, the heat of dissolution becomes negative ($-\Delta H$), and the process of dissolution becomes favourable. The polymer is then completely soluble in the solvent in all proportions. In the absence of such strong attraction forces, only weak dispersion forces act between the polymer and the solvent and the ΔH becomes positive. As stated earlier, dissolution can, however, occur even when ΔH is positive as long as its value is less than $T{\cdot}\Delta S$. Fortunately, in the case of macromolecules also, the magnitude of entropy gain in the dissolution process is high.

We know that when a polymer changes its state from solid to solution, not only the molecular mobility but the segmental mobility also is set free. Since macromolecules are composed of hundreds of segments (usually numbering 500 to 1,000) rotating freely around fixed angles (independent of each other, in a limited sense, at least), they assume innumerable *'conformations'* in the dissolved state. The freedom of molecular mobility and the segmental mobility give an entropy increase to the dissolved polymer. The increased entropy factor is able to still accommodate $+\Delta H$, giving rise to negative ΔF, thus making dissolution possible. To start with, the process of dissolution is highly favourable. As the concentration of the solution increases, the $T{\cdot}\Delta S$ value becomes progressively lower, as in the case of a low molecular weight dissolution. An equilibrium is finally attained when $+\Delta H$ becomes equal to $T{\cdot}\Delta S$. But what happens at the time of equilibrium?

As we have already seen, the polymer molecule as a whole does not become solvated and diffuse out of the solute phase abruptly. Solvation occurs at random, segment after segment, until all segments in the molecular coil are fully solvated, after which only, the molecule diffuses out of the polymer phase completely. Since the process of solvation of the polymer molecules and their subsequent diffusion into the solvent phase is slow and since the molecular size of the solvent is quite small as compared to that of the polymer, even before the polymer molecules at the surface of the solid polymer get completely solvated and detached, solvent molecules penetrate deep into the polymer matrix and the entire polymer sample becomes more or less fully swollen. Consequently, a situation where molecules at the surface of the solid polymer go into solution while those at the core remain unswollen does not exist. Instead, here, the entire polymer sample is in a fully swollen state with polymer molecules interpenetrating and retained within the swollen mass. No single polymer molecule within the matrix is fully solvated, nor is there any single molecule which is not at all solvated. In fact, in each molecule, some segments are solvated (unfolded) while others are unsolvated (aggregated). There is a constant diffusion of the solvent molecules between the solvated and the aggregated segments. Due to this diffusion, some of the solvated segments get aggregated, while some of the aggregated ones get solvated, and a dynamic equilibrium is set up between the solvated and the aggregated segments within the swollen mass (Fig. 12.4). The polymer molecules, therefore, never move out of the polymer phase and no separate solution phase is formed. When some fresh quantity of the polymer is added to this swollen mass, the solvent molecules penetrate into the newly added polymer and make it swell. The attack on the freshly added polymer sample by the solvent molecules and the resultant polymer swelling continue till a fresh equilibrium is reached between the solvated and the aggregated chain segments. The newly added polymer thus loses its identity and becomes one with the already existing swollen mass. The volume of the swollen mass naturally increases. As there is an increase in the polymer-to-solvent ratio, the ratio of the aggregated to the solvated chain segments in the swollen mass also increases. It can thus be seen that a

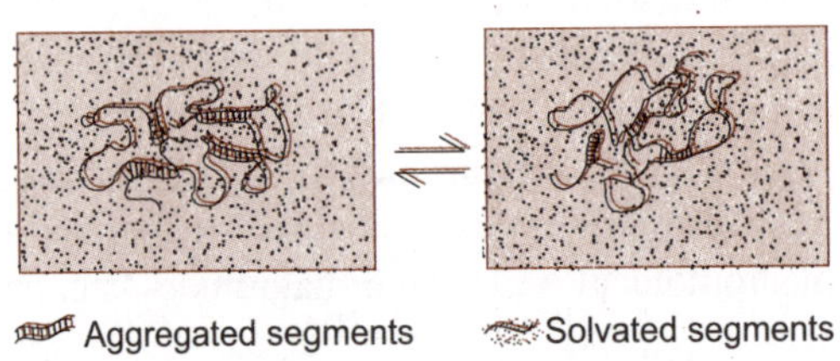

Fig. 12.4 Schematic picture showing existence of equilibrium between solvated and unsolvated (aggregated) segments of a polymer chain in swollen condition.

'saturation' point, where the unswollen solid polymer existing in equilibrium with the solution, is never reached.

In a polymer-solvent system, where $+\Delta H$ is very high and exceeds the $T \cdot \Delta S$ value, to start with itself, dissolution cannot occur at all and the polymer remains unaffected by the solvent. Such a solvent is known as a *'non-solvent'* or *'precipitant'* for that polymer.

12.2.2 Effect of Molecular Weight on Solubility

What we have seen so far corresponds to an ideal case of a monodispersed polymer system. In a polydispersed system (containing both low and high molecular weight components), low molecular weight components get fully solvated to start with and then they diffuse out, forming a solution, while higher molecular weight components are still in the mode of progressive solvation.

When ΔH is negative, the high molecular weight components also get fully solvated with time and go into solution. The higher the molecular weight, the longer the time taken for dissolution since the number of segments to be solvated is larger. On the other hand, when ΔH is positive, the point of equilibrium may be reached before the high molecular weight components also get fully solvated. At the point of equilibrium, $T{\cdot}\Delta S$ becomes equal to $+\Delta H$ and therefore, molecules with a molecular weight below a certain critical value will be in solution and those above it will be in a swollen state. Such a system, where equilibrium exists between the solution phase and the swollen solute phase, is referred to as a 'gel-sol' system ('gel' referring to the swollen polymer and 'sol' to the polymer solution) and the equilibrium as 'gel-sol' equilibrium. The critical molecular weight value, which determines the solubility (or otherwise) in such an equilibrium system, can be shifted by changing the *'solvent power'* using binary solvent mixtures in varying proportions. Such binary solvent systems usually comprise a *'good'* solvent (with $-\Delta H$ value) and a *'poor'* solvent or a *'non-solvent'* (with $+\Delta H$ values). All components of the polymer sample go into solution in a good solvent. When a non-solvent is progressively added to this, because of the $+\Delta H$ factor of the non-solvent, the equilibrium is shifted in such a way that some of the segments in the molecular coils of the highest molecular weight components are brought into aggregation (*i.e.*, they get desolvated) and hence, these molecules form the gel phase. This is to say that the highest molecular weight fraction of the polymer sample now precipitates out of the solution phase. When a further quantity of the non-solvent is added, the equilibrium is further shifted in such a way that the biggest molecules among the remaining molecules in solution precipitate out. As we progressively increase the amount of non-solvent, more and more polymer of comparatively lesser molecular weight precipitates out. This principle is utilised in the fractional precipitation method for separating a polymer sample into different fractions (see Section 13.3.1).

12.2.3 Solubility of Crystalline and Amorphous Polymers

At this stage an interesting question arises. Is the foregoing concept of dissolution true for both amorphous as well as crystalline polymers? Indeed, the thermodynamic feasibility is true for both. That is, the dissolution of a polymer in a solvent will occur, be it crystalline or amorphous, whenever the ΔH value is negative or, if positive, its magnitude is less than $T{\cdot}\Delta S$. Amorphous and crystalline polymers, however, differ in their phase state (cf. Section 6.2.2) and hence, it is but natural for them to behave differently as far as the physical process of their dissolution is concerned. In the gross sense of the term, dissolution of an amorphous polymer can be compared to the mixing of two liquids. Let us elaborate; two liquids, since both of them are in the same phase state, can freely mix in all proportions to form a single homogeneous phase, provided the mixing results in a negative ΔF. Likewise, an amorphous polymer, since it exists in the liquids phase state, can easily undergo dissolution in a solvent which is also in the liquid phase state, so long as ΔF is negative. On the other hand, dissolution of a crystalline polymer involves mixing of two substances existing in two different phase states. The crystalline polymer is in the crystalline phase state, while the solvent is in the liquid phase state. Since the two different phases cannot mix and form a single homogeneous system, mixing or dissolution is impossible unless both of them are brought into the same phase state. The first step involved in dissolving a crystalline polymer, therefore, is to bring the same into the liquid phase state, which can be done in two ways: (*i*) by heating the crystalline polymer to its melting point or (*ii*) by bringing in conditions so as to destroy the long-range order (which is responsible for the crystalline state). The reduced solubility on account of crystallinity or increased intermolecular cohesive forces is common with respect to both polymers as well as low molecular weight solutes. However, owing to the long chain-link nature of the polymer molecule and the presence of strong interchain interactions, the phenomenon is quite significant in the case of polymer dissolution. Crystalline polymers are not, as a rule, soluble in most solvents below their melting points. They are soluble below their melting points only in a few solvents where strong interaction between the polymer and the solvent (by way of hydrogen bonding or dipole interaction) overcomes the interchain cohesive forces within the polymer and destroys the crystalline lattice structure. For example, polyethylene can be dissolved in hydrocarbon solvents such as toluene or decaline only when heated. While atactic polystyrene is soluble in many solvents at ambient temperatures, isotactic polystyrene can be dissolved only at elevated temperatures. Nylon 6, in which the crystallinity is due to intermolecular hydrogen bonding, can be dissolved at ambient temperatures only in solvents such as formic acid or phenol, which are capable of forming strong hydrogen bondings. For the same reason, the fully aromatic polyamides can be dissolved only in concentrated sulphuric acid.

The free energy equation governing the solubility of crystalline polymers near their melting point can then be expressed as:

$$\Delta F = (\Delta H_m + \Delta H_f) - T\cdot(\Delta S_m + \Delta S_f) \tag{12.2}$$

where the subscripts m and f denotes mixing and fusion, respectively.

12.2.4 Heat of Dissolution and the Solubility Parameter

The heat of mixing (ΔH_m) of two liquids, whose dissolution involves only dispersion forces (endothermic mixing), has been expressed by Hildebrand and Scott as follows:

$$\Delta H_m = V_m \left[\left(\frac{\Delta E_1}{V_1} \right)^{1/2} - \left(\frac{\Delta E_2}{V_2} \right)^{1/2} \right]^2 \phi_1 \phi_2 \tag{12.3}$$

where H_m = heat of mixing; V_m = total volume of the two liquids;
ΔE_1 and ΔE_2 = energy of vaporisation of liquids 1 and 2; V_1 and V_2 = the molar volumes of the two liquids; and ϕ_1 and ϕ_2 = volume fractions of liquids 1 and 2.

The term ($\Delta E/V$) indicates the amount of energy required to vaporise a unit volume of the liquid. In the process of vaporisation, energy is spent to overcome the cohesive force holding the molecules together and to let the molecules completely separate out into the gaseous state. The term ($\Delta E/V$), therefore, is a measure of the cohesive force that holds molecules in the liquid together and has been designated as the *'cohesive energy density'* (CED). Hildebrand and Scott proposed the term 'solubility parameter' (δ), for the square root of the CED:

$$\delta = \left(\frac{\Delta E}{V}\right)^{1/2} \tag{12.4}$$

Now, ΔH_m can be related to the solubility parameters of the two liquids by the equation

$$\frac{\Delta H_m}{V_m \phi_1 \phi_2} = (\delta_1 - \delta_2)^2 \tag{12.5}$$

The square of the difference between the solubility parameters of two liquids decides the magnitude of the heat of mixing. Hence, for the positive heat of mixing to be at a minimum value so as to let mixing occur, the difference between the solubility parameters $(\delta_1 - \delta_2)$ of the two liquids must be as low as possible. When δ_1 equals δ_2, ΔH_m becomes zero and mixing will occur readily because of the entropy factor.

The above theory of mixing based on solubility parameters is valid also for polymer dissolution and Eqn. (12.5) can be simplified as

$$\Delta H = (\delta_p - \delta_s)^2 \phi_p \phi_s \tag{12.6}$$

where the subscripts p and s denote the polymer and solvent, respectively. Equation (12.6) predicts that a polymer will dissolve only in solvents having δ values very close to that of the polymer. If δ_p and δ_s values are very much different, dissolution will not take place. This has been verified in the case of amorphous polymers. The theory, however, fails when applied to crystalline polymers as well as to solvents with strong polar groups and hydrogen bonding. As already mentioned, the solubility parameter theory is based on dispersion forces and hence, does not apply to polymer-solvent systems with positive interaction where ΔH is negative. Dissolution cases occur in such cases even when δ_p and δ_s are quite apart.

12.3 THE FLORY-HUGGINS THEORY OF POLYMER SOLUTIONS

Notwithstanding some similarities with low molecular weight solutions, polymer solutions also exhibit some differences. They show considerable departure from Raoult's law and tend to obey the law only at extreme dilutions. In order to understand the reason for this departure, we have to look into the structure of the polymer and low molecular solutions as shown in Fig. 12.5, based on the quasi-crystalline lattice model for solutions. This figure shows a two-dimensional lattice model for a two-component solution where all the lattice sites (represented by squares) have the same area. The empty circles represent solvent molecules, and the shaded ones low molecular weight solute molecules or polymer chain segments. These two-dimensional lattice models, although oversimplifying the reality, do depict the ways of arranging the molecules in a two-component system. According to the Boltzmann's equation, the number of ways the solute and solvent molecules can be arranged (W) is defined by the following equation:

$$\Delta S_m = k \ln W \tag{12.7}$$

where ΔS_m is the entropy of mixing and k the Boltzmann's constant.

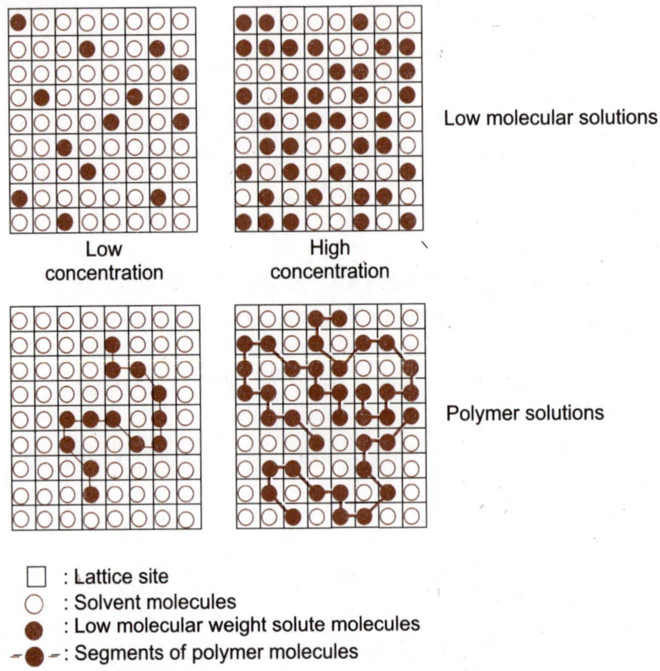

Fig. 12.5 Two-dimensional quasi-crystalline lattice model for solutions of low molecular weight compounds and polymers.

In the lattice model concept, it is assumed that the sizes of the polymer chain segments are comparable to those of the solvent molecules as well as of lattice sites and that the segments can occupy the lattice sites as shown in Fig. 12.5. Owing to the large size of the polymer molecule, it is evident that the number of ways of arranging the polymer segments will be relatively small. That is to say, the entropy of mixing for polymer dissolution is much less compared to that for low molecular weight solutes.

Through a series of theoretical arguments based on the lattice model, Flory and Huggins independently arrived at some interesting equations, which later came to be known as the Flory-Huggins theory. According to this theory, the entropy of mixing for polymer solutions may be written as

$$\Delta S_m = -k(N_s \ln \phi_s + N_p \ln \phi_p) \tag{12.8}$$

where the subscript s stands for the solvent and p for the polymer. The volume fractions ϕ_s and ϕ_p in the foregoing equations are given by

$$\phi_s = \frac{N}{(N_s + nN_p)} \tag{12.9a}$$

$$\phi_p = \frac{nN_p}{(N_s + nN_p)} \tag{12.9b}$$

where N_s denotes the number of solvent molecules; $N_p^{\cdot}$ the number of polymer molecules and n the number of segments in a polymer molecule.

The heat of mixing of polymer solutions could be written as

$$\Delta H_m = kT\chi N_s \phi_p \tag{12.10}$$

where χ is the Flory-Huggins interaction parameter which, in essence, measures the solvent power of the system and enables us to distinguish between a good solvent and a poor solvent. (A low value for the Flory-Huggins parameter is indicative of a good solvent, whereas a high value means a poor solvent). The free energy of mixing of polymer solutions is given in terms of χ as follows:

$$\Delta F_m = kT(\chi N_s \phi_p + N_s \ln \phi_s + N_p \ln \phi_p) \tag{12.11}$$

The Flory-Huggins theory arrives at the equation for total free energy of dilution based on the following trend of thought. There are two entropy factors which contribute to the free energy of dilution; one arising from contact between polymer segments and solvent molecules, and the other is called *conformational entropy.* Thus, the overall free energy of dilution will be the sum of the two terms as follows:

$$\Delta F_m = \Delta F_{\text{conform}} + \Delta F_{\text{contact}} \tag{12.12}$$

It is shown that

$$\Delta F_{\text{conform}} = RT\left[\ln(1-\phi_p) + \left(1-\frac{1}{n}\right)\phi_p\right] \tag{12.13}$$

and

$$\Delta F_{\text{contact}} = RT\,[\chi\phi_p^2] \tag{12.14}$$

so that

$$\Delta F_m = RT\left[\ln(1-\phi_p) + \left(1-\frac{1}{n}\right)\phi_p + \chi\phi_p^2\right] \tag{12.15}$$

This equation is derived for a monodispersed polymer. Further it has been shown that for a polydispersed system the n factor in the above equation has to be replaced with its average value denoted as $\bar{n}$. In that case, for a polydisperesed system Eqn. (12.15) becomes

$$\Delta F_m = RT\left[\ln(1-\phi_p) + \left(1-\frac{1}{\bar{n}}\right)\phi_p + \chi\phi_p^2\right] \tag{12.16}$$

In Eqn. (12.16), if the value of χ exceeds some critical value denoted as $(\chi)_c$, the system will tend to separate into two phases. In that case, the corresponding values in terms of n are mathematically shown to be as follows:

$$(\chi)_c = \frac{1}{2}n + \frac{1}{\sqrt{n}} + \frac{1}{2} \tag{12.17}$$

$$(\phi_p)_c = \frac{1}{1+\sqrt{n}} \tag{12.18}$$

Here $(\phi_p)_c$ represents the critical composition (*i.e.*, critical volume fraction of the polymer), corresponding to the critical value $(\chi)_c$.

Equation (12.16) has been derived based on many theoretical reservations. Therefore, attempts have been made to rectify the shortcomings of the theory. The most important correction has been done with respect to the term $\Delta F_{\text{contact}}$. For instance, Orofino and Flory argued that this term would be better represented by the equation

$$\Delta F_{\text{contact}} = RT[\chi\phi_p^2 + \chi'\phi_p^3 + \chi''\phi_p^4 + \cdots] \tag{12.19}$$

whereas, in Eqn. (12.14), only the first term is taken into account, neglecting the other terms. Hence, the ultimate free energy of dilution ΔF_m has also to be corrected accordingly, by using Eqn. (12.19) in place of Eqn. (12.14).

One interesting point that emerges now is that many parameters such as osmotic pressure, which could be experimentally determined, may be arrived at from this theory. Let us examine how this is possible. The osmotic pressure (π) is related to the free energy of dilution through the following equation:

$$\pi\overline{V}_s = -\Delta F_m \tag{12.20}$$

where $\overline{V}_s$ is the partial molar volume of the solvent.

On substituting for ΔF_m from this equation in Eqn. (12.16), we have

$$-\pi\overline{V}_s = RT\left[\ln(1-\phi_p) + \left(1-\frac{1}{\overline{n}}\right)\phi_p + \chi\phi_p^2\right] \tag{12.21}$$

$$\pi = \frac{-RT}{\overline{V}_s}\left[\ln(1-\phi_p) + \left(1-\frac{1}{\overline{n}}\right)\phi_p + \chi\phi_p^2\right] \tag{12.22}$$

In dilute solutions, the value of ϕ_p will be far less than unity and hence, $\ln(1 - \phi_p)$ may be written as follows:

$$\ln(1-\phi_p) = -\left[\phi_p + \frac{\phi_p^2}{2} + \frac{\phi_p^3}{3} + \cdots\right] \tag{12.23}$$

Now, Eqn. (12.22) may be rewritten as

$$\pi = \frac{RT}{\overline{V}_s}\left[\frac{\phi_p}{\overline{n}} + \left(\frac{1}{2}-\chi\right)\phi_p^2 + \frac{1}{3}\phi_p^3 + \cdots\right] \tag{12.24}$$

or

$$\pi = RT\left[\frac{\phi_p}{\overline{n}\overline{V}_s} + \left(\frac{1}{2}-\chi\right)\frac{\phi_p^2}{\overline{V}_s} + \frac{1}{3}\frac{\phi_p^3}{\overline{V}_s} + \cdots\right] \tag{12.25}$$

On appropriate substitutions, Eqn. (12.25) will take the following form [wherein the osmotic pressure is related to molecular weight ($\overline{M}_n$) and concentration (c) of the solution]:

$$\left[\frac{c}{\overline{M}_n}+\frac{\overline{v}_p^2}{\overline{V}_s}\left(\frac{1}{2}-\chi\right)c^2+\cdots\right] \tag{12.26}$$

Here, $\overline{v}_p$ is the partial specific volume of the polymer. Equation (12.26) may also be written as:

$$\left[\left(\frac{RT}{\overline{M}_n}\right)c+RT\left(\frac{\overline{v}_p^2}{\overline{V}_s}\right)\left(\frac{1}{2}-\chi\right)c^2+\cdots\right] \tag{12.27}$$

or

$$\frac{\pi}{c} \quad \left(\frac{1}{\overline{M}_n}+Bc\right) \tag{12.28}$$

where B is termed as the second virial coefficient. The force of interaction between the polymer molecules and the solvent is measured by this term. Equation (12.28) demonstrates the need to extrapolate the experimental osmotic pressure data to zero concentration at which the higher power concentration terms may be neglected and the equation gets reduced to:

$$\left(\frac{\pi}{c}\right)_{c\to 0} \quad \frac{RT}{\overline{M}_n} \tag{12.29}$$

The term on the left-hand side of Eqn. (12.29) indicates π/c at zero concentration or infinite dilution.

12.4 NATURE OF POLYMER MOLECULES IN SOLUTION

Staudinger's pioneering work led to the present-day knowledge of the macromolecular nature of polymers. Even prior to his work, starch and cellulose were known to be made up of glucose units. However, their abnormal behaviour as displayed, for instance, during fibre formation or due to the high viscosity of their solutions, was attributed to the formation of aggregates of oligomeric glucoside molecules (up to deca-saccharides) through intermolecular secondary valence forces. Starch solution in water was, therefore, thought of as a colloidal solution of aggregated molecules of oligomeric α-1, 4,—glucosides and was compared to a soap solution where molecule aggregates (micelles) form a colloidal solution.

By some simple and interesting experiments, Staudinger's demonstrated that starch or cellulose particles in solution are individual (and not aggregate) molecules and each molecule is made up of thousands of glucoside rings linked through primary covalent bonds to form long chain-like molecules. He took a cellulose sample and carefully measured its molecular weight in an acqueous copper-ammonium hydroxide solution. The sample was then converted into cellulose nitrate by treatment with nitric acid, when the —OH groups were replaced by —ONO_2 groups. The product formed had completely different solubility properties in that it was insoluble in copper-ammonium hydroxide but was soluble in acetone, methyl ethyl ketone, etc. The molecular weight of the nitro-ester derivative was measured from its solution in acetone. This nitro-ester derivative was then carefully regenerated by saponification and the cellulose thus obtained was again molecular weight from the copper-ammonium hydroxide solution. The molecular weight of the original cellulose before nitration and that of the cellulose after regeneration were found to be the same. Staudinger also found that the particle weights of cellulose in the copper-ammonium hydroxide solution and those of the cellulose nitrate in acetone solution led to the same degree of polymerisation. Based on these findings, he argued that the cellulose molecule is made up of a number of glucosidic rings connected through primary valence linkages and that these rings and linkages remain intact during conversions.

If the solutions were made up of aggregates of molecules of oligomeric glucosides, then: (*a*) the degree of aggregation of cellulose molecules with —OH groups in copper-ammonium hydroxide due to secondary valence forces would be quite different from that of cellulose nitrate containing $—ONO_2$ groups in the organic solvent and (*b*) the particle weights would not lead to the same degree of polymerisation. Staudinger further proved this by showing that fatty acid soaps, which can form a colloidal solution in water, when esterified, form only a normal solution, and not a colloid, in organic solvents. He also pointed out that substances that can form colloidal solutions by way of molecular aggregation or micelle formation do so only under specific conditions and in specific solvents, whereas polymers exhibit a colloidal nature in whichever solvents they undergo dissolution. He thus differentiated colloidal sols of silver and gold as well as soap solutions from polymer solutions. He called the former 'micellar colloids' and the latter 'molecular colloids' (Fig. 12.6), and postulated a macromolecular structure for polymer molecules. According to Staudinger, the large size of solute molecules in polymer solutions (although comparable in size to colloidal particles) is not due to the formation of the molecular aggregation by secondary valence forces as in the case of micellar colloids. Instead, these are due to the enormous size of the polymer molecules resulting from the primary valence linkages between a very large number of monomeric units.

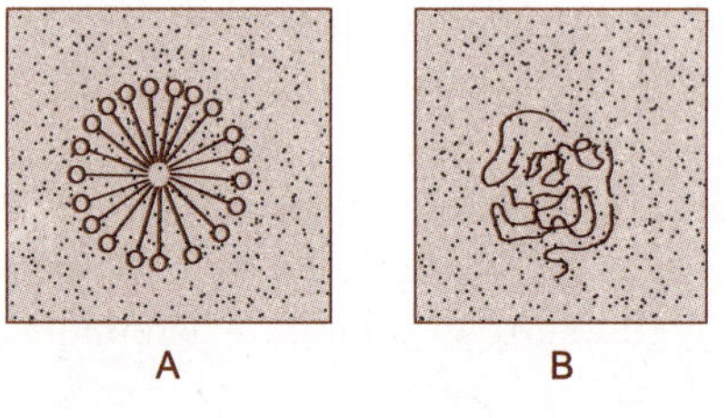

Fig. 12.6 Schematic representation of (A) aggregates of soap molecules forming 'micellar colloid' and (B) single polymer molecule forming 'molecular colloid'.

Although a polymer molecule can be visualised as a long chain or a thin long thread, in reality, it does not exist in solution as a stretched out chain or thread, but as a random coil. Such randomly coiled molecules: (*a*) can remain isolated and independent of their neighbouring molecules in the case of very dilute solutions; (*b*) can be interpenetrating and entangled with the neighbouring ones in the case of moderately concentrated solutions; and (*c*) can be aggregated into a uniform jelly-like mass in the case of swollen gels.

12.5 SIZE AND SHAPE OF MACROMOLECULES IN SOLUTION

In order to get a reasonably correct picture of the size and shape of a polymer molecule, we consider one which is free and not entangled with the neighbouring ones. Since polymers cannot be vaporised, a state in which the molecules can remain independent of each other, we can only visualise a system wherein a single polymer molecule is suspended within a space occupied by a large number of solvent molecules. In other words, we can consider an extremely dilute solution of a polymer in a solvent, and see what the polymer molecule looks like.

We have seen in Section 3.7.2 that a linear polymer molecule is made up of several segments linked to each other in a linear fashion and that each of these segments can rotate around fixed angles at the links between the successive segments.

Let us first assume that the polymer molecule suspended in a solvent medium is free from any interactive forces. Here, the absence of interaction means that there are no inter-segmental attraction forces nor any other interaction forces exerted by the solvent medium on the above mentioned segments. The segmental conformations are governed purely by random flight probability, restricted only by the angle between successive segments and by a constraint on their rotation due to steric hindrance. The molecule, over a period of time, assumes innumerable conformations (cf. Section 3.7.2). When all such conformations are statistically averaged over a period of time, the probability is that, most of the time, the two ends of the chain would be separated by a distance r_0 (Fig. 12.7), which is extremely small compared to the distance between the two ends of the molecule when it is fully

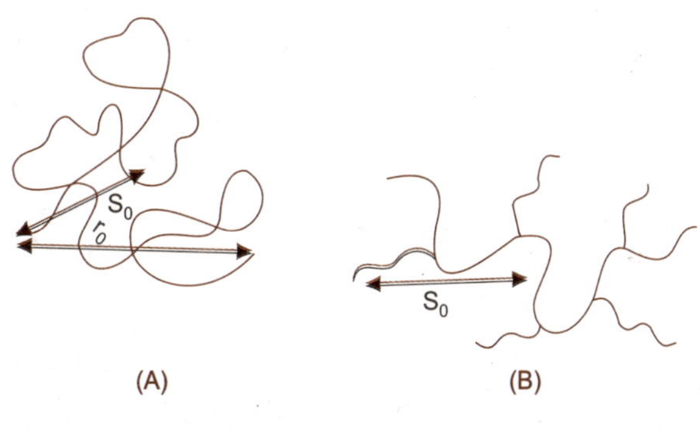

Fig. 12.7 Representation of one possible conformation of: (A) a random coil linear polymer molecule and (B) a branched molecule, showing end-to-end distance (r_0) and radius of gyration (S_0).

stretched out. This distance r_0 gives an idea of the size or the dimension of the chain coil and is known as the '*unperturbed dimension*' of the polymer chain. It is represented as:

$$\left(\overline{r_0^2}\right)^{1/2} \propto \bar{l}(n)^{1/2} \tag{12.30}$$

where $\left(\overline{r_0^2}\right)^{1/2}$ = root mean square end-to-end distance; $\bar{l}$ = average length of a segment; and n = number of segments in the chain.

CHAPTER 12

When we deal with branched (instead of linear) polymers with the same number (n) of segments of length $\bar{l}$ the main backbone length gets reduced as some of the segments are utilised in forming the branches. The random coil conformation of a branched polymer, therefore, results in a lower dimension as compared to the linear chain of the same molecular weight. Such a coil will contain more than two chain ends, and we cannot speak in terms of the end-to-end distance while denoting the coil dimension. The term radius of gyration, $\bar{S}_0$ defined as the average distance between the molecular coil centre of gravity and the chain ends, is used for this purpose (Fig. 12.7). Such a definition and concept of radius of gyration are true in the case of linear polymer molecules also and has been shown to be related with the end-to-end distance by the equation

$$\left(\overline{r_0^2}\right)^{1/2} \quad 6(\overline{S_0^2})^{1/2} \tag{12.31}$$

In actual polymer solutions, however, different types of interaction forces do exist. Cohesive forces act between the polymer chain segments and as a consequence, the segments try to come closer and tighten the coil. While this happens, strong interaction as well as dispersion forces act between the polymer segments and solvent molecules, pushing the segments apart and loosening the coil. The actual dimension of the polymer chain suspended in a solvent will, therefore, depend on the net result of these opposing forces and will be different from the *'unperturbed dimension'*. In thermodynamically *'good'* solvents, strong interaction forces are developed between the chain segment and the solvent. The polymer coil, therefore, opens up considerably increasing the coil dimension. If we can denote the end-to-end distance of a chain in a solvent as (r), the same can be related with (r_0) through an expansion factor α which describes the increase in size of the polymer coil in solution over the unperturbed dimension:

$$\left(\overline{r^2}\right)^{1/2} \quad \left(\overline{r_0^2}\right)^{1/2} \tag{12.32}$$

It should be noted that α will be large in the case of polymer molecules in a good solvent. In a thermodynamically *'poor'* solvent, where there are no strong interaction forces, but only dispersion forces acting between the solvent and polymer segments, the chain expansion is relatively less and the value of α is also low. When a *'non-solvent'* is added to a polymer solution, what happens is that the polymer molecule is surrounded by the *'non-solvent'* molecules and there are very weak, if at all any, dispersion forces acting between the polymer and the solvent. The intersegmental cohesive forces, therefore, take the upper hand and the segments aggregate, and the polymer coil tightens further, squeezing out the solvent. Under such conditions, the polymer molecule can no more remain in solution and hence, is thrown out as a swollen mass. In some other solvents bordering the area between *'poor'* solvent and *'non-solvent'*, at a particular temperature characteristic of the polymer/solvent system, the segmental cohesive forces are just balanced by the solvent dispersion forces and the molecular dimension remains unperturbed by these opposing interactions. The value of $\left(\overline{r^2}\right)^{1/2}$ will then be equal to $\left(\overline{r_0^2}\right)^{1/2}$ with α attaining a value of unity. For a given polymer and solvent, the temperature at which α becomes unity is called *Flory's theta temperature* (θ) and the solvent at that temperature is called the *'theta solvent'* for that polymer. The same solvent above the θ temperature behaves like a poor solvent and below this temperature acts as a non-solvent. The theta temperature can therefore, be defined as the temperature at which polymer molecules are just held in solution or are at the threshold of precipitation. The expansion factor of α polymer in a given solvent, however, varies slightly with the molecular weight—the higher the molecular weight, the higher the α value.

12.6 VISCOSITY OF DILUTE POLYMER SOLUTIONS

The viscosity of a dilute polymer solution depends on several factors, namely, the nature of the polymer and that of the solvent, their concentrations, the polymer molecular weight, the temperature and the shear rate. (The effect of the shear rate has been neglected in the following discussion for the sake of simplicity.)

At the molecular level, the viscosity of a polymer solution is a direct measure of the hydrodynamic volume of the polymer molecules. Hydrodynamic volume, as we have seen in Section 12.1, is the apparent volume occupied by the expanded or swollen molecular coil along with the imbibed solvent and can be defined in terms of the expansion factor α and unperturbed end-to-end distance $\left(\overline{r_0^2}\right)^{1/2}$. We have earlier seen the dependence of α on the solvent nature and of $\left(\overline{r_0^2}\right)^{1/2}$ on the polymer molecular weight. It, therefore, follows that a polymer will exhibit a higher viscosity in a good solvent than in a poor solvent and that in the same solvent, the viscosity will be directly proportional to the molecular weight.

12.6.1 Viscosity and Molecular Weight

Staudinger, during the early 1930s, used viscosity as a measure of the molecular weight of the polymer, to postulate his hypothesis about the long-chain nature of polymer molecules (see Section 12.4). He proposed the relationship:

$$\eta_{sp} = K_s c M \tag{12.33}$$

where K_s is a constant for a given polymer/solvent/temperature; c the concentration; M the molecular weight; and η_{sp} the specific viscosity denoting the increase of viscosity of a polymer solution over that of the pure solvent according to the relation

$$\eta_{sp} = \frac{\eta - \eta_0}{\eta_0} \tag{12.34}$$

where η = viscosity of the polymer solution and η_0 = viscosity of the solvent.

It can be seen from Eqn. (12.33) that the specific viscosity is dependent on the concentration. In order to quantify a viscosity function of a polymer in a solvent, which will be independent of the concentration, the limiting value of reduced viscosity (η_{sp}/c) or that of the inherent viscosity ($\ln \eta_r/c$) at infinite dilution is chosen and termed *'intrinsic viscosity'* or *'limiting viscosity number'*, [η]:

$$(\eta_{sp}/c)_{c \to 0} = [\eta] = (\ln \eta_r/c)_{c \to 0} \tag{12.35}$$

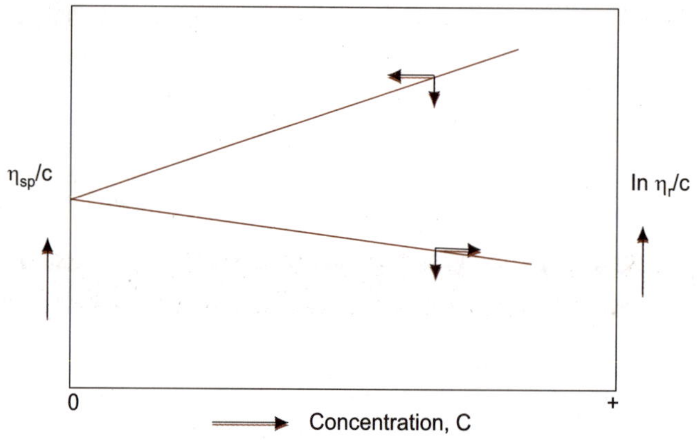

Fig. 12.8 Typical plots of viscosity data for computing intrinsic viscosity, [h], according to Eqns. (12.36) and (12.37). The ordinate intercept gives [h].

The term [η] has been related to the two viscosity functions through the following two equations by Huggins and Kraemer, respectively:

$$\frac{\eta_{sp}}{c} = [\eta] + k'[\eta]^2 c \tag{12.36}$$

$$\frac{\ln \eta_r}{c} = [\eta] - k''[\eta]^2 c \tag{12.37}$$

where k' and k'' are constants for a given polymer/solvent/temperature system. For many linear flexible polymer systems, k' often indicates the measure of the solvent power: the poorer the solvent, the higher the values of k'. The value of $k' + k''$ has been shown to be 0.5.

It should be noted that [η] is dependent on polymer/solvent/temperature and can be computed by measuring the viscosity of dilute solutions of the polymer in a given solvent at various concentrations (c less than 0.5 g/dl) and extrapolating the viscosity functions to zero concentration according to Eqns. (12.36) and (12.37). Representative plots are shown in Fig. 12.8 (see also Fig. 13.20).

Staudinger's equation was subsequently replaced by the famous Mark-Houwink equation:

$$[\eta] = K \cdot \overline{\overline{M}}^a \tag{12.38}$$

where K and a are constants for a given polymer/solvent/temperature system. The value of a varies from 0.6 to 0.8 and becomes 0.5 in theta solvents, while the value of K varies widely from one system to another. K and a values for some polymer/solvent systems are given in Tables 12.1 and 12.2.

Flory and Fox (based on Einstein's theory of viscosity, which states that the specific viscosity of a suspension of spheres in a liquid is proportional to the volume fraction of the spheres) deduced an empirical relationship connecting the hydrodynamic volume with the intrinsic viscosity. They showed that the hydrodynamic volume is proportional to the cube of root mean square end-to-end distance:

$$[\eta] = \Phi \frac{(\overline{r^2})^{3/2}}{M} \tag{12.39}$$

Table 12.1 Viscosity-molecular weight constants for polymer/solvent systems at 30°C*.

Polymer	*Solvent*	*K × 10³ (ml/g)*	*a*
Cellulose triacetate	Chloroform	4.5	0.90
	o-cresol	6.15	0.90
Cellulose trinitrate	Ethyl acetate	2.50	1.01
Natural rubber	Benzene	18.50	0.74
Polyacrylic acid	1, 4-dioxane	76.00	0.50
Polyacrylonitrile	γ-butyrolactone	34.2	0.70
	Dimethyl formamide	20.9	0.75
Polybutadiene—high cis	Benzene	33.7	0.715
	Toluene	30.5	0.725
Polybutadiene—high trans	Toluene	29.4	0.753
Polybutadiene—cis, trans mix	Toluene	39.0	0.713
Polybutyl methacrylate	Benzene	4.0	0.77
Polyethyl acrylate	Acetone	20.0	0.66
	Benzene	27.7	0.67
	Chloroform	31.4	0.68
	Ethyl acetate	26.0	0.66
Polyethylene terephthalate	Trifluoro acetic acid	43.3	0.68
	Phenol/tetrachloroethane (3/5 vol)	22.9	0.73
	Phenol/trichlorophenol (10/7 vol)	630.0	0.47
Polyisobutylene	Benzene	61	0.56
	Carbon tetrachloride	29	0.68
	Cyclohexane	27.6	0.69
	Toluene	20.0	0.67
Polyisoprene—cis	Toluene	20.0	0.728
Polymethyl acrylate	Acetone	28.2	0.52
	Benzene	3.56	0.798
	Toluene	7.79	0.697
Polymethyl methacrylate—atactic	Acetone	7.70	0.70
	Acetonitrile	39.30	0.50
	Benzene	5.20	0.76
	Toluene	7.0	0.71
Polymethyl methacrylate—isotactic	Acetone	23.0	0.63
	Benzene	5.2	0.76
Polypropylene—atactic	Benzene	33.8	0.67
	Cyclohexane	20.9	0.76
	Toluene	21.8	0.725
Polypropylene—syndiotactic	Heptane	31.2	0.71
Polystyrene—atactic	Benzene	11.5	0.73
	Chloroform	4.9	0.794
	Decaline (100% trans)	61	0.53
	Dichloro ethane	8.38	0.74
	Toluene	11	0.725

contd.

contd.

Polymer	Solvent	$K \times 10^3$ (ml/g)	a
Polystyrene—isotactic	Benzene	10.6	0.735
	Chloroform	25.9	0.734
	Toluene	11.0	0.725
Polyvinyl acetate	Acetone	10.2	0.72
	Acetonitrile	41.5	0.62
	Benzene	56.3	0.62
	Ethyl formate	32	0.65
	Methanol	31.4	0.60
Polyvinyl alcohol	Water	45.3	0.64
Polyvinyl butyrate	Benzene	11.15	0.735
Polyvinyl chloride	Chlorobenzene	71.2	0.59
	Cyclohexanone	16.3	0.77
	Tetrahydrofuran	83.3	0.83
Polyvinyl methyl ether	Benzene	76.0	0.60
Polyvinyl pyrrolidone	Methanol	23.0	0.65
	Water	39.3	0.59
SBR-rubber	Cyclohexane	31.6	0.70
	Toluene	37.9	0.71

*Data from *Polymer Handbook*, Eds. Brandrup, J. and Immergut, E.H., John Wiley & Sons, New York, 1975.

Table 12.2 Viscosity-molecular weight constants for polyisobutylene in toluene at different temperatures*.

Polymer	Solvent	Temp (°C)	$K \times 10^3$ (ml/g)	a
Polyisobutylene	Toluene	0	40	0.60
Polyisobutylene	Toluene	15	24	0.65
Polyisobutylene	Toluene	30	20	0.67
Polyisobutylene	Toluene	50	20	0.68
Polyisobutylene	Toluene	60	13.5	0.71
Polyisobutylene	Toluene	90	12.6	0.72

*Data from *Polymer Handbook*, Eds. Brandrup, J. and Immergut, E.H., John Wiley & Sons, New York, 1975.

where Φ is a universal constant independent of polymer type, solvent and temperature. When $[\eta]$ is expressed in dl/g, the value of Φ is 2.84×10^{21}.

Since $\overline{r^2} = \alpha^2 \overline{r_0^2}$ from Eqn. (12.32)], Equation (12.39) can be rewritten as:

$$[\eta] = \Phi\left\langle \frac{\overline{r_0^2}}{M} \right\rangle^{3/2} M^{1/2}\alpha^3 \tag{12.40}$$

The term $\Phi\left\langle \frac{\overline{r_0^2}}{M} \right\rangle^{3/2}$ is a constant for a polymer and is independent of solvent and temperature and can be denoted by K_θ.

Now, the final Flory-Fox equation for molecular weight determination can be written as

$$[\eta] = K_\theta M^{1/2} \alpha^3 \tag{12.41}$$

and at Flory's theta temperature, when a becomes 1, the equation is reduced to

$$[\eta]_\theta = K_\theta M^{1/2} \tag{12.42}$$

where $[\eta]_\theta$ is the intrinsic viscosity at the theta temperature in a theta solvent.

From Eqn. (12.38), we can see that the experimental determination of the molecular weight of a polymer from viscosity measurements requires a prior knowledge of the two constants K and a. However, if the measurements are made, in a theta solvent, the knowledge of one constant (K_θ) alone is necessary.

Theta solvents are known for limited polymers and even a very slight variation in the temperature condition makes the system shift from the theta conditions, making measurements less reliable. Attempts were, therefore, made to compute viscosity functions that would very closely resemble $[\eta]_\theta$ from measurements made in common solvents at ambient temperatures.

According to Bawn, visosity measurements carried out in a solvent/non-solvent mixture in which the polymer is just at the point of incipient precipitation will yield the intrinsic viscosity at the precipitation point, $[\eta]_p$ which is supposed to be identical with $[\eta]_\theta$. $[\eta]_p$ can be computed by measuring $[\eta]$ in solvent/non-solvent mixtures of varying proportions and extrapolating the data to that proportion where the polymer is at the threshold of precipitation, *i.e.*, precipitation point. This will avoid actual measurement at the precipitation mixture and the related uncertainties associated with such measurements. A typical plot is shown in Fig. 12.9.

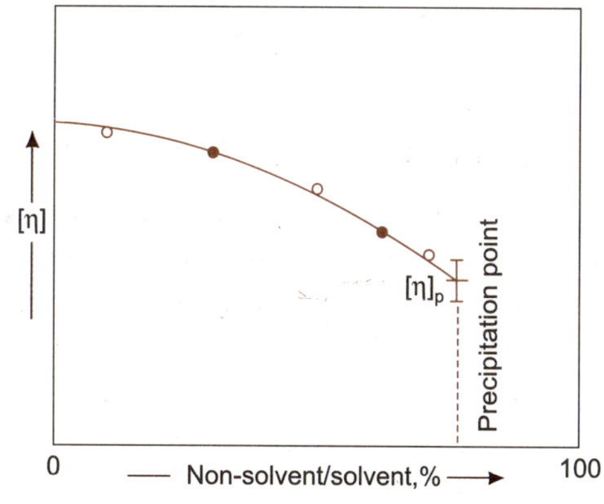

Fig. 12.9 Computation of $[h]_p$ from [h] measurements in a series of solvent/non-solvent mixtures of different proportions and extrapolating to precipitation point.

Cleverdone and Smith showed that when the η_{sp}/c versus c plots of a polymer sample in different solvents are extrapolated to negative concentrations, they converge at a point called the *'true intrinsic viscosity'*, $[\eta]_T$ (Fig. 12.10) and claimed that $[\eta]_T$, unlike $[\eta]$, should be independent of the solvent. They, however, found that $[\eta]_T$ was lesser than $[\eta]_P$ and $[\eta]_\theta$. Gundiah and Kapur argued

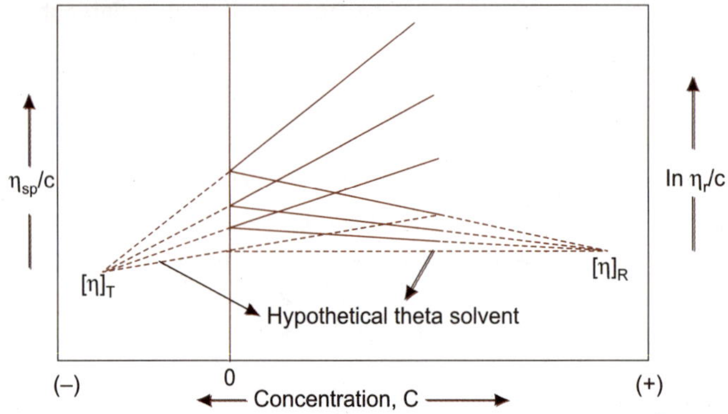

Fig. 12.10 Extrapolation of η_{sp}/c and ln η_r/c plots for computing $[\eta]_T$ and $[\eta]_R$ respectively. Note that $[\eta]_T$ is less than $[\eta]_R$.

that although a plot of η_{sp}/c versus c for a hypothetical theta solvent with a k' value of 0.5 would also converge at the same point (at$[\eta]_T$) at a negative concentration, such a plot would not be able to give any idea of the intrinsic viscosity in that solvent at zero concentration, *i.e.*, $[\eta]_\theta$, since k' alone cannot define the slope of the curve. They, nevertheless, put forward an alternative method of extrapolating the ln η_r/c versus c plots to a positive concentration and showed that such curves do meet at a common point of intersection. They termed the intrinsic viscosity at that point as $[\eta]_R$ (Fig. 12.10). According to them, the plots of a hypothetical theta solvent should meet at the same point and since the slope of that plot would be zero, $[\eta]_R$ should be identical with $[\eta]_\theta$.

Further, $[\eta]_P$ and $[\eta]_R$ are believed to be reasonably close to $[\eta]_\theta$ and hence, can be utilised to compute molecular weights according to the equation

$$[\eta]_R = [\eta]_P = [\eta]_\theta = K_\theta M^{1/2} \tag{12.43}$$

12.6.2 Unperturbed Dimension and Intrinsic Viscosity

In dilute solutions, we know that polymer molecules exist as expanded or swollen chain coils and that the degree of swelling depends on the nature of the solvent. The swelling is higher in a good solvent than in a poor solvent. In a theta solvent, the coil assumes the unperturbed dimension. It is possible to calculate the unperturbed dimension using the equation

$$[\eta]_\theta = \Phi \left\langle \frac{\overline{r_0^2}}{M} \right\rangle^{3/2} M^{1/2} \tag{12.44}$$

by knowing the molecular weight and the value of $[\eta]_\theta$. As already mentioned, the measurement of viscosities at theta conditions demands exacting experimental conditions. Moreover, theta temperatures may not lie in a convenient, experimentally realisable range.

Attempts have, therefore, been made to compute the unperturbed dimensions from viscosity measurements using any solvent under normal conditions. The most popular relationship is that of Stockmayer and Fixman, which is as follows:

$$[\eta] = K_\theta M^{1/2} + 0.51\Phi BM \tag{12.45}$$

where B is related to the Flory-Huggins interaction parameter χ according to the equation

$$B = \overline{v_p^2}(1-2\chi)/\overline{V}_s N_A \tag{12.46}$$

where N_A is Avogadro's number.

For easy graphical treatment, Eqn. (12.45) may be rewritten as

$$[\eta]/M^{1/2} = K_\theta + 0.51\Phi B\, M^{1/2} \tag{12.47}$$

If the molecular weights of a series of fractions of a polymer and their corresponding intrinsic viscosities in any single solvent at any one temperature are known, a simple plot of $[\eta]/M^{1/2}$ against $M^{1/2}$ as per Eqn. (12.47) will give K_θ as the ordinate intercept. The unperturbed root mean square end-to-end distance can then be calculated from the equation

$$K_\theta = \Phi\left\langle\frac{\overline{r_0^2}}{M}\right\rangle^{3/2} \tag{12.48}$$

The interaction parameter χ for the solvent used can also be calculated by knowing the slope of the plot and by using Eqn. (12.46).

12.7 VISCOSITY OF CONCENTRATED POLYMER SOLUTIONS

While the study of the viscosity behaviour of a dilute polymer solution is of significant theoretical importance, that of the flow behaviour of a concentrated polymer solution is of practical importance. Many industrial problems associated with adhesives, coatings, cast films and spun fibres cannot be fully appreciated without a proper knowledge of the properties of concentrated polymer solutions.

In a concentrated solution (generally concentration above 5% by weight), polymer molecules do not exist as isolated chain coils, but are interpenetrating and entangled with each other. Shear effects are, therefore, pronounced in the case of the viscosity behaviour of concentrated solutions. Any viscosity measurement will have to be made at different shear rates and extrapolated to zero shear so as to get the true picture of the solution behaviour.

The concentration dependence of viscosity of concentrated solutions is expressed by the equation

$$\frac{\eta_{sp}}{c(\eta)} = \exp\left\{\frac{k'[\eta]c}{1-bc}\right\} \tag{12.49}$$

where b is a constant. The values of k' and $[\eta]$ for a polymer-solvent system can be found out using Eqn. (12.36).

Similarly, the viscosity as a function of molecular weight and concentration can be expressed by the equation

$$\eta = K^*[c^{\alpha/\beta}M]^{3.4} \tag{12.50}$$

Here K^* is a characteristic constant for a polymer-solvent system and is temperature dependent. Its value is roughly of the order of 2.0×10^{-15} when c is expressed in g/ml and η in poise. The value of α/β ranges from 1.38 to 1.85 with an average of 1.5.

Books for Further Reading

Tompa, H., *Polymer Solutions,* Academic, New York, 1956.

Tanford, C., *Physical Chemistry of Macromolecules*, Wiley-Interscience, New York, 1961.

Morawetz, H., *Macromolecules in Solution,* Wiley-Interscience, New York, 1965.

Tveskov, V.N., Eskin, S.Y. and Frenkel, S.Y., *Structure of Macromolecules in Solution,* Butterworths, London, 1970.

Yamakawa, H., *Modern Theory of Polymer Solutions,* Harper and Row, New York, 1971.

Bohdanecky, M. and Kovar, J., *Viscosity of Polymer Solutions,* Elsevier, Amsterdam, 1982.

13

EXPERIMENTAL METHODS

13.1 POLYMER SYNTHESIS

Here, it is proposed to briefly discuss laboratory syntheses of a few polymers in order to get the feel of handling them. The experimental methods given in the forthcoming sections are preparatory and not meant for kinetic studies; the latter calls for rigorous control on the ingredient purity and polymerisation conditions, which are beyond the scope of this book*.

The experimental techniques are outlined essentially by way of guidelines. It must be stressed that many of the chemicals are toxic, corrosive or obnoxious and should be handled with extreme care, avoiding direct contact with skin, eyes, etc. Inhalation of vapours should be minimised and, wherever possible, experiments should be carried out in fume hoods. Necessary protective appliances such as hand gloves. safety spectacles, face shield, overcoat, etc., should be used. Many of the solvents are highly inflammable and should be handled appropriately. Peroxide initiators being shock- and thermal-sensitive may lead to explosions and should therefore be handled with utmost care. Polymerisation reactions are exothermic and can get out of control, leading to a fire or an explosion. All polymerisation reactions should be conducted behind a laboratory safety screen. Before starting an experiment, one should acquaint oneself with the potential hazards associated with the experiment and with appropriate precautionary measures to be adopted and observed.

13.1.1 Monomer Purification

Unsaturated monomers (vinyl monomers) are highly reactive and can polymerise during transport or storage. To avoid such polymerisation, they are transported and stored in an inhibited condition. Inhibitors such as hydroquinone or tertiary butyl catechol are used for this purpose.

Inhibited monomers will have to be freed of inhibitors as it may not be possible to initiate polymerisation otherwise. Vinyl monomers are purified (or freed of inhibitors) either by washing

---

*For experimental techniques on kinetic studies, reference may be made to *Laboratory Experiments in Polymer Synthesis and Characterisation* (by E.M. Pearce *et al.*) a publication of EMMSE Project, The Pennsylvania State University, Materials Research Lab., University Park.

away the inhibitor with dilute alkali or by the distillation of monomers, the latter being done carefully to avoid polymerisation due to excess heating. Distillation under a reduced pressure helps in overcoming this problem. In the washing process, the monomer is taken in a separating funnel and shaken with a 5% solution of potassium hydroxide in water (This method is useful only in the case of water insoluble monomers). The inhibitor reacts with the alkali, and the reaction product being water soluble, comes out of the monomer phase and can be removed along with the aqueous phase. On allowing to stand after thorough shaking, the monomer and alkali solutions separate. The alkali solution is removed, and washing with alkali is repeated several times.

In many cases, the alkaline solution of the reaction product is coloured and washing can be stopped when no more colouration of the aqueous phase remains. The monomer is then washed with distilled water to remove the alkali traces. It can then be dried using suitable drying agents such as calcium chloride, sodium sulphate, sodium bicarbonate or phosphorus pentoxide (all anhydrous). When monomers are needed for suspension or emulsion polymerisation, they need not be dried after washing with water.

Saturated monomers containing functional groups (used for step polymerisation reactions) should be purified to render them free of non-functional and monofunctional impurities. No generalised scheme can be given for this purpose and the procedure varies from one monomer to another, depending on their chemical nature (Reference should be made to standard books on organic chemistry for such purposes).

While carrying out the experiments described hereafter, it should be ensured that the monomers are purified prior to use and that the chemicals and solvents are of Technical Grade or Reagent Grade.

13.1.2 Bulk Polymerisation of Styrene

In this case, 50 mg of a freshly recrystallised dry initiator, Azobis Isobutyronitrile (AIBN)*, is taken in a clean and dry 100 ml round bottomed flask. To this is added 10 ml of styrene monomer. The flask is swirled gently till the AIBN dissolves in the monomer. Another 40 ml of styrene and 500 mg of Dodecyl Mercaptan (DDM) chain modifier are added to the flask and the contents are flushed with oxygen-free dry nitrogen for about 15 minutes. A water condenser is fitted to the flask and the flask is placed in a constant temperature bath maintained at 70 ± 1°C. The flask is swirled, say, every 5 minutes. It may be noticed that after 30 to 40 minutes, the viscosity of the contents in the flask increases steadily, indicating the progress of polymerisation. The viscosity increased is carefully observed and the flask is removed from the bath when the contents of the flask attain a high viscosity (say, the consistency of honey). The contents are emptied into 250 ml of toluene taken in a beaker. The flask is washed a few times with small quantities of toluene and the washings are added to the beaker. The contents of the beaker are now stirred with a glass rod till a homogeneous solution is formed. The polymer formed is now in the form of a solution in a mixture of toluene and the unpolymerised residual monomer. The polymer can be recovered by pouring the solution slowly into 2 litres of distilled methanol under constant agitation (see Section 13.2 for further details of polymer

*AIBN can be recrystallised from methanol. Due to the explosive nature of AIBN, the methanol solution should not be heated. A saturated solution at room temperature can be made and either cooled in a refrigerator or evaporated slowly at that temperature under reduced pressure to get crops of AIBN crystals.

purification). The polymer will precipitate out as a white fluffy solid. The solid polymer is washed further with methanol and dried under vacuum at 60°C.

The contents should, however, be emptied before the viscosity becomes abnormally high as otherwise the polymerisation might reach 100% conversion with the polymer solidifying inside the flask and rendering its removal very tedious.

Solution polymerisation of styrene can be carried out by taking 200 ml of toluene and dissolving the monomer, initiator and chain modifier in it. The rest of the procedure is the same. The viscosity increase in this case will, however, be gradual.

13.1.3 Precipitation Polymerisation of Acrylonitrile

Here, 50 mg of AIBN is dissolved in 50 ml of acrylonitrile monomer and 200 ml of toluene added to it. The contents of the flask are flushed with nitrogen and the polymerisation started by keeping the flask in a constant temperature bath of 70 ± 1°C. After some time, turbidity develops and a white precipitate starts collecting at the bottom of the flask. In the case of acrylonitrile, the polymer formed is insoluble in its monomer as well as in toluene. Hence, the polymer precipitates out as it is formed. The precipitated polymer can be filtered from the liquid phase, washed several times with methanol and dried in vacuum at 60°C.

13.1.4 Emulsion Polymerisation of Styrene

In this case, 180 ml of distilled water is taken in a 1- litre, three-necked round bottomed flask. To this is added 100 mg of potassium persulphate (initiator), 100 mg of sodium hydrogen phosphate (buffer) and 1.0 g of sodium lauryl sulphate (emulsifier). Further, 100 ml of styrene monomer is added and the flask is fitted with a water condenser, a high-speed stirrer and an inlet tube dipping into the contents on the flask. Nitrogen gas is flushed slowly for a few minutes taking care that undue frothing does not take place. After the flushing is completed, the nitrogen inlet tube is removed and the mouth closed with a stopper. The whole setup is placed in a constant temperature bath maintained at 70 ± 1°C (Fig. 13.1). The stirrer is now put into operation and the contents kept vigorously stirred. If the frothing makes the contents spill over the mouth of the flask, the stirring should be adjusted. The experiment is continued for 4 to 6 hours, at the end of which the contents of the flask attain the form of the latex. The assembly is removed and the flask taken out of the constant temperature bath. To the latex in the flask is added 50 ml of a 3% solution of aluminium sulphate in water and stirred gently when the emulsion is broken and the latex coagulated to give the polystyrene. The polymer can be washed with warm water and methanol and finally dried in vacuum at 60°C.

Fig. 13.1 Laboratory setup for carrying out emulsion polymerisation of styrene.

13.1.5 Suspension Polymerisation of Methyl Methacrylate

Here, 0.1 g of benzoyl peroxide (initiator) is dissolved in 100 g of methyl methacrylate monomer contained in a 500 ml round bottomed flask. To this is added 180 ml of distilled water containing 0.02 g of cetyl dimethyl benzyl ammonium chloride (or any other suitable quarternary ammonium salt type emulsifier) and 0.05 g of polyvinyl alcohol (protective colloid). The contents of the flask are flushed with nitrogen and placed in a constant temperature bath maintained at 75 ± 1°C. A medium-speed stirrer is fitted and the contents stirred uniformly. It can be seen at this stage that the monomer gets dispersed in the water phase as tiny droplets. The size of the droplets can be adjusted by controlling the stirrer speed. The experiment is continued for 8 hours at the end of which fine spherical beads of polymethyl methacrylate settle at the bottom of the flask. The beads are filtered and washed several times with warm water to free them from the emulsifier.

13.1.6 Polycondensation of Ethylene Glycol with Maleic Acid

Five hundred ml of xylene, 62 g of ethylene glycol, 116 g of maleic acid and 1 g of *p*-toluene sulphonic acid catalyst are taken in a 1 litre, three-necked round bottomed flask. One or two fragments of pumice stone are added to the flask and it is placed in an electrical heating mantle with an energy regulator. To the central neck of the flask is attached, a Dean and Stark apparatus fitted with a Liebig condenser and to the side necks are fixed a gas inlet tube and mercury well for the thermometer (Fig. 13.2).

The Dean and Stark apparatus is filled with xylene up to the graduated portion. A slow stream of carbon dioxide gas is bubbled through the contents of the flask to prevent any oxidative degradation of the polyester formed. The flask is heated gently until its contents become homogeneous. Heating is then adjusted such that the contents of the flask start refluxing gently. As the polyester is formed by the polycondensation reaction, the water eliminated is carried along with the xylene vapour in the form of an azeotropic mixture. When the azeotropic vapour mixture condenses at the receiver part of the Dean and Stark apparatus, the xylene and water separate out and only xylene returns to the reaction flask. The reaction is carried out until 18 ml of water is collected, when the polyester formation will be completed. Next, the Dean and Stark apparatus is removed and the xylene separated by distillation. The liquid polymer obtained is further purified by washing with methanol and drying in vacuum at 60°C. The product obtained is an unsaturated polyester, *i.e.*, a polyester containing double bonds in its backbone. This can be cured by cross-linking with styrene using a free-radical initiator.

Fig. 13.2 Laboratory setup for carrying out polycondensation of ethylene glycol with maleic acid.

13.1.7 Interfacial Polycondensation of Terephthaloyl Chloride and Ethylene Diamine

Twenty g of terephthaloyl chloride is dissolved in 200 ml of carbon tetrachloride taken in a 500 ml glass beaker. Separately, to 50 ml of water taken in a 400 ml beaker, 2 to 3 pallets of sodium hydroxide are added and dissolved. To this, alkaline water is added 5 g of ethylene diamine and mixed well. The diamine solution is very carefully added to the beaker containing the acid chloride solution and care is taken to ensure that the amine solution does not fall straight on the acid chloride solution, but slowly trickles down the sides of the beaker and forms a separate layer over the acid chloride solution.

The beaker is allowed to stand undisturbed. After a few minutes, an opaque film is formed at the interface between the organic and the aqueous layers. This is the polyamide formed by condensation between the acid chloride and the amine at the interface. If the film formed is held gently at its centre with a pair of forceps and slowly lifted out of the beaker, we can see fresh layers of the film being formed at the interface and a continuous stretch of the film can be pulled out of the beaker as if it contains an inexhaustible supply of the polymer. It may be noted that the experiment can also be carried out with 8 g of sebacoyl chloride and 10 g of hexamethylene diamine.

13.2 ISOLATION AND PURIFICATION OF POLYMERS

We know that solution, emulsion or suspension techniques are some of the methods available for preparing a polymer. Polymers formed will as such, however, be contaminated with the monomer initiator, solvent, emulsifier, etc., and hence, need to be isolated and purified.

Some polymers precipitate out of the monomer or monomer-solvent mixture after acquiring a particular molecular weight and hence, are straightaway obtained as fine powder. Polymers produced by the suspension method are obtained in the form of solid beads. Both these types of polymers can be filtered or centrifuged and washed several times with water or suitable non-solvents in order to free them from the contaminants.

In the case of bulk or solution polymerisation, the polymer is obtained as a solution either in the monomer or the solvent. Purification in these cases can be done by precipitating the polymer using a non-solvent. To get the best results, the polymer solution is diluted to 2% to 5% concentration and added as a thin stream to a large volume of non-solvent under vigorous stirring. One important point to be noted here is that the non-solvent should be freely miscible both with the solvent and the monomer in all proportions. As increasing quantities of the polymer, solutions are poured into the non-solvent, the solvent from the solution mixes with the non-solvent and slowly reduces the precipitating power of the non-solvent. If a sufficient quantity of non-solvent is not taken, the solvent may completely nullify the non-solvent power at some stage and no more polymer would precipitate out of the solution; hence the need for taking a large quantity of the non-solvent. A list of suitable solvents and non-solvents for some important polymers is given in Table 13.1.

On precipitation from the solution with a non-solvent, the polymer separates out as solid lumps. Redissolving and reprecipitating the polymer a couple of times will improve the purity of the polymer.

Table 13.1 Selected solvents and non-solvents for some polymers

Polymers	*Solvents*	*Non-solvents*
Cellulose triacetate	Methylene chloride, chloroform, trichloroethane, tetrahydrofuran, dioxane, acetone, ethyl acetate	Aliphatic hydrocarbons, benzene, dichloro ethane, chlorobenzene, ethanol
Cellulose trinitrate	Halogenated hydrocarbons, acetone, methyl amyl ketone, cyclohexanone, ethyl acetate, ethyl butyrate, ethylene glycol, ethyl acetate, nitrobenzene	Aliphatic hydrocarbons aromatic hydrocarbons, lower alcohols ethylene glycol, diethyl ether, water
Ethyl cellulose	Benzene, toluene, methylene chloride, alcohols, esters	Hydrocarbons, decaline, xylene, carbon tetrachloride, n-propyl ether.
Nylon 6	m-cresol, chlorophenol, formic acid, acetic acid, trichloroacetic acid, sulphuric acid, phosphoric acid, hexamethyl tris-phosphoric acid, hexamethyl tris-phosphoramide	Hydrocarbons, chloroform, alcohols, ethers, ketones, esters
Nylon 6, 6	Trifluoroethanol, trichloroethanol, phenol, cresols, formic acid, halogenated acetic acid, sulphuric acid, phosphoric acid	Hydrocarbons, aliphatic alcohols, chloroform, diethyl ether, aliphatic ketones and ethers
Polyacrylonitrile	o-, m- and p-phenylene diamines, N-nitroso piperidine, maleic anhydride, dioxanone, p-dioxanedione, ethylene oxolate, propylene carbonate, 2-oxazolidone, dimethyl formamide, cyanoacetic acid, N-methyl acetamide, dimethyl acetamide, hydroxy acetonitrile, adiponitrile, dimethyl sulphoxide, dimethyl sulphone, sulphuric acid, nitric acid	Hydrocarbons, chlorinated hydrocarbons, alcohols, diethyl ether, ketones, formamides, acrylonitrile
Polybutadiene/ polyisoprene/natural rubber	Aliphatic, cycloaliphatic and aromatic hydrocarbons, chlorinated hydrocarbons, tetrahydrofuran, higher ketones, higher aliphatic esters	Alcohol, lower ketones and esters, nitromethane, propionitrile, water
Polybutyl acrylate	Aromatic hydrocarbons, chlorinated hydrocarbons, tetrahydrofuran, esters, ketones, butanol	Aliphatic hydrocarbons, diethyl ether, methanol, ethanol, ethyl acetate
Polyethyl acrylate	Aromatic hydrocarbons, chlorinated hydrocarbons, tetrahydrofuran, esters, ketones, methanol, ethanol, butanol, glycol ether	Aliphatic alcohols of more than 5 C-atoms, cyclohexanol, aliphatic hydrocarbons, diethyl ether
Polyethylene		
(i) High density	*Above 80°C*: Aliphatic, cycloaliphatic and aromatic hydrocarbons, halogenated aliphatic, cycloaliphatic and aromatic hydrocarbons, higher aliphatic esters and ketones di-n-amyl ether	All common organic solvents at room temperature; highly polar organic solvents at elevated temperatures
(ii) Low density	As above but dissolve between 40°C and 60°C	As above
Polyethylene terephthalate	Chloral hydrate, phenol, chlorophenol nitrobenzene, dimethyl sulphoxide (hot), halogenated aliphatic carboxylic acids	Hydrocarbons, chlorinated hydrocarbons, aliphatic alcohols, ketones, ethers
Polyisobutylene	Aliphatic, cycloaliphatic and aromatic hydrocarbons, chlorinated aliphatic, cycloaliphatic and aromatic hydrocarbons, tetrahydrofuran, dioxane, aliphatic ethers, higher esters, higher alcohols, carbon disulphide, ethyl sulphide	Lower ketones and esters, lower alcohols, nitromethane, propionitrile

Polymers	Solvents	Non-solvents
Polymethyl acrylate	Aromatic hydrocarbons, chlorinated hydrocarbons, tetrahydrofuran, esters, ketones	Aliphatic hydrocarbons, diethyl ether, methanol, ethanol, carbon tetrachloride
Polymethyl methacrylate	Benzene, toluene, xylene, methylene chloride, chloroform, chlorobenzene, dioxane, methyl ethyl ketone, cyclohexanone, acetic acid, methyl formate, ethyl acetate	Hexane, cyclohexane, mineral oil, methanol, ethylene glycol, glycerol
Polypropylene		
(i) Atactic	Hydrocarbons and chlorinated hydrocarbons, isoamyl acetate, diethyl ether	More polar organic solvents with a small hydrocarbon group at elevated temperature
(ii) Isotactic	See polyethylene	
Polystyrene	Cyclohexane (above 35°C), benzene, toluene, ethyl benzene, lower chlorinated hydrocarbons, tetrahydrofuran, dioxane, methyl ethyl ketone, diisopropyl ketone, cyclohexanone, ethyl acetate, butyl acetate, carbon disulphide	Saturated hydrocarbons, alcohols, diols, ethylene chlorohydrine, diethyl ether, acetone
Polyvinyl acetate	Benzene, toluene, chloroform, chlorobenzene, methanol, allyl alcohol, benzyl alcohol, tetrahydrofuran, dioxane, glycol ethers, acetone, methyl ethyl ketone, acetic acid, lower aliphatic esters, acetonitrile, nitromethane, dimethyl formamide, dimethyl sulphoxide, chloroform, chlorobenzene	Saturated hydrocarbons, ethanol, cyclohexanol, diethyl ether, carbon disulphide
Polyvinyl alcohol	Glycols (hot), glycerol (hot), triethylene diamine, formamide, dimethyl formamide, dimethyl sulphoxide (hot), water	Hydrocarbons, chlorinated hydrocarbons, lower alcohols, tetrahydrofuran, dioxane, ketone, esters
Polyvinyl chloride	Tetrahydrofuran, methyl ethyl ketone, cyclopentanone, cyclohexanone, dimethyl form-amide, nitrobenzene, dimethyl sulphoxide	Aliphatic hydrocarbons, mineral oils, aromatic hydrocarbons, alcohols, vinyl chloride, esters, carbon disulphide

The probability of losing some low molecular weight species of the polymer is, of course associated with the process of reprecipitation.

In the case of emulsion polymerisation, the polymer can be isolated by breaking the emulsion with an electrolyte, such as hydrochloric acid, sodium chloride or alum, or by freezing. The polymer crumbs obtained can be washed with water to free them from the contaminant emulsifier or can be purified by dissolution and precipitation. Polymers obtained in the solid form (as in the case of precipitation or suspension polymerisation) can also be purified by the precipitation method so that a pure sample for analytical purposes can be obtained.

Purified polymers in all the above cases can be freed from the non-solvent either by drying in a vacuum at a suitable temperature or by employing infrared lamps. Care should be taken to ensure that polymer degradation does not occur. This can be achieved by keeping the drying temperature as low as possible, say, around 60°C, or by drying in an inert atmosphere of, say, nitrogen.

Low molecular weight polymers obtained as viscous liquids can be purified by precipitation using a non-solvent. The liquid polymer contaminated with the non-solvent can be dried at 60°C to

80°C by employing a flash/rotary evaporator. Here, as the flask containing the liquid polymer is rotated, the viscous liquid polymer will spread and form a thin layer on the inner surface of the flask, resulting in the rapid and effective removal of the non-solvent under the influence of heat and vacuum. In most cases, the liquid polymer containing the non-solvent appears turbid and as the latter is removed, the polymer will progressively become clearer. When the liquid polymer becomes completely clear, it indicates that the non-solvent has been fully removed. When all the non-solvent is removed, the liquid polymer can be transferred to the appropriate container. Generally, it is better to transfer the dried liquid polymer in the hot condition as the viscosity of the liquid polymer will be low at high temperature and hence, relatively free flowing.

13.3 POLYMER FRACTIONATION

We have seen that in a polymer sample, all the molecules do not have the same molecular weight and that the sample, as such, has a certain average molecular weight. The whole sample can, however, be separated into several smaller samples, each having a narrower molecular weight distribution and different average molecular weights of their own. Each of the smaller samples is known as a fraction of the original unfractionated sample. The process of separating the unfractionated sample into several fractions of progressively increasing (or decreasing) average molecular weight is known as *polymer fractionation*. From the knowledge of the weights of the individual fractions and their corresponding average molecular weights, we can compute the distribution curve of the original unfractionated sample. The pattern of the distribution curve for an unfractionated polymer sample and those for a set of fractions obtainable from it are shown in Fig. 13.3.

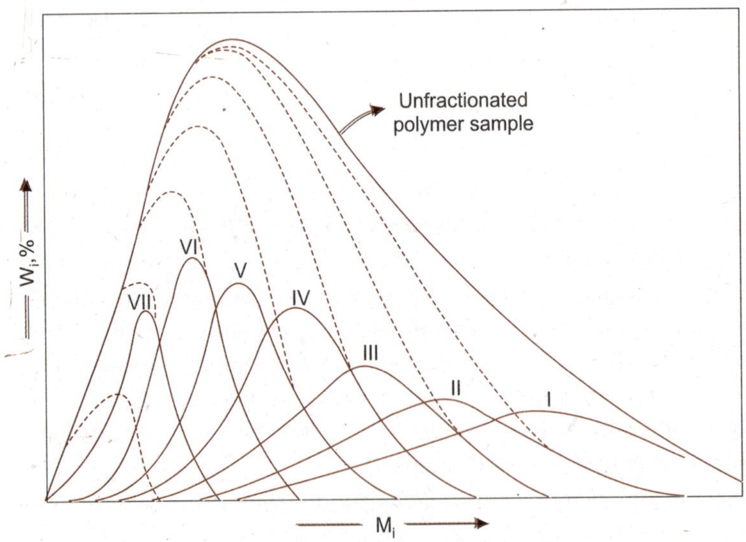

Fig. 13.3 Pattern of molecular weight distribution curve for an unfractionated polymer sample and those for a set of fractions (I to VII) obtainable from it (The curves shown in broken lines represent the distribution for the polymer remaining in solution after successive fractions are removed.).

However, before knowing the procedure for computing the molecular weight distribution curve, we should know how to separate the original sample into several fractions. There are, in fact, several methods available for this, the most important being fractional precipitation, partial dissolution, gradient elution and gel permeation chromotographic techniques.

13.3.1 Fractional Precipitation Technique

When a small quantity of a non-solvent (miscible with the solvent) is added to a polymer solution in a poor solvent*, the non-solvent decreases the solvent power further to such an extent that the molecular species of the highest molecular weight can no longer be held in solution and hence, are thrown out of the solution as swollen gel. If the swollen gel is removed and a further quantity of the non-solvent is added to the remaining solution, the solvent power is further decreased and the molecular species of the next higher molecular weight are thrown out. In this way, each time some quantity of the non-solvent is added to a polymer solution, a partial precipitation of the polymer occurs. The fractional precipitation technique makes use of this principle.

The major difference between fractionation by precipitation (also called *partial precipitation*) and purification by precipitation lies in the quantities of the polymer solution and the non-solvent taken. In the former case, small quantities of non-solvents are added to a large volume of the solution, whereas, in the latter case, only a small quantity of the solution at a time is added to a large volume of the non-solvent.

The experimental procedure for the fractional precipitation is as follows. A 1% solution of the polymer in a poor solvent is taken in a sufficiently large container with provision for stirring and dropwise addition of the non-solvent. The whole setup is thermostated so as to maintain the contents at a constant temperature within ±0.01°C. Since dissolution and precipitation are equilibrium processes and sensitive to temperature change, the whole process of fractional precipitation should be carried out at a constant temperature for effective fractionation. After allowing a sufficient time for the contents to attain the required temperature, the non-solvent is added drop-wise, with continuous stirring. As the non-solvent is added progressively, turbidity appears, indicating the precipitation of the highest molecular weight species. When sufficient turbidity has been developed, the addition of the non-solvent is stopped and the contents are slowly heated till the turbidity disappears. In this heating process, the precipitated polymer gets redissolved at a slightly higher temperature. At this stage, the stirring is stopped and the setup is allowed to cool down slowly to the original precipitation temperature. Now, the turbidity reappears. A proper equilibrium between the low molecular weight species in the dissolved state and the high molecular weight species in the precipitated state is established, resulting in an effective fractional precipitation. The whole setup is then allowed to stand at a constant temperature for a sufficient period of time (say, 8 to 10 hours), so that the precipitated polymer fraction settles down in the container as a gel-like mass. The gel is carefully separated by siphoning without disturbing the supernatant clear solution. After the gel has been removed, a fresh lot of non-solvent is added to the clear solution till turbidity appears and the whole sequence of operations repeated till the second fraction is separated. By repeating the process several times, we

*The importance of taking a poor solvent lies in the fact that precipitation of the polymer can be achieved with lesser volumes of the non-solvent.

can collect several fractions. The last fraction, however, is generally recovered by simply evaporating off the solvent/non-solvent mixture.

There are several modifications to the precipitation technique. One such method of fractional precipitation involves the cooling of the polymer solution (in a single solvent). The method is very simple. The solution is made at an elevated temperature, say, 80°C, and then slowly cooled and maintained at a lower temperature, say, 75°C, for some time. Now, the highest molecular weight species are thrown out of the solution and are carefully removed. The solution is further cooled slowly and maintained at a still lower temperature, say, 70°C, when the second fraction gets precipitated. The process is continued till several fractions are obtained. The drawback of the process is that in the case of some polymers simple cooling alone does not precipitate the polymer.

Yet another way of fractional precipitation is by solvent volatilisation. The polymer is dissolved in a binary mixture of solvent and non-solvent. The solvent/non-solvent system is chosen in such a way that the solvent is highly volatile compared to the non-solvent. By inducing a slow suction of the vapour phase above the polymer solution, the solvent slowly evaporates. The solution, therefore, becomes progressively richer in non-solvent. Hence, the polymer starts precipitating out, the high molecular weight species being thrown out first. By controlled evaporation of the solvent and intermittent collection of the polymer precipitated, several fractions can be separated.

13.3.2 Partial Dissolution (Extraction) Technique

The partial dissolution technique is based on the fact that the polymer solubility in a solvent is restricted, depending on the solvent power, by its molecular weight. A poor solvent can, therefore, dissolve only low molecular weight species, whereas a good solvent can dissolve higher molecular weight species too. If we have a series of solvents with increasing solvent powers, by using each solvent separately and starting from the one with the lowest solvent power, we can dissolve the fractions of different molecular weights one after another starting from the lowest fraction.

The experimental procedure can be outlined as follows. The polymer sample to be fractionated is taken in the form of fine powder or as small pieces (as small as possible). A series or solvent mixtures with increasing solvent powers is developed by using binary mixtures, comprising the solvent and non-solvent in different ratios. It should be noted that the mixture having the lowest solvent to-non-solvent ratio will have the least solvent power and the one having the highest ratio will have the highest solvent power. Next, the binary mixture of the lowest solvent power is added to the powdered polymer sample. The polymer is allowed to swell fully and then the entire content is stirred well to enhance the process of dissolution of the lowest molecular weight species in the solvent mixture. Depending on the quantity of the polymer taken, the stirring time should be adjusted to achieve the complete dissolution of all the soluble components. At this stage, the stirring is stopped, the solution filtered off and the filtrate collected. The first fraction is recovered by completely evaporating the solvent from the filtrate. To the swollen mass left undissolved by the first lot of the binary mixture (*i.e.*, to the solid obtained by filtration), the binary mixture with the next higher solvent power is added and stirred and the whole process repeated till the second fraction is recovered. In this way, by successive dissolution with binary mixtures of increasing solvent powers, a series of fractions with increasing average molecular weights can be separated.

13.3.3 Gradient Elution Technique

The gradient elution technique too works on the partial dissolution principle explained above. In this technique, the polymer is precipitated on glass wool packed at the top of a column filled with silica or alumina, as shown in Fig. 13.4. The polymer is eluted successively with solvent/non-solvent mixtures of increasing solvent power. The lowest molecular weight fraction will be eluted first and the highest one will emerge last from the column. The efficiency of the method is increased by maintaining a temperature gradient in the column. The upper end of the column is kept at a slightly higher temperature than the lower end. The temperature gradient makes each molecular species undergo a series of dissolutions and precipitations at different levels of the column and hence, improves the efficiency of fractionation.

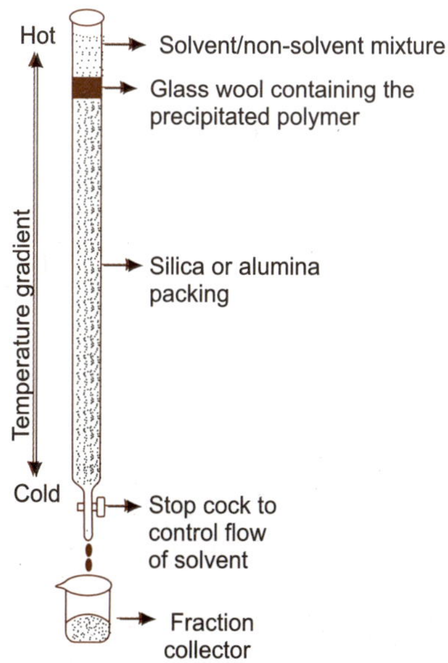

Fig. 13.4 A simple laboratory setup used for gradient elution method of fractionation.

13.3.4 Gel Permeation Chromatographic Technique

A rather quick and efficient method for polymer fractionation and simultaneous computation of the molecular weight distribution curve is provided by the Gel Permeation Chromatography (GPC).

The principle of GPC is as follows. In any column chromatographic process, as the solute travels inside the column from one end to another, it is transferred between two phases, one remaining stationary and the other mobile. In solid-liquid chromatography, the stationary phase is a solid (such as silica, alumina or sintered glass powder packed in the column) and the mobile phase is a liquid (*i.e.*, the solvent used to elute the solute). In liquid-liquid chromatography, both the phases are liquids, the two being immiscible with each other.

In GPC, however, a single liquid, *viz.*, the solvent itself acts as both stationary as well as mobile phases. But how can this be possible? The column in GPC is filled with a material, called *'gel'**. The latter is in the form of fine spherical beads (usually 100 μ in diameter and made of cross-linked styrene-divinyl benzene copolymer or glass) containing a large amount of micropores of uniform size. The gel beads are hard and incompressible and are prepared in different porosity grades, *i.e.*, having a pore size ranging form 50 Å to 1,000 Å. The column can be packed either with gel beads of the same pore size or with beads of different porosities. When the GPC column is filled with a suitable solvent, the solvent occupies not only the free volume between the beads (void volume) but also the volume of the pores (pore volume) inside the gel beads. The solvent phase occupying the void volume acts as the mobile phase, while that occupying the pore volume acts as the stationary phase (Fig. 13.5).

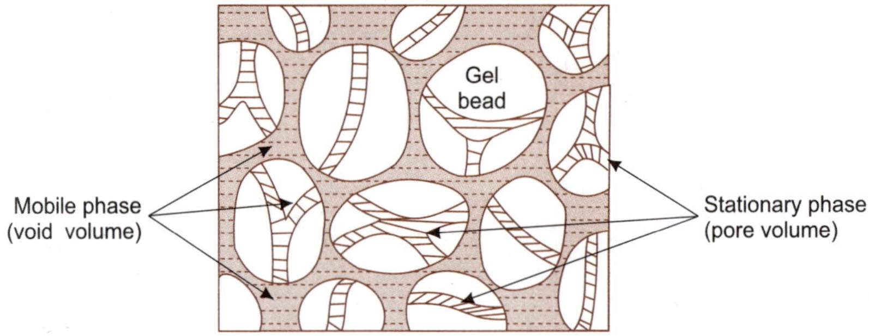

Fig. 13.5 Schematic diagram showing gel beads, the pore volume and void volume in a GPC column.

The GPC operates on the principle that polymer molecules in solution separate according to their size and not chemical interaction or chemical retention. The separation is based on the difference in the hydrodynamic volumes of polymer molecules of varying molecular weights, in their dissolved state. When a dilute solution (say, 0.5% concentration) of the polymer sample in a suitable solvent is injected at the top of the column and the column is eluted with the solvent, the polymer solution along with the solvent starts streaming down the column. The mobile phase around the gel beads now contains polymer molecules, whereas the stationary phase within the gel beads does not contain any. Due to this difference in polymer concentration between the mobile and the stationary phases, the polymer molecules start diffusing into the stationary phase so as to equalise the concentration. Although all the polymer molecules try to enter the pore volume, the pore size restricts the entry and allows molecules up to a certain hydrodynamic size only to enter the pores. Molecules of bigger size are thus 'excluded'** from entering the pores and are washed down the column by the solvent. The highest molecular weight fraction therefore elutes out of the column first. Among the molecules that can enter the pores the lowest molecular weight species, due to their lowest hydrodynamic size, are able to diffuse faster into the pores and hence, occupy the maximum part of the pore volume; while those of intermediate sizes diffuse slowly and are able to occupy only a part of the pore volume. As

*The term 'gel' used here is different from the one used earlier in Section 12.3 and other sections.
**For this reason, GPC is termed size exclusion chromatography.

the column is eluted further with a pure solvent, the mobile phase becomes devoid of polymer molecules, whereas the stationary phase contains a large quantity. This difference in concentration drives the polymer molecules in the opposite direction, which now start diffusing out of the pores. In this diffusion process (in and out of the pore volume), the smallest molecular weight species are retained in the column for the maximum period and hence, elute last. The molecules of intermediate sizes elute inbetween.

If we denote the pore volume by V_p and the void volume by V_0, then the largest molecules, which are totally excluded from the pores, elute out of the column when a solvent volume equal to V_0 elutes out after the injection of the sample. The smallest molecules which permeate the maximum volume of the pores elute out at a volume of $(V_0 + V_p)$. The intermediate size molecules elute between V_0 and $(V_0 + V_p)$. Hence, the elution volume directly gives an estimate of the molecular size. In practice, however, no attempt is made to actually compute V_0 or V_p. Calibration curves in the form of log$M$ versus elution volume V_E are drawn using a series of *'standards'* of known molecular weights and narrow molecular weight distribution. Chromatograms of these standards are run and the elution volume corresponding to the peak position of bands obtained for each '*standard*' sample (Fig. 13.6) is read. From the knowledge of the molecular weight of the standards and their corresponding elution volume (see Table 13.2), a calibration curve of log$M$ against V_E is drawn (Fig. 13.7). This peak position calibration method has been found to give reliable results, provided the standards used are of narrow molecular weight distribution (*i.e.*, $\overline{M}_w/\overline{M}_n < 1.2$) and the chemical types of the *'standard'* and the sample are the same.

Table 13.2 GPC data for calibration standards (polystyrene) in toluene solutions at 25°C

Polystyrene standards, serial no.	*Average molecular weight of the standard, $\overline{M}$*	*log $\overline{M}$*	*Elution volume corresponding to the standard, V_E (ml)*
1	450,000	5.6532	31.80
2	240,000	5.3802	33.00
3	100,000	5.0000	34.80
4	50,000	4.6990	36.00
5	19,000	4.2788	38.00
6	9,000	3.9542	40.00
7	4,000	3.6021	41.50
8	2,000	3.3010	43.10

*Data computed from Fig. 13.6

The GPC apparatus (Figs. 13.8*a* and *b*) consists of two columns: one for injecting the sample and other serving as a reference column. The solvent elutes through both the columns. A differential refractometer is used to measure the refractive index difference between the polymer solution and the pure solvent eluting out of the sample and reference columns, respectively. This refractive index difference directly gives the concentration of the polymer being eluted at different elution volumes. From a knowledge of elution volume and refractive index and by comparison with the calibration curve, we can directly get the weight and the molecular weight of different fractions eluting at

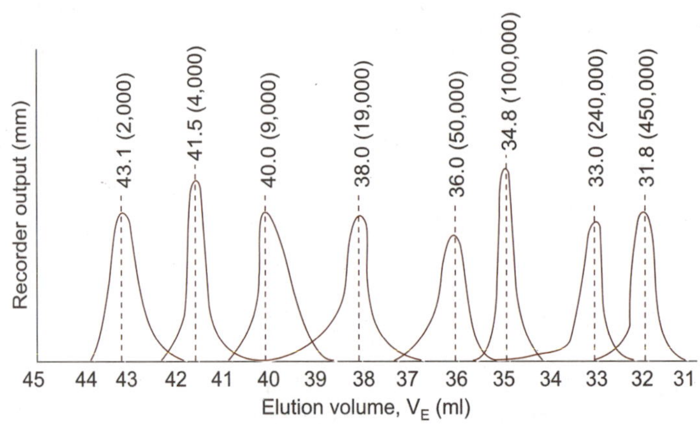

Instrument: Waters ALC/GPC-244
Column: μ Styragel 10^5, 10^4, 10^3, 500 and 100 Å
Sensitivity: (RI) 2x
Pressure: 700 psi

Solvent: Toluene
Concentration: 0.5%
Injection volume: 100 μl (each)
Temperature: 25°C

Fig. 13.6 Gel permeation chromotograms of a number of 'standard' polystyrene samples showing elution volumes corresponding to different molecular weights (figures given in brackets are molecular weights).

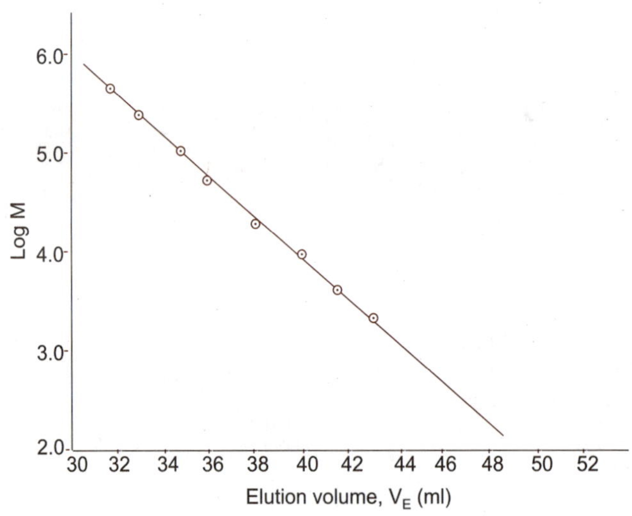

Fig. 13.7 GPC calibration curve showing the relation between log M and V_E (Data from Fig. 13.6).

different elution volumes. In the case of completely automated equipments, the data are processed and recorded in the form of a plot of elution volume versus recorder output of the differential refractometer. The elution volume corresponds to the molecular weight and the recorder output to the weight concentration of the polymer. This plot directly gives the molecular weight distribution. Such

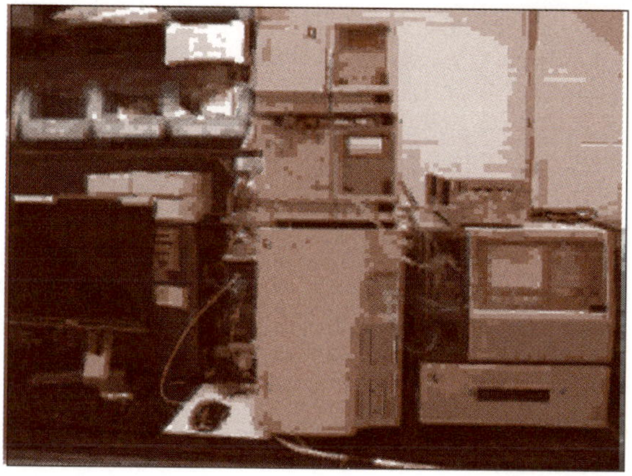

(a)

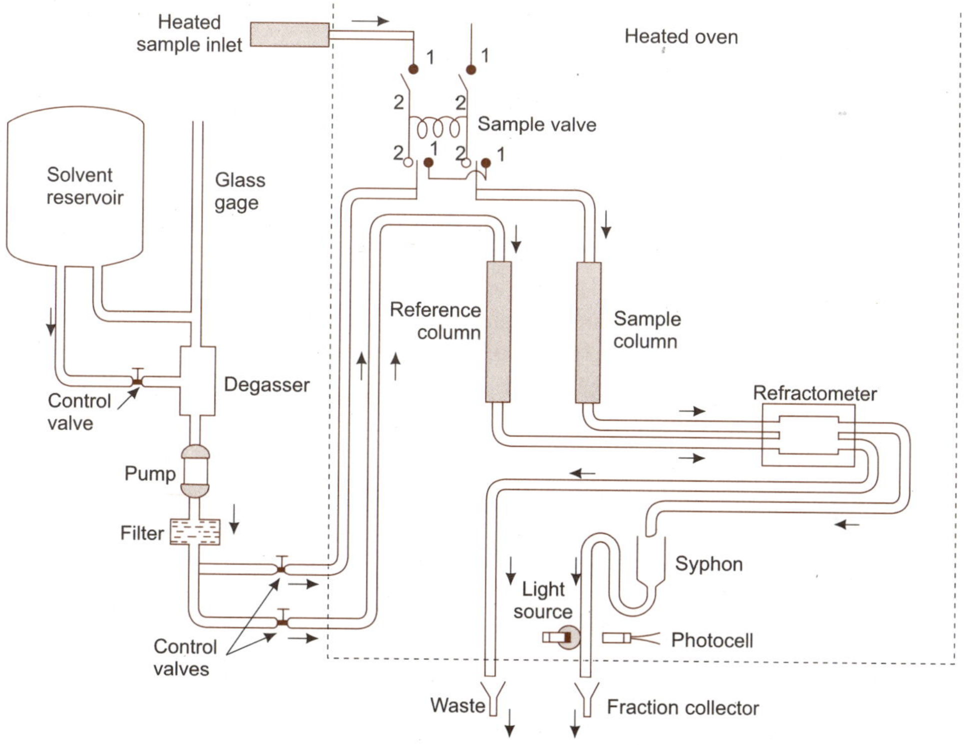

(b)

Fig. 13.8 (a) GPC apparatus. (b) Flow scheme of a gel permeation chromatograph.

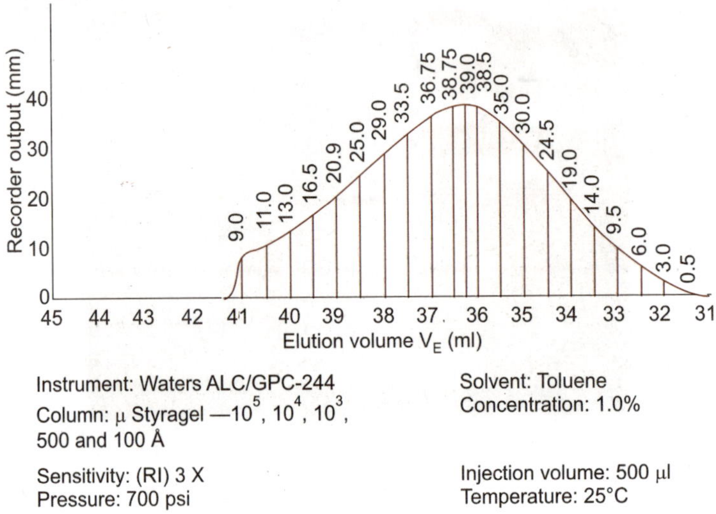

Fig. 13.9 GPC chromatogram of a polystyrene sample.

a distribution curve for a polystyrene sample is given in Fig. 13.9. This curve can also be used to find out the $\overline{M}_w$ and $\overline{M}_n$ values of the polymer sample as now explained. From the curve, the heights (h) of the recorder output axis corresponding to different elution volumes (V_E) are read and from the calibration curve, molecular weights corresponding to these elution volumes are noted. The values are tabulated as shown in Table 13.3. The height of the curve in Fig. 13.9 at any particular elution volume is a measure of the concentration of molecules of the corresponding molecular weight. This height h can, therefore, be equated to n_iM_i and the molecular weight M to M_i of the various equations we have come across in Chapter 3. The $\overline{M}_n$ and $\overline{M}_w$ of the sample can be calculated as follows:

From Table 13.3, We have

$$\Sigma h = 452.40,\ \Sigma(h/M) \times 10^3 = 16.2107 \text{ and } \Sigma(h\cdot M) = 29{,}552{,}925$$

Now,

$$\overline{M}_n \quad \frac{\Sigma n_i M_i}{\Sigma n_i} \quad \frac{\Sigma h}{\Sigma(h/M)} \quad \frac{452.40\times 10^3}{16.2107} = 27{,}907 \text{ (or, say) } 28{,}000$$

$$\overline{M}_w \quad \frac{\Sigma n_i M_i^2}{\Sigma n_i M_i} \quad \frac{\Sigma(h.M)}{\Sigma h} \quad \frac{29{,}552{,}925}{452.40} = 65{,}325 \text{ (or, say) } 65{,}500$$

13.4 MOLECULAR WEIGHT DETERMINATION

In Chapter 3, we have referred to various molecular weights *viz.*, number-average ($\overline{M}_n$), weight-average ($\overline{M}_w$), viscosity-average ($\overline{M}_v$), etc. Let us now see, how they are experimentally determined.

Table 13.3 GPC data for a polystyrene sample*

Elution volume, V_E (ml)	Recorder output height, h (mm)	Molecular weight, $\bar{M}$	(h/M) × 10^3 (mm)	(h.M) (mm)
31.5	0.5	500,000	0.0010	250,000
32.0	3.0	395,000	0.0075	1,185,000
32.5	6.0	310,000	0.0193	1,860,000
33.0	9.5	240,000	0.0395	2,280,000
33.5	14.0	190,000	0.0736	2,660,000
34.0	19.0	150,000	0.1266	2,850,000
34.5	24.5	120,000	0.2041	2,940,000
35.0	30.0	94,000	0.3191	2,820,000
35.5	35.0	74,000	0.4729	2,590,000
36.0	38.5	58,000	0.6637	2,233,000
36.25	39.0	52,000	0.7500	2,028,000
36.5	38.75	45,500	0.8516	1,763,125
37.0	36.75	36,000	1.0208	1,323,000
37.5	33.5	29,000	1.1551	971,500
38.0	29.0	22,500	1.2888	652,500
38.5	25.0	17,500	1.4285	437,500
39.0	20.9	14,000	1.4928	292,600
39.5	16.5	11,000	1.5000	181,500
40.0	13.0	8,600	1.5116	111,800
40.5	11.0	6,800	1.6176	74,800
41.0	9.0	5,400	1.6666	48,600

*Data computed from Figs. 13.7 and 13.9.

The type of average obtained depends on the experimental technique and the property of the polymer molecules being measured. Any experiment, in which the measured property is dependent on the total number of molecules present regardless of their masses, will yield the number-average molecular weight. Precisely for this reason, physical methods based on cryoscopy, ebulliometry and osmometry which are colligative in nature give $\overline{M}_n$ Some chemical methods are also used to determine $\overline{M}_n$, wherein the chemical reactivity is a direct function of the number of molecules present in the sample analysed. End group analysis is a chemical method employed to evaluate the number-average molecular weight of a polymer having reactive functional end groups. Experiments based on viscosity measurements yield the viscosity-average molecular weight $\overline{M}_v$. There are a few measurements where polymer molecules contribute to the measured property according to their size. Here, bigger molecules contribute substantially more to the measured property than the smaller ones do. These measurements give the weight-average molecular weight $\overline{M}_w$. Light scattering and ultracentrifuge techniques are examples of such types of measurements.

The important point to be noted is that all these measurements are made on polymer molecules in solution and that they are concentration dependent. Moreover, the mathematical equations (relating the properties measured to the molecular weight) used in the following sections are applicable to solutions of very low concentration only, say, less than 1.0 g/dl. All measurements should, therefore, be done with solutions below this concentration.

13.4.1 Cryoscopy

We know that the freezing point of a solvent is lowered when a non-volatile solute is dissolved in it. For example, water which freezes at 0°C will no longer freeze at this temperature if common salt or calcium chloride is dissolved in it. With the dissolution of a solute in it, the freezing point of water is *'depressed'*. In general, solutions freeze at a temperature lower than that at which corresponding pure solvents do. This phenomenon of depressing the freezing point or a liquid by the addition of a solute is made use of in the technique known as cryoscopy. The interesting feature to be noted is that the extent of the freezing point depression depends on the number of solute molecules dissolved per unit volume of solution and is independent of the size (mass) and nature of the solute molecule.

This means that, when the cryoscopic method is applied to determine the molecular weight of polymers, what we get is the number-average molecular weight $\overline{M}_n$. If the depression of the freezing point is denoted by ΔT_f, it is related to the number-average molecular weight and concentration through the following equation:

$$\frac{\Delta T_f}{C} \quad \frac{RT_f^2}{\rho \Delta H_f}\left(\frac{1}{\overline{M}_n} + BC\right) \tag{13.1}$$

where, C is the concentration of the solution; ρ the density of the solvent; R the universal gas constant; T_f the freezing point of the solvent; B a constant called *second virial coefficient;* and ΔH_f the heat of fusion of the solvent. The freezing points of the pure solvent and of very dilute solutions of the polymer sample in different concentrations are determined using standard freezing point techniques. In one method, the solvent or the solution is carefully supercooled to a temperature that is about 5°C lower than the freezing point. The supercooled liquid, when stirred, starts freezing rapidly with a rise in temperature which attains a maximum and then drops down (Fig. 13.10) (The maximum temperature attained is the freezing point of the liquid). Once the freezing points of the solvent and solutions of different concentrations are known, ΔT_f can be calculated for all these concentrations. Using these data, a graph of ΔT_f versus C is plotted and extrapolated to infinite dilution, *i.e.*, zero concentration (Fig. 13.11). From the value of the ordinate intercept of the plot at zero concentration, $\overline{M}_n$ can be computed from the equation

$$\left(\frac{\Delta T_f}{C}\right)_{C\to 0} = \frac{RT_f^2}{\rho \Delta H_f} \cdot \frac{1}{\overline{M}_n} \tag{13.2}$$

One of the serious drawbacks of this method is the error introduced in the measurement of temperatures. Many times, the difference in temperature between the freezing points of the solvent and solution may be of the order of 0.001°C; hence, the need for high-precision temperature sensors. The advent of thermistors and the high amplification effected by sophisticated electronic devices have made temperature measurements more accurate and meaningful. Now, temperature changes as

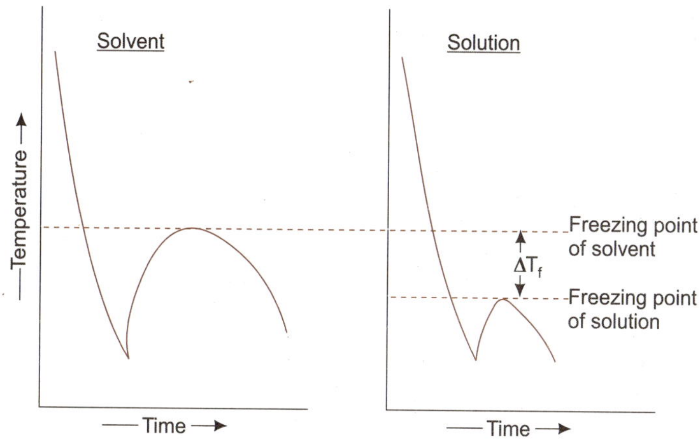

Fig. 13.10 Freezing point curves for a solvent and a solution.

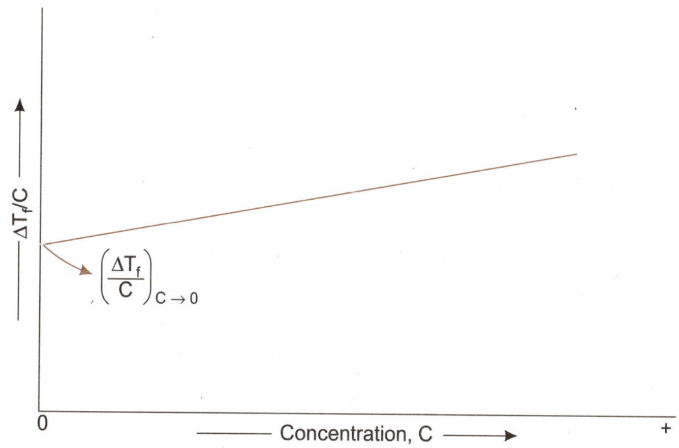

Fig. 13.11 Representative plot of $\Delta T_f/C$ against C.

small as 0.0001°C can be measured with precision. Nevertheless, the cryoscopic technique has an inherent limitation in that the measurement of a molecular weight beyond a certain figure starts becoming unreliable. This is evident from Eqn.(13.1) itself. As the molecular weight ($\overline{M_n}$) increases, the depression in freezing point (ΔT_f) assumes smaller and smaller values, which become increasingly difficult to measure with accuracy. Molecular weights up to 30,000 can be determined accurately by this method. For this molecular weight range, however, vapour phase osmometry (see Section 13.4.4) provides a more reliable and easily adoptable method of measurement, as compared to cryoscopy and ebulliometry.

13.4.2 Ebulliometry

Like cryoscopy, ebulliometry evolved itself into a useful technique for the determination of molecular weights of solutes. It is based on a very simple observation that the boiling point of a solution (of a non-volatile solute) is higher than that of the pure solvent. For example, a salt solution in water boils at a temperature higher than 100°C. This phenomenon is made use of in determining the molecular weight of polymers.

The basic equations, which relate the boiling point elevation ΔT_b to the solute molecular weight and concentration, are as follows:

$$\frac{\Delta T_b}{C} \quad \frac{RT_b^2}{\rho \Delta H_v}\left(\frac{1}{\overline{M}_n} + BC\right) \tag{13.3a}$$

$$\left(\frac{\Delta T_b}{C}\right)_{C\to 0} \quad \frac{RT_b^2}{\rho \Delta H_v} \cdot \frac{1}{\overline{M}_n} \tag{13.3b}$$

where ΔH_v and T_b are the heat of vapourisation and the boiling point of the solvent, respectively.

As far as the experimental part of ebulliometry is concerned, there are two general types of ebulliometers. In the first type, the solution of the polymer (whose molecular weight is to be determined) is heated in a boiler. The solution is then brought in contact with a sensor. Side by side, another temperature sensor measures the condensation temperature of the solvent. The boiling point of the solution is thus compared directly with that of the solvent. In the second type, the solvent and the solution are kept in identical containers and uniform boiling is achieved through mild rocking. The containers are provided with thermistors for measuring temperatures. Number-average molecular weights up to 30,000 can be determined fairly accurately by ebulliometry. This is because, like in cryoscopy, the problem relating to the accurate measurement of extremely small temperature changes exists in the case of ebulliometry also.

13.4.3 Membrane Osmometry

Membrane osmometry is a very widely used technique to determine the number-average molecular weight of polymers. It is based on the phenomenon of osmosis, which can be described as follows. If a pure solvent is separated from the solution by a semipermeable membrane*, on account of the difference in chemical potentials between the solvent and the solution, the solvent molecules diffuse into the solution through the membrane till the solution is rendered infinitely dilute. The flow of solvent into the solution could, nevertheless, be practically stopped by applying an appropriate pressure on the solution side. Osmotic pressure of a solution can thus be defined as the pressure that must be applied to the solution so as to totally prevent the flow of the solvent through the semipermeable membrane into the solution.

The osmotic pressure (π) of a polymer solution is related to the number-average molecular weight of the polymer by the relations

---

*A semipermeable membrane is one which allows only one species to pass through but not the others.

$$\frac{\pi}{RTC} = \frac{1}{\overline{M}_n} + BC \tag{13.4}$$

$$\left(\frac{\pi}{RTC}\right)_{C \to 0} = \frac{1}{\overline{M}_n} \tag{13.5}$$

A plot of (π/RTC) versus C therefore, gives a straight line, the ordinate intercept of which will give 1 $\overline{M}_n$. Such plots for a series of cellulose acetate fractions in acetone are given in Fig. 13.12. This figure shows that all the plots are straight lines with different intercepts. The striking feature is that [as predicted by Eqn.(13.4)] all these plots have approximately the same slope. Since the solvent and the polymer types are the same, the slopes for the plots of different molecular weight fractions will almost be the same. However, each fraction of cellulose acetate has a different molecular weight and hence, the plots will have different ordinate intercepts. The nature of the plots will be different if the osmotic pressure of a single fraction is measured in different solvents. Here, all the plots will have the same ordinate intercept since the molecular weight is the same. But as we have used different solvents, the plots will have different slopes as shown in Fig. 13.13.

Different types of membrane osmometers are now commercially available. For instance, the Fuoss-Mead block osmometer, the Pinner-Stabin glass osmometer, the Stabin-Shell automatic osmometer and the Reiff osmometer. These osmometers basically fall into two categories, depending on whether the static or the dynamic equilibrium method has been used for osmotic pressure measurement. The static equilibrium osmometers are less preferred these days. However, description of one variety would not be out or place here as it can provide an understanding of the principle of osmometry.

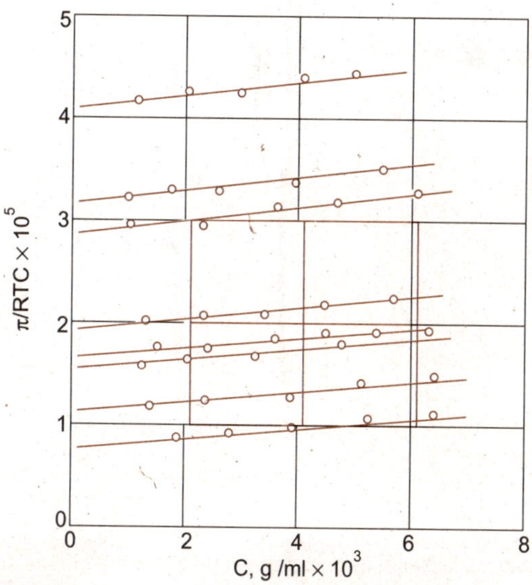

Fig. 13.12 Plots of π/RTC versus C for different fractions of cellulose acetate in acetone solutions.

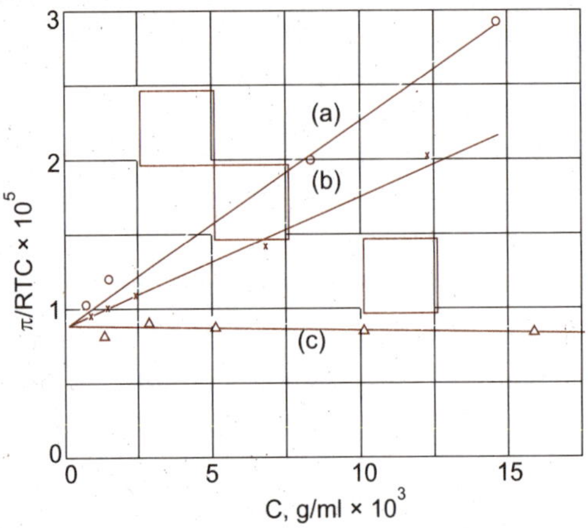

Fig. 13.13 Plots of π/RTC versus C for one sample of nitrocellulose in (a) acetone, (b) methanol and (c) nitrobenzene solutions.

A simple schematic diagram of a static equilibrium osmometer is shown in Fig. 13.14. Essentially, the apparatus consists of two metallic blocks madeup of stainless steel, one face of each block having a series of concentric grooves. These grooves, in turn, are interconnected by vertical-horizontal channels. In between these grooved faces of the two metallic blocks, a semipermeable membrane is placed. On bolting the apparatus, the two internal grooved faces of the metallic blocks come in intimate contact through the semipermeable membrane. The solvent and the solution are kept

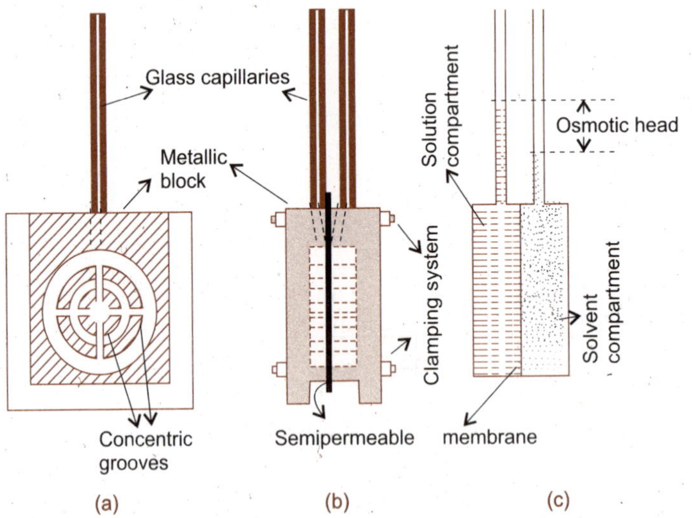

Fig. 13.14 Schematic diagram of static equilibrium osmometer: (a) details of the internal face of one of the two metallic blocks; (b) side view cross-section of the assembled osmometer; and (c) details of the solution and solvent compartments.

in the space formed between the grooves on the two blocks and are prevented from freely mixing by the membrane separating them. The vertical channels of the blocks are connected to glass capillaries of a suitable bore through a drill in the metallic blocks of the solvent and solution compartments. The actual experiment is done as follows. The two compartments are assembled with a membrane in between, both being filled with solvent and made leak-proof. One of the compartment is then emptied of the solvent and filled with the solution of known concentration. The assembly is placed in a thermostat bath maintained at a constant temperature. Due to the osmotic pressure, the solvent will start passing through the membrane to the solution side and this will be reflected in a steady rise of the liquid level in the capillary attached to the solution compartment. The process will continue until the hydrostatic pressure developed in the capillary due to the increased level of the solution exactly balances the osmotic pressure. Once the osmotic equilibrium is achieved, the solvent stops entering the solution side. When this steady state is achieved, the height difference between the solution and the solvent levels in the two capillaries is measured using a cathetometer. The height gives the *'osmotic head'* of the solution for that particular concentration. This procedure is repeated with solutions of different concentrations and the corresponding osmotic heads are measured. Typical data are given in Table 13.4. The calculation of molecular weight from these data is done as follows.

Table 13.4 Osmotic head data for a polymethyl acrylate sample in benzene solution at 30°C *

Concentration, C(g/dl)	*Osmotic head, h (cm)*	*h/C (cm dl/g)*
0.260	0.140	0.538
0.540	0.516	0.955
0.755	0.966	1.279
0.920	1.389	1.510
1.025	1.730	1.688

*Data (unpublished) from Gundiah, S. *et al.*, National Chemical Laboratory, Pune, India.

Let C be the concentration of the solution, g/dl; h the osmotic head, cm; d the density of the solution, g/ml (this can be taken as the density of the solvent since the solution is very dilute); T the temperature at which the measurements are done, °K; h/C the reduced osmotic head, cm dl/g; and $(h/C)_{C\to 0}$ the limiting osmotic head, cm dl/g (this can be computed by plotting h/C against C and extrapolating to zero concentration, see Fig. 13.15).

Let the limiting osmotic head be denoted by L for simplicity.

Now,

$$1 \text{ atmosphere} = 1033 \text{ g. wt/cm}^2$$

$$\text{Therefore } \pi \text{ in atmosphere} = \frac{\text{osmotic head} \times d}{1033}$$

$$(\pi/C)_{C\to 0} = \frac{L.d}{1033} \text{ atmosphere}$$

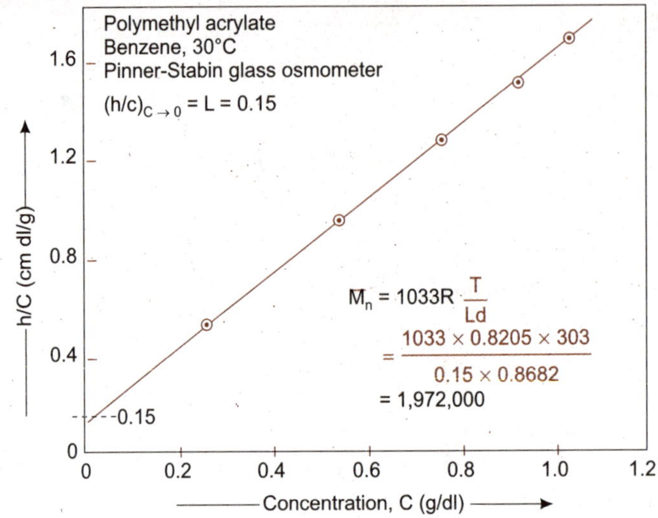

Fig. 13.15 Plot of *h*/*C* versus *C* for a polymethyl acrylate sample in benzene solution.

We know that

$$(\pi/C)_{C\to 0} = \frac{RT}{\overline{M}_n} = \frac{L\cdot d}{1033}$$

$$\overline{M}_n = \frac{1033\cdot RT}{L\cdot d} = \frac{1033\times 0\cdot 8205T}{L\cdot d}$$

The static equilibrium method discussed above is relatively simple, but suffers from one major drawback. Since the diffusion of solvent through the membrane is a very slow process, the time taken to attain equilibrium is extremely high. A single molecular weight determination may take several hours or even a few days. This is the main reason why static equilibrium osmometers have become obsolete these days. This disadvantage is overcome in some of the present-day osmometers which operate on the dynamic equilibrium principle. The High-Speed Membrane Osmometer (HSMO) schematically represented in Fig. 13.16 is one such apparatus and is very commonly used in all modern polymer characterisation laboratories.

The HSMO consists essentially of two cells, the upper cell holding the solution and the lower one, the solvent. They are separated by a horizontal semipermeable membrane. The movement of an air bubble inside the capillary immediately below the solvent cell indicates the solvent flow to the solution cell. Such movement is immediately detected by a photocell. This setup, in turn, is coupled to a servo mechanism. If any movement of the air bubble is detected, the servo mechanism will be stimulated to move the solvent reservoir in the upward or downward direction, and to adjust the hydrostatic pressure such that the solvent flow is completely arrested. The pressure head of the reservoir gives the osmotic head. The main advantage of the high-speed membrane osmometer is that

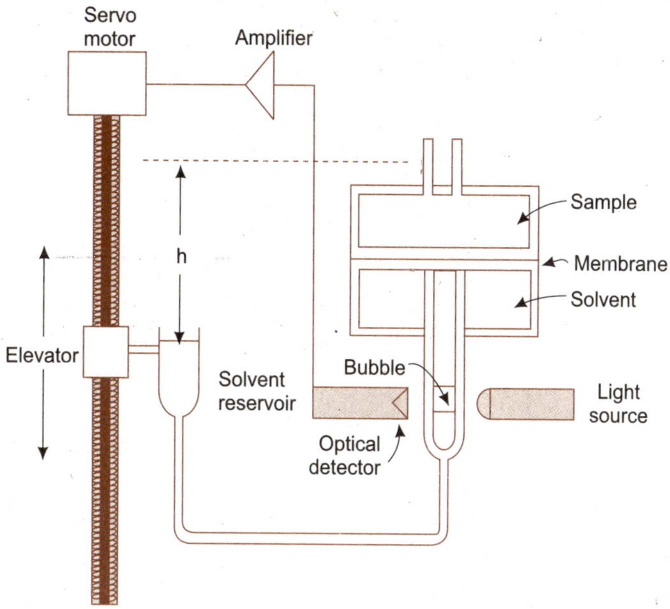

Fig. 13.16 Schematic diagram of essential components of a high-speed membrane osmometer.

it takes hardly 5 to 7 minutes to attain the osmotic equilibrium which, in turn, permits rapid determination of polymer molecular weights.

A limitation of membrane osmometry is the non-availability of true semipermeable membranes. Commercially available membranes such as *'gel-cellophane'* membrane, *'bacterial-cellulose'* membrane and *'denitrated collodion'* membranes are permeable to very low molecular weight polymer species and hence, the average molecular weight measured will always be higher than the actual. Membrane osmometry is useful in the molecular weight range of 30,000 to 1,000,000.

13.4.4 Vapour Phase Osmometry

This technique is based on the principle that at a given temperature the vapour pressure of a solution is less than that of the pure solvent. Now, keep a drop each of a pure solvent and the solution in an atmosphere saturated with the solvent vapour (continuously supplied from a large reservoir of solvent). Since the vapour pressure of a solution is lower than that of the pure solvent, condensation of the solvent takes place from the saturated vapour phase on to the solution droplet, whereas practically nothing happens as far as the solvent droplet is concerned. The solution droplet, therefore, starts getting diluted as well as heated up by the latent heat of condensation of the solvent condensing on it. Owing to the temperature rise and increased concentration of the solvent, the vapour pressure of the solution droplet increases steadily. The process of condensation and the resultant temperature rise continues till the vapour pressure of the solution droplet at the new elevated temperature becomes equal to that of the pure solvent at the original temperature. The total rise in temperature, ΔT, will be proportional to the mole fraction of the solute, n, in the solution, as per the following equation:

$$\Delta T = \frac{RT^2 n}{\Delta H_v} \tag{13.6}$$

where, ΔH_v is the heat of vapourisation of the solvent. In an actual Vapour Phase Osmometer (VPO), also called vapour pressure osmometer, the solution and the solvent droplet are placed directly on two thermistors arranged in a Wheatstone bridge circuit in such a way that the temperature rise can be measured very accurately as a function of the bridge imbalance output voltage, ΔV. The factor ΔV is related to the molecular weight or the solute by the equation

$$\frac{\Delta V}{C} = \frac{K}{\overline{M}_n} + KBC \tag{13.7}$$

where, K is the calibration constant. It may be noted that molecular weights up to 30,000 can be reliably measured using a VPO.

A general description of a VPO is as follows. A solvent vessel madeup of glass is placed in an anodised aluminium block and is sealed by a stainless steel lid through a Teflon gasket (Fig. 13.17). This vessel provides the space for the atmosphere of solvent vapour wherein the thermistors are

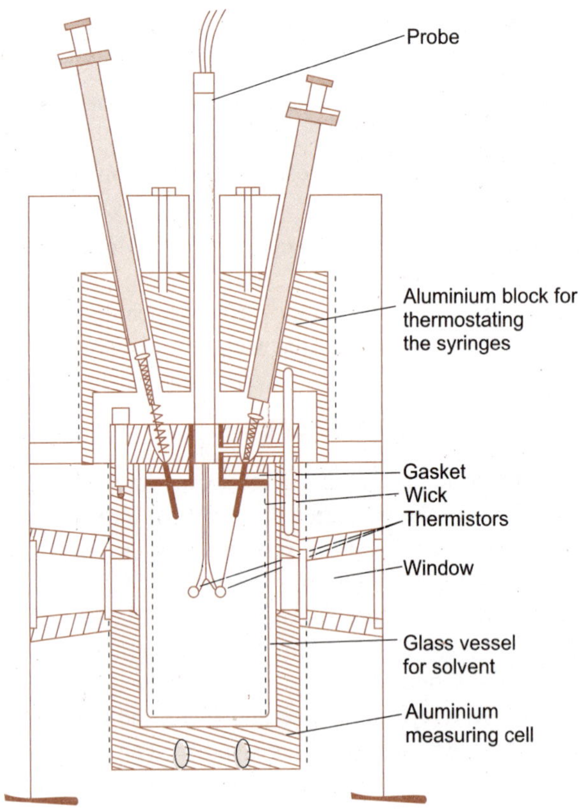

Fig. 13.17 Schematic diagram of a vapour pressure osmometer. (*Courtesy*: Knauer & Co., Berlin)

suspended. The stainless steel lid is covered by another metal block through which several syringes for placing the solution/solvent drops on the thermistors are introduced. The entire assembly is thermostated to maintain a constant temperature with an accuracy of ±0.001°C. At the start of the experiment, the solvent is added to the solvent vessel through a syringe and solvent drops are also placed on each thermistor. Now, since there will not be any condensation on either of the thermistors, there is obviously no temperature difference and hence, the instrument should read zero. If it does not do so, the bridge is adjusted till a steady zero reading is obtained. The next step is to calibrate the instrument with a standard sample of a known molecular weight. Solutions of a standard sample in four to five known concentrations are made. One of the thermistors is flushed with the solution of one particular concentration and a drop of the same is placed on the thermistor, while the other thermistor still contains a drop of the pure solvent. Now, the solvent starts condensing on the solution droplet and the rise in temperature is directly measured as the bridge imbalance output value, ΔV, after the output attains a steady state. The process is repeated with solutions of different concentrations and ΔV values corresponding to each concentration are noted (see Table 13.5). A plot of $\Delta V/C$ versus C is

Table 13.5 Vapour phase osmometry data for a calibration standard (polystyrene) and an experimental sample (HTPB) in toluene solutions at 70°C*

Polymer	Concentration, C (g/kg)	Bridge output, ΔV (μV)				ΔV/C (μV.kg/g)
		Reading 1	Reading 2	Reading 3	Average	
Standard polystyrene of $\overline{M}_n$ = 1790	8	148.4	140.6	140.6	143.2	17.9
	12	218.8	218.8	218.8	218.8	18.2
	16	296.8	304.7	296.8	299.4	18.7
	20	382.8	382.8	382.8	382.8	19.1
HTPB of unknown molecular weight	8	113.3	113.3	113.3	113.3	14.1
	12	171.9	171.9	171.9	171.9	14.3
	16	234.4	234.4	234.4	234.4	14.6
	20	300.8	300.8	300.8	300.8	15.0

*Data (unpublished) from Chemicals Group, Vikram Sarabhai Space Centre, Trivandrum, India.

made and extrapolated to zero concentration, as shown in Fig. 13.18. The ordinate intercept will then be $(\Delta V/C)_{C\to 0}$. The calibration constant K can be computed using the equation

$$K = M\cdot(\Delta V/C)_{C\to 0} \tag{13.8}$$

where, M is the molecular weight of the known standard sample. For determining the molecular weight of an unknown sample, solutions of the same are made in different concentrations in the same solvent used for the standard sample and the whole procedure is repeated to get the ordinate intercept $(\Delta V/C)_{C\to 0}$. The molecular weight of the unknown sample will then be given by

$$\overline{M}_n = \frac{K}{(\Delta V/C)_{C\to 0}} \tag{13.9}$$

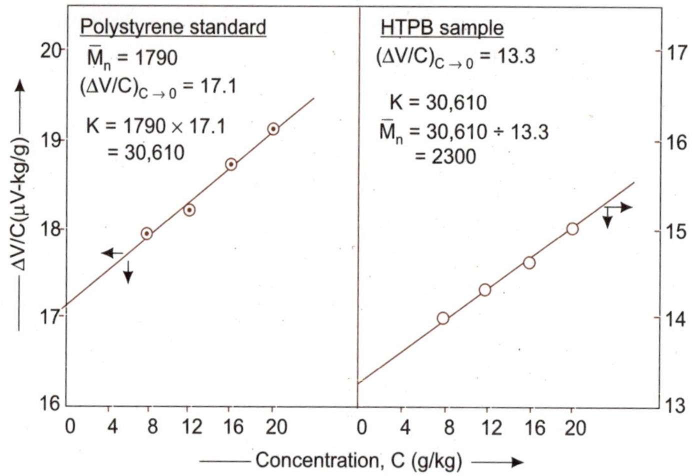

Fig. 13.18 Plots of $\Delta V/C$ versus C for a polystyrene 'standard' and a HTPB sample.

A set of typical data for an HTPB sample is included in Table 13.5 and the corresponding graph in Fig. 13.18.

13.4.5 End Group Analysis

The end group analysis is a chemical method used for calculating the number-average molecular weight of polymer samples whose molecules contain reactive functional groups at one end or both ends of the molecule. For example, we know the structures of Carboxyl-Terminated Polybutadiene (CTPB) and Hydroxyl-Terminated Polybutadiene (HTPB). These molecules have carboxyl or hydroxyl groups at their respective ends. Each molecule theoretically possesses two functional groups. In other words, for every two functional groups identified in a sample of the polymer, one molecule has to be present. We are already familiar with the term *'functionality'* used in polymer chemistry. It is defined as the number of reactive functional groups present in each macromolecule. Thus, the functionality of HTPB and CTPB is two, as each molecule possesses two reactive functional groups.

Apart from hydroxyl and carboxyl, several other end groups (such as aldehyde, amino, ester or methyl) can possibly be present. Usual chemical methods are employed to find out the total number of functional groups present in a given weight of the sample and this is expressed as a functional group equivalent/100 g. From a knowledge of the functional group equivalent and the functionality, the molecular weight can be calculated using the equation

$$\bar{M}_n = \frac{\text{Functionality}}{\text{Functional group equivalent}} \tag{13.10}$$

For instance, the carboxyl value of CTPB can be determined by dissolving a known quantity of the polymer in a mixture of toluene and ethanol (1:3) and titrating against alcoholic potassium hydroxide. The procedure is as follows. Commercial grade ethanol is distilled and neutralised with a dilute solution of alcoholic potassium hydroxide. Then, a 1:3 mixture of ethanol and toluene is made.

A precisely weighed quantity (less than 1 g) of the CTPB sample is dissolved in the ethanol-toluene mixture. This is titrated against a 0.1N alcoholic potassium hydroxide solution using phenolphthalein as the indicator. The end point is the appearance of a slightly pink colour.

Let the volume of KOH consumed = V ml.

Normality of KOH solution = N

Weight of the sample taken = Wg

$$\text{Carboxyl value} = \frac{VN(56.1)}{W}$$

$$\text{Carboxyl equivalent/100 g.} = \frac{VN(56.1)}{W(1000)} \times \frac{100}{56.1} \quad \frac{VN}{10W}$$

In a particular measurement, if 0.8734 g of the sample has consumed 5.1 ml of 0.1242 N alcoholic potassium hydroxide solution, then,

$$\text{Carboxyl equivalent/100 g} = \frac{5.1 \times 0.1242}{10 \times 0.8734}$$

Assuming a functionality of 2, we get

$$\overline{M}_n \quad \frac{2 \times 100}{0.0725}$$

The end group analysis method can be used reliably only for samples consisting of linear molecules with determinable end groups and for those obtained by a known polymerisation mechanism without side reactions. This is because side reactions during polymerisation (such as branching or chain transfer) may introduce errors in the assumed functionality and hence, lead to erroneous results.

The method could, however, be conveniently used to find out the average functionality of any polymer sample. For this purpose, the functional equivalent of the sample is obtained by chemical analysis and the molecular weight determined by any other method such as osmometry. Then the functionality of the polymer sample can be given by the equation

$$\text{functionality} = \text{molecular weight} \times \text{functional equivalent} \tag{13.11}$$

13.4.6 Viscometry

Viscometry is an useful technique for determining the polymer molecular weight. The molecular weight obtained by this technique is the viscosity-average molecular weight, $\overline{M}_v$. The viscosity of a polymer solution is considerably high compared to that of the pure solvent. The increase in viscosity imparted by the macromolecules in solution is a direct function of the hydrodynamic volume and hence, the molecular weight of the macromolecule.

We are already familiar with different relationships between the viscosity of a polymer solution and the molecular weight (see Section 12.6). In this section, we shall recapitulate only the famous Mark-Houwink equation [Eqn. (12.38)]:

$$[\eta] = K\overline{M}^{a} \tag{13.12}$$

where $[\eta]$ is the intrinsic viscosity; $\overline{M}$ the molecular weight, and a and K are constants for a particular polymer/solvent/temperature system. (K and a values are available for many of the known polymers; some such values are given in Tables 12.1 and 12.2). For a polymer type of known K and a values what is required is the determination of the intrinsic viscosity and computation of the molecular weight using Eqn. (13.12). For an unknown polymer system, the K and a values have to be generated. This is done by fractionating the polymer sample into several fractions and for each fraction, the molecular weight is determined by osmometry or by the light-scattering technique and the corresponding intrinsic viscosities are also measured. Typical data are given in Table 13.6.

Table 13.6 Molecular weight and viscosity data for a polymethyl acrylate sample in benzene solution at 30°C*.

Fraction number	*Molecular weight, $\overline{M}_n \times 10^{-5}$*	*$[\eta]$, (dl/g)*
1	8.237	1.65
2	11.390	2.19
3	13.890	2.49
4	15.120	2.65
5	19.720	3.20

*Data (unpublished) from Gundiah, S, *et al.*, National Chemical Laboratory, Pune, India.

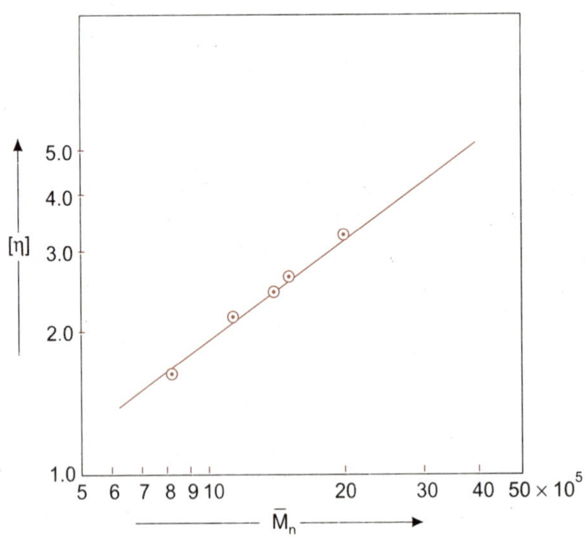

Fig. 13.19 Plot of $[\eta]$ versus $\overline{M}_n$ for polymethyl acrylate in benzene solution.

A log-log plot of [η] against $\overline{M}$ gives a straight line as shown in Fig. 13.19. The ordinate intercept and slope of this line will give K and a respectively. For the data given in Table 13.6, the K and a values are 8.630×10^{-5} (dl/g) and 0.725, respectively.

Now, let us see how intrinsic viscosity can be measured. Assume that a liquid is flowing through a capillary tube. The time required for the liquid of volume V to pass through the capillary of radius r and length l is related to its absolute viscosity η by the famous Poiseuille equation as follows:

$$\eta = \frac{3.14 Pr^4 t}{8Vl} \tag{13.13a}$$

$$t = \frac{8Vl\eta}{3.14 Pr^4} \tag{13.13b}$$

where P is the pressure head under which the liquid flow takes place. Equation (13.13a) thus correlates the flow time with the absolute viscosity of the liquid. If η and η_0 are the absolute viscosities of a solution and the pure solvent, respectively, and t and t_0 are their corresponding flow times, we can derive the following relationships:

$$t = \left(\frac{8Vl}{3.14 Pr^4}\right)\eta \tag{13.14}$$

$$t_0 = \left(\frac{8Vl}{3.14 Pr^4}\right)\eta_0 \tag{13.15}$$

For the same capillary tube, the quantities within parenthesis have the same value. Hence, dividing Eqn. (13.14) by Eqn. (13.15), We have

$$\frac{t}{t_0} = \frac{\eta}{\eta_0} \tag{13.16}$$

The term η/η_0 is known as the *'relative viscosity'*. This term and other terms related to viscosity measurements are given as follows:

Relative viscosity = $\eta/\eta_0 = t/t_0 = \eta_r$.

Specific viscosity = $(\eta - \eta_0)/\eta_0 = (t - t_0)/t_0 = \eta_r - 1 = \eta_{sp}$.

Reduced viscosity = $\eta_{sp}/C = \eta_{red}$, dl/g.

Inherent viscosity = $\ln \eta_r/C = \eta_{inh}$, dl/g.

Intrinsic viscosity = $(\eta_{sp}/C)_{C\to 0} = (\ln \eta_r/C)_{C\to 0} = [\eta]$, dl/g.

Intrinsic viscosity is also known as the Staudinger *index* or *limiting viscosity number.* As specific viscosity is, by definition, a dimensionless quantity, the intrinsic viscosity has the reciprocal of concentration as its dimension. It may be noted from the above discussion that for calculating the intrinsic viscosity of a polymer sample in solution, we need not to know the absolute viscosities of the solvent and the solutions but only the flow times of a constant volume of solvent and the solutions through a particular capillary tube. This principle is used in the viscometric technique of molecular weight determination.

In an actual experiment, solutions of known concentrations of the polymer sample are made. The solvent flow time (t_0) and the solution flow times (t) for different concentrations are measured using the same viscometer. For each concentration, the corresponding reduced viscosity and the inherent viscosity are calculated. Next, double extrapolation plots of reduced viscosity against concentration and inherent viscosity against concentration (both extrapolated to zero concentration) are prepared. The common ordinate intercept of these graphs gives the intrinsic viscosity [see Eqns. (12.36) and (12.37)]. The data and plots for a typical measurement are given in Table 13.7 and Fig. 13.20, respectively.

Of all the methods available for the molecular weight determination of polymers, viscosity measurement may perhaps be the least complex and the most convenient. The solution viscosities of

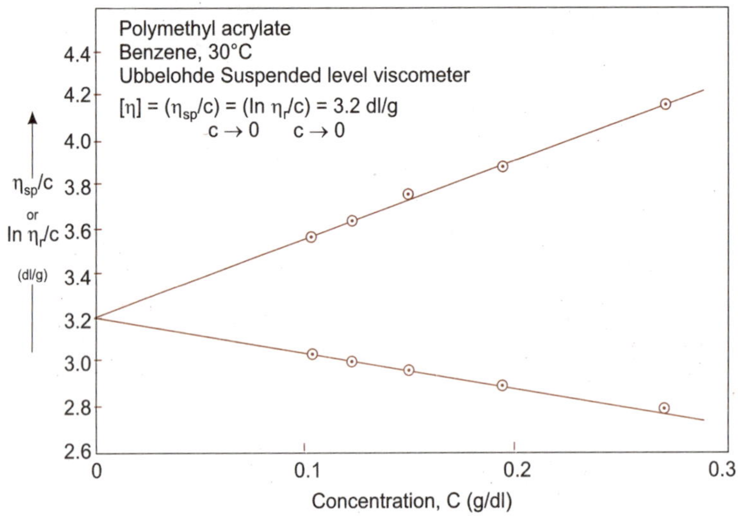

Fig. 13.20 Plots of η_{sp}/C versus C and ln η_r/C versus C for a sample of polymethyl acrylate in benzene solution.

Table 13.7 Viscosity data for a polymethyl acrylate sample in benzene solution at 30°C*
(Flow time for solvent, t_0 = 216.0 sec.)

Concentration C (g/dl)	Flow time, t (sec)	$t/t_0 = \eta_r$	$\eta_r - 1 = \eta_{sp}$	η_{sp}/C (dl/g)	ln η_r	ln η_r/C (dl/g)
0.2716	459.8	2.1287	1.1287	4.156	0.7556	2.782
0.1940	378.2	1.7509	0.7509	3.870	0.5601	2.887
0.1509	337.9	1.5644	0.5644	3.740	0.4475	2.966
0.1235	312.8	1.4481	0.4481	3.629	0.3702	2.998
0.1045	296.4	1.3722	0.3722	3.562	0.3164	3.028

*Data (unpublished) from Gundiah S. *et al.*, National Chemical Laboratory, Pune, India.

polymers are determined by viscometers. The Ostwald viscometer and the Ubbelohde suspended level viscometer (USLV) are the two popular viscometers used for this purpose (Fig. 13.21). As can be seen from the figure, these viscometers are simple glass capillary devices. While using the Ostwald viscometer, care must be taken to see that all the measurements are made using a constant volume of the solution. Otherwise, the effective pressure head will vary from one solution to another and Eqn. (13.16) will no longer be applicable. However, in the USLV, the design is such that the measurement is unaffected by the volume of the solution taken. The main advantage of the USLV is that only a single solution of a known concentration is required to be made to start with. A known volume of this solution is taken in the viscometer and the flow time measured. Subsequent concentrations can be achieved by adding known volumes of pure solvent and mixing inside the viscometer itself. This, in turn, yields sufficient data for computing flow times at different concentrations. In contrast to this, the experimental procedure with respect to the Ostwald viscometer is rather tedious, as, each time, we have to empty, clean and refill the viscometer with a fresh solution before measuring the flow time.

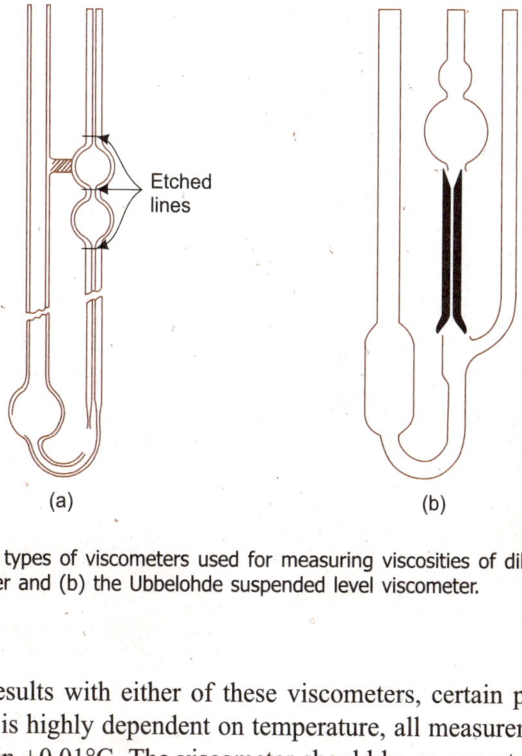

Fig. 13.21 The two common types of viscometers used for measuring viscosities of dilute polymer solutions: (a) the Ostwald viscometer and (b) the Ubbelohde suspended level viscometer.

To get reproducible results with either of these viscometers, certain precautions are absolutely essential. As the viscosity is highly dependent on temperature, all measurements should be done at a constant temperature within ±0.01°C. The viscometer should be so mounted as to hold the capillary perfectly vertical all the time as, otherwise, even with the same volume of the solution the pressure head will vary. Another source of error lies in the measurement of the flow time. The flow time for a particular volume of the solution or solvent must be sufficiently long so that the difference in flow time between two successive concentrations is at least of the order of 3 to 5 seconds. All solutions should be free from suspended impurities as, otherwise, the latter may obstruct the free flow of the solution within the capillary, leading to erroneous flow times.

13.4.7 Ultracentrifugation

Of all the existing methods for the molecular weight determination of polymers, the technique based on ultracentrifugation is perhaps the most complex. Usage of a centrifugal force to separate suspended particles from liquids is well known. Old-time housewives, for instance, separated butter from curds in their *'kitchen laboratory'* by churning, which was nothing but a crude form of centrifugation. A sophisticated version of this is found in large milk-bottling plants, where huge quantities of milk are centrifuged to separate the butter from it. Polymer molecules in solution, when subjected to a very high centrifugal force, exhibit similar behaviour and settle down. The sedimentation rate of polymer molecules settling down under the influence of a constant centrifugal force is related to their molecular weight. Making use of this principle, two different methods are adopted to determine the molecular weight of a polymer. They are the sedimentation velocity method (or sedimentation transport method) and the sedimentation equilibrium method.

Sedimentation Velocity Method

The earth's gravitational force makes any particle in suspension settle down. Macromolecules in a polymer solution are also affected by gravity in a similar way. However, the majority of polymers forms stable homogeneous solutions, and the polymer molecules present in them do not settle down. This is because of a counter-balancing diffusion effect which always tends to keep the macromolecules dispersed all over the solution. This opposing effect is due to the fast random movement of the solute molecules which possess a fairly good amount of kinetic energy at ordinary temperatures. If we can increase the gravitational force acting on macromolecules by any means, the latter can be made to sediment. This is precisely what is done in ultracentrifugation. The sedimentation velocity of the polymer molecules depends on the molecular weight.

In the sedimentation velocity method, the solution is subjected to a very high gravitational fields. This process is effected by an ultracentrifuge which can provide spins as high as 65,000 revolutions per minute. Due to the high centrifugal force, the polymer molecules will start sedimenting. By applying Stoke's law, it is possible to correlate the sedimentation coefficient S (a function derived from the sedimentation velocity) with molecular weight $\overline{M}$ by the following mathematical expression:

$$\overline{M} = \frac{S_0 RT}{D_0 (1 - \rho \overline{V})} \tag{13.17}$$

where S_0 and D_0 are the sedimentation and diffusion constants, respectively, obtained by extrapolating the sedimentation coefficient (S) and the diffusion coefficient (D) at different concentrations to zero concentration: $\overline{V}$ is the specific volume of the polymer in solution; and ρ the density of the solvent Determination of D_0 and S_0 will be dealt with later in this section. Equation (13.17) which enables us to evaluate the molecular weight by ultracentrifugation is known as the *Svedberg equation*.

In an actual experiment, a homogeneous solution of a known concentration is taken and subjected to ultracentrifugation. As the sedimentation commences, a boundary between the solvent phase and the solution phase will be formed, as represented in Fig. 13.22. As centrifugation proceeds,

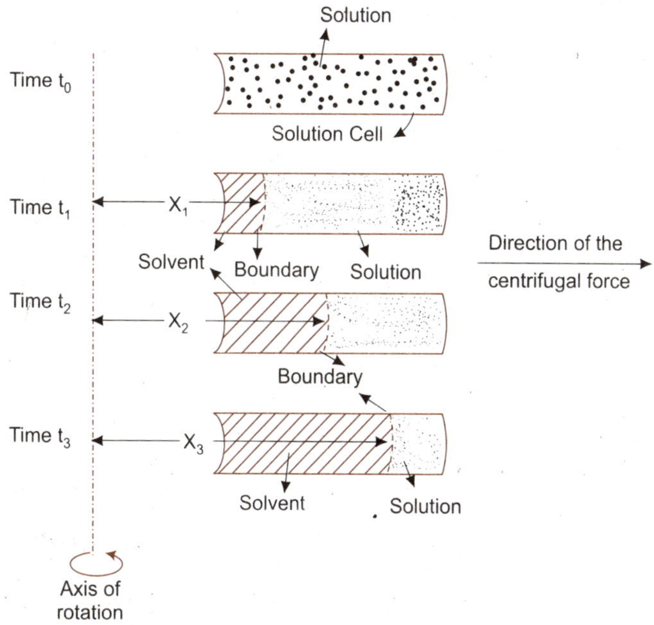

Fig. 13.22 Boundary between solvent and solution phases receding away from axis of rotation as ultracentrifugation proceeds.

this boundary between the solvent and solution phases slowly recedes away from the axis of rotation and by carefully following its displacement with respect to time, the sedimentation velocity can be computed. Usually, the movement of the boundary is monitored by taking photographs or by ultraviolet detection at suitable intervals of time. The rate at which the boundary moves is the sedimentation velocity, U. It can be calculated using the equation

$$U = \frac{X_2 - X_1}{t_2 - t_1} \tag{13.18}$$

where X_1 and X_2 are the distances of the boundary from the axis of rotation at times t_1 and t_2, respectively (see Fig. 13.22). The sedimentation coefficient S is related to the sedimentation velocity by the equation

$$S = \frac{U}{\omega^2} \cdot \frac{2}{X_1 + X_2} \tag{13.19}$$

where ω is the angular velocity.

The sedimentation constant S_0 used in Eqn. (13.17) can be computed by extrapolating S values for different concentrations to zero concentration according to the equation

$$\frac{1}{S} = \frac{1}{S_0} + k_s C \tag{13.20}$$

The diffusion coefficient D can be similarly calculated by measuring the concentration gradient at different times using the refractive index as a measure of concentration. Diffusion coefficients for different concentrations are measured and extrapolated to zero concentration to obtain D_0 (diffusion constant) as per the equation

$$D = D_0 + k_d C \tag{13.21}$$

The terms k_s and k_d appearing in Eqns. (13.20) and (13.21) are constants for a given polymer-solvent system.

Sedimentation Equilibrium Method

In this method, a thermodynamic equilibrium is achieved by practically equalising the sedimentation and diffusion velocities. For attaining the equilibrium state, the ultracentrifuge has to be spun at a lower rate for a longer period. Usually, the experiment takes several days to reach the equilibrium state. During this period, the centrifuge spin is kept at around 15,000 revolutions per minute (rpm). Under this centrifugal force, an equilibrium concentration gradient is established along the cell, and the macromolecules get distributed according to their molecular weights, as shown in Fig. 13.23.

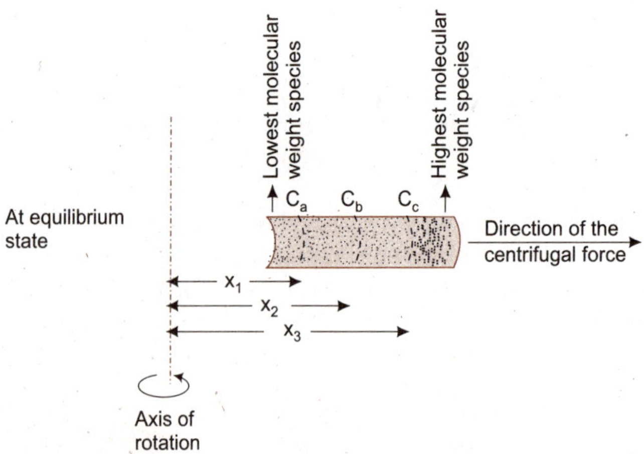

Fig. 13.23 Distribution of polymer molecules at various distances from axis of rotation under influence of equilibrium centrifugal force.

At equilibrium, the molecular tendency to diffuse into the less concentrated zones (*i.e.*, towards the axis of rotation) is balanced by the opposing tendency of polymer molecules to sediment. Thus, at this stage, the sedimentation and diffusion balance out and the molecular weights is given by the equation

$$\overline{M} = \frac{2RT \ln (C_a/C_b)}{\omega^2 (1 - \rho\overline{V})(X_1^2 - X_3^2)} \tag{13.22}$$

In this equation, the molecular weight obtained is the weight-average molecular weight $\overline{M}_w$. Here, C_a and C_b represent the concentration values in the cell at distances X_1 and X_2 from the axis of rotation. For a polydisperse polymer system, X_1 and X_2 would represent points at the meniscus and bottom of the cell, respectively.

Theoretically speaking, since the macromolecules are distributed in the centrifuge cell according to their molecular size, the sedimentation equilibrium method can be used to get a molecular weight distribution of a heterogeneous system. The drawback of this method is that it is rather laborious and time consuming.

Figure 13.24 shows the schematic representation of a commonly used analytical ultracentrifuge set-up. It has an aluminium or titanium alloy rotor. This rotor is capable of giving extremely high-speed rotations. A modern ultracentrifuge can operate from 2,000 rpm to 65,000 rpm. Usually, it is driven electrically though, sometimes, air turbines and similar methods are also employed. The polymer solution whose molecular weight is to be found is kept in the solution cell of the rotor. A camera is used to photograph the solution cell at different times, from which, the sedimentation rate can be computed, as explained earlier. The refractive index can also be measured at different levels of the solution. For very accurate results, it is advisable to use dust-free solvents having a relatively low viscosity. The density and refractive index of the solvent must be substantially different from those of the polymer. In fact, maximum density difference between the solvent and polymer is required for effective sedimentation.

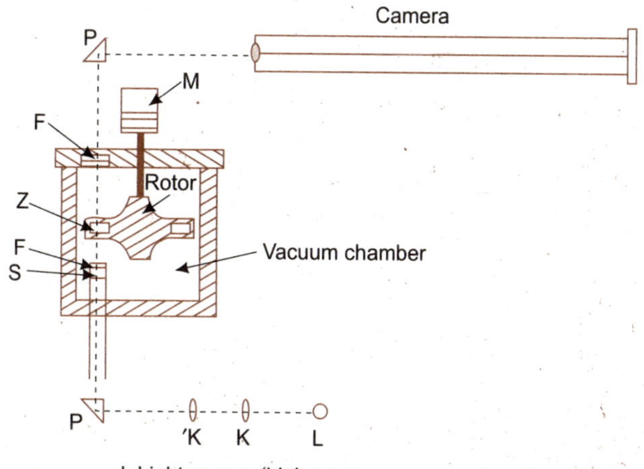

L Light source (high pressure mercury lamp)
K Condensor lens
P Reflecting prism
S Scale
F Quartz window
Z Solution cell with quartz windows
M Electromotor

Fig. 13.24 Schematic representation of an ultracentrifuge measurement set-up.

13.4.8 Light Scattering

As it happens, many aesthetic phenomena of nature, such as the rainbow, colourful dawn and dusk, the blue of the sky and the sea, are in one way or another connected to a phenomenon called *scattering of light*, which occurs when light beam encounters matter.

The light-scattering phenomenon is used to measure the weight-average molecular weight, $\overline{M}_w$, of polymers. This is possible because not only a suspended particulate matter can scatter a light beam but even solute molecules in a solution can do so. In fact, it has been recognised that in solutions and liquid mixtures, scattering of light occurs due to changes in density or refractive index within the system arising from compositional variations.

In all light-scattering phenomena, the amplitude of scattering is found to be proportional to the mass, M, of the particle which scatters the beam of light. However, in practice, it is the intensity of the scattered light that is experimentally determined and which, as we know, is the square of the amplitude. Debye, in 1944, arrived at an expression, as given in Eqn. (13.23), to relate the molecular weight of the solute particle to the intensity of scattered light. This equation is known as the *Debye equation* and holds good only for particles which are smaller than the wavelength of light used for the scattering experiment

$$\frac{KC}{R_{90}} = \frac{HC}{\tau} = \frac{1}{M} + 2BC \qquad (13.23)$$

where B is the second virial coefficient; C the concentration of the solution; and R_{90} is what is called the *Rayleigh ratio* at 90° observation angle. This ratio in a generalised case is represented as R_θ. In Eqn. (13.23) the Rayleigh ratio is determined at an observation angle of 90°. Hence $R_\theta = R_{90}$.

Again, according to Rayleigh

$$R_\theta = \frac{i_\theta r^2}{I_0 V} \qquad (13.24)$$

where i_θ is the intensity of the scattered light per unit volume V of scattering material which is observed at a distance r and at an angle of θ with reference to the incident beam, and I_0 is the intensity of the incident beam.

Now, coming back to Eqn. (13.23), τ is the turbidity of the medium which results from the scattering of light. Let us now see in what way this quantity is related to the intensity of the incident light. While passing through thickness X of the medium, if the intensity of the incident beam is reduced from I_0 to I, then the turbidity of the medium is given by the following equation

$$\frac{I}{I_0} = e^{-\tau X} \qquad (13.25)$$

K and H in Eqn. (13.23) are light-scattering calibration constants and are defined by Eqns. (13.26) and (13.27), respectively, as follows:

$$K = \frac{2\pi^2 n^2 (dn/dC)^2}{\lambda^4 N_A} \qquad (13.26)$$

where π is the mathematical symbol with a numerical value of 3.14; n the refractive index of the solution; (dn/dC) the specific refractive index increment, *i.e.*, the change of refractive index with concentration; λ the wavelength of the incident light; and N_A Avogadro's number. Similarly, H is written as

$$H = \frac{32\pi^3 n^2 (dn/dC)^2}{3\lambda^4 N_A} \tag{13.27}$$

It should be noted that for polymer solutions in which the solute molecules are larger in size compared to the wavelength of light, Eqn. (13.23) needs modification.

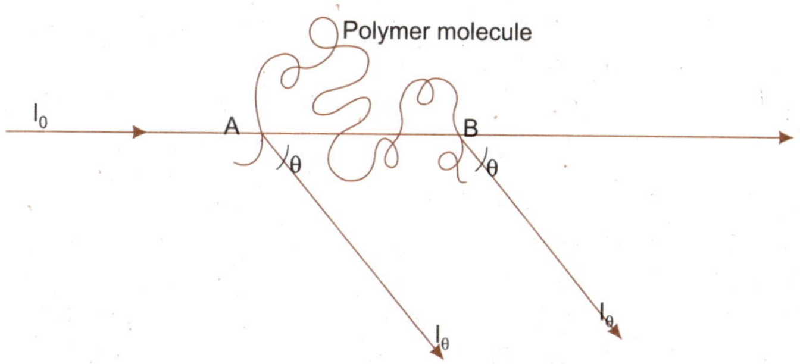

Fig. 13.25 Scattering of light by a polymer molecule (for details see text).

In the case of polymer solutions, the polymer molecules are in the form of randomly coiled swollen chains and different regions of the macromolecular coil (shown as A and B in Fig. 13.25) simultaneously scatter the incident light beam. Thus, different regions of the same particle get exposed to incident light having a different phase. The ultimate result is that the scattered beam consisting of light waves emerging from different regions of the same particle interfere with one another and hence, the intensity of the scattered light diminishes when observed at an angle θ. In other words, we can say that the value of R_θ varies with θ. Debye showed that under such conditions Eqn. (13.23) gets modified as follows:

$$\frac{KC}{R_\theta} = \frac{HC}{\tau} = \frac{1}{\overline{M}_w\, P(\theta)} + 2BC \tag{13.28}$$

where $P(\theta)$ is the particle-scattering factor describing the angular dependence of the scattered light according to the relationship

$$\frac{1}{P(\theta)} = 1 + \frac{16\pi^2}{3\lambda^2} \langle \overline{S}^2 \rangle \sin^2\left(\frac{\theta}{2}\right) \tag{13.29}$$

where $\langle \overline{S}^2 \rangle$ is the mean square radius of gyration of the polymer random coil. Substituting this value of $P(\theta)$ in Eqn. (13.28), we get

$$\frac{KC}{R_\theta} = \frac{HC}{\tau} = \frac{1}{\overline{M}_w} + \frac{16\pi^2}{3\lambda^2 \overline{M}_w} <\overline{S}^2> \sin^2\left(\frac{\theta}{2}\right) + 2BC \tag{13.30}$$

$P(\theta)$ can be considered as a correction factor to the scattered intensity at different values of θ, such that $P(\theta) = 1$ at $\theta = 0$. Similarly, $BC = 0$ at $C = 0$. Now, Eqn. (13.30) can be rewritten in the following forms:

$$\left(\frac{KC}{R_\theta}\right)_{C\to 0} = \frac{1}{\overline{M}_w} + \frac{16\pi^2}{3\lambda^2 \overline{M}_w} <\overline{S}^2> \sin^2\left(\frac{\theta}{2}\right) \tag{13.31}$$

$$\left(\frac{KC}{R_\theta}\right)_{\theta\to 0} = \frac{1}{\overline{M}_w} + 2BC \tag{13.32}$$

$$\left(\frac{KC}{R_\theta}\right)_{C,\theta\to 0} = \frac{1}{\overline{M}_w} \tag{13.33}$$

There are two methods (namely, the Debye method and the Zimm method) by which we can make use of various aforementioned equations to determine the molecular weight of a polymer. The Debye method which involves the determination of the particle-scattering factor $P(\theta)$ needs a prior knowledge of the exact shape of the macromolecule in solution. The Zimm method, however, eliminates this requirement by adopting a double extrapolation of the plots of C/R_θ versus $\sin^2(\theta/2) + kC$ (where k is an arbitrary constant) to both zero angle of scattering ($\theta = 0$) and zero concentration ($C = 0$), when $P(\theta) = 1$ and $BC = 0$. The ordinate intercept will then give the value of $1/(K\overline{M}_w)$. The Zimm plot for an anionic polystyrene sample in toluene is shown in Fig. 13.26. From the Zimm plot, the molecular weight, the second virial coefficient and the mean square radius of gyration can be calculated using Eqns. (13.33), (13.32) and (13.31), respectively.

The light-scattering technique offers a convenient method for measuring the molecular weights of polymers in the range of 10,000 to 10,000,000. This technique requires the experimental determination of either R_θ or τ using a light-scattering photometer, and of n and dn/dC using a differential refractometer.

The schematic set-up of a light scattering photometer is shown briefly in Fig. 13.27. L is the source of light (mercury lamp). The light, after leaving the source, passes through condensing lenses K and also through a monochromatising filter IF. It then passes through the slit B and enters an achromatic system A. Next, it strikes against a beam splitter H and, after passing through the polariser N, falls on the sample (*i.e.*, the polymer solution) kept in the solution cell *Ce*. After leaving the cell, the incident (primary) beam is absorbed by a light trap shown as LA; the scattered light both from the cell and calibrated turbidity standard S_t reaches Ph1 and Ph2, the moveable and the reference photocells, respectively. Theoretically, the turbidity can be determined from the scattered intensity. But this requires precise information on the geometry of the light-scattering photometer, the distance of the scattering centre from the photometer, etc. The measurement of all these factors is

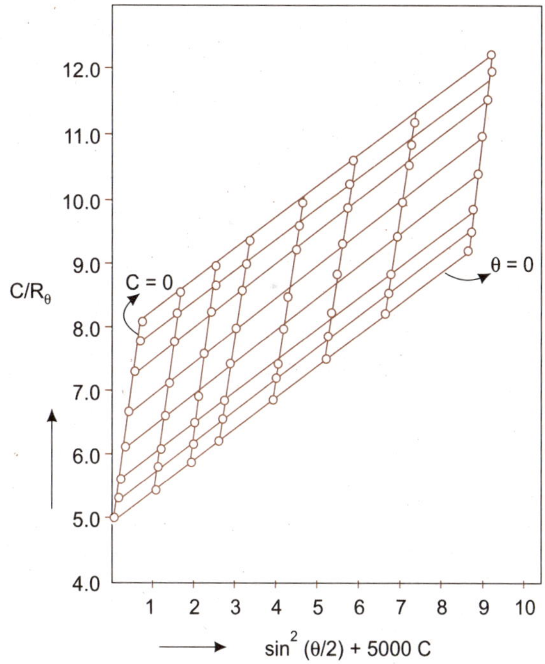

Fig. 13.26 A typical Zimm plot.

cumbersome. The idea of using a standard is to overcome these problems. The standards usually used are organic liquids, colloidal suspensions or even well-calibrated polymeric substances. The scattering values of these substances are obtained under conditions identical to those of the sample. A comparison of these results dispenses with the need for having to know the details mentioned above. For determining if $\overline{M}_w$ apart from the scattering values, we need to know the polymer solution concentrations, the refractive index of the solvent and also the refractive index difference between the polymer solutions at different concentrations and the solvent. One major precaution to be taken while conducting light-scattering experiments is that the solution should be made completely dust-free. A high-speed centrifuge is generally used to remove the dust. The solution should be maintained at a constant temperature throughout the experiment.

13.5 MOLECULAR WEIGHT DISTRIBUTION CURVE

Computing the molecular weight distribution curve is quite easy once the polymer sample has been fractionated into several fractions and the actual weight as well as the average molecular weight of each fraction have been determined. It is observed that the larger the number of fractions separated, the more accurate is the distribution curve. In theory, if we can separate a sample into completely monodisperse fractions, *i.e.*, each fraction containing molecules of only one particular degree of

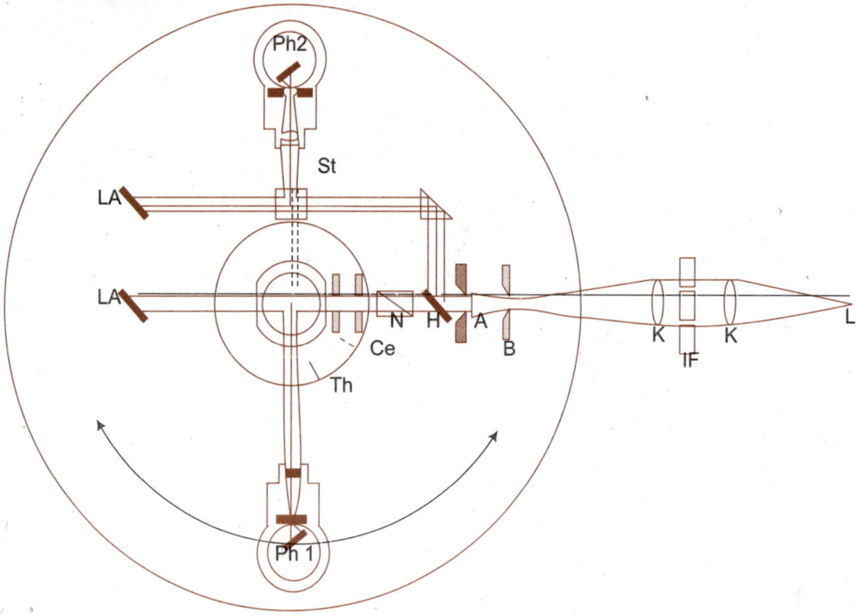

Ph1	Moveable photocell
Ph2	Reference photocell
St	Calibrated turbidity standard (translucent glass)
Ce	Solution cell
Th	Thermostat
N	Polariser
H	Light beam splitter
A	Achromatic system
B	Slit
K	Condensing lenses
IF	Interference filter
L	Light source (high-pressure mercury lamp)
LA	Light absorber (beam stop)

Fig. 13.27 Schematic diagram of an experimental set-up for light-scattering measurements.

polymerisation (such as $DP = 1$, $DP = 2$, $DP = 3$, ..., $DP = 10^8$ or $DP = 10^9$), we can get a 100% representation of the molecular weight distribution. In practice, however, such a fractionation is not possible and the maximum that could be done is to separate the sample into as many fractions as possible and to subfractionate each of them into further fractions. In this way, we can get fractions of better homogeneity.

Let us now consider a hypothetical case where a sample has been separated into 10 fractions, whose weights and molecular weights are as given in Table 13.8. A simple curve can be developed, using these data by plotting the weight percentage of each fraction w_i against the molecular weight $\overline{M}_i$ as shown in Fig. 13.28. This is a very simple way of expressing the distribution pattern and can

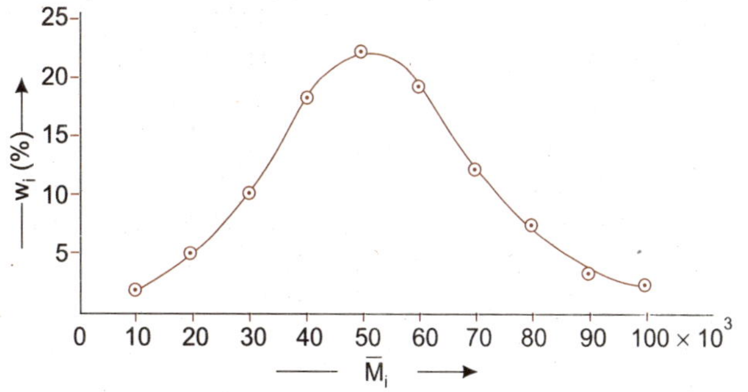

Fig. 13.28 Plot of w_i versus $\overline{M}_i$ depicting a simple molecular weight distribution curve.

Table 13.8 Fractionation data for a hypothetical polymer sample wherein the difference in molecular weight between adjacent fractions is the same.

Fraction number	*Weight per cent of fraction, w_i*	*Cumulative weight per cent, Σw_i*	*Molecular weight, $\overline{M}_i$*
1	2	2	10,000
2	5	7	20,000
3	10	17	30,000
4	18	35	40,000
5	22	57	50,000
6	19	76	60,000
7	12	88	70,000
8	7	95	80,000
9	3	98	90,000
10	2	100	100,000

Table 13.9 Fractionation data for a hypothetical polymer sample wherein the difference in molecular weight between adjacent fractions is not the same.

Fraction number	*Weight per cent of fraction, w_i*	*Cumulative weight per cent of fraction, Σw_i*	*Molecular weight, $\overline{M}_i \times 10^{-3}$*
1	2.0	2.0	80
2	7.5	9.5	150
3	2.5	12.0	240
4	8.0	20.0	310
5	5.5	25.5	350
6	12.0	37.5	425
7	20.0	57.5	505
8	24.0	81.5	580
9	5.5	87.0	720
10	8.0	95.0	770
11	3.0	98.0	825
12	2.0	100.0	900

more or less represent the overall distribution pattern, provided the difference between the molecular weights of the successive fractions are the same, as in the case of the hypothetical example. But in actual fractionation, it is very difficult to get fractions of this type, and usually, the difference between the molecular weights of successive fractions will be spaced quite apart as given in Table 13.9. In such cases, a simple curve of w_i against $\overline{M_i}$ can give an erroneous picture of the distribution pattern, as shown in Fig. 13.29. To overcome this difficulty, the cumulative weight percentage Σw_i

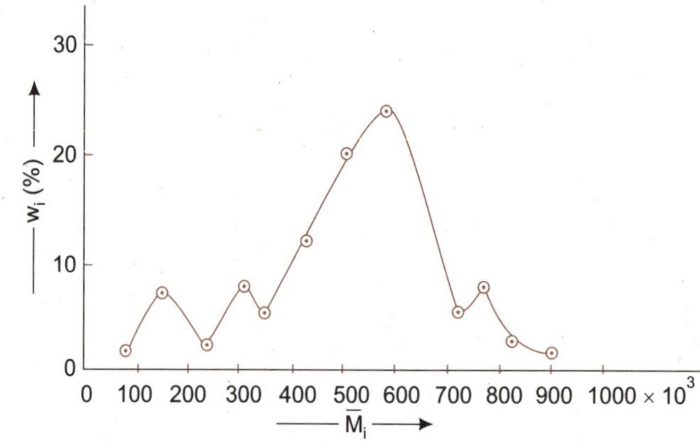

Fig. 13.29 Plot of w_i versus $\overline{M_i}$ for a hypothetical polymer sample.

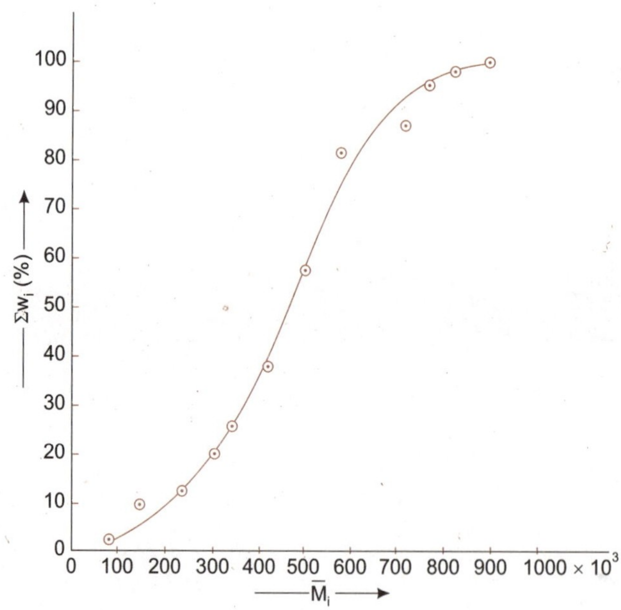

Fig. 13.30 Plot of Σw_i versus $\overline{M_i}$, depicting an integral distribution curve for molecular weight distribution of hypothetical polymer sample represented in Fig. 13.29.

(i.e., the weight percentage of the sum of all the fractions containing molecules up to a particular molecular weight), is plotted against the molecular weight $\overline{M}_i$, as shown in Fig. 13.30. The curve is called an integral distribution curve. Graphical differentiation of this curve, by determining tangents at as many points as possible on the integral curve, leads to the computation of a differential distribution curve (Fig. 13.31) from which the weight percentage of molecules of a particular molecular weight present in the whole sample can be read out.

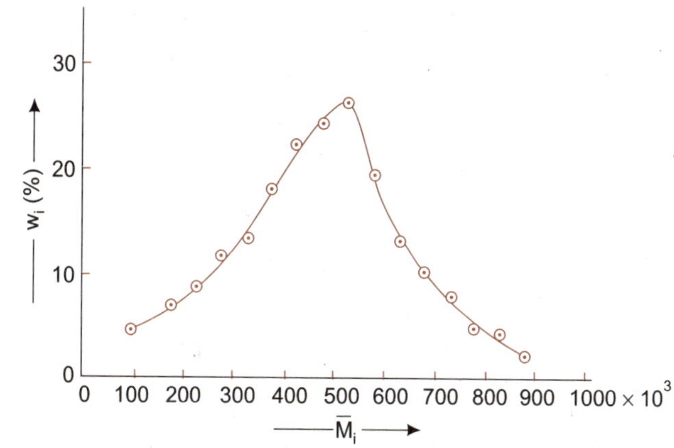

Fig. 13.31 A differential distribution curve for molecular weight distribution of hypothetical polymer sample. This curve has been computed by graphical differentiation of the plot in Fig. 13.30.

With the advent of GPC, however, there is no need for the cumbersome procedure of separating the sample into fractions, determining their molecular weights and computing the distribution curve. As already explained under Section 13.3.4, the GPC straightaway gives a plot of the distribution curve of the sample.

13.6 DETERMINATION OF GLASS TRANSITION TEMPERATURE

We have seen in Chapter 6 that many properties of the polymeric material undergo a change near the glass transition region. The coefficient of expansion and heat capacity are two such important properties. By measuring one of these properties as a function of temperature, we can determine the $\overline{T}_g$ of polymers.

13.6.1 Dilatometric Method

A dilatometer can be used to measure the expansion coefficient of the sample. As already explained in Section 6.3, the dilatometer measures the change in volume (ΔV) of the sample under a constant rate of heating. From a plot of ΔV against temperature T, the T_g can be computed as shown in Fig. 13.32. It may be mentioned here that this method is seldom used these days as the thermomechanical and calorimetric methods give fast and more reliable data.

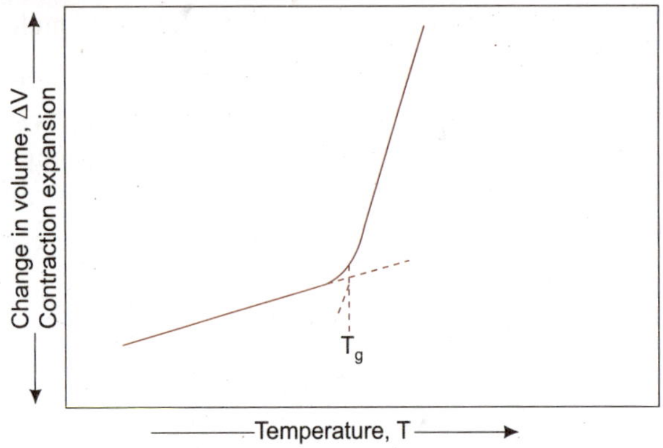

Fig. 13.32 Representative plot of ΔV against T used for computing T_g of a polymer.

13.6.2 Thermomechanical Method

Thermomechanical analysis (TMA) is a technique used for making accurate measurements of dimensional changes of polymeric samples as a function of temperature. TMA instruments operate on the float suspension principle (which allows reproducible measurements of minute dimensional changes) and have provision for programmed heating. The polymeric sample in the form of a disc is loaded into the instrument as schematically shown in Fig. 13.33. As the material expands with increase in temperature, it actuates an expansion probe and the change in dimension ΔL is directly recorded against temperature T as given in Fig. 13.34. The T_g value can be read directly from the plot where the expansion coefficient undergoes a sudden change.

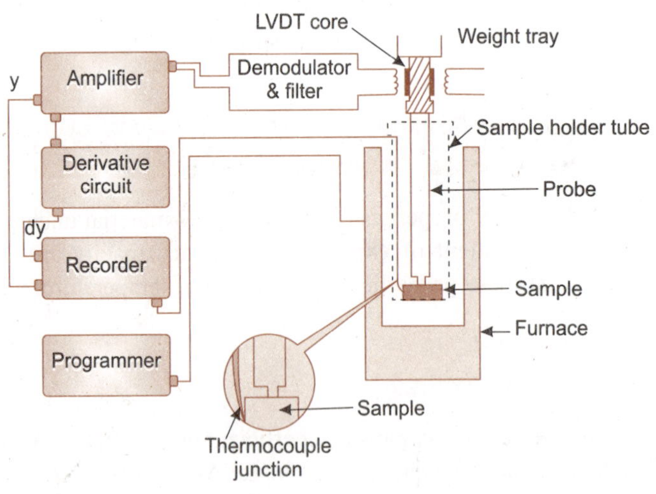

Fig. 13.33 Schematic details of a thermomechanical analyser. (*Courtesy*: E I Du Pont De Nemours & Co. Willmington, USA).

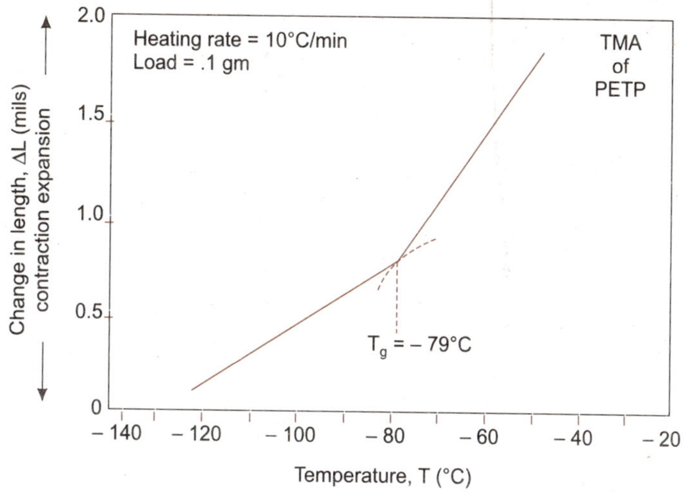

Fig. 13.34 Plots of ΔL versus T for a sample of polyethylene terephthalate.

13.6.3 Calorimetric Method

The calorimetric method is used to measure the heat capacity of the polymer as a function of temperature. In an instrument called the differential scanning calorimeter (Fig. 13.35), the sample and a reference material which does not undergo any transition over a wide range of temperature are heated at a programmed rate of heating and the heat capacity of the sample is measured as a function of the differential heat flow rate between the sample and the reference material. The instrument directly gives a recording of heat flow rate (C_p) against temperature (T), as shown in Fig. 13.36. From

Fig. 13.35 Differential scanning calorimeter.

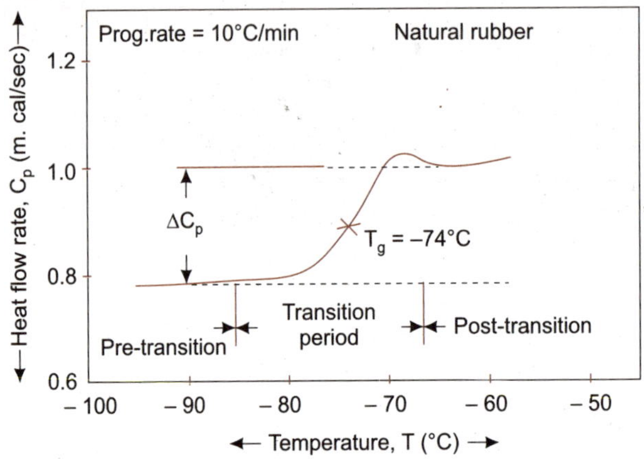

Fig. 13.36 Plot of C_p versus T for a natural rubber sample.

this figure it can be seen that the heat flow rate of the sample undergoes a change during transition. The T_g value can be computed from the plot by extrapolating the pre-transition and post-transition lines and by calculating the temperature where the heat flow rate is exactly at the middle of pre- and post-transition rates.

Books for Further Reading

Svedberg, T. and Pedersen, K.D., *The Ultracentrifuge,* Clarendon, Oxford, 1940.

Bonnar, R.V., Dimbat, M. and Stross, F.H., *Number-average Molecular Weights,* Wiley-Interscience. New York, 1950.

Allen, P.W., Ed., *Techniques of Polymer Characterisation,* Academic, New York, 1959.

Pinner, S.H., *A Practical Course in Polymer Chemistry,* Pergamon, New York, 1961.

Tranford, C., *Physical Chemistry of Macromolecules,* Wiley-Interscience, New York, 1961.

Chien Jen-Yuen, *Determination of Molecular Weights of High Polymers*, Old bourne, London, 1963.

Williams, J.W., Ed., *Ultracentrifugational Analysis in Theory and Experiment*, Academic, New York, 1963.

McIntyre, D. and Gornick, F., Eds., *Light Scattering from Dilute Polymer Solutions*, Gordon and Breach, New York, 1964.

Ke, B., *Newer Methods of Polymer Characterisation,* John Wiley, London, 1964.

Cantow, M.J.R., Ed., *Polymer Fractionation*, Academic, New York, 1967.

Sorenson, W.R. and Campbell, T.W., *Preparative Methods of Polymer Chemistry*, 2nd edn., Interscience, New York, 1968.

MacCaffery, E.M., *Laboratory Preparation for Macromolecular Chemistry,* McGraw-Hill, New York, 1970.

Braun, D., Cherdon, H. and Kern, W., *Techniques of Polymer Synthesis and Characterisation*, Wiley-Interscience, New York, 1971.

Altgelt, K.H. and Segal, L., *Gel Permeation Chromatography*, Dekker, New York, 1971.

Huglin, M.B., Ed., *Light Scattering from Polymer Solutions*, Academic, London, 1972

Collins, E.A., Bares, J. and Billmeyer, F.W., *Experiments in Polymer Science*, Wiley-Interscience, New York, 1973.

Pearce, E.M., Wright, C.E. and Bordoloi, K., *Laboratory Experiments in Polymer Synthesis and Characterisation*, a publication of the EMMSE Project, The Pensylvania State University, Materials Research Laboratory, 1982.

14

ELASTOMERIC, FIBRE-FORMING AND PLASTIC MATERIALS

14.1 INTRODUCTION

In Chapter 1, we came across some versatile applications of polymers which compete well with metals, alloys, ceramics, etc. In packaging, textiles, household articles, construction materials, etc. polymers are used in the form of fibres, films, coatings, foams or moulded and fabricated articles and exhibit a wide range of properties in terms of strength, rigidity, toughness, flexibility, elasticity, resilience, optical clarity, chemical and solvent resistivity, etc. These properties are inherently built into polymeric materials and can be further modified by adding plasticisers, fillers, colouring agents, stabilisers, etc.

Depending on their ultimate properties, polymeric materials can be classified into three broad categories: elastomeric, fibre-forming and plastic. In a simplistic manner, we can say that elastomeric materials exhibit rubber elasticity; fibre-forming materials possess rigidity and stiffness; and plastic materials are generally strong and tough. We know that all these materials are polymeric materials with a degree of polymerisation (DP) in the useful property range (refer to Fig. 3.6). But then what makes them behave differently? To answer this question: It is the chemical and geometrical structure of the individual polymer chains as well as the type of interchain arrangement within the chain aggregates at the '*use*' temperature that decide the property of a polymeric material. Let us discuss these aspects with respect to elastomeric, fibre-forming and plastic materials.

14.2 ELASTOMERIC MATERIALS

Take a rubber band. It is a typical elastomeric material. When pulled, it is elongated to many times its length. When the force is released, it snaps back to its original length. This is the characteristic behaviour of an elastomer. Now, take some putty or coal-tar and pull it. They can also be stretched considerably. But they are not elastomeric and differ from the rubber band in two aspects. First, they do not return to their original shape on release of the force and secondly, continued pulling can break them without much effort. On the contrary, a rubber band rebounds instantaneously on release of the force and can be broken only with some effort. It is soft, exhibits reversible deformation and has strength, whereas the putty exhibits only a '*cold flow*' without any strength. We can summarise the properties of an elastomeric material by a simple term, '*rubber elasticity*', which represents a combination of softness, stretchability, resilience and toughness. Rubber elasticity can be qualitatively compared to the behaviour of a metallic spring. When we apply some force to a spring, it elongates by virtue of its uncoiling and when we release the force, the spring regains its original shape owing to recoiling. Something similar happens in the case of the rubber band also. The chain molecules undergo uncoiling and recoiling on the application and release of a small force. In order to undergo coiling and uncoiling, the molecular chains in the rubber band should comprise flexible segments capable of attaining free rotation. In the unstrained state, they should tend to take up the more probable randomly coiled conformation such that the entropy factor is the highest at its normal state. When strained, the chains should be able to get extended and be brought to more ordered conformations. An ordered arrangement of the chain molecules gives rise to partial crystallinity and also a decrease in the entropy factor. It is thus the entropy factor that favours recoiling on release of the force. There should also be sufficient interchain free volume to provide for unhindered segmental mobility during coiling and uncoiling. Further, the interchain cohesive force should also be low enough so as not to hinder the free segmental mobility.

Another important requirement is that the individual chains should not, under prolonged stretching, slip past each other and get separated (resulting in the breakage of the material). Something should be present to hold the individual chains together without permitting them to go apart, while, at the same time, allowing sufficient segmental mobility. We can fulfil all these requirements and prepare an elastomeric material by what is known as '*molecular engineering*'. Chain flexibility and segmental mobility can be obtained by selecting repeat units made of the C—C and C—O linkages and by avoiding bulky side groups on the repeat units. By avoiding aromatic and cyclic structures in the chain backbone, chain stiffening can be eliminated. In the case of diene polymers, an increase in the cis-content will lead to a better chain flexibility. Branched molecules will contribute towards an increased interchain free volume on account of the fact that such molecules cannot assume close packing. Interchain cohesive forces can be minimised by avoiding polar groups in the repeat units. Cross-links provided at suitable intervals will hold the molecules together and will not allow them to slip past each other on stretching. In fact, it has been established that rubber elasticity can only be exhibited by such polymeric systems consisting of long flexible chain coils with very weak interchain cohesion and interconnected here and there through cross links to form a three-dimensional network. This, in summary, is the molecular engineering of elastomeric materials. These materials exihibit a combination of local mobility and overall rigidity.

We know that for a polymeric material to exhibit localised segmental mobility without any overall molecular mobility, it should be amorphous in nature and should exist in the rubbery state. The amorphous nature is taken care of by molecular engineering. But what about the rubbery state? For the material to be in the rubbery state, its glass transition temperature (T_g) should be very much below, and the flow temperature (T_f) much above, the '*use*' temperature. The polymeric material can be made to possess the elastomeric property over a wide range of temperature, or its '*use*' temperature range as an elastomeric material can be widened by pushing down the T_g as low as possible and pushing up the T_f as high as possible. This is what is done in rubber technology. We can push down the T_g by plasticising the material using either internal or external plasticisation. We know that the T_g of a polymer can be lowered by copolymerisation with small quantities of a suitable comonomer. This is internal plasticisation. On the other hand, external plasticisation involves compounding the polymer with a mutually compatible high boiling liquid, called a *plasticiser*. The plasticiser reduces the interchain cohesion in the polymer and favours segmental motion. Milling or mastication of rubber, by which the molecular weight and hence the T_g of the rubber is reduced considerably, also provides a method of plasticisation. The elevation of the T_f is accomplished by cross-linking or vulcanisation. In a cross-linked or vulcanised material, the T_f is generally so high that the material on heating starts decomposing before its T_f is reached. All rubber technology is essentially a combination of plasticisation and cross-linking.

14.3 FIBRE-FORMING MATERIALS

These are polymeric materials from which man-made fibres are formed. Nylon 6, 6 is a typical fibre-forming polymer. Now, take a monofilament made of this nylon and try to pull it. It can be noticed that unlike a rubber band, the nylon filament does not stretch, nor does it get deformed so easily. It offers considerable resistance before it can be deformed to a limited extent. Now release the force. Does the filament snap back like the rubber band? No. It just retains its slightly elongated shape and does not go back to its original shape. Pull it again and try to break it. It would be almost impossible to break it just by pulling. It may also be noticed that after elongation, the nylon filament becomes even stiffer. These are the characteristic properties of fibre-forming polymers. They exhibit high tensile strength and very high rigidity or stiffness and undergo irreversible deformation. For imparting these properties, the molecules in these materials should have a very high DP. They should neither exhibit localised segmental mobility nor overall chain mobility. They should be packed very close to each other in a highly ordered manner and should be held together by strong interchain cohesion forces which resist deformation and do not allow any relative movement between the chains. In other words, the material should have a very high crystallisability. This property, as we have seen earlier, can be increased through structural regularity, polar groups in the repeat units and stiffening units such as aromatic and cyclic rings in the chain backbone. Since branching reduces crystallisability, it should be avoided. This, in brief, is the molecular engineering of fibre-forming materials.

Now, how about the '*use*' temperature? In a fibre-former such as nylon, we would naturally like to emphasise the solid state where the segmental as well as molecular mobilities are forbidden. A 100% crystallinity, although desirable, is difficult to achieve and hence, we should take care of the crystalline as well as amorphous components in the material. The melting temperature (T_m) should be much above the '*use*' temperature, such that the crystalline component is in the solid state. Its T_g should also be much higher than its '*use*' temperature, so that whatever little amorphous components may be present in the material do not exhibit segmental mobility. At the same time, the T_m should not be so high that the material becomes untenable for processing. A good compromise would be to have the T_g and T_m in the range of 200–300°C. In fibre technology, we cannot afford to go in for plasticisation or cross-linking since both these will reduce the crystallinity. How do we then proceed? We know that the molecular engineering takes care of the crystallisability aspect of the material. A high crystallisable polymer need not, however, always p ess a high degree of crystallinity, which depends entirely on factors such as the crystallisa temperature, rate of cooling, rate of solvent evaporation and flow orientation. The fibre techn gy, therefore, aims at achieving a maximum degree of crystallinity when the fibres are formed er from a solution or from a melt. When the fibre is formed, we try to orient the molecules and them together from the beginning in order to facilitate efficient crystallisation. After fibre formation too, we draw the fibres further under tension so as to orient the molecules more unidirectionally. Such unidirectional orientation of the molecules increases the strength and rigidity of the fibre in the direction in which it is drawn.

14.4 PLASTIC MATERIALS

The third category of polymeric materials exhibits properties intermediary to elastomeric and fibre-forming polymers. Plastic materials possess good tensile strength and rigidity with or without elongation or impact strength. These materials can either be fully amorphous or partially crystalline. A fully amorphous polymer will be a plastic material when its T_g is above the '*use*' temperature. Polystyrene with a T_g of 100°C is a good example. Since these polymeric materials do not possess any crystallinity and are in the glassy state at the '*use*' temperature, amorphous plastics are brittle in nature. Polystyrene, in fact, is a fairly brittle material. Partially crystalline polymers, on the other hand, will be tough plastics above T_g (and below T_m). Polyethylene with a T_g of –125°C and a T_m of 146°C is a good example of a partially crystalline plastic. At the '*use*' temperature, which is between the T_g and T_m, the crystalline component in these materials gives rigidity and strength while the amorphous component existing in the rubbery state provides for sufficient resilience. A combination of these properties leads to the material exhibiting toughness and impact strength.

The brittleness of a fully amorphous polymer can be improved by physically blending some elastomeric material with it. The elastomer part of the blend improves the resilience and thus reduces the brittleness. For example, if we polymerise a styrene monomer containing a certain percentage of butadiene rubber dissolved in it, we get a polystyrene material called '*high impact polystyrene*', possessing good strength and impact properties.

In plastics technology, the desirability of plasticisation or cross-linking is guided by the nature of the polymeric material and the property of the end product aimed at. No plasticisation is required when: (*a*) the T_m or T_f is not very high; (*b*) the material can be melted and moulded without decomposition; and (*c*) the unplasticised material itself can exhibit the properties demanded of the end product. Polystyrene and polyethylene, for example, are used as such without plasticisation. If, on the other hand, the T_g or T_f is too high, posing consequent processing difficulties or if improved flexibility and properties such as softness are called for in the end product, then plasticisation is resorted to PVC, for example, has a T_m of 310°C and is a tough horn-like material. It is, therefore, invariably plasticised to a varying degree, and products ranging from rigid tubes to soft flexible toys are made from it.

Cross-linking of plastic materials, as in the case of elastomerics, improves the structural rigidity and the resistance to swelling by solvents and also increases the upper range of the '*use*' temperature. When products do not demand high rigidity, high temperature serviceability and solvent resistance, uncross-linked systems are used. Polystyrene and polyethylene, again, can be cited as examples of such systems. On the other hand, phenol-formaldehyde, urea-formaldehyde and epoxy systems involve cross-linking. Generally, the uncross-linked plastic materials are thermoplastic and the cross-linked systems are thermosetting.

Summarising the foregoing concepts, we can say that the amorphous and crystalline nature of the polymeric material together with the actual '*use*' temperature decides whether the material will be elastomeric, fibre-forming or plastic. When the material is fully amorphous, and the '*use*' temperature is below its T_g, the polymer will be in its glassy state and will behave like a brittle plastic. When the '*use*' temperature is above its T_g, the same polymer will be in the rubbery state and will exhibit an elastomeric nature. A crystallisable polymer whose T_m is very much above the '*use*' temperature, when allowed to crystallise, attains a very high degree of crystallinity and becomes a fibre-forming material. Partially crystalline polymers neither behave fully like an elastomer nor can they have enough strength to form fibres. In fact, they become plastic. If the '*use*' temperature is below their T_g, partially crystalline polymers contain components existing both in glassy as well as crystalline states. While the crystalline component increases the tensile properties, the glassy component lowers the impact properties. When the '*use*' temperature is above their T_g, such polymers contain components in the rubbery and crystalline states and by combining them appropriately we can achieve good tensile and impact properties from the plastic materials. As crytallinity increases, they tend to become fibre-forming materials. Nylon, for example, can be plastic as well as fibre-forming, depending upon the degree of crystallinity.

14.5 RHEOLOGY OF POLYMERIC MATERIALS

Rheology is the study of deformation or flow behaviour exhibited by materials. Low molecular weight solids and liquids exhibit their characteristic flow behaviour according to Hooke's equation or Newton's equation, which are now discussed individually.

14.5.1 Hooke's Equation

The behaviour of an ideal solid obeying Hooke's law can be represented by the elongation versus time plot given in part A of Fig. 14.1. Application of a load at a certain point of time t_1 to the solid material elongates the material by a certain length and the extended length remains as long as the load is retained. We remove the load at time t_2 and the material goes back to its original length instantaneously. This is the deformation behaviour of an ideal solid. In other words, an ideal solid elongates instantaneously upon loading and retracts instantaneously upon unloading. This behaviour is called *elasticity* or *elastic deformation* and is time-independent. The stress-strain (or force-elongation) plot for such a deformation will be as represented in part (A) of Fig. 14.1.

From the diagrams, it can be seen that the applied stress (S) is proportional to the resultant strain or elongation (E) and is independent of the rate of strain (dE/dt). This is *Hooke's law*, and is represented as follows:

$$S = GE \tag{14.1}$$

where G is the modulus of elasticity, a quantity that is representative of the material resistance to deformation. Differentiation of Eqn. (14.1) gives

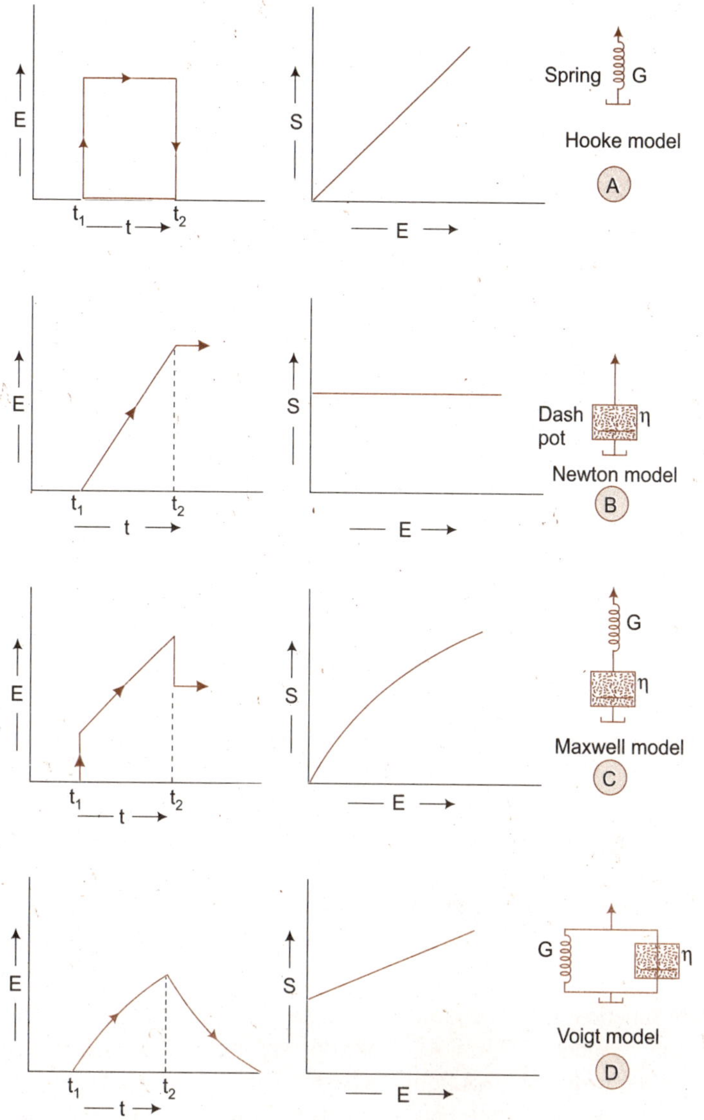

E = elongation, t = time, S = stress, G = modulus, η = viscosity

Fig. 14.1 Plots showing elongation-time behaviour (*E* vs. *t*) and stress-strain behaviour (*S* vs. *E*) of solids and liquids, and different models explaining such behaviours (for details see text).

$$\frac{dS}{dt} = G\frac{dE}{dt} \tag{14.2}$$

[Equation (14.2) will be useful later while deriving Eqn. (14.8)]. Now, elongation is given by the stress divided by modulus:

$$E = \frac{S}{G} \tag{14.3}$$

Thus, the larger the modulus, the more rigid will be the material. The behaviour of an ideal solid can be compared to that or a steel spring which elongates readily upon loading and goes back to its original shape upon unloading.

14.5.2 Newton's Equation

In the case of an ideal liquid, the elongation versus time plot will be given in part B of Fig. 14.1. At time t_1 we apply a force to a viscous liquid (say, coal-tar) which then starts elongating and continues to do so until the force is released at time t_2, when the flow stops but the material never recovers its original shape. The deformation here is time-dependent and is permanent. This type of deformation is called the '*viscous flow*'.

The stress-strain curve for viscous flow is given in part B of Fig. 14.1. It can be seen that the applied stress (S) is proportional to the rate of strain (dE/dt), but is independent of strain (E). This is Newton's law, and is represented as follows:

$$S = \frac{dE}{dt}\eta \tag{14.4}$$

where η is the viscosity of the liquid, a quantity which is representative of the material resistance to flow. Elongation in this case is given by

$$E = \frac{S}{\eta}t \tag{14.5}$$

The viscosity function in the denominator indicates that the higher the viscosity of a liquid, the lesser is its flowing tendency. This flow behaviour can be compared to the movement of a piston inside a cylinder filled with a viscous liquid (a dash pot or a shock absorber). Here, the viscous liquid does not allow the piston to move fast, even if an attempt to pull the piston forcibly is made. The piston, therefore, moves slowly. On releasing the force, the viscous liquid in the cylinder does not allow the piston to move back and hence, there is a permanent displacement of the piston.

14.5.3 The Maxwell and Voigt Models for Visco-elasticity

A steel spring with a modulus G represents the behaviour of an elastic solid (See part A in Fig. 14.1). A dash pot containing a liquid of viscosity η represents the behaviour of a viscous liquid. (See part B in Fig. 14.1). The spring and the dash pot form the basic models, which help us in studying the deformation behaviour of many materials. The former is called *Hooke's model* and the latter *Newton's model*.

Maxwell, while studying the behaviour of many metals, found that certain metals, such as copper or lead, exhibited a behaviour as represented in part C of Fig. 14.1. A copper wire when loaded immediately elongates to a certain length and then continues to elongate slowly, and on unloading, retracts immediately just as much as it had elongated originally, but remains permanently set at that length. Part C of Fig. 14.1 (if we look carefully) is just a linear superpositioning of parts A and B. The total strain behaviour of the copper wire is an additive function of the behaviour of an elastic solid which is recoverable (S/G) and that of a viscous liquid which is not recoverable $[(S/\eta)t]$:

$$E_{\text{Maxwell}} = E_{\text{elastic}} + E_{\text{viscous}} = (S/G) + (S/\eta)t \tag{14.6}$$

Maxwell explained this behaviour by connecting a spring and a dash pot in series (Fig. 14.1, part C). On applying a force, there is an instantaneous elongation of the spring followed by a slow movement of the piston in the dash pot and on releasing the load, the spring jumps back instantaneously, but the dash pot remains permanently in the displaced state. So, while describing the behaviour of a metal such as lead or copper, we cannot do away with either the modulus or the viscosity, since the material behaves both like a viscous liquid and an elastic solid. This behaviour is known as '*visco-elasticity*'. The model in which a spring and a dash pot are connected in series is called the *Maxwell model* for visco-elasticity. The stress-strain plot for such a model is given in part C of Fig. 14.1.

About the same time, Voigt proposed a different model to describe visco-elasticity. He studied the behaviour of a rubber band on straining (which we have discussed earlier). It elongates on loading and regains its original shape on unloading. This may appear to be comparable with the behaviour of a steel spring, but it is not so. The actual behaviour is as given in part D of Fig. 14.1. On applying a small load at time t_1, instead of an instantaneous elongation as exhibited by the spring, we get a slow but continuous elongation, till the load is released at time t_2. On releasing the load, again, the recovery is not instantaneous but slow and continuous, and also complete. Voigt explained this behaviour by using a spring and dash pot connected in parallel (Fig. 14.1, part D). On loading, the spring tries to elongate instantaneously, but the piston in the dash pot moves rather slowly and delays the response of the spring. On releasing the load, the spring tries to pull the system back, but, again, the dash pot delays the motion. Here, we have a phenomenon of '*delayed elasticity*' which can be described, again, by both the modulus and the viscosity. This model is called the *Voigt model* for visco-elasticity. The stress-strain curve for the Voigt model is shown in Fig. 14.1, part D. Here, the total stress is equal to the sum of the elastic and viscous stresses:

$$S_{\text{Voigt}} = S_{\text{elastic}} + S_{\text{viscous}} = GE + \eta\frac{dE}{dt} \tag{14.7}$$

14.5.4 Deformation Behaviour of Polymeric Materials

Now, what about the behaviour of polymeric materials? We know that most of the polymeric materials at the '*use*' temperature are partially crystalline, *i.e.*, they contain both crystalline and amorphous components. In the glassy state both these components exhibit elasticity, while, in the rubbery state, the amorphous components exhibit visco-elasticity. Uncross-linked elastomeric materials too exhibit viscous flow. Polymeric melts, of course, exhibit viscous flow. A majority of the polymers give elongation versus time curves as represented in Fig. 14.2, exhibiting a combination of

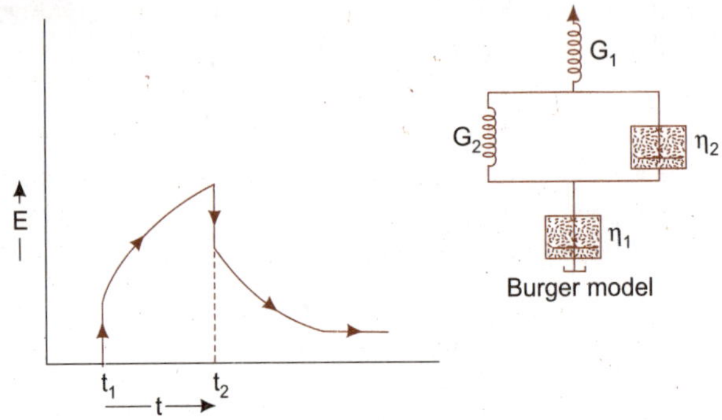

Fig. 14.2 Plot showing generalised elongation versus time behaviour exhibited by many common polymers (for details see text).

instantaneous elongation, delayed elongation, instantaneous recovery, delayed recovery and permanent set. The magnitude of these responses in any individual polymeric material depends on the crystallinity and the extent of cross-linking. To describe this combination of behaviours as exihibited by polymeric materials, we need a Hooke model, a Voigt model and a Newton model combined as given in Fig. 14.2. This model is called the *Burger model*. Here, we have two moduli G_1 and G_2 and also two viscosities η_1 and η_2. When we start from the bottom of the Burger model, the dash pot with viscosity η_1 describes the viscous flow and the irrecoverable deformation (or permanent set). This is caused in polymeric materials by individual molecules of the uncross-linked elastomeric components slowly moving past each other under the influence of the load and taking up new positions (see Section 14.5.5). Next, in the middle part of the Burger model, we have the spring and dash pot with modulus G_2 and viscosity η_2 together describing the delayed elongation and delayed recovery. The cross-linked molecules of the elastomeric component in the polymeric material are responsible for viscous flow (η_2) coupled with the elasticity (G_2). And, finally, we have the spring at the top of the Burger model describing the elasticity (G_1). The crystalline and glassy components of the polymeric material are responsible for this.

The stress-strain curves of polymeric materials vary, depending upon certain factors, *viz*., the crystallinity, T_g and the degree of cross-linking. On the basis of these factors, polymeric materials can be broadly classified as hard and brittle, hard and strong, hard and tough, soft and weak, and soft and tough, as per the scheme given in Table 14.1. The representative stress-strain curves of these categories are shown in Fig. 14.3.

Table 14.1 Broad classification of polymeric materials based on mechanical behaviour

Classification	*Crystallinity*	*'Use' temp.*	*Typical example*
Hard and brittle	Amorphous	Below T_g	Brittle plastics (polystyrene)
Hard and strong	Highly crystalline	Below T_g	Strong fibres (nylon)
Hard and tough	Partially crystalline	Above T_g	Strong plastics (polyethylene)
Soft and weak	Amorphous	Above T_g	Uncross-linked elastomers (polyisobutylene)
Soft and tough	Amorphous	Above T_g	Cross-linked elastomers (vulcanised rubber)

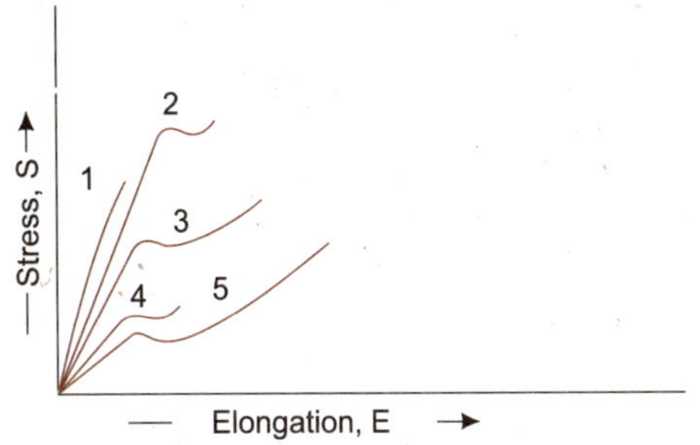

1. Hard and brittle
2. Hard and strong
3. Hard and tough
4. Soft and weak
5. Soft and tough

Fig. 14.3 Plots of stress versus elongation for different types of polymeric materials.

Now, the spring and dash pot models can also be used to explain the *Brownian motion* (BM) within the polymeric materials. The spring element characterises the modulus and the absence of motion and hence, can represent the negative BM. The dash pot, on the other hand, characterises viscosity and flow (motion) and can represent the positive BM. In the solid state, both IBM and EBM* are absent and there is stress transferring and storing of energy. This state can be compared to the spring element with a high modulus (G_1). The polymer melt is characterised by the onset of both IBM and EBM, *i.e.*, both segmental as well as molecular motions and no storing of energy. This corresponds to the dash pot with a low viscosity (η_1). In the rubbery state, we have the positive IBM and the negative EBM, and the material possesses both rigidity and flow (*i.e.*, visco-elasticity). This state can be represented by a spring with a low modulus (G_2) and a dash pot with a high viscosity (η_2). The solid, rubbery and melt states of polymeric materials can now be represented as shown in Fig. 14.4.

SOLID POLYMERS (Crystalline and glassy state)	ELASTOMERICS (Rubbery state)	POLYMER MELTS (Liquid state)
IBM – EBM – Spring (G_1)	IBM + EBM – Dash pot (η_2) and Spring (G_2)	IBM + EBM + Dash pot (η_1)

Fig. 14.4 Behaviour of solid, elastomeric and melt states of polymers explained by presence and/or absence of IBM and EBM (for details see text).

14.5.5 Relaxation and Retardation

We have seen that the visco-elasticity of polymeric materials is a time-dependent phenomenon. This leads to two associated behaviour types called *relaxation* and *retardation*. When an uncross-linked elastomeric polymer is stretched and released, it always gains on its original length. That is, it suffers

*Please see Section 6.2.1 for details of IBM and EBM.

a certain amount of permanent deformation, which gets worse as the extent of stretching and its duration increase. When the material is stretched and maintained in the stretched condition for a long time, nothing apparently happens to it. This is true at the macroscopic level, but not at the microscopic or molecular level. In a stretched condition, individual molecules of the materials are extended and are subjected to tension. They try to relieve the tension by changing their position with respect to each other such that they attain a more probable random conformation in their displaced positions (Fig. 14.5). The same phenomenon occurs in liquids too. On applying a force, the individual molecules which are small in size are able to take up new positions very quickly and relieve the strain or tension. This process whereby the molecules take up new positions in order to relieve the strain, is known as '*relaxation*'. The relaxation process in liquids is so fast that they exhibit only flow and not elastic recovery. In elastomeric materials, however, because of the long chain nature and interpenetrating coil structure of the molecules and also because of the restricted molecular mobility at lower temperatures, the individual molecules take a sufficiently long time to slip into new positions. When we stretch such a material rapidly and release it back immediately, the molecules do not get any time to relax and hence, the material exhibits elasticity and regains its original shape. On the other hand, if we stretch it slowly and keep it stretched for a longer time, we are in fact allowing sufficient time for the relaxation process and the material exhibits more of viscous flow characteristics and undergoes a greater degree of permanent deformation. As the temperature is increased, the molecular mobility becomes increasingly activated. Hence, the relaxation process becomes quicker and the polymeric material becomes increasingly liquid in character.

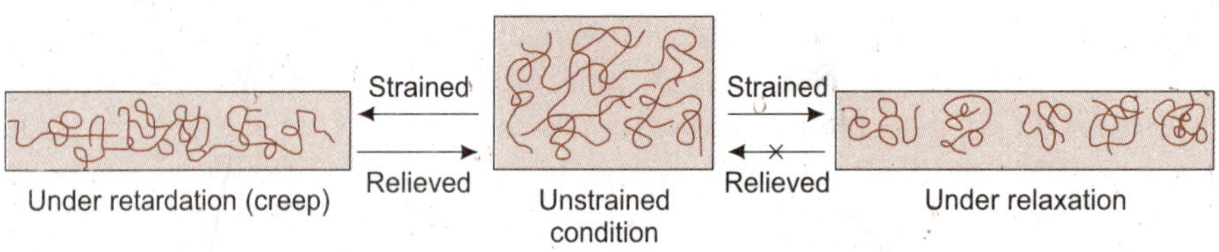

Fig. 14.5 Simplified representation of conformations assumed by polymer molecules present in an elastomer undergoing creep or relaxation on straining (for details see text).

The strain-relieving phenomenon during the process of relaxation can be understood by the following simple experiment. Take a piece of uncross-linked elastomer, say, polyisobutylene, load it on a tensile-testing machine, elongate it to a certain length and maintain it at that length. Now, measure the stress at different time intervals. In doing so, we are actually measuring the stress that is necessary and just sufficient to keep the material elongated in that constant length. It will be noticed that, as time passes, the measured stress becomes lesser and lesser (Fig. 14.6). This clearly indicates that the tension inside the material is progressively relieved. Relaxation can now be defined as the decay in stress with time under constant deformation. The Maxwell model describes this phenomenon well. When the Maxwell model is pulled, the spring in it immediately elongates to a certain length, but the dash pot, which is unable to respond quickly to the rapidly applied stress, remains undisturbed. Now, keep the model at that elongated position. The spring now slowly transfers the tension set in it on to the dash pot, which now starts responding. As the piston in the dash pot

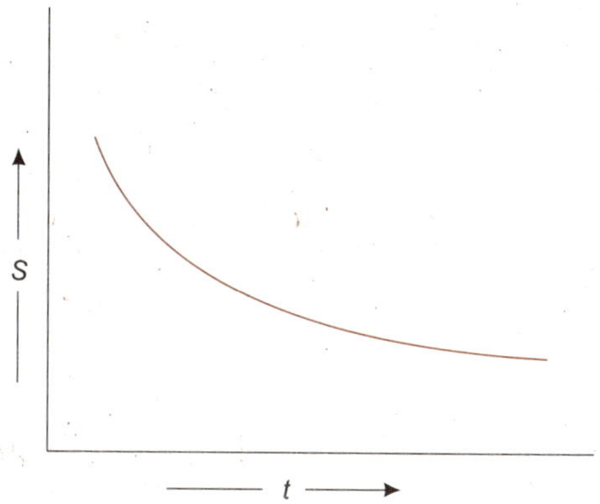

Fig. 14.6 Representative plot of stress (S) versus time (t) (relaxation curve) for an uncross-linked elastomer kept under constant elongation.

moves slowly in the direction of the applied stress, a corresponding amount of strain in the spring is relieved, and the elongation undergone by the spring element is reduced correspondingly although the model as a whole still remains in the same elongated position. At this stage, we release the pull. The spring jumps back to its original shape, but the piston of the dash pot does not move back and hence, the recovery of the model as a whole is not complete. The longer we keep the model under constant deformation, the more will be the displacement of the dash pot piston and hence, the greater will be the permanent deformation of the model.

From Eqn. (14.6), we know that the total deformation of a Maxwell model is additive of the elastic and viscous deformations. The rate of deformation is also additive. By combining Hooke's law and Newton's law [(*i.e.*, by combining Eqns. (14.3) and (14.6)], we can get the total rate of deformation as

$$\frac{dE_{\text{total}}}{dt} = \frac{dE_{\text{elastic}}}{dt} + \frac{dE_{\text{viscous}}}{dt} = \frac{1}{G}\cdot\frac{dS}{dt} + \frac{S}{\eta} \tag{14.8}$$

When the deformation is constant, *i.e.*, when $dE/dt = 0$, Eqn. (14.8) becomes

$$\frac{1}{G}\cdot\frac{dS}{dt} = -\frac{S}{\eta} \tag{14.9}$$

On integrating Eqn. (14.9), we get

$$S = S_o \exp\left(\frac{-G}{\eta}t\right) = S_o \exp\left(\frac{-t}{t_{\text{rel}}}\right) \tag{14.10}$$

where S_o is the original stress required to bring about the deformation; S the stress at a time t after the material has been kept under constant deformation; and t_{rel} $(= \eta/G)$ the relaxation time. Here, t_{rel} indicates the time taken for the stress to fall to a value $1/e$ times, *i.e.*, 1/2.7 or 0.37 times S_o.

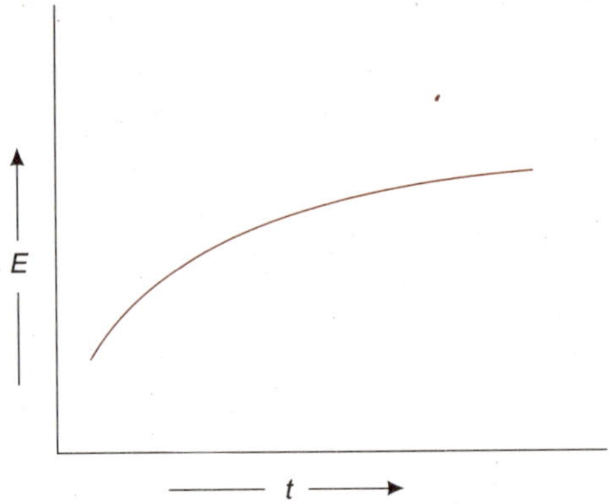

Fig. 14.7 Representative plot of elongation (E) versus time (t) (creep curve) for a cross-linked elastomer kept under constant stress.

Now, let us take up retardation. What precisely is retardation? Take a rubber band and load it with a small weight that is just sufficient to keep the band erect, but not stretched. Allow some time to pass and then observe. It will be noticed that the rubber band now slowly starts elongating and continues to do so with time. After it has elongated to the maximum, the weight is removed. The rubber band starts contracting slowly and on allowing sufficient time to elapse, it recovers its original dimension completely. The rubber band has behaved like an elastic solid in one sense, *i.e.*, it was elongated upon loading and fully recovered on unloading. The process of elongation and recovery is, however, not instantaneous but delayed. This delayed elasticity is called '*retardation*'. It can be defined as the increase in deformation with time under constant stress. Retardation is also referred to as '*creep*'. A typical creep curve is given in Fig. 14.7. In the process of retardation, the position of molecules within the material does not change but only their conformation does. The molecules, under the influence of a constant stress, without moving away from their original position, get slowly elongated from their most probable random conformation. When the stress is released, the molecules again go back slowly to their original most probable conformation. The retardation behaviour can be explained with a Voigt model. When a constant force is applied, the spring tries to elongate, but is allowed to do so only slowly by the dash pot. Given a sufficient time, the spring elongates to its full possible length. On releasing the force just the reverse takes place. As we have already seen, the total stress is equal to the sum of the elastic and viscous stresses [Eqn. (14.7)]:

$$S_{\text{total}} = S_{\text{elastic}} + S_{\text{viscous}} = GE + \frac{dE}{dt}\eta \tag{14.11}$$

On integration, we get

$$E = \frac{S}{G}\left[1 - \exp\left(-\frac{G}{\eta}t\right)\right] = E_{\infty}\left[1 - \exp\left(\frac{-t}{t_{\text{ret}}}\right)\right] \tag{14.12}$$

where E_{∞} is the elongation attained at time t after the stress has been applied and t_{ret} (= η/G) the retardation time. It indicates the time taken for the model to recover (after the release of the stress) to 1 – (l/e) or 1 - (1/2.7) or 0.63 times the E_{∞} value.

Books for Further Reading

Alfrey, T., *Mechanical Behaviour of High Polymers*, Interscience, New York, 1948.

Frith, E.M. and Tuckett, R.F., *Linear Polymers*, Longmans Green, London, 1951.

Tobolsky, A.V., *Properties and Structure of Polymers*, John Wiley, New York, 1960.

Ferry, J.D., *Visco-elastic Properties of Polymers*, John Wiley, New York, 1961.

Bueche, F., *Physical Properties of High Polymers*, Wiley-Interscience, New York, 1962.

Severs, E.T., *Rheology of Polymers*, *Reinhold*, New York, 1962.

Hearle, J.W.S. and Peters, R.H., *Fibre Structure,* Butterworths, London, 1963.

Meares, P., *Polymers: Structure and Bulk Properties*, Reinhold, London, 1965.

Nielsen, L.E., *Mechanical Properties of Polymers*, Reinhold, New York, 1965.

Lenk, R.S., *Plastics Rheology*, Maclaren, London, 1968.

Christensen, R.M., *Theory of Visco-elasticity: An Introduction*, Academic, New York, 1970.

Deanin, R.D., *Polymers: Structure, Properties and Applications*, Cahners, Boston, 1972.

Aklonis, J.J., MacKnight, W.J. and Shen, M., *Introduction to Polymer Visco-elasticity*, Wiley-Interscience, New York, 1972.

15

POLYMER PROCESSING

15.1 INTRODUCTION

Whether we want to make a toy or fabricate a spacecraft, polymers would often have some crucial role to play. But how does one convert a polymer into a broad spectrum of useful shapes and structures? To answer this question, we need to look at another branch of polymer technology called '*polymer processing*' which is the theme of this chapter.

In a very broad sense, polymer processing may be defined as an engineering speciality used to convert polymeric materials into useful end products. Most of the available techniques now employed in processing technology of polymers are basically refined versions of those used in ceramic or metal industries. In fact, the need to understand finer aspects of polymer processing surfaced as conventional materials were required to be substituted by those with better finish and elegance. The black ebony or the costly ivory keys in a harmonium or a piano gave way to wooden keys with a covering of celluloid.

15.2 PLASTICS, ELASTOMERICS AND FIBRES

In Chapter 14, we have defined the terms '*plastic*', '*elastomeric*' and '*fibre-forming*' based on the properties exhibited by polymeric materials. We now elaborate these three categories further in the following paragraphs.

15.2.1 Plastics

The word '*plastic*' originated from Greek, meaning a material which can be moulded or formed into any shape of one's choice. This etymology would allow even clay to be called a plastic, but in practice, only articles made of synthetic materials are referred to as '*plastics*'. The American Society for Testing and Materials (ASTM) has defined a plastic as "any one of the large and varied group of materials, wholly or primarily organic in composition, which may be formed into useful shapes by the application, singly or together, of heat and pressure."

There are hundreds of plastics they can, however, be grouped into a dozen or so major types. Table 15.1 gives the major types of plastics and some representative members belonging to each type. There is, however, no comprehensive way of describing these varieties of plastics as they are too numerous.

Table 15.1 Major types of plastics

Type	*Typical members*
Acrylic plastics	Polymethyl methacrylate (PMMA) Polyacrylonitrile (PAN)
Amino plastics	Urea-formaldehyde resin (UF) Melamine-formaldehyde resin (MF)
Cellulosics	Ethyl cellulose (EC) Cellulose acetate (CA) Cellulose nitrile (CN)
Epoxide plastics	Epoxy resins (EP) Epoxy novalac resins (EPN)
Fluoroplastics	Polytetrafluoro ethylene (PTFE) Polyvinylidene fluoride (PVDF)
Phenoplastics	Phenol-formaldehyde resin (PF) Phenol-furfural resin (PFF)
Polyamide plastics (nylons)	Polycaprolactam (Nylon 6) Polyhexamethylene adipamide (Nylon 6, 6)
Polyesters	Unsaturated polyester resins (UPER) Polyethylene terephthalate (PETP) Polyethylene adipate (PEA)
Polyolefines	Polyethylene (PE) Polypropylene (PP)
Styrene plastics	Polystyrene (PS) Styrene-acrylonitrile copolymer (SAN) Acrylonitrile-styrene-butadiene terpolymer (ABS)
Vinyl plastics	Polyvinyl chloride (PVC) Polyvinyl butyral (PVB) Vinyl chloride-vinyl acetate copolymer (PVCAC)

15.2.2 Elastomerics

Elastomerics are popularly known as '*rubbers*'. Balloons, shoe soles, tyres, surgeons' gloves, garden hoses are examples of articles made of elastomerics. Natural rubber is the most classical example of an elastomeric. There are several synthetic elastomerics now in use. These include polybutadienes, styrene-butadiene (SBR) and acrylonitrile-butadiene (nitrile rubber) copolymers, polyisoprene, polychloroprene (neoprene), ethylene-propylene copolymer, isoprene isobutylene copolymer (butyl rubber), polyfluorocarbon, polyurethane and silicone rubbers.

15.2.3 Fibres

We are all familiar with fibres of natural origin such as cotton, wool, flax and silk. We are also familiar with man-made or synthetic fibres made of nylon, polyesters, polypropylene and acrylics. One general definition of a fibre is that its length is at least hundred times its diameter. While natural fibres (except silk) occur as staples, synthetic fibres can be produced as continuous filaments or cut to form staples.

From the consumer's point of view, fibres can be classified into three types, *viz*., comfort fibres, safety fibres and industrial fibres. Comfort fibres are those used for making undergarments and garments. Fibres used for making underwears, socks, shirtings, suitings, ladies garments, etc., come under this category. Comfort fibres should have adequate strength and extensibility, softness, good moisture regain, and preferably some flame retardancy, and should be capable of being dyed. Cotton, silk, wool, nylon, polyesters, acrylics are common examples of comfort fibres.

Safety fibres are those used for making carpets, curtains, seat covers, draperies and so on. Such fibres should be strong, tough, durable and abrasion-resistant. The most important feature is that they should be difficult to ignite, have as low a flame spread as possible, liberate minimum heat, smoke and toxic gases if involved in a fire. Comfort fibres can be rendered flame-retardant and made to behave as safety fibres by adding small quantities of substances containing atoms such as B, N, Si, P, Cl, Br or Sb. While these substances serve as flame-proofing agents and delay the ignition and flame spread, they cannot reduce smoke and toxic gases if these treated fibres get involved in a fire. Recent investigations have shown that aromatic polyamides, polyimides, polybenzimidazoles and polyoxy diazoles are useful as safety fibres. These fibres have, however, a tendency to evolve toxic gases on burning owing to the presence of nitrogen in their molecules. Fully aromatic polyesters do not suffer from this disadvantage.

Industrial fibres are used as reinforcing materials in composite structures. These fibres are also called structural fibres as they possess very high modulus, strength, thermal stability, toughness and durability. Structural fibres are used to reinforce articles such as rigid and flexible tubes, pipes and tyres and also in composite structures called fibre-reinforced plastics (FRP) used in the construction of boats, cars, planes and even buildings. Uniaxially oriented fibres of aromatic polyamides and polyesters, carbon fibres and silica fibres come under this category.

15.3 COMPOUNDING

Polymers in their pure form, as obtained from the manufacturing plants after isolation and purification, are called '*virgin*' polymers or '*virgin*' resins. Barring a few (such as polystyrene, polyethylene or polypropylene), virgin polymers, as such, may not be amenable for processing straightaway. Virgin PVC, for example, is a material of horny texture and cannot be moulded without making it soft by the addition of a plasticiser. Natural rubber similarly requires a vulcanising agent to render it mouldable. Most polymers are protected from thermal, oxidative and photo degradation by incorporating appropriate stabilisers. Many articles are obtained in attractive colours by adding suitable dyes and pigments to the polymeric material prior to its moulding. Lubricants and process aids are added to most of the polymers to reduce friction and improve flow within processing machineries. Fillers are added to impart special properties and reduce the finished product cost.

The process involving incorporation of ingredients such as plasticisers, vulcanising agents, curatives, stabilisers, fillers, colouring agents, flame retardants and lubricants into the virgin resin is known as '*compounding*' of the resin.

Virgin resins of plastic materials such as polystyrene, polyethylene, polymethyl methacrylate and PVC are usually available as free flowing fine powders. Compounding ingredients in the form of fine powder or liquids are blended with fine powder of the virgin resin using pianatary mixers, V-blenders, ribbon-blenders, Z-mixers or tumblers. Blending can be done at ambient or elevated temperatures which should, however, normally be much below the softening temperature of the polymer. Liquid prepolymers are compounded using simple high-speed stirrers.

Virgin resins of elastomerics such as natural rubber, SBR and nitrile rubber are obtained in the form of crumbs physically compacted into thick slabs called '*bales*'. They are generally compounded with vulcanising agents, accelerators, fillers, antioxidants and lubricants. Since elastomerics are not free flowing powders (like virgin plastic materials are), they cannot be blended with a compounding ingredient using devices that are employed for virgin plastics. This is understandable as compounding of virgin plastic resins with other ingredients requires only uniform mixing to be achieved, whereas compounding of virgin elastomeric resins involves milling down the crumbs into pliable sheets and then incorporating the compounding ingredients. Compounding of elastomerics is done either on a two-roll rubber mill or in a Banburry internal mixing mill. Elastomerics in the form of latex or low molecular weight liquid resins, however, can be compounded by simple mixing using a high-speed stirrer.

Fibre-forming resins usually do not require compounding. Ingredients such as lubricants, stabilisers and fillers are straightaway incorporated into the polymer melt or polymer solution just prior to spinning.

15.4 PROCESSING TECHNIQUES

The very fact that polymeric materials are used in many forms such as rods, tubes, sheets, foams, coatings or adhesives and also as moulded and fabricated articles implies that there must be a variety of ways in which the compounded resins can be processed and converted into finished products. A majority of the articles are either moulded or fabricated, while many others are made by casting liquid prepolymers into a mould and allowing them to cure or cross-link. Fibres are made by a process known as '*spinning*'.

The process of moulding can be compared to making, for instance, a toy horse out of clay; the fabrication process can perhaps be compared to making the toy by carving it out of a soap cake. In the moulding process, the compounded material in the form of powder, chips or granules is placed in a mould of the required shape and subjected to heat and pressure, after which, the final product is obtained. Fabricated items are made of simple forms such as sheet, rod or tube, using processes such as cutting, shaping, gluing, bonding and welding.

About 50 years ago, there existed only a few processes for converting polymeric materials into finished products. Today, there are many processes and automatic machines for this purpose, the important ones being calendering, casting, compression moulding, injection moulding, extrusion moulding, blow moulding, cold forming, thermoforming, foaming, reinforcing, melt spinning, dry spinning and wet spinning. The last three techniques are used for making fibres from fibre-forming materials, while the rest are used for converting plastic and elastomeric materials into moulded and fabricated articles. In the forthcoming sections, we shall give an outline of these important processes. For a more detailed account of these and other processes such as dip coating, fluidised bed coating, heat sealing, electronic sealing and welding reference may be made to specialised textbooks on '*polymer processing*'. Similarly, details regarding coatings and adhesives are also beyond the scope of this book.

Before taking up various processing techniques, let us recapitulate that polymeric materials can either be thermoplastic or thermosetting. Once thermoplastic materials are moulded or shaped under heat and pressure, they are required to be cooled much below their softening temperature before being released out of the mould as otherwise they may get deshaped. In the case of thermosetting materials, however, there is no need to cool the article before releasing it out of the mould, since once set under the influence of heat and pressure, the article does not change its shape even when handled hot.

15.4.1 Calendering

The calendering process, in its simplest form, is employed to produce continuous films and sheets. The main part of a calendering machine (Fig. 15.1) is a set of highly polished metal rollers rotating in opposite directions with provision for precise adjustment of the gap between them. The gap between the rollers determines the thickness of the sheet calendered out. Compounded polymeric material is fed between the rollers which are maintained at an elevated temperature, and the sheet emerging from the rollers is cooled by passing through cold rollers. The sheets are finally wound up in rolls as shown in Fig. 15.1. If, however, thin films are required instead of sheets, a series of rollers with a

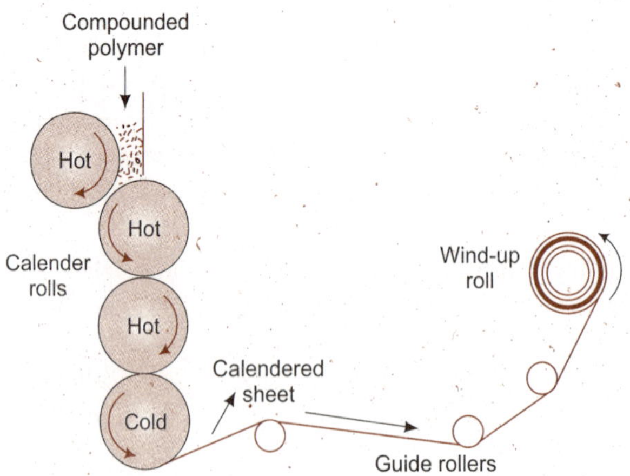

Fig. 15.1 Schematic diagram of a calendering machine.

successively diminishing gap between them is employed. Polyvinyl chloride, polyethylene, acrylonitrile-butadiene-styrene copolymer (ABS) and rubbers are among the main polymers which are usually calendered into sheets.

By incorporating a suitably engraved embossing roller in the calendering machine, embossed sheets of a suitable design could be produced. Besides this, special decorative effects (like marbleisation) could be achieved by feeding the calender with a mixture of compounded materials of different colours. The marbleisation technique is usually employed in the production of vinyl floor tiles made of PVC.

15.4.2 Die Casting

Die casting is a relatively low cost process which consists of converting a liquid prepolymer to a solid object with a desired shape. Sheets, tubes, rods and the like in limited lengths, can be produced by the casting process. Figure 15.2 illustrates the simplest version of the casting process. Here, the prepolymer compounded suitably with a curative and other ingredients is poured into a petridish, the latter representing the die. The dish is then kept in an oven at an elevated temperature for a few hours to complete the cure reaction. On cooling to room temperature, the solid product from the petridish is pulled out. The solid thus cast will have a shape identical to the interior of the petridish. Instead of a petridish, if a cylindrical glass tube closed at one end is used, we get the product in the form of a cylindrical rod. Instead of a prepolymer and a curative, a mixture of monomer, catalyst and other ingredients can be heated to the polymerisation temperature and poured into the die. Polymerisation is allowed to continue inside the die till the solid product is formed. Acrylics, epoxies, polyesters, phenolics and urethanes are suitable for die casting.

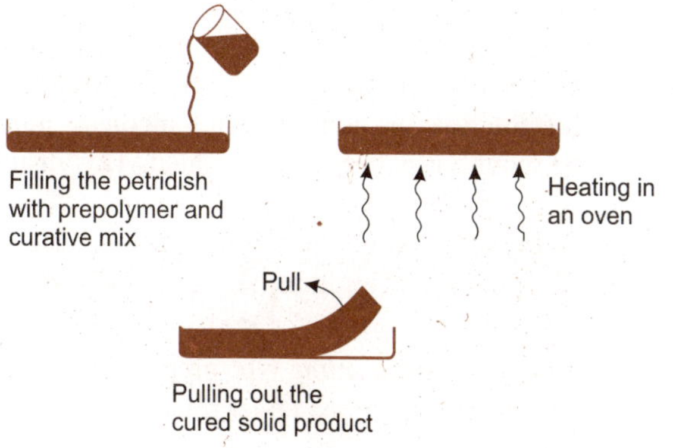

Fig. 15.2 Simple illustration demonstrating the casting process.

Depending on the convenience and availability, the dies for casting are made of plaster of Paris, lead or glass. During curing, the resin block shrinks in size which, in fact, helps in its removal from the die with ease.

15.4.3 Rotational Casting

Hollow articles such as balls and dolls are produced by a process called '*rotational casting*'. The apparatus used for this process is shown in Fig. 15.3.

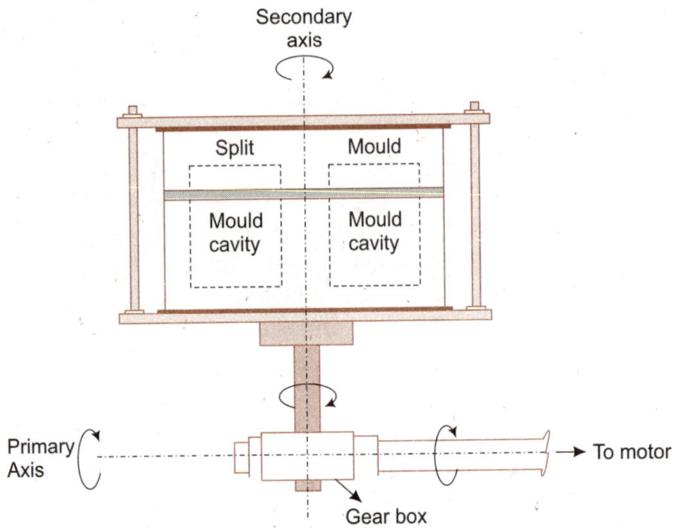

Fig. 15.3 Hollow moulds filled with the polymeric material being rotated simultaneously along the primary as well as the secondary axis during rotational moulding.

The compounded thermoplastic material in the form of a fine powder is taken in a hollow mould. The apparatus has provision for rotating the mould simultaneously along the primary and the secondary axes. After closing the mould, it is heated and rotated. This distributes the molten plastic uniformly along the entire surface of the inside cavity of the mould. After a while, the mould, still under rotation, is chilled with cold water. Now, the molten plastic material uniformly distributed inside the mould surface cools down and solidifies in the shape of that surface. The mould can now be opened and the product removed.

Instead of thermoplastic materials, a thermosetting type prepolymer and curative mixture can also be fed into the mould in liquid form and curing done under rotation at an elevated temperature, when the product is formed.

To a large extent, PVC articles such as rainboots, hollow balls of doll heads are made by rotational casting. Curing of the PVC is done by a physical gelation between the PVC and a liquid plasticiser. Gelation takes place at temperatures around 150–200°C. Fine particles of PVC resin are dispersed uniformly in the liquid plasticiser together with stabilisers and colouring agents to get a pasty substance of a relatively low viscosity. This pasty material called '*plastisol*' is fed into the mould and deaerated. Subsequently, the mould is rotated and heated for the required temperature and time, when gelation of the PVC occurs. The time of gelation controls the wall thickness of the product formed. After achieving the required wall thickness, the excess plastisol is removed for recycling. The gelled product is further heated inside the mould, which will ensure the final fusion of PVC resin particles with the plasticiser. The finished product is taken out of the mould after being cooled with a water spray. This method of rotational casting using a liquid material is referred to as '*slush moulding*'.

15.4.4 Film Casting

A casting technique is also used to produce polymeric films. In this technique, the solution in an appropriate concentration of the polymer in a suitable solvent is allowed to fall at a precalculated rate on an endless metallic belt of high finish moving at a constant speed (Fig. 15.4). A continuous sheet of the polymer solution is thus formed on the surface of the metallic belt. When the solvent is subsequently evaporated under controlled conditions, a thin film of the polymer is formed on the surface of the belt. The film could be removed simply by stripping. Most of the commercially available varieties of cellophane sheets and photographic films are prepared by this process.

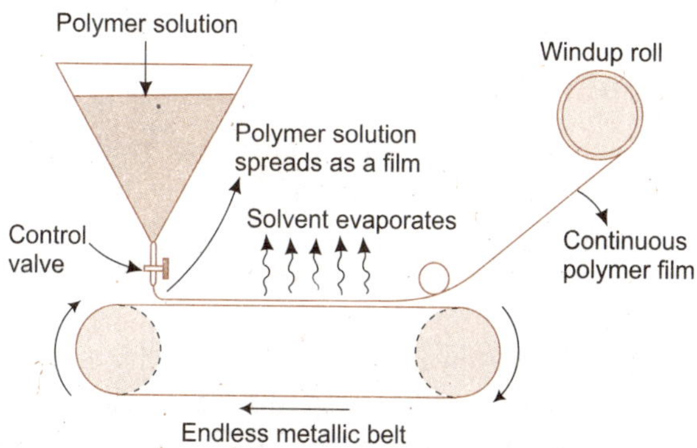

Fig. 15.4 Schematic diagram of a film-casting equipment.

15.4.5 Compression Moulding

The compression-moulding process is very widely used to produce articles from thermosetting materials. Figure 15.5 shows a typical mould employed for compression moulding. The mould is made of two halves—the upper and the lower halves or the male and the female. The lower half usually contains a cavity and the upper half has a projection which fits into the cavity when the mould is closed. The gap between the projected upper half and the cavity in the lower one gives the shape of the moulded article.

In compression moulding, the thermosetting material is subjected to heat and pressure in a single stroke. This is accomplished by using a hydraulic press with heated plattens. Moulding temperature and pressure can be as high as 200°C and 70 kg/cm^2, respectively. The actual temperature and pressure depends on the rheological, thermal and other properties of the plastic material to be moulded. The compounded material is placed in the cavity of the mould so as to fully fill the cavity. As the mould closes down under pressure, the material is squeezed or compressed between the two

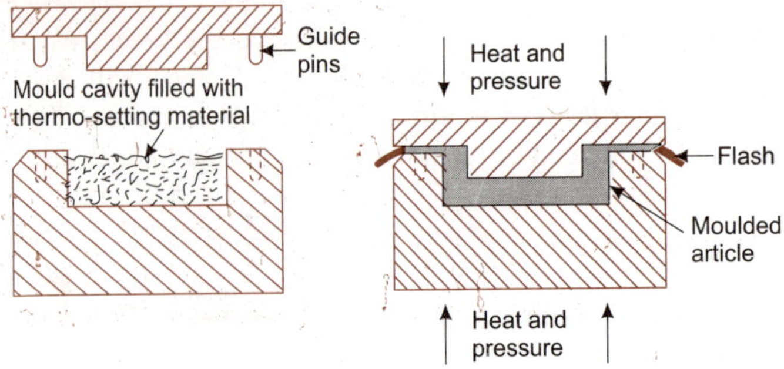

Fig. 15.5 Schematic diagram of a mould employed for compression moulding.

halves and compacted to shape inside the cavity. The excess material flows out of the mould as a thin film. This film expelled out of the mould is known as the '*flash*'. Under the influence of heat, the compacted mass gets cured and hardened to shape. The mould can be opened while it is still hot to release the moulded product.

15.4.6 Injection Moulding

The injection-moulding process is best suited for producing articles made of thermoplastic materials. Here, the equipment cost is relatively high but the main attraction is the amenability of the injection-moulding process to a high production rate. In injection moulding, a definite quantity of molten thermoplastic material is injected under pressure into a relatively cold mould where it solidifies to the shape of the mould.

The injection-moulding machine is shown in Fig. 15.6. The process consists of feeding the compounded plastic material as granules, pellets or powder through the hopper at definite time intervals into the hot horizontal cylinder where it gets softened. Pressure is applied through a

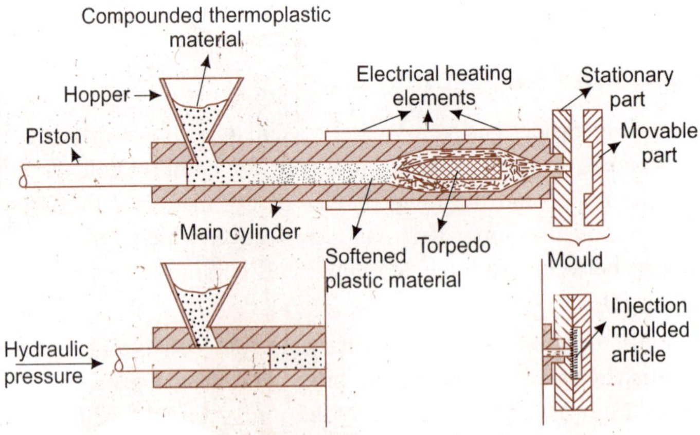

Fig. 15.6 Schematic diagram of an injection-moulding machine.

hydraulically driven piston to push the molten material through a cylinder into a mould fitted at the end of the cylinder. While moving through the hot zone of the cylinder, a device called '*torpedo*' helps spread the plastic material uniformly around the inside wall of the hot cylinder and thus ensures uniform heat distribution. The molten plastic material from the cylinder is then injected through a nozzle into the mould cavity.

The mould used, in its simplest form, is a two-part system. One is a movable part and the other stationary (see Fig. 15.6). The stationary part is fixed to the end of the cylinder while the movable part can be opened or locked on to the stationary part. By using a mechanical locking device, the mould is properly held in position as the molten plastic material is injected under a pressure as high as 1500 kg/cm^2. The locking device has to be very skillfully designed in order to withstand high operating pressures. Furthermore, a proper flow of the molten material to the interior regions of the mould is achieved by preheating the mould to an appropriate temperature. Usually, this temperature is slightly lower than the softening temperature of the plastic material undergoing moulding.

After the mould is filled with the molten material under pressure, then it is cooled by cold water circulation and then opened so as to eject the moulded article. The whole cycle could be repeated several times either manually or in an automated mode.

15.4.7 Blow Moulding

Most of the hollow plastic articles are produced by the blow-moulding technique. Containers (see Fig. 15.7), soft drink bottles and numerous other hollow articles are produced by this process. Thermoplastic materials such as polyethylene, polycarbonate, PVC, polystyrene, nylon, polypropylene, acrylics, acrylonitrile and ABS polymer can be blow moulded; however, high-density polyethylene tops the list in terms of annual consumption.

Fig. 15.7 Blow-moulded containers made of high-density polyethylene.

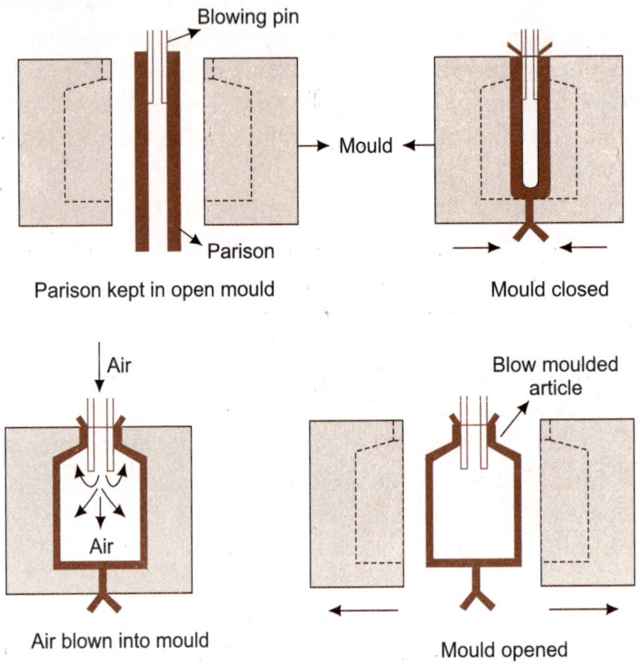

Fig. 15.8 Schematic diagram explaining the steps involved in blow-moulding process.

Blow moulding, in fact, basically belongs to the glass industry. The essence of the process is explained in Fig. 15.8. A hot, softened thermoplastic tube, usually called '*parison*', is properly placed inside a two-piece hollow mould. When the two halves of the mould are closed, it pinches and closes one end of the parison and encloses a blowing pin at the other end. The parison is now blown by pressurising from within by blowing compressed air through the blowing pin. The hot parison is inflated like a balloon and goes on expanding until it comes in intimate contact with the relatively cold interior surface of the hollow mould. Under pressure, the parison ultimately assumes the shape of the hollow cavity of the mould. The mould is allowed to cool and the rigid thermoplastic article formed is removed by opening the mould.

The parison needed for blow moulding can be made either by the injection or extrusion process, and the technique can accordingly be called '*injection* or *extrusion blow moulding*'.

15.4.8 Extrusion Moulding

Extrusion is one of the cheapest methods available for producing many common plastic products such as films, filaments, tubes, sheets, pipes, rods, hoses and straps, all in continuous lengths. The profile of a continuous extrudate is obtained by employing a suitable die. The molten plastic under suitable conditions is forced through the die orifice which imparts the desired profile to the extrudate. A simple extrusion machine is illustrated in Fig. 15.9.

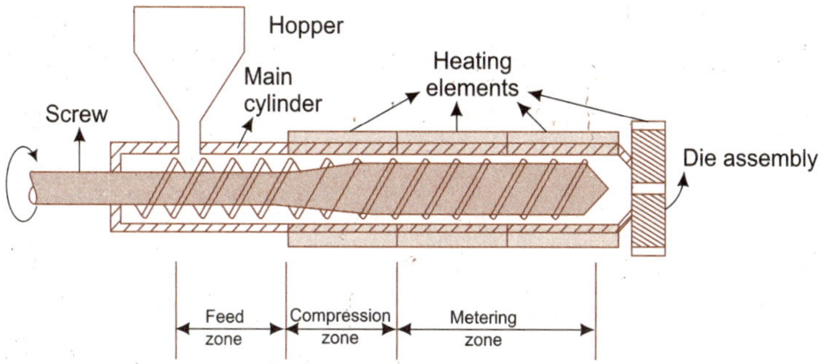

Fig. 15.9 Schematic diagram of a simple extrusion machine.

In this machine, the compounded plastic material is fed through the hopper as either powder or granules into a cylinder having provision for electrical heating for softening the material. The hot plastic charge is further worked through the cylinder by a helically flighted revolving screw. The temperature of the plastic material rises owing to the frictional heat produced by the compression of the charge between the rotating screw and the cylinder surface. During its journey from the hopper to the die, the plastic material passes through distinct zones named feed zone, compression zone and metering zone, as shown in the figure. Each zone contributes in its own way to the overall extrusion process. The feed zone, for instance, receives the charge from the hopper and sends it over to the compression zone. No heating takes place in the feed zone. In the compression zone, however, the powdered charge melts due to heat conducted from the heating elements and is compressed by the working of the screw. The pasty molten plastic material is then sent to the metering section, where it acquires a constant flow rate, imparted by the helical flight of the screw. The pressure built up in this section enables the polymer melt to enter the die and emerge out of it with the desired profile. Due to the highly viscous nature of some polymers, however, one more region called the working zone may also be required, wherein the polymer could be subjected to high shearing forces for effective mixing. The extruded material of the required profile emerging from the die is quite hot (usually 125°C to 350°C) and has to be cooled rapidly to avoid deshaping. The extrudate is accordingly carried over a conveyor belt through a tub containing cold water. During this process, the material dissipates heat to the water and gets set. Cold air blast and cold water spray can also be used for cooling. The product formed is cut to the desired length or wound on to rolls.

The extrusion process is also used for coating wires and cables with PVC or rubber and for coating roll-formed metal strips with suitable thermoplastics.

15.4.9 Thermoforming

Thermoforming is a highly useful process for fabricating three-dimensional articles from plastic sheets. Even some large products, such as submarine hulls are made of ABS sheets by thermoforming techniques.

The technique can be outlined as follows. The thermoplastic sheet is heated to its softening temperature. The warm flexible sheet is then pressed into the female half of a matched metal die duly

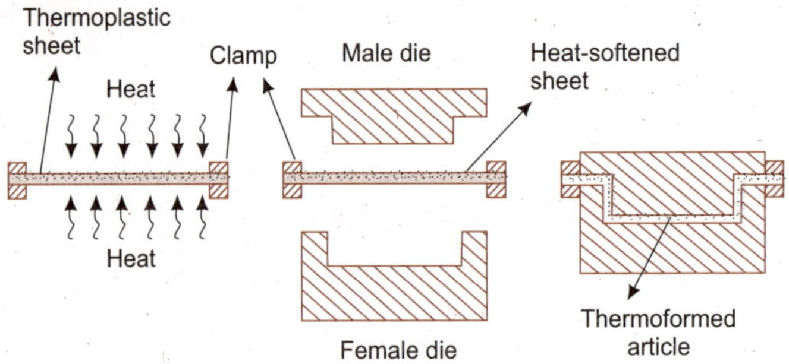

Fig. 15.10 Schematic diagram explaining thermoforming process.

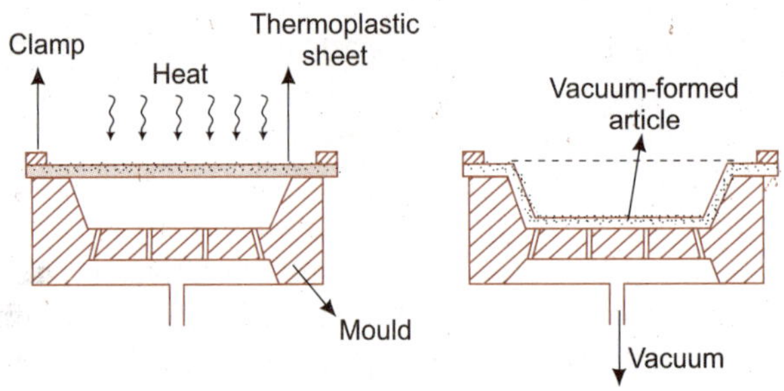

Fig. 15.11 Schematic diagram explaining vacuum-forming process.

assisted by the male half (Fig. 15.10). Now, the sheet assumes the required shape of the mould. On cooling, the shaped article becomes rigid and can be removed from the mould.

In a modified method, the hot plastic sheet is sucked into the cavity of the female mould under the influence of vacuum to give the desired shape (Fig. 15.11). This method is called '*vacuum forming*'.

15.4.10 Foaming

Foaming is an ingenious process for producing expanded or spongy materials. Special properties of this class of materials with cushioning ability, light weight and low thermal conductivity make them eminently suitable for several applications. The usual foamable polymers are polyurethanes, polystyrene, polyethylene polypropylene silicones, epoxy, PVC and the like. The foam structure consists either of discrete unit cells or of interconnecting cells. The discrete unit cells are closed cells and may enclose gases while the interconnecting cells are of the open type. The open type and closed type cell structures are schematically shown in Fig 15.12.

There are different methods for producing foamed or cellular plastics. One method is to melt a compounded thermoplastic material and blow air or nitrogen in such a way that the whole matrix foams up. Addition of the surface active agents helps the foaming up process. When the required degree of foaming is achieved in the midst or blowing air or nitrogen the matrix is cooled to the ambient temperature. In this process, the material is allowed to solidify in the foamed-up condition.

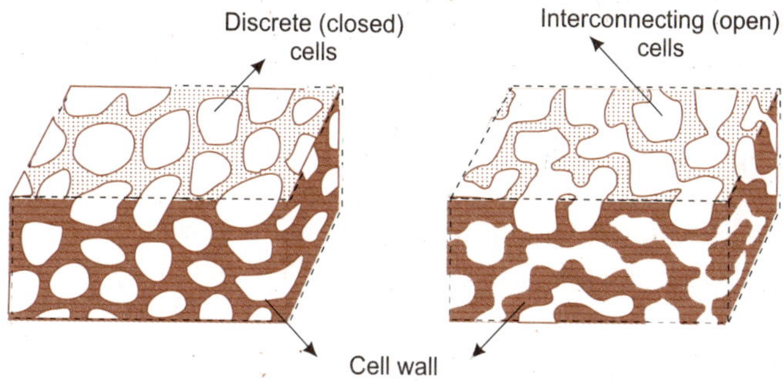

Fig. 15.12 Schematic representation of open type and closed type cell structures formed during foaming process.

Thermosetting liquid propolymers can be blown up cold and then heated to complete the cure in the foamed up condition. Usually, however, foaming is achieved by the addition of a foaming agent or blowing agent. Such agents are solvents with a low boiling point or certain chemical compounds. Solvents such as *n*-pentane or *n*-hexane, under normal cure temperatures of polymeric materials, boil off and liberate large volumes of vapours. Certain chemicals, on the other hand, can be thermally decomposed to generate inert gases. Azobis isobutyronitrile can decompose thermally and liberate large volumes of nitrogen. Carbon dioxide gas, produced as an *in situ* blowing agent by the reaction between the isocyanate and water can also be used for producing foams, for instance, polyurethane foams:

$$R-N=C=O + H-O-H \longrightarrow R-\underset{\underset{H}{|}}{N}-\underset{\underset{O}{\|}}{C}-OH \longrightarrow CO_2\uparrow + R-NH_2$$

Polyurethane is produced by a polyol reacting with a diisocyanate, and we have to only add some additional isocyanate and water to get the foam.

The large volumes of vapours or gases liberated by these blowing agents or foaming agents make the polymer matrix foam up. The polymer matrix in a foamed-up condition is cooled below the softening temperature (in the case of thermoplastic materials) or made to undergo the cure reaction or cross-linking (in the case of the thermosetting type) so that the matrix as a whole in the foamed condition attains structural rigidity to maintain the foam structure. This is called the '*stabilisation of the foam*'. If the polymer matrix is not cooled down to below its softening point or is not sufficiently cross-linked, the gases and vapours may escape out of the system and the foam may collapse.

Foamed plastics could be made in flexible, rigid or semi-rigid forms. When foaming is carried out inside the mould cavity, articles made of foamed plastic can be directly obtained. Foamed sheets and rods can also be used for fabricating articles of different shapes. Depending on the nature and extent of foaming, foamed plastics can have a density anywhere between 20 kg/m^3 and 1000 kg/m^3. Foamed plastics are put to several uses. The automobile industry consumes large quantities of flexible PVC and polyurethane foams as upholstery. These foams find a prominent place in the furniture industry also. Polystyrene rigid foams are widely used for packaging, and thermal insulation of buildings. For matresses and similar other applications, rubber and urethane foams are used to a large extent.

Urethane rigid foams also find application in thermal insulation of buildings and even in the making of artificial limbs.

15.4.11 Reinforcing

Reinforcing a plastic matrix with a high-strength fibre material results in the formation of what are called '*fibre-reinforced plastics*' (FRP). FRPs have outstanding properties such as high strength-to-weight ratio and excellent corrosion resistance and are easy to fabricate. A wide variety of articles are made by the fibre-reinforcing process. A remarkably high strength-to-weight ratio is the main feature attracting spacecraft designers to use reinforced plastics in satellites. Elegance, light weight and corrosion resistance qualifies reinforced plastics for use in boat hulls which are required to rough through marine salt water. The corrosion resistance property of reinforced plastics is fully utilised when they are employed as huge storage tanks for storing acids.

Now, let us examine the physical nature and chemical composition of these extremely useful materials. As has been said earlier, they are basically polymeric materials which have gained special properties through the use of reinforcing fibres. The main reinforcing fibres (both short-chopped and continuous) are those of glass, graphite, alumina, carbon, boron and beryllia. The latest trend in this field is to use fully aromatic polyamide fibres which, it is claimed, give, in some cases, over 50% weight reduction compared to reinforced plastics made of conventional fibres. Natural fibres such as sisal, asbestos, etc. are also used for reinforcement. The selection of the fibre, of course, depends on the property one expects the final reinforced composite to possess. Among these, however, glass fibre is the most extensively used and continues to be the workhorse of the FRP industry. Properties such as the low coefficient of thermal expansion, high dimensional stability, low cost of production, good tensile strength, low dielectric constant, non-flammability and chemical resistance constitute the attractive features of glass fibre. The other fibres are used mainly when some specific added advantage or service conditions are to be fulfilled. They are generally costlier than glass fibres.

Reinforced plastics are produced by suitably bonding a fibre material with a resin matrix and curing the same under pressure and heat. The reinforcement could be in different forms. For instance, it could be short-chopped fibres, continuous filaments or woven fabrics. The common resin matrix used in FRPs includes polyesters, epoxy, phenolic, silicone, melamine, vinyl derivatives and polyimides. A large proportion of the reinforced plastics are made of polyester resins taking advantage of their low cost. Phenolic resins are used wherever resistance to high temperatures is demanded. Epoxy resins impart extremely high mechanical properties to the FRP. Excellent electrical and thermal properties are obtained by using silicone resins.

There are several methods available for the production of reinforced plastics. Three of the most commonly used techniques are: (1) The hand lay-up technique; (2) The filament-winding technique; and (3) The spray-up technique.

The Hand Lay-up Technique

This is probably the simplest method for producing reinforced plastic articles. The quality of the end product depends to a large extent on the skill of the operator. The whole process consists of the following steps. To begin with, the mould is given a thin coating of a release agent such as polyvinyl

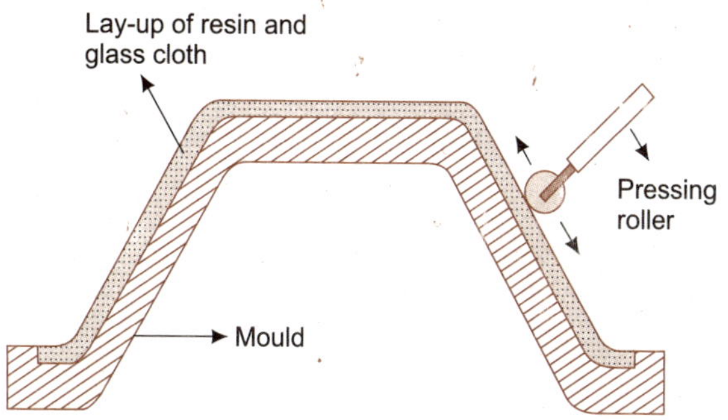

Fig. 15.13 Schematic diagram showing hand lay-up technique.

alcohol, silicone oil or wax. (This is done to prevent the final fabricated article from sticking to the mould.) The mould is then coated with a resin matrix. A precut glass cloth or mat is then laid over the resin layer. Now, another layer of the resin coating is given over the glass cloth. Rollers are used to press the glass cloth on the resin uniformly and also to remove the entrapped air bubbles. Alternate layers of resin and glass cloth are laid in a similar sequence until the required thickness is built up (Fig. 15.13). The whole set-up is then cured either at the ambient or elevated temperature. After the curing is completed, the reinforced plastic material thus formed is removed from the mould and subjected to trimming and finishing. Sheets, auto body parts, boat hulls, ducts and building components are produced by this technique.

Filament-winding Technique

This is a very widely used method for producing reinforced plastic articles such as high-pressure cylinders, storage tanks and rocket motor bodies. In this process, a continuous length of strand, roving or woven tape of the fibre is passed through a bath of resin and curative. As the strand comes out of the bath the excess resin is squeezed out. The resin-dipped filament or strand is then wound over a mandrel of the required shape and subsequently cured under the influence of heat. The

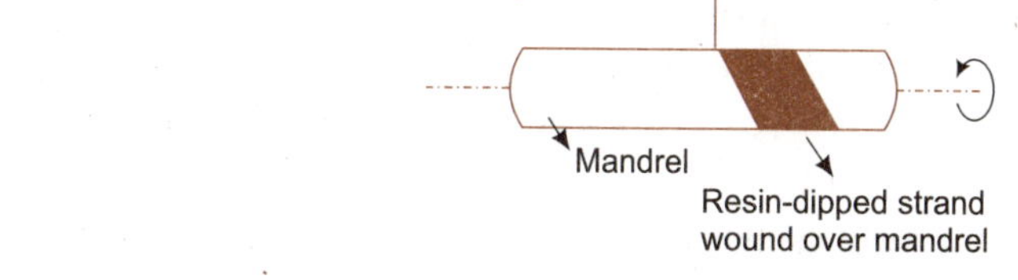

Fig. 15.14 Schematic diagram of a filament-winding machine.

winding machine (Fig. 15.14) is designed in such a fashion that the fibres can be wound in a predetermined pattern to suit the job. The tension of the fibre and the pattern of winding are very important factors as they influence the ultimate tensile property of the finished product.

Spray-up Technique

This technique employs a multiple headed gun. A spray of resin, a curative and chopped fibres are discharged simultaneously from a spray gun on the surface of a mould (Fig. 15.15) where they get deposited to a uniform thickness. The chopped fibre of a suitable length is obtained by continuously feeding rovings to the chopping head of the apparatus. Once the required thickness is built up by spraying, it is cured under heat. The spray-up technique is a quick method to cover large surface area moulds. Many present-day reinforced plastic articles such as truck bodies, storage vessels, lorry cabs and boat hulls are produced by the spray-up technique.

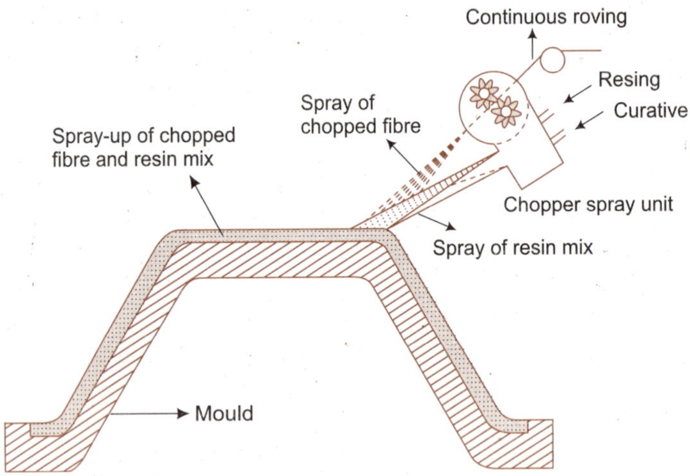

Fig. 15.15 Schematic diagram showing spray-up technique.

Other Techniques

Apart from the techniques described above, there are several other techniques available in the ever-growing reinforced plastic industry, each having a specific use. For example, a continuous laminating technique is used for producing continuous sheets of reinforced plastic laminates with varying thicknesses. In this process, individual layers of woven fabric are fed from individual rolls and impregnated with the resin and curative, and then pressed into a single layer by passing through a set of hot laminating rollers. The single layer coming out is cured by applying heat, which results in a laminate of a desired thickness (Fig. 15.16). The thickness can be adjusted by selecting the number of layers.

In yet another technique, known as '*pultrusion*', articles such as tubing or fishing rods are produced from continuous strands of a fibre. The process is relatively simple. The continuous strands

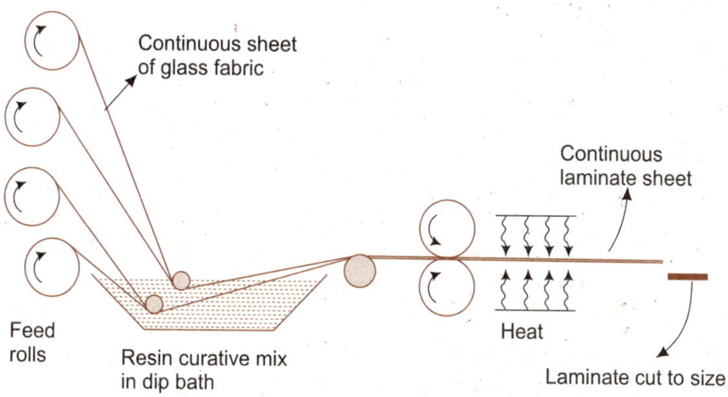

Fig. 15.16 Schematic diagram showing continuous lamination technique.

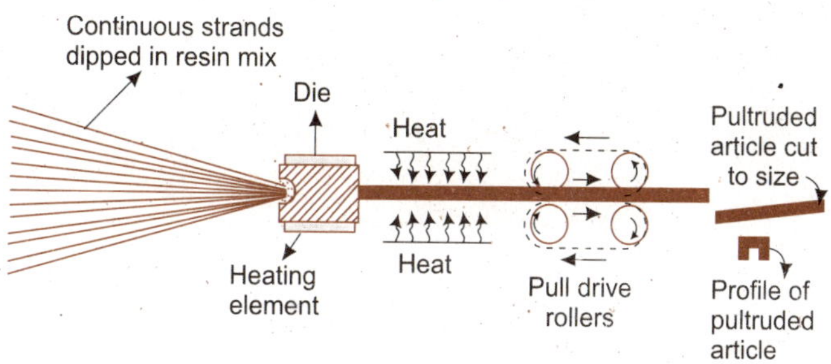

Fig. 15.17 Schematic diagram showing pultrusion technique.

are pretreated by passing through a resin-curative bath and then pulled through a die of a suitable profile (Fig. 15.17). The die is heated and the compacted profile coming out of the die is also heated externally. The cured profile is continuously pulled out of the die which provides the driving force for the impregnated strands to be forced through the die. This process is similar to extrusion except that, in extrusion, the polymeric material is forced through the die from inside by a rotating screw, while in pultrusion, the material is actually pulled through the die orifice from the outer side.

Apart from these methods, a mix comprising chopped fibres, resin and curative can be moulded by any of the conventional methods such as compression moulding technique. Thermoplastic materials filled with chopped fibres can also be moulded by compression, injection or extrusion to get a product with improved mechanical properties.

15.4.12 Fibre Spinning

Fibres are made from polymers by a process called '*spinning*'. There are three principal spinning methods, *viz*., melt spinning, dry spinning and wet spinning. In the melt spining process, the polymer is used in the molten state, while, in the other two cases, the polymer is used as a solution in an appropriate solvent. In all the three cases, however, the polymer (either in the molten or in the

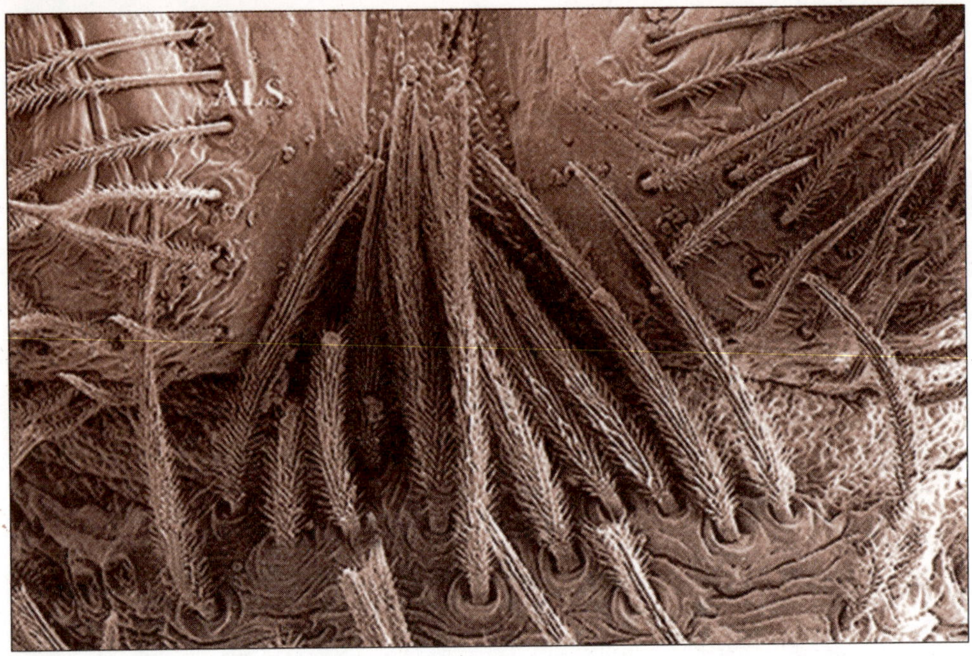

Fig. 15.18 Fibres emerging out of a spinneret.

solution form) is streamed through a spinneret which is a special kind of plate with extremely fine holes for the fibres to emerge (Fig. 15.18).

Melt-Spinning

In its simplest form the melt-spinning process may be described as follows. Polymer chips are electrically heated and melted in a heating grid. This converts the solid polymer into a viscous mobile liquid. Sometimes, during this heating process, cross-linking or thermal degradation of the polymer may occur leading to formation of lumps. These lumps can be easily removed from the hot molten polymer by using a filter pack. Furthermore, the hot molten polymer should be protected from the surrounding oxygen to avoid oxidative degradation. This is effected mostly by creating an inert environment around the molten polymer with nitrogen, carbon dioxide or steam. The molten polymer then enters a constant rate pump (known as the metering pump) which in turn, pumps the polymer melt through the spinneret at a specified rate. The polymer melt is thus forced through the fine holes of the spinneret from where continuous and very fine filaments emerge. The filament emerging from the spinneret solidifies as it comes in contact with cold air. The cooling and solidification process may be enhanced by a cold air blast. The solid filaments emerging from the spinneret are subsequently wound upon spindles (Fig. 15.19).

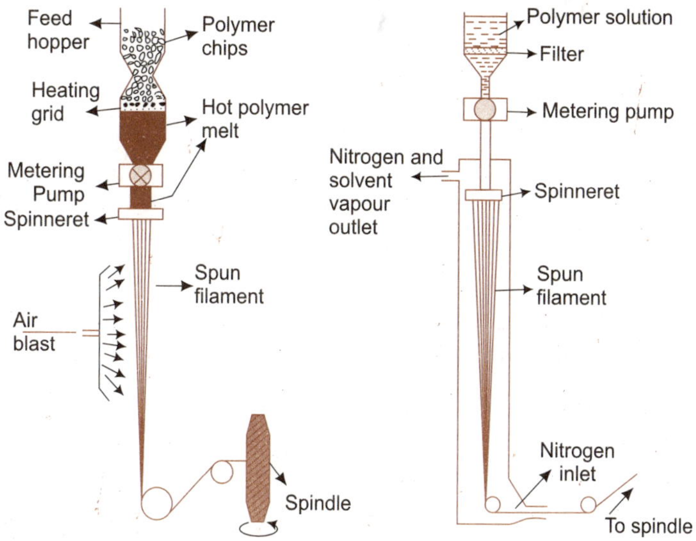

Fig. 15.19 Schematic diagrams of melt-spinning and dry spinning processes respectively.

One important point to remember in the melt-spinning process is that the diameter of the filament depends as much on the rate at which the molten polymer streams through the holes as on the rate at which the filament is pulled and wound up as on the hole diameter.

Dry Spinning

Many common polymers such as polyacrylonitrile or polyvinyl chloride are converted to fibres on a large scale by using the dry spinning process. The process is also shown in Fig. 15.19. The polymer is dissolved in an appropriate solvent to get a solution of high concentration. The viscosity is controlled by increasing the temperature of the solution. The hot viscous solution is pumped through the spinneret when fine continuous jets of the solution come out of the orifices. The filament is formed from these jets by the simple evaporation of the solvent. Solvent evaporation can be enhanced by passing dry nitrogen in a counter-current manner. Filaments formed from the polymer solution ultimately reach the spindle where they are wound. The rate at which the fibres are spun could be as high as 1000 m/min. Manufacture of cellulose acetate fibres from a 35% solution of the polymer in acetone at 40°C could be cited as a typical example of the utility of the dry spinning process.

Wet Spinning

Wet spinning, like dry spinning, employs a fairly concentrated polymer solution. The increase in polymer solution viscosity is controlled by spinning the solution at an elevated temperature. The process details for wet spinning are shown in Fig. 15.20. Wet spinning also converts a viscous polymer solution into fine jets through a spinneret. These jets, however, are led into a coagulation bath containing a large volume of a non-solvent which can precipitate the polymer from its solution. When continuous jets of the polymer solution come in contact with a non-solvent, they precipitate in

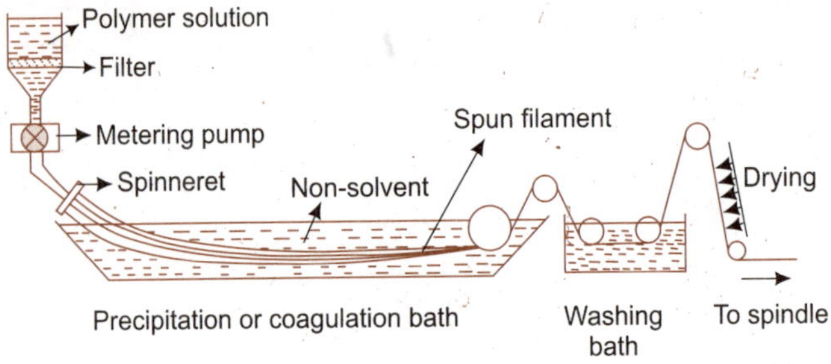

Fig. 15.20 Schematic diagram of wet spinning process.

the form of fine filaments. The filaments formed are gathered on a spindle after undergoing washing, drying, etc. One important point to be borne in mind while carrying out wet spinning is that sometimes droplets are formed instead of continuous filaments. This happens as a result of the breaking of continuity in the jet as it emerges from the spinneret hole and is caused by the forces of interfacial tension acting on the newly formed jet. This could be avoided by increasing the viscosity of the polymer solution.

Coagulation, an essential step in wet spinning, is a rather slow process. Hence, compared to the other two processes, the rate of wet spinning may be as low as 50 m/min. The wet spinning process is employed commercially to make fibres from cellulose, viscose rayon, polyacrylonitrile, etc.

Uniaxial Orientation

During the spinning of filaments from a polymer melt or solution, the polymer molecules in the filament are not properly oriented and hence, the percentage crystallinity is relatively low. This will be reflected in poor physical properties of the fibre. To get better physical properties, the filaments are subjected to a process called uniaxial '*drawing*'. This is done by employing a particular kind of stretching machine.

The main feature of the machine are two rollers (A and B shown in Fig. 15.21), revolving at different speeds. Roller B may be revolving 4 to 5 times faster than roller A. The spun yarn is sent in

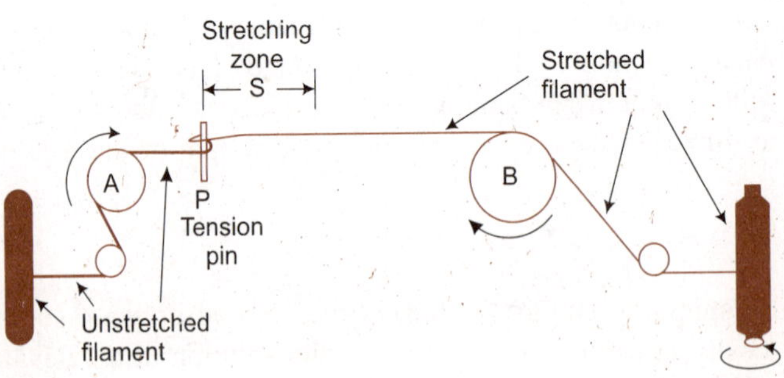

Fig. 15.21 Schematic diagram of uniaxial orientation done in a stretching machine.

through roller A and reaches roller B by passing over a tension pin *P.* Since roller B is rotating at a speed higher than that of roller A, the filament is actually pulled and stretched by roller B under tension imparted by the pin *P*. Stretching takes place at zone *S*. After passing through roller B, the properly drawn filament is collected on a metal bobbin. During the drawing process, the filament diameter gets reduced as it is stretched under tension, but, at the same time, the molecules get oriented parallel to the axis of the filament. This orientation improves the tensile properties very significantly.

Post Treatments for Fibres

Many special post treatments are given to fibres to make them more useful. The treatments include scouring lubricating, sizing, dyeing, etc.

Soap and synthetic detergents are used for '*scouring*'. Scouring, in fact, is nothing but the removal of dirt and other impurities from the fibre surface. '*Lubrication*' is a treatment given to fibres to protect them from external friction that may be imparted on the fibre by neighbouring fibres or by a rough metallic surface with which the fibres may come in contact during processing. Mostly, vegetable oils are used as lubricants. Lubrication also helps to reduce static electricity build-up on the fibres.

'*Sizing*' is a process by which a protective coating is given to fibres. For many fibres, polyvinyl alcohol or gelatin is employed as a sizing material, which holds the projecting fibres on the main axis of the yarn and this ensures smooth weaving. Weaving is generally followed by dyeing and hence, before dyeing the '*size*' should be removed by washing.

Customers are attracted by providing an increasing variety of colours to the fabrics, and this is done by dyeing. Dyeing is carried out by immersing fibres into a die solution till the dye penetrates into the fibres. Hydrogen bonds or van der Waals' type of bonds play a very significant role in the absorption of the dye by the fibre. Penetration usually takes place at the amorphous regions of the fibre.

Acid dyes are absorbed quickly by fibres made from cellulose or proteins. This is understandable as acid dyes can easily bond with amino or hydroxyl groups present in such polymers. The process of dyeing synthetic fibres such as polyester, acrylics or polyamides is rather slow. Elevated temperatures, of course enhance the dyeing process. Colouring of fibres made from polyvinyl chloride, polyethylene, etc., is practically impossible unless sites are made available for the absorption of dyes through copolymerisation or by chemical oxidation.

Books for Further Reading

Mokelvey, J.A., *Polymer Processing,* Wiley-Interscience, New York, 1962.

Hearle, J.W.S. and Peters, R.H., *Fibre structure,* Butterworths, London, 1963.

Kobayashi, A., *Machining of Plastics*, McGraw-Hill, New York, 1969.

Elder, R.E and Swan, A.D., *Calendering of Plastics*, Iliffe, London, 1971.

Fisher, E.G., *Blow Moulding of Plastics,* Butterworths, London, 1971.

Blow, C.M., *Rubber, Technology and Manufacture,* Butterworths, London, 1971.

Ritchie, P.D., Ed., *Plasticisers, Stabilisers and Fillers,* Butterworths, London, 1972.

CHAPTER 15

Penn, W.S., *PVC Technology,* MacLaren, London, 1972.
May, C.A. and Tanaka, Y., Eds., *Epoxy Resins—Chemistry and Technology*, Dekker, New York, 1973.
Rubin, I.I., *Injection Moulding Theory* and *Practice,* Wiley-Interscience, New York, 1973.
Milby, V., *Plastics Technology,* McGraw-Hill, New York, 1973.
Catherall, J.A., *Fibre Reinforcement*, Mills and Boon, London, 1973.
Moncrieff, R.W., *Man-made Fibres,* Halsted, New York, 1975.
Seymour,W.B., *Modern Plastics Technology*, Reston Publishing Co., Reston, 1975.
Mascia, L., *The Role of Additives in Plastics*, Halsted, New York, 1975.
Baird, R.J., *Industrial Plastics,* Goodheart-Willeox, South Holland, 1976.
Levy, S. and Davis, J.H., *Plastics Product Design Engieering Handbook*, Reinhold, New York, 1976.
Powell, P.C., *The Selection and Use of Thermoplastics*, Oxford University Press, Oxford, 1977.
Dubois, J.H. and John, F.W., *Plastics*, Reinhold, New York, 1981.

BIBLIOGRAPHY*

Billmeyer, F.W., *Textbook of Polymer Science*, Wiley-Interscience, New York, 1971.

Bolkev, H.I., *Natural and Synthetic Polymers,* Marcel Dekker, New York, 1974.

Elias, H.G., *Macromolecules,* Vol. 1 and 2, Plenum, New York, 1977.

Flory, P.J., *Principles of Polymer Chemistry*, Cornell University Press, Ithaca, 1953.

Ivin, K.J. and Rose, J.B., *Advances in Macromolecular Chemistry*, Vol. 1, Academic, New York, 1968.

Jenkins, A.D., Ed., *Polymer Science*, Vols. 1 and 2, North Holland, Amsterdam, 1972.

Kaufman, H.S. and Falcetta, J.J. Eds., *Introduction to Polymer Science and Technology,* Wiley-Interscience, New York, 1977.

Kennedy, J.P. and Tornquivst, E.G.M., Eds., *Polymer Chemistry of Synthetic Elastomers*, Wiley-Interscience, New York, 1969.

Kumar, A. and Gupta, S.K., *Fundamentals of Polymer Science and Engineering,* Tata McGraw-Hill, New Delhi, 1978.

Lenz, R.W., *Organic Chemistry of Synthetic High Polymers*, Wiley-Interscience, New York, 1971. Mark, H., *Physical Chemistry of High Polymeric Systems*, Interscience New York, 1940. Mark, H., Gaylord, N.G. and Bikles, N.M., Eds., *Encyclopaedia of Polymer Science and Technology*, Wiley-Interscience, New York, 1971.

Pinner, S.H., *A Practical Course in Polymer Chemistry*, Pergamon, New York, 1961.

Ravve, A., *Organic Chemistry of Macromolecules*, Marcel Dekker, New York, 1967.

Rosen, S.L., *Fundamental Principles of Polymeric Materials*, John Wiley, New York, 1982.

Seymour, R.B., *Introduction to Polymer Chemistry*, McGraw-Hill, New York, 1971.

Seymour, R.B., and Carraher, C.E. in Lagowski, J.J., Ed., *Polymer Chemistry: An Introduction,* Marcel Dekker, New York, 1981.

Strepikheyev, A., Derevitskaya,V. and Slonimsky, G., *A First Course in Polymer Chemistry*, Mir Publishers, Moscow, 1971.

Tager, A., *Physical Chemistry of Polymers*, Mir Publishers, Moscow, 1978.

Van Krevelen, D.W., *Properties of Polymers,* Elsevier, Amsterdam, 1972.

Vollmert, B., *Polymer Chemistry*, Springer-Verlag, New York, 1973.

Young, R.J., *Introduction to Polymers*, Chapman and Hall, London, 1981.

*Books dealing with specific aspects of polymer Science have been listed at the end of each chapter. This bibliography gives a list of books covering various aspects of polymer science under a single title.

Appendix A

Table of Physical Constants

Symbol	Name	Value
Å	Angstrom unit	1 Å = 10^{-10} m (not a preferred unit)
e	Base of natural logarithms	2.7183
h	Planck's constant	6.6255×10^{-27} erg-sec
k	Boltzmann's constant	1.3805×10^{-16} erg/deg
N_0	Avogadro's number	6.0226×10^{23} mole^{-1}
R	Gas constant	8.314×10^{7} erg/mole-deg
		8.314 J/mole-deg
		82.1 cm^3 atm/mole-deg
K	Absolute temperature	0°C = 273.16 K (Kelvin)
μm	Micrometer	1 μm = 10^{-3} mm = 10^{-6} m
nm	Nanometer	1 nm = 10^{-3} μm = 10^{-6} mm = 10^{-9} m

Appendix B
Polymer in Daily Life

S. No.	Polymer	Used in
1.	Polyethylene	
	(a) LDPE	Packing and wrapping frozen food, textile products.
	(b) HDPE	Toys and other household articles.
2.	Polypropylene	Appliances such as refrigerators, radios and TVs. It is also used for producing package films, pipes, storage tanks, seat covers, monofilaments and ropes.
3.	Polystyrene	Manufacture of articles such as moulded containers, lids, jars, bottles, radio and television cabinets, toys, foamed plastics and many other household items. It is most suited for injection-moulded articles of household items and packing containers.
4.	Polyacrylonitrile	Producing PAN fibres.
5.	Polymethyl Methacrylate	Buildings for decorative purposes.
6.	Polyesters	Manufacture of magnetic recording tapes, aluminised sheets for certain special applications and so on. Also used as the resin matrix in fibre-reinforced plastic (FRP) structures and as decorative coatings.
7.	Polyformaldehyde	Engineering plastic.
8.	Polyparaphenylene	Hoses, drive belts, ropes and cords.
9.	Polycarbonates	Many useful articles such as safety goggles, safety shields, telephone parts and machinery housings.
10.	Polysulphone	Making injection-moulded items and films.
11.	Polyimides	Surface coating in supersonic aircraft.
12.	Polyamides	As a good substitute for metals in gears and bearings, textile fibre, brushes and bristles.
13.	Polyurethanes	Preparing polyurethanes commercially which is used to make foams, coatings, adhesives and elastomers.

14.	Polyureas	Providing corrosion and abrasion protection of pipelines that carry abrasive and corrosive meterial. It is applied as a base coating to prevent cathodic disbonding.
15.	Polyethylene Glycols (PEG)	Pharmaceutical preparations and cosmetics
	and Polypropylene Glycols (PPG)	As the diols in the preparation of unethane polymers
16.	Polyacrylic Acid	Adhesives as a thickneing agent.
17.	Polymethacrylic Acid	PMMA brushes
18.	Polyvinyl Acetate	Preparing polymer and manufacture of polyvinyl alcohol. Also used in adhesives, lacquers and for the large-scale manufacture of gramophone records.
19.	Polyvinyl Alcohol	Manufacture of textile fibres.
20.	Polyvinyl Acetals	Adhesive compositions and manufacture of shatter-proof glass.
21.	Polyvinyl Carbonate	Protecting tissue cells from high energy ionising radiations.
22.	Polyvinyl Chloride	Emulsion systems, while azo compounds are employed in bulk and suspension polymerisation. It is also used in certain formulations called plastisols and organosols. The main use is in the manufacture of adhesives, coatings and fibres.
23.	Polyvinylidene Chloride	Emulsion technique, thermoplastic materials and making fibres.
24.	Polyvinyl Fluoride	Paints and other surface coatings.
	and Polyvinylidene Fluoride	Coating compositions.
25.	Polytetrafluoroethylene	Making articles such as pump valves and pipes where chemical resistance is required. It is also used in non-lubricated bearings and its fibre is used to form belts, filter cloth, and similar products where resistance to acid and alkalies is needed.
26.	Polyvinyl Carbazole, Polyvinyl Pyrrolidone and Polyvinyl Oxazolidinone	Complexing agents to a number of inorganic as well as organic compounds.
27.	Polybutylene and Polyisobutylene	Rubber-based industries. Also used as an insulation material in the electrical industry and added to motor oils to supress sharp fall in viscosity with increasing temperature.
28.	Polyisoprenes	Latex form or the solid rubber.
29.	Polybutadienes	Chain modifier to control molecular weight and oil seals and other oil resistant applications.
30	Polychloroprene	Oil-resistant insulation coatings to wires and cables and for producing shoe soles, solid tyres, gloves and industrial hoses.
31.	Polysulphides	Making of sealants, gaskets, balloons, fabric coatings and gasoline hoses.

32.	Phenol-Formaldehyde Resin	Making of laminates, paper, fabric or asbestos cloth are impregnated with the resols, electrical circuits and core binders in foundries.
33.	Urea-Formaldehyde and	Paper industry to improve strength and in plywood industry as adhesives.
	Melamine-Formaldehyde Resins	Melamine resin is used in manufacture of decorative laminates and preparation of lacquers.
34.	Epoxy Polymers	Industrial floorings, foams, potting materials for electrical insulations, etc.
35.	Silicone Polymers	Surface coatings, to manufacture laminates. Light foam made from silicone resin is used in aeroplanes, and missiles. Silicon polymers are used as elastomers. Silica in the form of fine powder is used as a filler to reinforce silicone rubbers.
36.	Organotin Polymers	Film for food packaging and in PVC articles.
37.	Organotitanium Polymers	Major industrial processes.
38.	Cellulose and its Derivatives	
	(a) Cellulose	Cotton
	(b) Rayon	As regenerated cellulose
	(c) Cellophane	Package and wrapping industries
	(d) Cellulose Nitrate	Making moulded decorative articles, fountain pens, frames for spectacles, etc.
	(e) Cellulose Acetate and other related polymers	Manufacture of fibres, non-flammable safety cinematographic films, moulded articles and fibres and automobile components.
	(f) Ethyl and Methyl Cellulose	Textile industry as finishing material and as paint bases.
	(g) Carboxymethyl Cellulose	Textile additives, cosmetics, cakes, ice-creams, etc.

Appendix C

A Representative List of Universities/ Institutions Offering Courses in Polymer Science and Technology

Part A: List of Universities/Institutions*

I. Argentina

1. Faculty of Chemical Engineering; National University of Litoral; (Universidad Nacional del Litoral); Santiago del Estero 2829; 3000 Santa Fe (Republica Argentina).

   'Synthetic polymers' forms one out of the 21 'themes' covered in the postgraduate course in organic chemistry.

II. Australia

1. Department of Polymer Science; School of Chemical Technology; The University of New South Wales; PO Box 1, Kensington; New South Wales, 2033 (Australia).

   Graduate course leading to Master of Applied Science (M. App. Sc.) degree with a complete programme in the polymer area is available at the School of Chemical Technology.
2. Department of Chemistry; The University of Newcastle; New South Wales, 2308 (Australia).

   Polymer chemistry is covered, in 13 lectures and 21 hours of laboratory work, during the second year of the B.Sc. course in chemistry.
3. Department of Industrial Science; University of Melbourne; Parkville, Victoria 3052 (Australia).

   (*a*) Polymer work is offered as a part of a four-year Bachelor of Applied Science course in applied chemistry or industrial materials.

*This is not an exhaustive list.

(*b*) At the postgraduate level, conventional research in the polymer area leading to the degree of Master of Applied Science and Ph.D. is undertaken.

4. Department of Chemical Engineering; The University of Adelaide; GPO Box 498; Adelaide 5001 (Australia).

   No formal course work leading to graduate degrees is offered in the polymer area. Postgraduate research work is undertaken in the Materials Engineering Group.

III. Austria

1. Institut fur Physikalische Chemie; der Universitat Graz; Heinrichstrasse 28; A– 8010 Graz (Austria).

   High polymers form part of the graduate course.

2. Institut fur Chemische und Physikalische; Technologie der Kunstoffe; Montan universitat; Frauz—Josef-Strasse 18; A 8700 Leoben (Austria).

   A special study of 'polymer engineering' over five years leading to the degree 'Diploma-Ingenieur' is offered, which can be followed by another special study for three years leading to the Doctor's degree.

IV. Bangladesh

1. Department of Chemistry; University of Dacca; Dacca–2 (Bangladesh).

   Polymer chemistry forms a part of course leading to degrees like B.Sc. (Hons.) and M.Sc.

V. Belgium

1. Universite Catholique de Louvain; Laboratorie de Physique et de Chimie des Hauts Polymeres; Place Croix du Sud, 1; 1348, Lauvain-la-Neuve; (Belgium). Offers graduate courses covering high polymers as part fulfilment.

VI. China (Peopleís Republic of)

1. University of Nanking; Chemistry Department; Nanjing Jainso (People's Rebulic of China), Lecture courses in 'polymer science' are offered to graduate students.

VII. China (Taiwan)

1. Department of Chemical Engineering and Technology; National Taiwan Institute of Technology; 43, Keeling Road, Section 4; Taipei (Taiwan).

   (*a*) A two-year and a four-year undergraduate course leading to a BS degree in engineering with 'Polymer Chemistry Major' are offered.

   (*b*) A two-year programme leading to a MS degree in engineering includes polymer chemistry as part credit.

2. National Taiwan Normal University; College of Science; 88 Sec., 5 Roosevelt Road; Taipei (Taiwan 117).

   A one-semester introductory course in polymer chemistry is offered at the graduate level.

VIII. Federal Republic of Germany

1. Institut fur Chemische Technik; der Universitat Karlsruhe; Post Fach 6380; Kaiserstrasse 12; 7500 Karlsruhe 1 (West Germany).
2. Universitat Fridericiana Karlsruhe: Fakultat fur Chemie; Karlsruhe (West Germany).

   At both the places, polymer chemistry is covered as a selective subject at the 'Diploma-Hauptprufung' level.
3. Technische Universitat Munchen; Institut fur Technische Chemie; Lichtenbergstrasse 4; 8046, Garching (West Germany).

   Offers research opportunities with themes in macromolecular chemistry at the graduate level.

IX. Finland

1. Department of Wood and Polymer Chemistry; University of Helsinki; Meritullinkatu lA; Helsinki 17 (Finland).
   - (*a*) Courses in polymer chemistry are offered leading to degrees like Candidate of Science degree (B.Sc.), Master of Philosophy diploma (M.Sc.), Licentiate of Philosophy and Doctor of Philosophy (Ph.D).
   - (*b*) Research programmes concerned with theoretical and experimental topics in polymer science and with the technological applications of polymers leading to Licentiate of Philosophy and Doctor of Philosophy degrees are offered.

X. France

1. Université des Sciences et Techniques de Lille; 59650 Villeneuve d' Ascq; Cedex (France).

   Offers research themes in physical chemistry of macromolecules leading to the doctoral degree.

XI. Hong Kong

1. Department of Chemistry; The Chinese University of Hong Kong; Shatin, New Town (Hong Kong).

   A course on 'polymer chemistry and physics' is offered at the graduate level.

XII. India

1. Faculty of Science; Sardar Patel University; Vallabh Vidyanagar; Gujarat (India).
   - (*a*) At the undergraduate level, polymer chemistry forms part of a three-year B.Sc. (industrial chemistry) course.
   - (*b*) At the graduate level, polymer chemistry forms one of the majors in a two-year M.Sc. degree course in 'petrochemicals and polymer chemistry.' Some aspects of the subject are also covered as part credit in M.Sc. (physical chemistry) and M.Sc. (organic chemistry).
2. Chemical Engineering Education Development Centre; Indian Institute of Technology; Madras 600 036 (India).

   Polymer Chemistry forms an elective in an 'eight-semester first degree course in chemical engineering.'

3. Department of Polymer Science; University of Madras; A.C. College of Tech. Campus, Madras 600 025 (India).

   A four-semester (two-year) course in M.Sc. (polymer chemistry) and a two-semester (one-year) course in Diploma in Polymer Science are offered.

4. Rubber Technology Centre; Indian Institute of Technology; Kharagpur 721 302 (India).

   A three-semester (two-year) course in M. Tech. (rubber technology) is offered.

5. Department of Plastics and Rubber Technology; University College of Science and Technology; 92, Acharya Prafulla Chandra Road; Calcutta 700 009 (India).

   'Polymer science and technology' form a part of a composite course for a B. Tech. degree in 'plastic and rubber technology'.

6. Centre for Materials Science and Technology; Indian Institute of Technology; Hauz Khas; New Delhi 110 016 (India).

   A three-semester M. Tech. programme in 'polymer science and technology' is offered.

7. Department of Polymer Science and Rubber Technology; University of Cochin; Thrikkakara Campus; Cochin 682 022 (India). A three-year B. Tech. course in 'polymer science and rubber technology' is offered.

8. Department of Chemistry; Marathwada University; University Campus; Aurangabad 431 004 (India).

   Polymer chemistry forms an optional paper at the M.Sc. final level.

9. Department of Chemical Technology; University of Bombay; Matunga Road; Bombay 400 019 (India).

   A three-year course in B.Sc. (Tech.) in 'technology of plastics' and a two-semester course in M. Sc. (Tech.) in 'technology of plastics' are offered.

10. Plastic Technology Department; Harcourt Butler Technological Institute; Kanpur 208 002 (India).

    A three-year B. Tech. (Chem. Tech.) course in 'plastic technology' is offered by this institute which is affiliated to the University of Kanpur.

11. Government Polytechnic; Kottayam; Kerala (India).

    Offers a full-time three-year Diploma course in 'polymer technology'. The minimum educational qualification for admission to this Diploma course is a pass in SSLC.

XIII. Israel

1. Bar-Ilan University; Department of Chemistry; Ramat Gan (Israel).

   A course in polymer chemistry for graduate students is offered.

2. Technion-Israel Institute of Technology; Technion City; Haifa 32000 (Israel).

   Lectures covering some aspects of 'high polymers' are given in the Department of Chemistry and Department of Chemical Engineering to graduate students.

3. Ben Gurion University of the Neger; Faculty of Natural Sciences; PO Box 2053; Beersheva 84120 (Israel).

   A course on polymers is given to research students in the second year of their graduate study.

XIV. Japan

1. University of Osaka Prefecture; Dept. of Applied Chemistry; Sakai City, Osaka (Japan).
2. Faculty of Engineering; Doshisha University; Karasuma Imedegawa; Kamikgo-ku, Kyoto 602 (Japan)

   Polymer chemistry and polymer physics are covered in two-year Master of Engineering and three-year Doctor of Engineering courses at both these universities.

XV. Norway

1. University of Trondheim; The Norwegian Institute of Technology; Department of Chemistry; Norges Tekniske Kogskole; Kjemiavdelingen (Norway).

   Lectures on high polymers form part of the graduate course.

XVI. Rhodesia (now Zimbabwe)

1. University of Rhodesia; Department of Chemistry; PO Box MP 167; Salisbury (Rhodesia).

   Short courses in polymer chemistry (about five lectures each) form part of undergraduate courses in chemistry.

XVII. South Africa

1. University of Natal; Department of Chemistry; PO Box 375; Peitermaritzburg 3200 (South Africa).

   Polymer chemistry is taught at the undergraduate level.

XVIII. Turkey

1. Ege University; Faculty of Chemistry; Bornova, Izmir (Turkey). A postgraduate course of 40 credits leading to MS degree is available at the Polymer Chemistry Department.
2. Technical Institute of Istanbul; Faculty of Chemical Engineering; Taksim, Taskisla; Istanbul (Turkey).

   Basic polymer science lectures together with laboratory work are offered at the postgraduate level in the Organic Chemistry Department.

XIX. United Kingdom

1. The University of Manchester Institute of Science and Technology; PO Box 88, Sackville Street; Manchester M 60 I QD (UK).
   - (*a*) Three-year B.Sc. and B.Sc. (Hons.) courses are provided in 'polymer science and technology (chemistry),' 'polymer science and technology (physics)' and 'polymer engineering'.
   - (*b*) Research themes on various aspects of polymer science and technology leading to M.Sc. and Ph.D. degrees are available.

2. Institute of Polymer Technology; Loughborough University of Technology; Loughborough, Leicestershire LE 11.3TU (UK).

   Postgraduate degree course leading to M.Sc. in 'polymer technology' is offered.
3. Department of Physical Sciences and Technology; Polytechnic of the South Bank, Borough Road; London SEI OAA (UK).

   'Polymer science' and 'plastic technology/surface coating' are covered in a four-year B.Sc. (Hons.) course in chemical and polymer technology.
4. The University of Aston in Birmingham; Gosta Green; Birmingham B 4 TET (UK).
   - (*a*) One-year M.Sc. degree course in 'chemistry and technology of polymers' is offered.
   - (*b*) Research opportunities in 'polymer science and technology' are available at the Chemistry Department.
5. Department of Materials; Crossfield Institute of Technology; Cranfield, Bedfordshire MK 43 OAL (UK).

   One-year or two-year M.Sc. degree course in 'polymeric engineering' offered.
6. Lanchester Polytechnic; Department of Applied Chemistry; Priory Street; Coventry C VI, 5 FB (UK).

   Polymer chemistry is covered in the final year of a 'B.Sc. applied chemistry' degree course.
7. Manchester Polytechnic; Department of Polymer Technology; Chester Street, Manchester MI 5 GD (UK).

   A four-year sandwich course leading to B.Sc. degree/Degree with Honours offered in 'polymer science and technology'.

XX. USA

1. Case Western Reserve University; The Department of Macromolecular Science; Cleveland, Ohio 44106 (USA).

   Programmes are offered leading to BS, MS, Ph.D. and MD/Ph.D. degrees.
2. North Dakota State University; State University Station; Fargo, North Dakota 58102 (USA).

   The Department of Polymers and Coatings offers graduate work leading to MS and Ph.D. degrees in chemistry with a major in polymers and coatings.
3. Princeton University; Princeton, NJ 08544 (USA).

   Interdepartmental postgraduate research themes are offered in 'polymer science and materials' leading to the Ph.D. degree.
4. University of Southern Mississippi; Southern Station, Box 8276; Hattiesburg, Mississippi 39401 (USA).

   Polymer science is covered both in the graduate and postgraduate courses offered at this university.
5. The University of Michigan; Institute of Science and Technology; 3200 Bonisteel Boulevard; Ann Arbor, Michigan 48109 (USA).

   Offers an interdepartmental 'macromolecular science programme' leading to MS and Ph.D. degrees.

6. University of Massachusetts; Materials Research Laboratory; Polymer Science and Engineering Department; Amherst, Massachusetts 01003 (USA).

   M.Sc. and Ph.D. degree courses are available as interdisciplinary graduate level programmes.

XXI. USSR

1. Kharkov State University of A.M. Gorkov; Kharkov, Ploshchad Dzerzhin; Skogo-4; Ukraine Republic (USSR).

   A course on 'high molecular mass compounds' is offered to students of the chemistry faculty.

Part B: Samples of Typical Course Content

1. Department of Chemistry; The University of Newcastle; New South Wales, 2308 (Australia).

   Lecture courses are as follows:

   (1) Introduction—common terms; description of common polymers.

   (2) Polymerisation reaction types. Condensation (step growth) and chain polymerisation. Characteristics of reaction and of polymers produced.

   (3) Molecular weights and their determination.

   (4) Structure of solid polymers.

   (5) Protein and related molecules.

   (6) Melting, crystallisation and the glassy state.

   (7) Homogeneous and heterogeneous polymerisations. Low density polyethylene—chain branching, chain transfers.

   (8) Kinetics of free-radical chain polymerisation.

   (9) Kinetics of condensation polymerisation. Molecular size distribution.

   (10) Non-radical chain polymerisations. Ziegler-Natta catalysis. High density polyethylene, polybutadiene.

   (11) Copolymers. The copolymer equation.

   (12) Elastomers.

   (13) Polymers for high temperature applications.

2. Department of Chemistry; The Chinese University of Hong Kong; Shatin, New Town (Hong Kong).

   Polymer chemistry:

   (1) Introductory survey:

      (*a*) Concept of polymers.

      (*b*) Classification of polymers.

      (*c*) Common terms used in polymers.

   (2) Polymerisation:

      (*a*) Step polymerisation.

(*b*) Chain polymerisation:

Radical polymerisation.

Anionic polymerisation.

Cationic polymerisation.

Coordination polymerisation.

(*c*) Copolymerisation.

(3) Structure and properties of polymers:

(*a*) Molecular weights and conformation in solution.

(*b*) Glass state and glass transition.

(*c*) Rubber elasticity.

(*d*) Crystallinity.

(4) Polymer technology:

(*a*) Plastic.

(*b*) Fibre.

(*c*) Rubber.

(*d*) Paint and coating.

(*e*) Adhesives.

Polymer physics: Physical states of amorphous polymers; crystallinity; orientation. Linear viscoelasticity, mechanical models; time-temperature equivalence; mechanical anisotropy. Heat capacity, thermal conductivity. Electrical conduction; dielectric constant and loss. Optical properties, nuclear magnetic resonance.

3. Department of Polymer Science; University of Madras; A.C. College of Tech. Campus, Madras 600 025 (India).

(*a*) *M.Sc. (polymer chemistry)*

(1) Types of polymeric substances—classification of polymers—condensation polymers—addition polymers—polymerisation of cyclic compounds—inorganic polymers.

(2) Principles of condensation polymerisation—chemical reactivity of homologous series of monomeric compounds—theory of the reactivity of large molecules—kinetics of condensation polymerisation—kinetics of degradation of condensation polymers— ring formation versus chain polymerisation.

(3) Polymerisation of unsaturated monomers by free-radical mechanisms—kinetic scheme for polymerisation—initiator efficiencies—autoacceleration—chain length—kinetics of chain transfer—absolute values for the rate constant—inhibition and retardation.

(4) Copolymerisation, emulsion polymerisation and ionic polymerisation—the composition of addition copolymers—monomer reactivity ratios—theory of emulsion polymerisation—cationic polymerisation—anionic polymerisation—ionic copolymerisation—coordination polymerisation—electropolymerisation—graft copolymerisation.

(5) Determination of molecular weights: osmotic methods—theory—the number average molecular weight—experimental methods—light scattering theory for particles—the weight average molecular weight—ultracentrifuge—diffusion constants— sedimentation equilibrium—intrinsic viscosity—molecular weight relationship—the viscosity average molecular weight and the influence of molecular heterogeneity—IR and NMR of polymers—gel permeation chromatograph.

(6) Molecular weight distributions in linear polymers: the most probable distribution—experimental methods—condensation polymers—addition polymers—polymer fractionation.

(7) Configuration of polymer chains—statistical distribution of end-to-end dimensions—the freely jointed chain in one dimension—the freely jointed chain in three dimensions—the distribution of high extensions—influence of bond angle restrictions—calculation of average dimensions for various polymer chain structures—hindered polymer chains—configuration of polymer molecules in dilute solution—long-range intramolecular interactions—the unperturbed polymer chain.

(8) The thermodynamics of polymer solutions: the theory of mixing—liquid lattice theory—the heat and free energy of mixing—partial molar quantities—the excluded volume.

(9) Phase equilibria in polymer systems: theory of binary systems—ternary systems—theory of polymer fractionation—the nature of the crystalline state in polymers—theory of swelling—swelling of non-ionic network systems—swelling of ionic networks.

(10) Mechanical (rheological) properties of polymers in the viscofluid state: parameters characterising flow regions—simplest cases of deformation—dependence of viscosity on shear rate and stress—initial Newtonian viscosity of polymeric systems—apparent (effective) viscosity—normal stresses in polymeric systems—dynamic properties and relaxation spectrum.

(11) Plastics—thermosetting and thermoplastics—phenolics—novolaks—extrusion—moulding—properties.

(12) Plasticisation: effect of plasticisers on glass transition temperature and fluidity—effect on mechanical properties—effect on dielectric properties—compatibility of plasticisers and polymers—mechanism of plasticisation—theory of plasticisation.

(13) Rubber: the composition of rubber latex—the processing of rubber—softeners—additives—accelerators—fillers—synthetic rubber—polybutadiene—polystyrene copolymers—carbon black.

(14) Natural and synthetic fibres: criteria for fibre formation—orientation of molecules on drawing—spinning processes—additives—dyes—testing of fibres—modification of fibres.

(15) Adhesives: the thermodynamic condition for wetting and spreading—surface tension—polymer—substrate interface—surface treatments—surface morphology—adhesive composition, selection and evaluation—manufacture of adhesives—metal

processing—food industry—paper coating—glass industry—classes of adherent materials.

(16) Medical and synthetic applications of polymers—polymeric drugs—polymer supports in organic synthesis—artificial organs—other polymeric substitutes in biology.

(*b*) *Diploma in Polymer Science*

(1) Analysis and testing of polymers: analytical chemistry of polymers—infrared spectroscopy—X-ray diffraction analysis—nuclear, magnetic and electron paramagnetic resonance spectroscopy—thermal analysis—microscopy—physical testing.

(2) Structure and properties of bulk polymers: configurations of polymer chains—crystallisation process—orientation and drawing—rheology and the mechanical properties of polymers—viscous flow—kinetic theory of rubber elasticity—visco elasticity—the glassy state and the glass transition—the mechanical properties of crystalline polymers.

(3) Polymerisation process: classification of polymers—chemistry of step-wise polymerisation—addition polymerisation—kinetics of vinyl polymerisation—molecular weight and its distribution—thermochemistry of chain polymerisation—ionic polymerisation—copolymerisation—composition of copolymers—block and graft copolymers—polymerisation in homogeneous systems—polymerisation in heterogeneous systems—degradation of polymers—radiation chemistry of polymers.

(4) Properties of commercial polymers: LDPE, HDPE—polypropylene—natural rubber—polybutadiene—synthetic elastomers—polystyrene—acrylic polymers—chlorine-containing polymers—fiuorine-containing polymers—polyamides—cellulosic polymers—aromatic heterochain thermoplastics—ladder and inorganic polymers—phenolic resins—epoxy resins—urethane foams—silicone polymers.

(5) Polymer processing: moulding—fillers—plasticisers and other additives—textile and fabric properties—compounding and elastomer properties—vulcanisation—reinforcement.

4. Technion—Israel Institute of Technology; Technion City; Haifa 32000 (Israel).

Introduction and definitions. Mechanisms of polymerisation; Kinetics of chain polymerisation. Kinetics of copolymerisation. Kinetics of stage-wise polymerisation. Industrial polymerisation process in bulk, suspension, emulsion and solution. Chemical structure of useful polymers.

Condensation polymerisation; phenolic resins; amino resins;. polyamides: polyesters; high-temperature stable polymers; relation between the chemical structure of polymers and their physical properties.

Thermodynamics of polymer solutions; methods for molecular weight determination; rheology of polymeric solutions and melts; physical properties of polymers in the solid state; amorphous and crystalline structures; rubber elasticity; fabrication methods of thermoplastics.

5. Lanchester Polytechnic; Department of Applied Chemistry; Priory Street; Coventry C VI, 5 FB (UK).

   Chain growth polymerisation in bulk and solution by radical and ionic mechanisms, the control of microstructure, molecular size and its distribution. Polymerisation carried out under heterogeneous conditions. Copolymerisation leading to random and sequence copolymers. Step growth polymerisation under melt, solution and interfacial conditions; reaction variables controlling heterogeneity. Cyclopolymerisation.

   Solution thermodynamics of non-ideal systems and their application to such factors as molecular mass determination, molecular shapes, polymer solubility, polymer fractionation and network swelling. Consideration of synthetic and naturally occurring macromolecules. Viscous flow of dilute solutions, Huggin's constant, the Mark-Houwink relationships, hydrodynamic volume, aggregation of macromolecules, shear effects.

   A detailed study of autooxidative degradation processes in polymeric materials, chemistry of antioxidants and their application. Mechanism of thermal degradation and the importance of chemical structure. Chemical modification to macromolecules, *e.g.*, in enzymic reactions. Biological polymer molecules. Importance of polymer morphology in controlling polymer reactions, diffusion of gases and vapours.

   Use of spectroscopic methods and X-ray diffraction studies in the elucidation of structure.

6. University of Southern Mississippi; Southern Station, Box 8276; Hattiesburg, Mississippi 39401; (USA).

   Introduction to polymer concepts; step-growth polymerisation; polymer reactions; ring-opening polymerisation, free-radical reaction mechanisms; homogeneous chain growth polymerisation of vinyl compounds; homogeneous free-radical copolymerisation of vinyl compounds; degradation mechanisms of polymers; methods of studying stereoregular polymers; anionic polymerisation; cationic polymerisation; stereoregular polymerisation via homogeneous and heterogeneous catalysts.

   Fundamental theoretical concepts in physical chemistry important to polymer science, conformation of isolated macromolecules in a vacuum; conformation of macromolecules in solution; Solution thermodynamics; molecular weight averages and molecular distributions; determination of molecular weight and molecular weight distribution.

Appendix D

Structures of Some Common Polymers

S.No.	Name of Polymer	Structure
1.	Acrylonitrile-butadiene-styrene terpolymer (ABS)	$-[CH_2-CH(CN)-CH_2-CH=CH-CH_2-CH_2-CH(C_6H_5)]_n-$
2.	Butyl rubber	$-[CH_2-C(CH_3)_2-CH_2-CH=C(CH_3)-CH_2]_n-$
3.	Cellulose	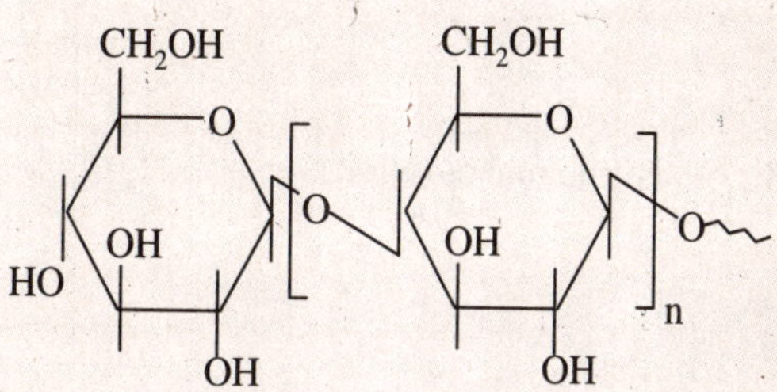
4.	Cellulose ethers	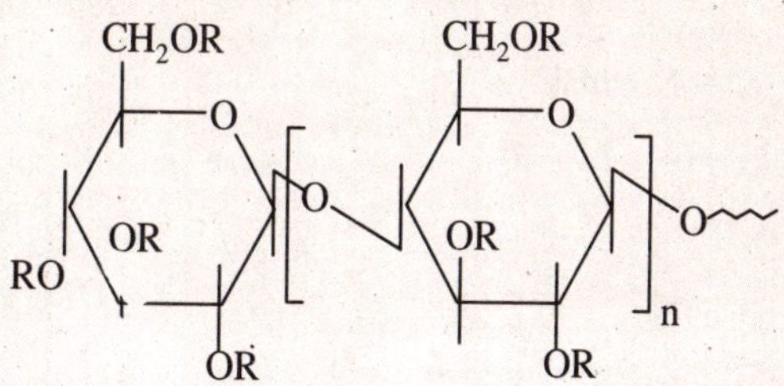

S.No.	Name of Polymer	Structure
5.	Cellulose nitrate	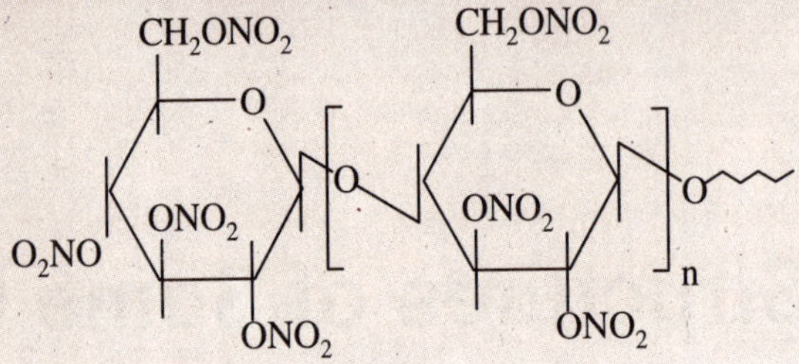
6.	Ethylene-propylene elastomers	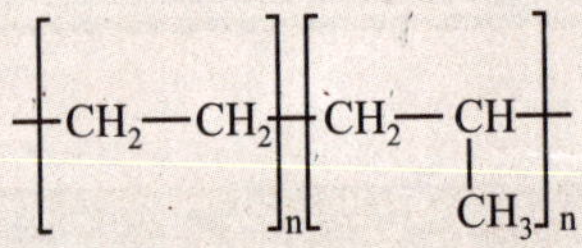
7.	Melamine-formaldehyde resins (MF)	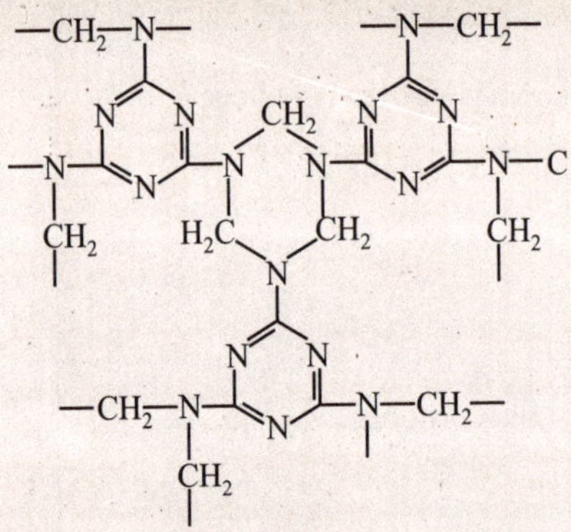
8.	Nitrile rubber (NBR)	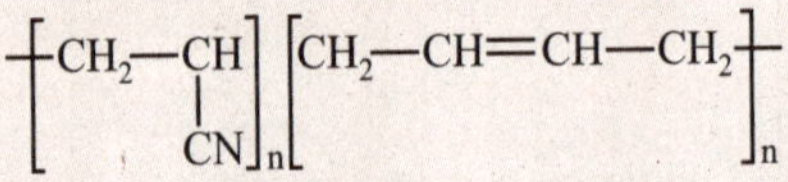
9.	Novolak	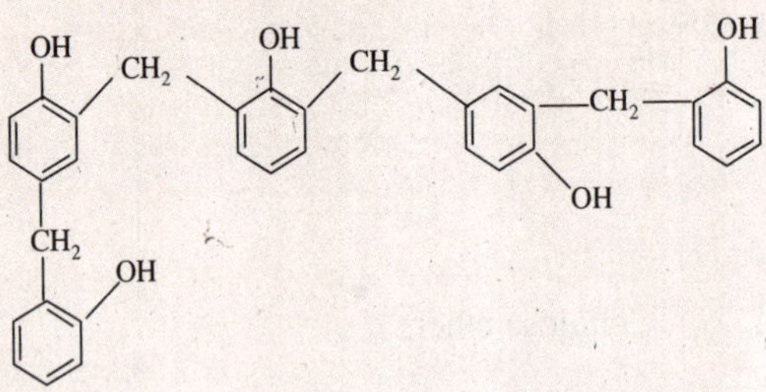
10.	Nylon-6	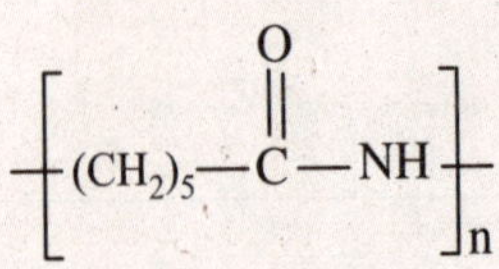

S.No.	Name of Polymer	Structure
11.	Nylon-11	$\left[(CH_2)_{10}-\overset{\overset{\displaystyle O}{\|}}{C}-NH \right]_n$
12.	Phenol-formaldehyde resins (PF)	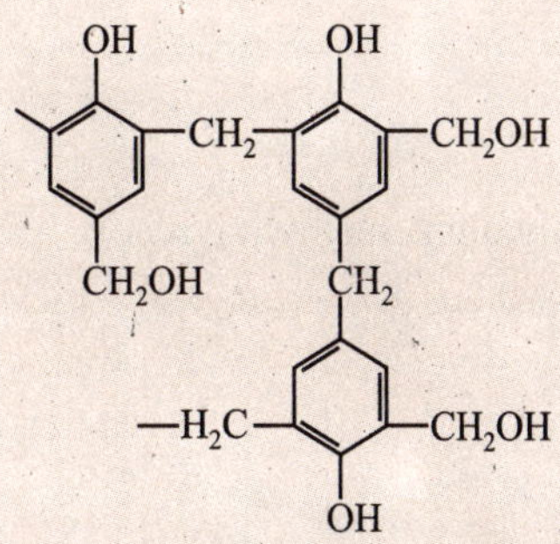
13.	Polyacrolein	$\left[CH_2-\underset{\underset{\displaystyle CHO}{\vert}}{CH} \right]_n$
14.	Polyacrylamide	$\left[CH_2-\underset{\underset{\displaystyle CONH_2}{\vert}}{CH} \right]_n$
15.	Poly (acrylic acid)	$\left[CH_2-\underset{\underset{\displaystyle COOH}{\vert}}{CH} \right]_n$
16.	Polyacrylonitrile (PAN)	$\left[CH_2-\underset{\underset{\displaystyle CN}{\vert}}{CH} \right]_n$
17.	Poly(allyl alcohol)	$\left[CH_2-\underset{\underset{\displaystyle CH_2OH}{\vert}}{CH} \right]_n$

S.No.	Name of Polymer	Structure
18.	Poly(alpha-cyanomethyl acrylate)	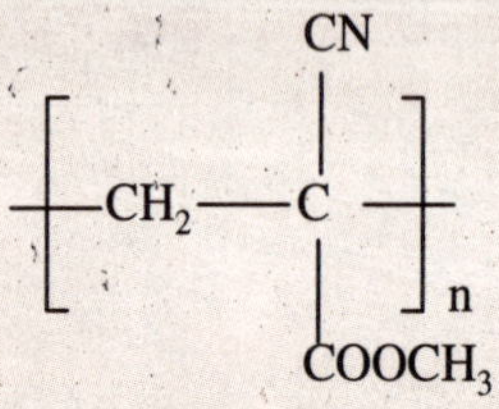
19.	Poly(alpha-methyl styrene)	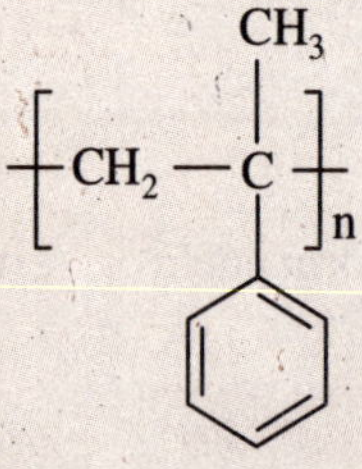
20.	Poly(β-alanine) (3-Nylon)	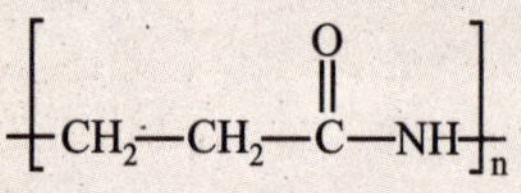
21.	1, 2-Polybutadiene	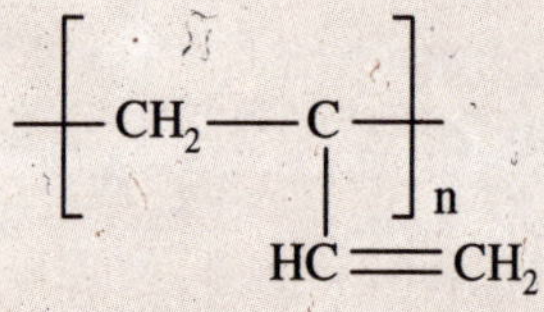
22.	trans-1, 4-Polybutadiene	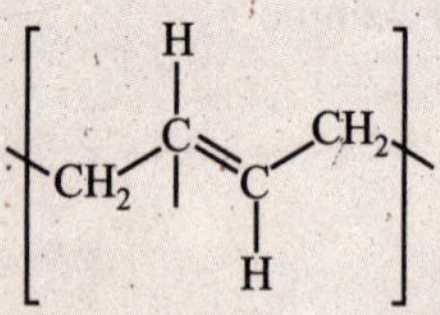
23.	Polybutadiene (Butadiene rubber, BR)	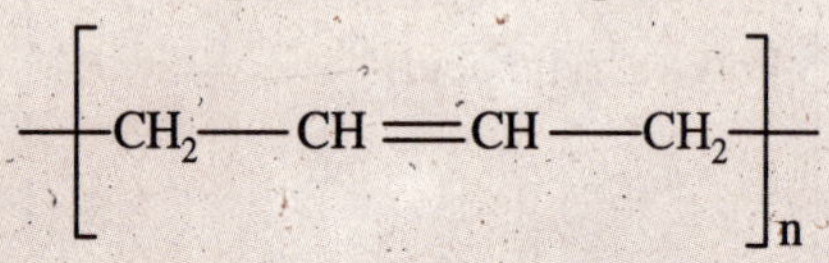
24.	Polybutylene	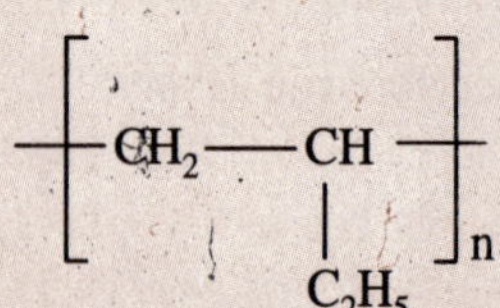
25.	Poly(butylene terephthalate) (PBT)	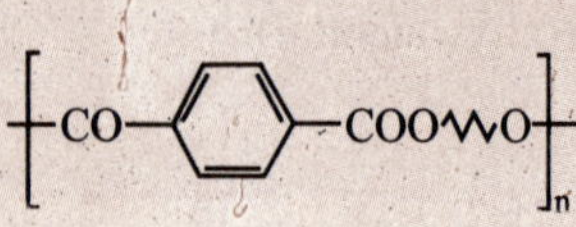

S.No.	Name of Polymer	Structure
26.	Polycarbonate (PC)	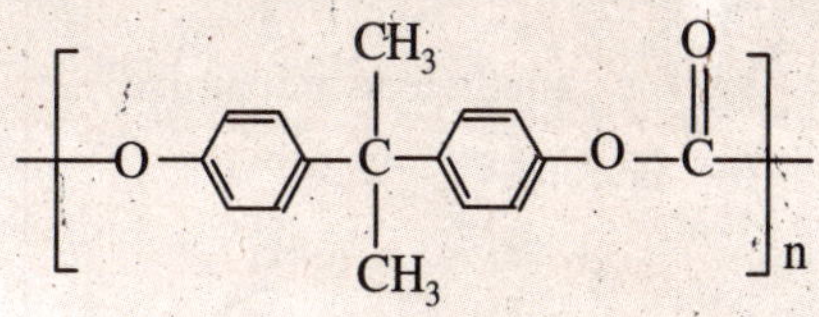
27.	Polychloroprene	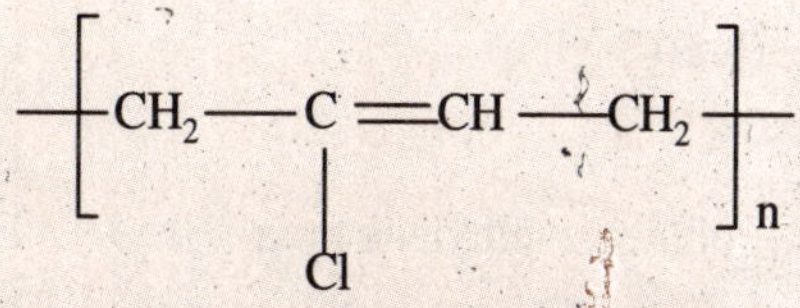
28.	Polyethylene (PE)	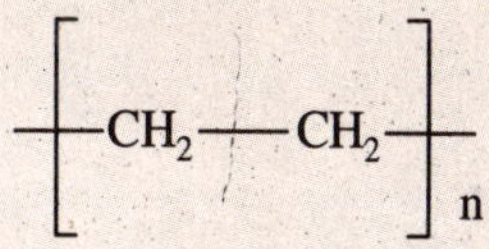
29.	Polyethylene adipate	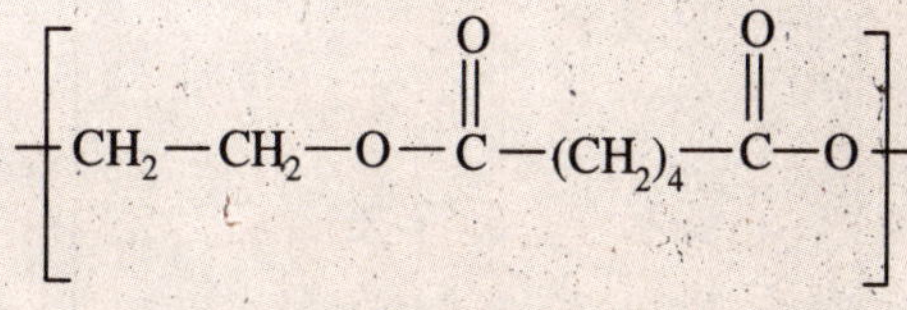
30.	Poly(ethylene glycol) (PEG)	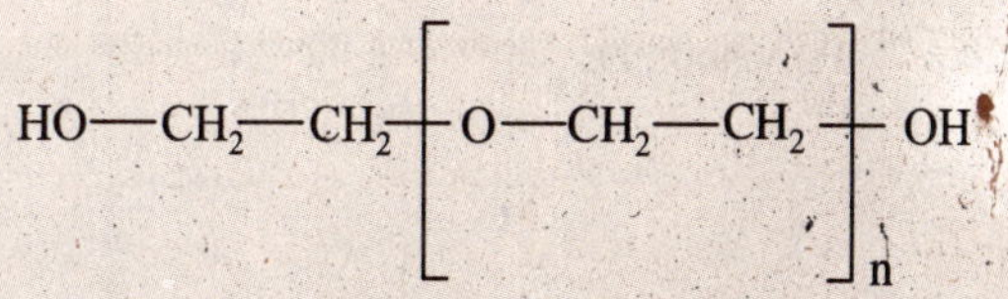
31.	Poly(ethylene terephthalamide)	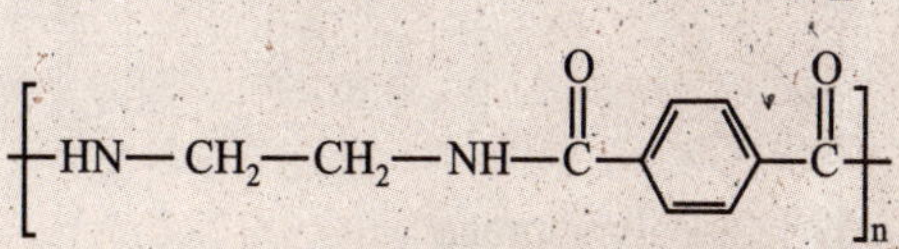
32.	Poly (ethylene terephthalate) (PET)	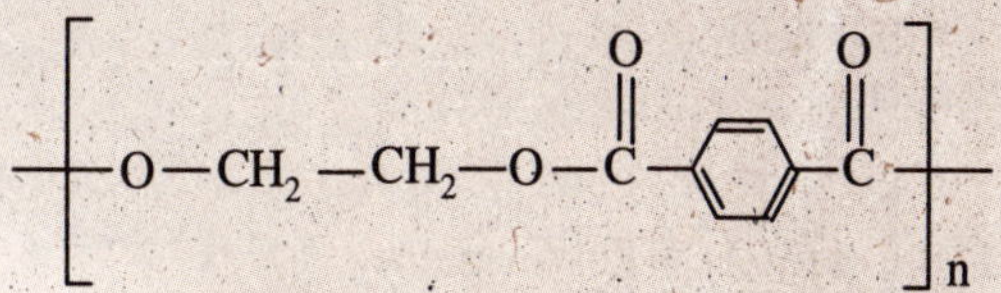
33.	Poly(ethylene-co-propylene adipate)	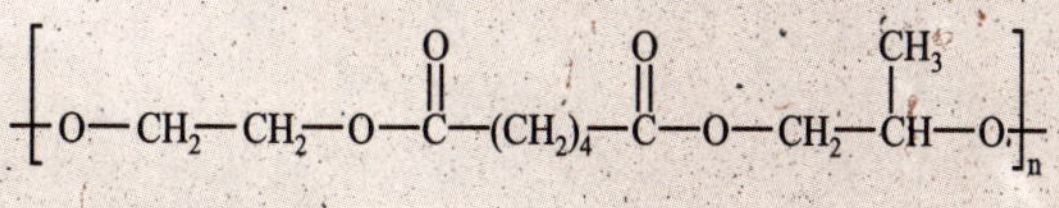
34.	Poly(hexamethylene thioether)	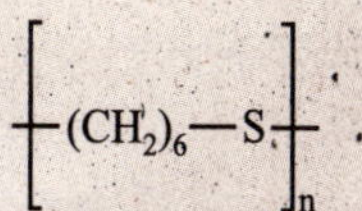

S.No.	Name of Polymer	Structure
35.	Poly(hexamethylene adipamide) (6-6 Nylon)	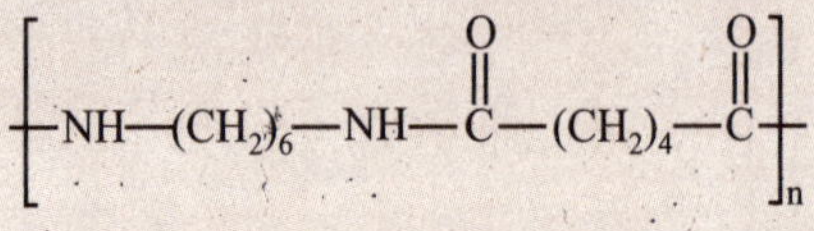
36.	Polyhexamethylene sebacamide (6-10 Nylon)	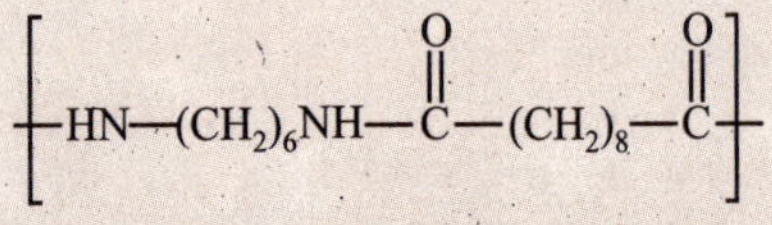
37.	Polyhexamethylene urea	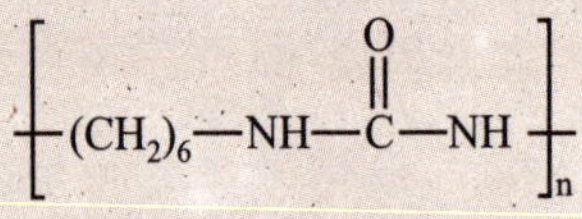
38.	Polyisobutylene (PIB)	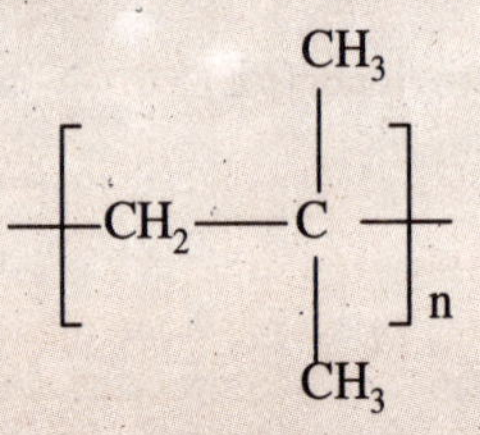
39.	Polyisoprene	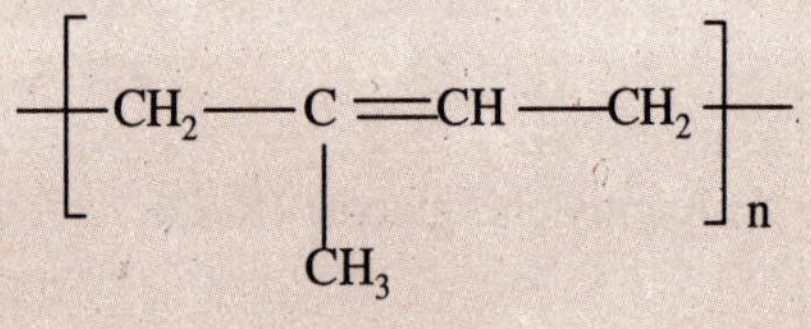
40.	3, 4-Polyisoprene	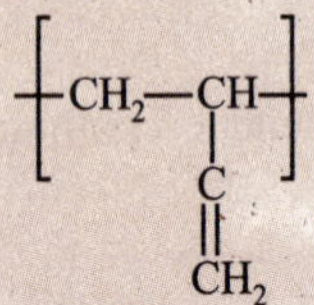
41.	trans-1, 4-Polyisoprene	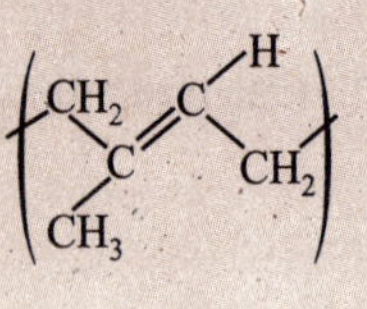
42.	Poly(methyl methacrylate) (PMMA)	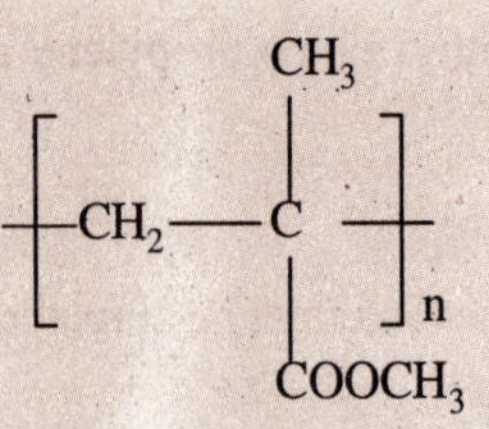

S.No.	Name of Polymer	Structure
43.	Poly(methyl acrylic acid)	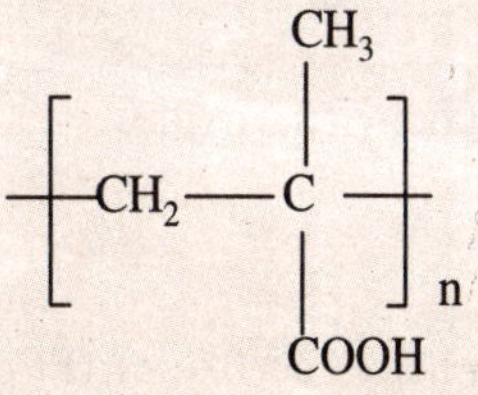
44.	Poly(methyl vinyl ketone)	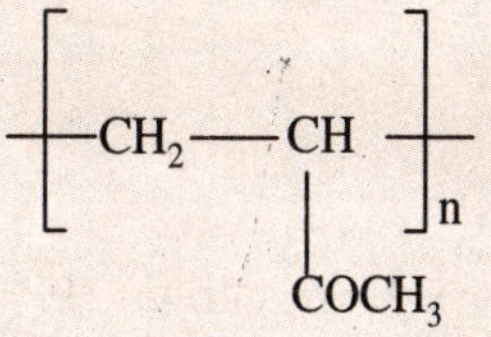
45.	Poly methyl acrylate (PMA)	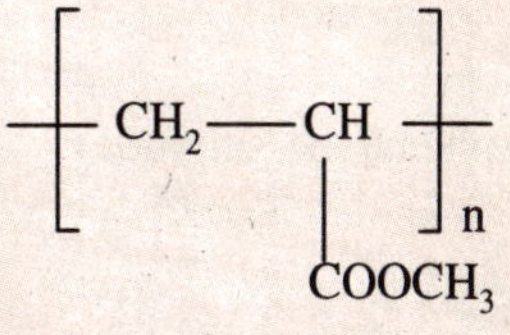
46.	Polymethylene adipamide (1-6 Nylon)	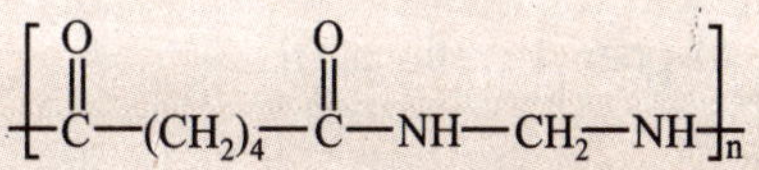
47.	Poly(m-phenyl carboxylate)	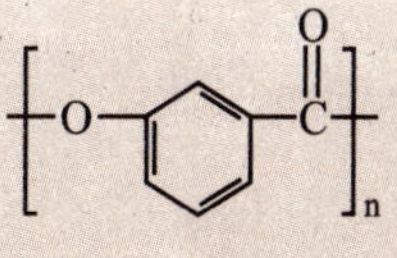
48.	Poly[(2,2'(m-phenylene)-5,5' (6,6'-benzimidazole)]	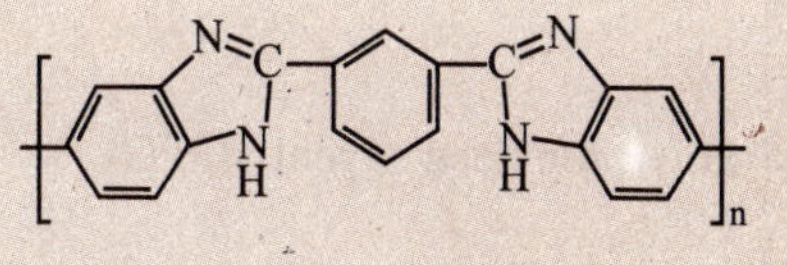
49.	Polyoxymethylene (polyacetal)	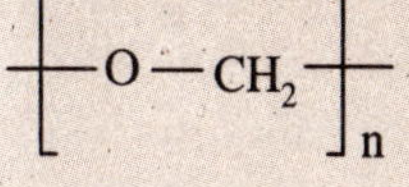
50.	Poly(1, 4-phenylene adipate)	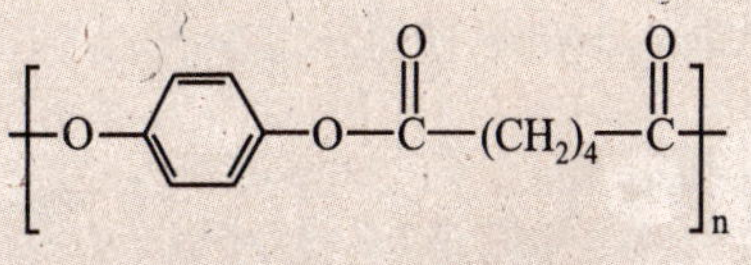

S.No.	Name of Polymer	Structure
51.	Poly(phenylene oxide)	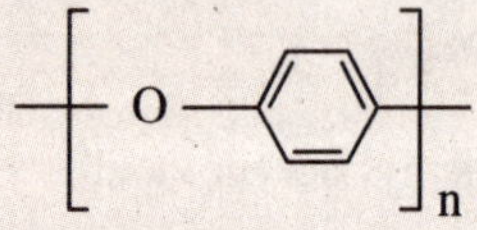
52.	Poly(phenylene oxide) (PPO) Poly (2,6-dimethyl-p-phenylene ether)	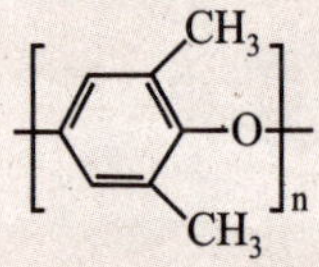
53.	Poly(phenylene sulfide) (PPS)	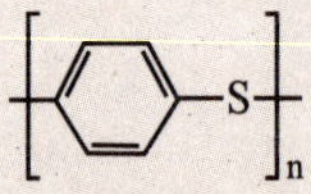
54.	Poly(phenylene sulphide)	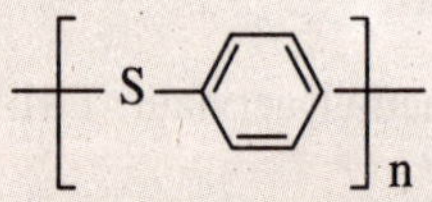
55.	Poly[2,2-propanebis (4-phenyl carbonate)]	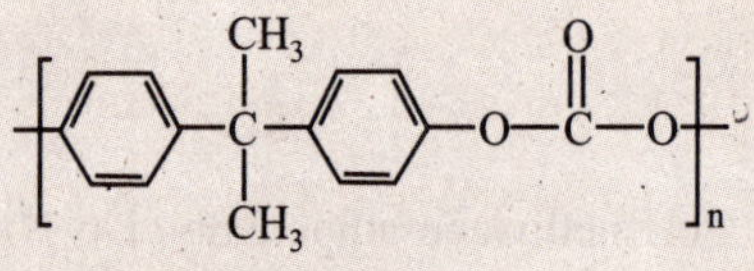
56.	Polypropylene (PP)	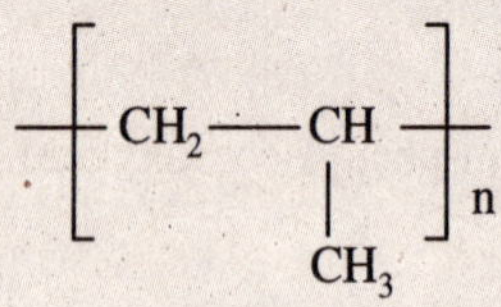
57.	Poly(propylene glycol) (PPG)	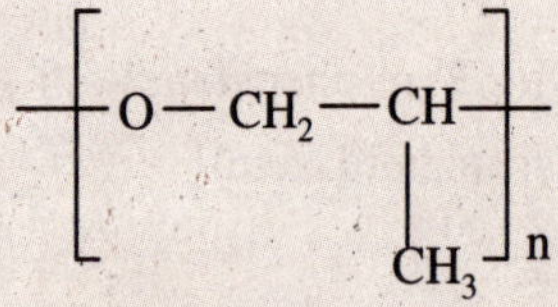
58.	Polystyrene (PS)	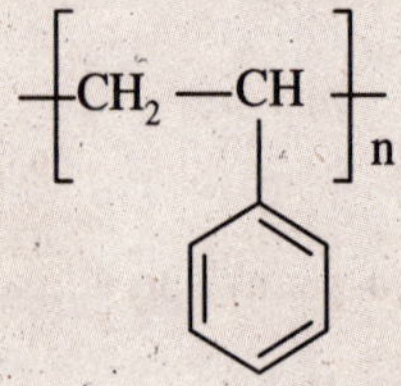

S.No.	*Name of Polymer*	*Structure*
59.	Polysulphide	$\left[-CH_2-CH_2-S-S-\right]_n$
60.	Polytetrafluoroethylene (PTFE)	$\left[-CF_2-CF_2-\right]_n$
61.	Poly(11-undecanoamide)	$\left[-(CH_2)_{10}-\overset{O}{\overset{\Vert}{C}}-NH-\right]_n$
62.	Poly(vinyl acetate) (PVAc)	$\left[-CH_2-\underset{OCOCH_3}{\underset{\vert}{CH}}-\right]_n$
63.	Poly(vinyl alcohol) (PVA)	$\left[-CH_2-\underset{OH}{\underset{\vert}{CH}}-\right]_n$
64.	Poly(vinyl butyrate)	$\left[-CH_2-\underset{OCOCH_2CH_2CH_3}{\underset{\vert}{CH}}-\right]_n$
65.	Poly(vinyl chloride) (PVC)	$\left[-CH_2-\underset{Cl}{\underset{\vert}{CH}}-\right]_n$

S.No.	Name of Polymer	Structure
66.	Poly(vinyl ether)	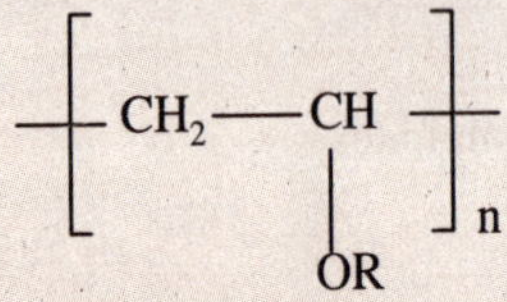
67.	Poly(vinylidene chloride)	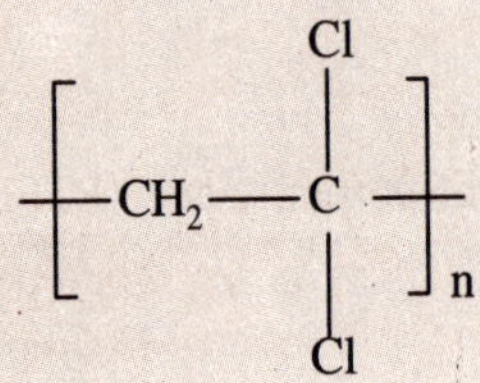
68.	Resol	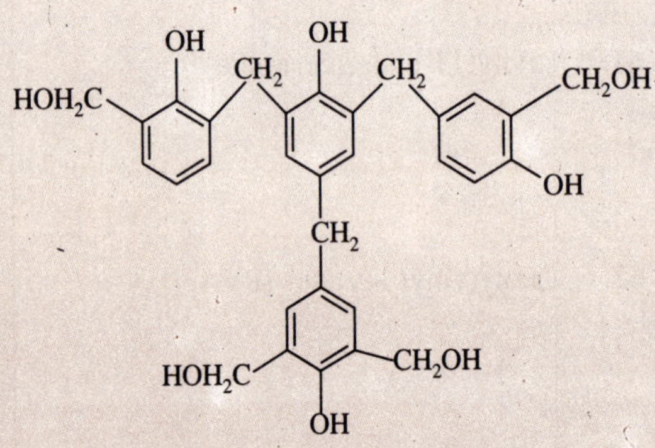
69.	Styrene-acrylonitrile copolymer (SAN)	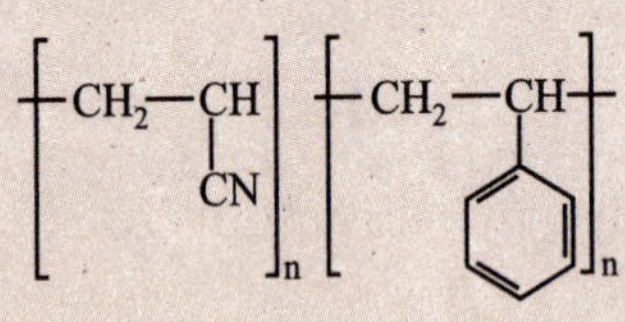
70.	Styrene-butadiene rubber (SBR)	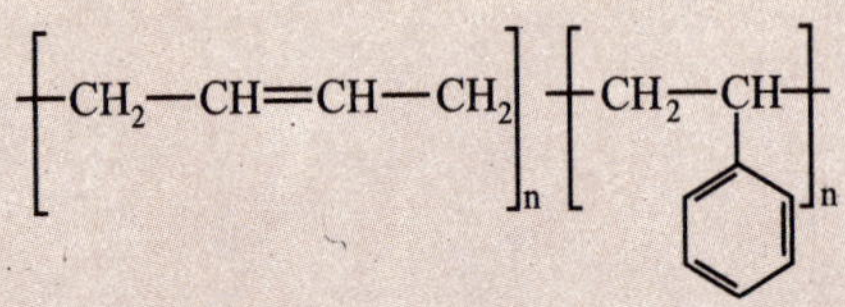
71.	Urea-formaldehyde resins (UF)	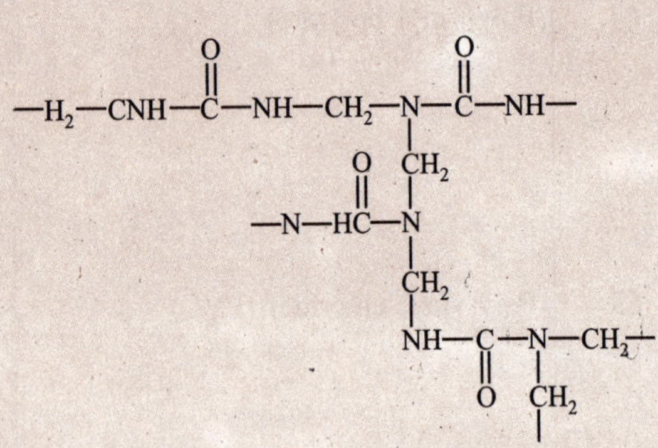

INDEX

ABS Polymer 357
Accelerators 184, 351
Acidolysis 225, 228
Acrylic acid 16, 149, 151
Acrylic plastics 349
Acrylonitrile 16, 29, 53, 118, 149
Active centre 30, 32, 82
Addition polymerisation 8, 41, 43
Alkali cellulose 197, 199
Allyl compounds 15, 16
Alternating copolymer 50, 107, 138, 153, 228
Amide interchange 45
Aminolysis 225, 228,
Amino plastics 349
Amino resins 192
Amorphous 102, 105, 113, 116, 121
 components 133
 polymer 1, 4, 11, 20
 regions 117, 123, 133, 196
 solid 1, 31, 57, 105, 183
Aniodic polymerisation 28, 52
Antioxidants 222, 351
Arrhenius equation 78
Asymmetric carbon atom 108
Atactic 99, 109, 137, 163, 279
Autoacceleration 52, 70
Average molecular weights: 15, 69, 292, 304
 concept of 69, 119, 157, 267, 276
 number ($\bar{M}_n$) 8, 11, 57, 63, 124
 viscosity ($\bar{M}_v$) 7, 52, 68, 71, 91
 weight ($\bar{M}_w$) 3, 11, 17, 39, 53
 Z-($\bar{M}_z$) 67
Azeotropic composition 154, 157
Azeotropic copolymerisation 153, 157
Azo-bis-isobutyronitrile 17

Bakelite 2, 190
Banburry mill 351
Bead 15, 53, 288
Bead polymerisation 53
Benzoyl peroxide 17, 21, 158, 182, 242
Bimetallic mechanism 32,
Birefringent 140, 141
Bisphenol 118, 168, 193
Block copolymers 30, 107, 154, 252
Blowing agents 179, 361
Blow moulding 352, 369
Boltzmann equation 269
Bond 4, 21, 76, 108, 123
 double 13, 21, 41, 98, 179
 single 8, 22, 37, 104, 110
 triple 13
Bond angles 76, 108
Bond energy 205
Bond stability 204
Bound solvent 263
Branched polymers 135
Branching 40, 70, 84
Brownian (movement) motion 114, 115, 343
Bulk polymerisation 52, 161, 178, 286
Buna rubber 185
1, 3-Butadiene 16, 110, 111
Butyl rubber 183, 349

Calendering 129, 352
Caprolactam 44, 146, 202
Carbanion 28, 31, 41, 98, 256
Carbon black 186, 213
Carbonium ion 26, 41, 238
Carboxymethyl cellulose 199
Carbowax 174
Catalyst 11, 21, 91, 162, 178
Cationic polymerisation 28, 47, 53, 86, 158, 167, 183
Ceiling temperature 86, 202
Cellophane 2, 197, 355
Celluloid 2, 198, 348
Cellulose 1, 72, 118, 125, 145
  alkali 29, 43, 163, 181, 227
  regenerated 35, 196, 273
Cellulose acetate 198, 305, 349, 367
Cellulose acetate butyrate 198
Cellulose ethers 199
Cellulose nitrate 125, 198, 273
Cellulose xanthate 197
Cellulosics 349
Chain aggregation 135, 136
Chain backbone 10, 103, 123, 159, 168
Chain coiling 135
Chain end 24, 38, 51, 73, 127
Chain entanglement 135, 262
Chain flexibility 123, 335
Chain folding 143, 144
Chain growth 15, 27, 41, 89, 256
Chain length 24, 50, 74, 111, 252
Chain modifier 50, 83, 185, 287
Chain polymerisation 15, 38, 58, 201
Chain rigidity 159
Chain scission 203, 209, 227
Chain segment 76, 105, 123, 263, 276
Chain transfer 24, 52, 82, 162, 313
  intermolecular 39, 102, 121, 162, 262
  intramolecular 39, 102, 162, 238, 262
  termination by 15, 23, 44, 78, 215
Chain transfer constants 83, 85
Chlorinated polyethylene 231
Chlorinated PVC 163, 177, 182, 231, 290
Chlorinated rubber 230
Chlorination 179, 230
Chloroprene 16, 57, 186, 242, 349
Chlorosulfonated polyethylene 231
Cis-configuration 111, 184
Cis-isomer 137
Co-catalyst 27, 49, 87
Coefficient of expansion 121
Cohesive energy density 268
Coil dimension 276
Coil volume 263
Cold flow 335
Cold rubber 185
Collodion 309
Colloids 53, 274
  micellar 274
  molecular 5, 39, 43, 61, 63, 78
Colourants 179
Comonomer 104, 148, 164, 252, 336
Complex 1, 30, 63, 140, 213
  coordination 15, 53, 162
  monomer-catalyst 30
Compounding 336, 350, 351
Compression moulding 352, 355, 365
Condensation polymerisation 8
Configuration 43, 105, 137, 163, 186
Conformation 105, 134, 265, 335
Conjugated double bonds 168, 179, 207
Continuous lamination 365
Contour length 73, 76, 77
Coordination polymerisation 30, 70, 162
Copolycondensation 159
Copolymer(s) 30, 107
  alternating 35, 50, 107, 138, 153, 157, 252
  block 30, 51, 107, 154, 202
  definition 133, 175, 276, 315, 350
  graft 107, 252, 254
  random 49, 74, 105, 128, 147
Copolymer composition 153, 157
Copolymer equation 148, 152, 157

Copolymerisation 129, 146, 153, 252, 336
azeotropic 153, 157, 288
free-radical 15, 25, 46, 78, 86
ideal 121, 153, 197, 267, 338
ionic 15, 44, 86, 89, 157, 185
random 49, 63, 105, 128, 201
Copolymerisation behaviour 153
Cotton 1, 9, 177, 197, 350
Counterion 27, 31, 145, 157, 264
Coupling termination by 253
Creep 180, 344
Crepe rubber 184
Critical micelle concentration 54
Cross-linked polymers 13, 105
Cross-link(ing) 26, 190, 213, 238, 239, 242, 243
chemical 1, 9, 19, 47, 63
physical 9, 47, 72, 116, 136
Cryoscopy 301, 302
Crystal growth 135
Crystal lattice 34, 114, 133, 261
Crystalline 58, 102, 113, 121
components 2, 16, 31, 52, 128
phase 26, 31, 52, 116, 220
polymers 1, 8, 31, 49, 67
regions 117, 133, 208, 323
solid 1, 11, 31, 114, 261
Crystallinity 11, 121, 133, 159, 231
degree of 8, 57, 81, 93, 121
Crystallisability 123, 134, 137, 336
Crystallisation 134, 136, 143, 238, 337
Crystallite 117, 121, 136
Curative(s) 243, 244, 246, 249, 351, 353, 363, 364, 365
Cure reactions 186, 190, 243, 249
Cyclic macromolecule 49
Cyclic products 39
Cyclisation 219, 236
Cyclised rubber 238
Cyclodienes 50
Cycloolefines 49, 50

DNA 63, 140
Dash pot 340, 345
Deactivated chain 24
'Dead' polymer 24
Debye equation 322
Debye method 324
Deformation 115, 171, 335, 341, 346
elastic 115, 129, 183
irreversible 238, 336
permanent 209, 340
reversible 86, 202, 238, 335
viscoelastic 113
viscous 2, 6, 52, 113, 129
Degradation 57, 145, 161, 171, 195
Degradation, types of 201
chain end 201, 255, 276
chemical 1, 10, 21, 47, 63
enzymatic 203
hydrolytic 203, 225
mechanical 9, 72, 129, 145, 163
oxidative 215, 219, 225, 288, 350
photo 58, 350
random 49, 63, 75
thermal 9, 15, 58, 85, 105
ultrasonic 209,
Degree of polymerisation (DP) 8, 9, 50, 57, 68, 71, 81
number average 64, 69, 95, 300, 312
weight average 64, 95, 301, 321
Density 47, 118, 127, 161, 183
Depolymerisation 164, 201
Depropagation 86, 202
Dianion 89, 185
Diaziridine 248
Die 353, 358
Die casting 353
Dienes 15, 30
Diepoxide 245
Differential distribution curve 329
Differential refractometer 297, 324
Differential scanning calorimeter 331
Diffraction grating 132
Diffraction pattern 132, 142
electron 15, 17, 33, 98, 142
x-ray 131, 133, 145, 213

Diffusion coefficient 318, 320
Diffusion constant 318
Diisocyanates 249
Dilatometer 120, 329
Diol 42, 91
Diradical 59
Dispersity 50, 68, 70
Disproportionation termination by 24, 81, 85, 255
Dissolution 7, 196, 261, 267
Divinyl benzene 239
Dodecyl mercaptan 84, 185, 286
Droplets monomer 56
Dry spinning 352, 365

Ebulliometry 301, 304
Elasticity 334, 346
  delayed 129, 341
  rubber 1, 10
Elastomer(s) 10
Elastomeric(s) 337, 348, 349, 351
Electrochemical polymerisation 46
Electron pair 16, 21, 44, 98
Ellipsoid 75, 263
Elution volume 297, 300
Emulsion polymerisation 54, 56, 185, 186, 199, 286, 287
End group(s) 24, 37, 40, 84, 109, 168, 187, 301, 312
End group analysis 301, 312
End-to-enddistance 73, 74, 76, 77, 275, 276
Entangled 134, 262, 283
Enthalpy 134, 264
Entropy 264, 335
Epichlorohydrin 193,
Epoxy (resins) 193, 194, 247, 349, 362
Equilibrium sedimentation method 318, 320, 321
Esterification reaction 13, 91, 232
Ethyl cellulose 118, 199, 290, 349
Ethylene 8, 37
Ethylene glycol 12, 37, 39, 57, 166
Ethylene oxide 44, 48, 174, 185, 231, 255
Expansion factor 121, 276, 329, 362
Extended (chain) length 76, 77
Extent of reaction 40, 91
Extrudate 358
Extrusion machine 358
Extrusion moulding 352, 358

Fibre(s) 10, 167, 172, 178, 334
  comfort 350
  industrial 26, 52, 83, 163, 172
  reinforcing 186, 350, 362
  safety 169, 198, 285
Fibre forming (polymers) 180, 336
(Fibre) reinforced plastics (FRP) 194, 350, 362
Fibrils 140
Filament 197, 336
Filament winding technique 362, 363
Fillers 179, 186, 334
Film casting 355
Fineman-Ross method 152
Flame retardant(s) 351
Flash 356
Flexible (Polymer) chain 105, 155, 262
Flexural strength 144
Flory-Huggins interaction parameter 271
Flory-Huggins theory 269
Flow-behaviour 67, 283, 338, 340
Flow temperature 113, 335
Fluorocarbon 11, 180
Fluoroplastics 349
Foaming 195, 352, 360
Foaming agents 195, 361
Foams 4, 54, 173, 194, 334
Forces 35, 77, 102, 123, 209
  attraction 105, 127, 138, 261, 275
  cohesive 123, 262
  dispersion 56, 102, 262
  interaction 77, 102, 128, 210
  intermolecular 39, 102, 117, 238
  van der Waals' 123, 261
Formaldehyde 2, 58, 167, 188
Fractional precipitation 267, 293
Fractionation 292, 327

Fractionation techniques
fractional precipitation 267, 293
gel permeation chromotagraphic 293, 295
gradient elution 293
partial dissolution 293, 294
partial precipitation 293
solvent volatiliasation 294
Free energy 135, 264, 271
of dilution 271
of mixing 268, 271
Free-radical(s) 15, 17, 20, 78, 147, 252
Free-radical copolymerisation 147, 157
(Free-) radical polymerization 158, 164, 180, 185
Free-radical site 22, 79, 147, 206
Free rotation 73, 108, 124, 335
Free solvent 263, 321
Freely jointed chain 73, 76
Freon 11
Friedel-Crafts reaction 235
Functional group 8, 36, 91, 224, 243
carboxyl 8, 12, 37, 91, 167
hydroxyl 8, 37, 42, 91, 168
Functional group equivalent 312
Functionality 11, 243, 312
bi 8
mono 13
tri 13

Gas phase polymerisation 58
Gel 40
Gelation 40, 354,
Gel permeation chromatography 295
Gel point 40
Gel-sol equilibrium 267
Giant molecules 5, 261
Glass transition 113, 123, 329
Glass transition temperature 113, 126, 335
Glassy solid 113, 116, 136
Gradient elution 293, 295
Graft copolymers 107
Group transfer polymerisation (GTP) 50, 52
Gun cotton 2, 198
Gutta-percha 137, 183

Half-life, of initiators 20
Hand lay-up technique 362
Hardness 144, 162, 192
Head-to-head addition 23
Head-to-tail addition 23
Heat distortion temperature 130, 164
Heat of mixing 268, 271
Helical 105, 139, 359
chains 15, 52, 63, 98, 123
conformations 105, 134, 265, 275
structure 9, 32, 61, 102, 137
Helix 139, 181
Heterochain polymer 103
Heterotactic 109
Hevea brasiliensis 184
Hexamethylene tetramine (Hexa) 251
Homochain polymer 103
Homolytic decomposition 17, 78
Homopolymer 51, 104, 128, 146, 164
Homopolymerisation 146
Hooke's (equation) law 338, 345
Huggins equation 283
Hydrodynamic volume 263, 278, 296, 313
Hydrogen bond(ing) 102, 123, 138, 196, 238
Hydrogen peroxide 18, 56, 181, 255
Hydrogenation 161, 228
Hydrolysis 145, 176, 185, 225
Hydroperoxides 17, 82, 161, 222
Hydroxy acid 8, 39

Ideal copolymerisation 153
Industry, polymer 25, 26, 130, 224
Inhibitors 25, 285
Initiation 15, 21, 78, 147
in anionic polymerisation 29,
in cationic polymerisation 29
in free-radical polymerisation 82
in ring-opening polymerisation 172
in Zeigler-Natta polymerisation 31, 183, 185
Initiators(s) 15, 21, 53, 82, 161
anionic 28, 44, 89, 158, 164, 255

cationic 27, 47, 86, 158
free-radical 15, 41, 78, 146, 185
decomposition rate 19, 78, 99
efficiency 79, 127, 295
group-transfer 50
ionic 15, 26, 44, 86, 157, 183
redox 21, 56, 178, 181
Injection moulding 352, 356
Injection moulding machine 356
Inorganic polymers 10, 103
Insertion polymerisation 31
Integral distribution curve 329
Interaction 77, 102, 210, 262, 296
dipole 102, 123, 262
intermolecular 39, 102, 162, 238
intramolecular 39, 102, 162, 238
solvent-solute 77, 262
Interfacial (poly) condensation 57, 169
Interpenetrating 105, 134, 262, 266, 283, 347
Intrinsic viscosity 68, 277, 281, 314
Ion-exchange resins 54
Ion pair 27, 86
Ion pair formation 86, 89
Ion pair precipitation 28
Ionic copolymerisation 157
Ionic polymerisation 26, 52, 86, 89, 158, 185
Isobutylene 16, 28, 88, 146
Isocyanates 12, 43, 232
Isomerism 108, 111
geometric 102, 105, 110, 134, 334
optical 108, 165, 334
Isoprene 16, 29, 137, 151, 183, 208
Isotactic 35, 109, 137, 183, 268

Kevlar 172
Kinetic chain length 53, 80, 87
Kraemer equation 278

Lactams 43, 172
Lactones 43
Ladder polymer 126, 237
Lamella 142
Lamellar thickness 142
Laminates 190, 364
Latex 56, 184, 287, 351
Lead dioxide 187, 244
Light scattering 68, 301, 322
Light scattering photometer 324
Limiting viscosity number 277, 315
Linear polymers 135, 243, 276
Liquid polymers 29, 51
Liquid supercooled 10, 174, 187, 243, 351
Long-range order 114, 131, 268
Lubricants 350, 369

Macromolecule 5, 49, 102, 213, 275
Macrostructure 102
Man-made fibres 336
Mark-Houwink equation 68, 278, 314
Mastication 208, 336
Mayo equation 85
Mayo-Lewis method 152
Melamine-formaidehyde (MF) resins 190
Melamine resins 192
Melt polycondensation 57, 171
Melt spinning 352, 366
Melt viscosity 71
Melting point (T_m) 57, 114, 121, 162, 168, 180
Melting temperature 117, 336
Membrane osmometry 304, 309
Mercaptans 83, 241
Metathetical polymerisation 48, 50
Methacrylic acid 16, 151, 175
Methyl acrylate 149, 156
Methyl cellulose 199
Methyl methacrylate 10, 47, 146
α-methyl styrene 88, 211
Methylol phenols 189
Micelle 54, 57, 274
Microscope 140, 143
electron 15, 142, 213, 302, 352
polarising 140
Microstructure 102, 140
Milling 208, 336, 351

Mixed polyamides 159
Mixed polycondensates 159
Mixed polyester 159
Mobile phase 295, 297
Mobility (or motion) 58, 115
  molecular 5, 37, 61, 83, 91
  segmental 115, 121, 136, 262, 275
Model 73, 269, 340
  burger 342
  hooke 338, 342
Maxwell 340, 345
  Newton 338, 340
  voigt 340, 346
Modulus (of elasticity) 338
Molecular engineering 335, 336
Molecular weight 5, 11, 40, 63, 92
Molecular weight averages 68,
  (see also average molecular weights)
Molecular weight distribution (curve) 293, 295, 325
Monodisperse 68, 267, 325
Monodispersity 68
Monomer(s) 8, 22, 52, 79
  list of 17, 289
  purification of 289
  Q-e values of 156
  reactivity ratios of 148, 154, 155
  vinyl 13, 41, 53, 72, 104
Monomer feed ratio 149, 153
Monomer pair 41, 146, 152
Monomer(ic) component 128, 152
Monomeric composition 152
Monomeric unit 8, 23, 31, 102, 147
Monometallic mechanism 32
Mould 348, 350, 351, 354

Natural polymer(s) 1, 9
Natural rubber 10, 118, 137, 183, 230, 279
Neoprene (rubber) 186, 349
Network polymers 107
Newton's (equation) law 340, 345
Nitrile rubber 165, 185, 349, 351
Nitrocellulose 198, 306
Non-solvent 50, 266, 276, 281, 367
  for polymers 71
Novolak 190, 251
Nylons 1, 40, 171, 349
Nylon-6, 44, 57, 124, 138, 146, 159, 172
Nylon-6, 6, 146, 159
Nylon 10, 172
Nylon 11, 172

Olefines 15, 30, 50
Olefine metathesis 48
Oligomers 43, 169
Organosols 179
Organotin polymers 195
Orientation 31, 123, 337, 368
  chain 5, 15, 29, 44, 73
  uniaxial 368
Osmometers 305, 308
Osmometry 301, 304, 313
Osmotic head 307, 308
Osmotic pressure 68, 272, 304
Ostwald viscometer 317
Oxirane 43
Ozonide 218, 229
Ozonolysis 218

Parison 358
Partial dissolution 293, 295, 295
Partial precipitation 293
Particle-scattering factor 323, 324
Peracids 17
Peresters 17
Permeability 144, 162, 180, 188
Peroxides 21, 178, 222
Persulphates????
Phenol 2, 13, 168, 172, 189
Phenol-formaldehyde (PF) resin 188, 191
Phenolic resin 192, 362
Phenoplastics 349
Photoabsorber 211
Photodecomposition 20
Photo (initiated polymerisation) 20
Photostabilisers 211

INDEX

Plastic(s) 2, 10, 129, 164, 194
  types of 9, 39, 103, 136, 147
  plasticisation 336
  external 23, 105, 129, 205, 336
  internal 115, 129, 142, 306
Plasticiser 127, 162, 179, 198, 336
Plasticiser efficiency 127
Plastisol 179, 354
Platelets 142
Polar groups 123, 138, 335
Polarity 31, 88, 99, 127, 158
Polyacrolein 232
Polyacrylamide 235
Polyacrylates 125, 213, 254
Polyacrylic acid 174, 233, 279
Polyacrylonitrile 164, 254
Polyaddition polymerisation 41, 43
Polyalkenamers 49
Polyamic acid 170, 171
Polyamides 40, 57, 105, 159, 171
  aliphatic 159, 166, 184, 290
  aromatic 21, 57, 105, 125, 205
Polybutadienes 184, 185, 349
Polybutylene 182
Polycaprolactam 202, 349
Polycarbonate 4, 118, 168, 205, 357
Polychloroprene 186, 242, 349
Polycondensation 37, 57, 91, 147, 166
  AA-BB type 39, 40
  A-B type 39
  interfacial 57, 169, 289, 368
  intermolecular 39, 102, 117, 121, 162
  intramolecular 39, 102, 162, 238, 262
  melt- 366
  self- 15, 39, 172, 187
Polydispersed system 68, 267, 271
Polydispersity ($\bar{M}_w/\bar{M}_n$) 68, 70, 121
Polyelectrolyte 175, 234
Polyenanthoamide 39
Polyester(s) 1, 40, 105, 159, 166
  aliphatic 159, 166, 172, 184, 290
  unsaturated 57, 215, 243, 285, 349
Polyethers 40, 193
Polyethylene 6, 9, 39, 57, 103
Polyethylene glycol 174
Polyethylene terephthalate 39, 118, 159, 279, 349
Polyformaldehyde 58, 167
Polygermanes 103
Polyimides 170, 350, 362
Polyisobutylene 53, 88, 146, 182, 204
Polyisoprene 137, 183, 202, 228, 349
Polymer definition of 5, 6
Polymers classification of 9
Polymer analogue reactions 227
Polymer backbone 166, 202, 243
Polymer bead 54
Polymer melt 119, 134, 163, 343, 366
Polymer molecular shape 74–76
Polymer molecular size 73
Polymer single crystals 142
Polymerisation 8, 11, 26, 36, 41
  chemistry of 2, 11, 235, 286, 312
  definition 133, 175, 276, 315, 350
Polymerisation, kinetics of 78
  anionic 28, 44, 70, 89, 158
  cationic 27, 47, 58, 86, 158
  free-radical 15, 41, 78, 100, 146
  polycondensation 37, 42, 57, 91, 147
Polymerisation mechanism (reactions) 313
  addition 8, 20, 33, 41, 56
  anionic 28, 47, 89, 100, 158
  cationic 27, 47, 58, 86, 100
  chain 5, 10, 22, 30, 39
  coordination 15, 30, 70, 162
  condensation 8, 15, 37, 45, 57
  electrochemical 46
  (free-) radical 15, 17, 20, 21, 23
  group transfer 50, 52
  insertion 31, 35
  ionic 15, 26, 44, 86, 147
  metathetical 48, 50
  photo initiated 20, 185
  polyaddition 41, 70, 170
  polycondensation 37, 57, 91, 147, 170

redox 21, 56, 178
ring-opening 36, 43, 172
salient features of 60, 98, 100
step 15, 30, 38, 74, 187
Polymerisation techniques 52, 70, 161
bead 15, 53, 288
bulk 47, 70, 99, 134, 161
emulsion 54, 70, 99, 176, 199
gas phase 58
high pressure 161, 208, 363
industrial 26, 52, 83, 157, 185
interfacial 172, 289, 368
melt 6, 57, 99, 117, 134
salient features of 60, 98
solid phase 31, 58, 261
solution 50, 53, 55, 57, 67
suspension 53, 70, 99, 164, 185
pearl 53
precipitation 28, 53, 267, 281, 291
Polymethacrylic acid 175
Polymethyl acrylate 118, 279, 307
Polymethyl methacrylate 10, 50, 118, 146, 165
Poly α-methyl styrene 88, 118
Poly-4-methyl pentene-1 143
Polyol 361
Polyolefines 349
Poly (para) phenylene 168
Polypentenamer 49
Polypropylene 11, 118, 140, 163, 204
Polypropylene glycol 174
Polystyrene 11, 54, 83, 85, 105, 124, 163, 202, 208
Polysulphide 10, 186, 244
Polysulphone 169, 170
Polyterephthalamide 172
Polytetrafluoroethylene 181
Polytitanoxanes 196
Polyurea 43, 173
Polyurethanes 43, 173, 232, 360
Polyvinyl acetals 177
Polyvinyl acetate 54, 104, 175, 227, 280
Polyvinyl alcohol 7, 54, 118, 138, 176
Polyvinyl amine 234
Polyvinyl benzene 163
Polyvinyl butyral 177, 237, 349
Polyvinyl carbazole 118, 125, 181
Polyvinyl carbonate 178
Polyvinyl chloride (PVC) 104, 118, 178, 280, 349
Polyvinyl cyanide 164
Polyvinyl ethers 231
Polyvinyl fluoride 138, 180
Polyvinyl formal 177, 236
Polyvinyl methyl ketone 233
Polyvinyl pyrrolidone 181, 227, 280
Polyvinylidene chloride 118, 179
Polyvinylidene fluoride 180
Poly-p-xylylene 59
Pore volume 296
Precipitation polymerisation 287
Prepolymers 219, 243, 351
Primary homopolymer 252
Processing 57, 129, 181, 336, 348
Processing techniques 129, 184, 351
blow moulding 352, 357
calendering 129, 352, 353
compression moulding 352, 355, 365
continuous lamination 365
die casting 353
extrusion moulding 352, 358
fibre spinning 365
filament winding 362, 363
film casting 355
foaming 195, 352, 360
hand lay-up 362, 363
injection moulding 352, 356
pultrusion 364
reinforcing 186, 350, 362
rotational casting 354, 355
spray-up 362, 364
thermoforming 352, 359
vacuum forming 360
anionic 28, 44, 70, 89, 100
cationic 27, 47, 70, 86, 158
free-radical 15, 21, 41, 70, 100
Zeigler-Natta 31, 59, 163, 183, 185

Propylene 13, 48, 137, 156, 174
Pultrusion 364, 365
Pyromellitic anhydride 170

Q-e scheme 157
Q-e values 156, 157
Quasi crystalline lattice model 270
Quenching 117, 136

Radius of gyration 275, 323
Random coil 105, 134, 262, 323
Random copolymers 51, 128, 252
Random flight 75, 275
Random walk 74
Rayleigh ratio 322
Rayon 9, 197, 368
Reactive site 8, 11
Reactivity 26, 41, 79, 148, 226
Reactivity ratios 148, 152, 158
  of monomer pairs 155
Redox initiators 21, 56, 178, 185
Redox reaction 21
Redox polymerisation 21
Reduced viscosity 210, 277, 315
Reinforcing 186, 350, 362
Relative viscosity 315
Relaxation 343, 345
Release agent 362
Repeat unit 8, 14, 24, 42, 68
Resol 190, 250
Retardation 343
Rheology 338
Rigid (polymer) chain 105
Ring-opening polymerisation 43, 70, 172
Rotational casting 354, 355
Root-mean-square end-to-end 76, 77, 275
  distance 73, 116, 132, 275, 319
Rubber 1, 9, 83, 121, 137
  buna 185
  butyl 16, 29, 89, 125, 183
  chlorinated 163, 177, 230, 290
  cold 176, 185, 335, 354, 361
  cyclised 238
  natural 1, 9, 118, 137, 183
  neoprene 186, 349
  nitrile 156, 165, 186, 349
  silicone 10, 103, 194, 242, 349
  styrene-butadiene (SBR) 83, 146, 229, 256, 349
  synthetic 1, 9, 67, 146, 177
  warm 134, 185, 287, 359
Rubber elasticity 334
Rubber mill 351

SAN copolymer 164
SBR rubber 280
Saponification 178, 273
Saturation point 7, 266
Second virial coefficient 273, 302, 322
Sedimentation 67, 318, 320
  coefficient 121, 212, 273, 302, 362
  constant 64, 79, 85, 98, 127
  equilibrium 49, 86, 98, 263, 293
  rate 19, 52, 78, 91, 130
  velocity 318,
Self (poly) condensation 39, 40, 46, 172, 187
Semipermeable membrane 304, 309
Shape of polymer molecules 275
Shatter-proof glass 177
Short-range order 116
Short stops 26
Silicone polymer 103, 194
Single crystal 58, 142
Sintering 181
Size of polymer molecules 73
Size exclusion chromatography 296
Slurry 53, 58
Slush moulding 355
Smoked sheet 184
Solid phase polymerisation 58
Solubility 7, 54, 127, 261, 294
  of amorphous polymers 269
  of crystalline polymers 140, 268
Solubility parameter 127, 262, 268
Solubility behaviour/pattern 7

Solution polycondensation 57
Solution polymerisation 53, 161, 176, 287
Solvents for polymers 263, 270, 271
Solvent power 267, 293
Solvent volatilization 294
Specific viscosity 277, 315
Specific volume 119, 134, 273, 318
Spherulites 140
Spinneret 366, 368
Spinning 57, 351, 365, 368
  dry 197, 286, 352, 367
  melt 6, 57, 71, 117, 134
  wet 176, 192, 352, 367
Spray-up technique 362, 364
Spring 111, 335, 344
Stabilisers 179, 207, 211, 334
State 33, 52, 73, 105, 116
  amorphous 102, 113, 130, 164, 267
  change of 114, 227, 264, 323
  condensed 52, 114, 166, 238
  crystalline 58, 102, 114, 116, 134
  glassy 113, 121, 337
  liquid 6, 10, 52, 84, 121, 145
  liquid phase 58, 116, 133, 267
  molten 6, 47, 116, 123, 181
  visco-elasticity 340, 341, 343
  viscous fluid 115
States of aggregation 113, 114
States of phase 115
Stationary phase 295, 297
Staudinger equation 277, 278
Staudinger index 315
Step polymerisation 15, 36, 286
Stereoregular polymers 99
Stereo-regularity 31, 35, 196
Stockmayer-Fixman equation 282
Stoichiometric imbalance 97, 100
Stoke's law 318
Stress-strain behaviour 339
Structure 1, 9, 23, 47, 84
  aromatic 21, 57, 105, 123, 159
  chemical 1, 9, 21, 47, 63
  coiled 75, 135, 262, 323
  cross-linked 13, 105, 183, 213, 239
  cyclic 39, 43, 123, 217, 335
  foam 4, 54, 173, 334, 361
  geometrical 102, 109, 134, 334
  helical 105, 139, 359
  isomeric 109
  lattice 34, 114, 132, 268
  macro 115
  micro 115
  network 244, 245, 246, 249
  planar zig-zag 108, 109, 110
  rigid 102, 114, 125, 159, 168
  spherical 15, 53, 140, 288, 296
  stereoregular 99
Structural identity 8
Structural regularity 108, 137, 336
Styrene 13, 28, 41, 82, 146
Styrene-Butadiene rubber 83, 146, 229
Styrene copolymers 164
  ABS polymer 164
  SAN copolymer 164
  SBR rubber 280
Styrene plastics 349
Surfactants 54
Suspension polymerisation 53, 178, 288
Svedberg equation 318
Swelling 196, 262, 282, 337
Swollen polymer 56, 262, 267
Syndiotactic 35, 109, 111, 137, 163

Tail-to-head addition 23
Tail-to-tail addition 23
Technology 4, 185, 240, 336
  fibre 4, 10, 141, 165
  plastic 2, 10, 129, 168, 198
  rubber 1, 9, 83, 113, 137
Teflon 11, 180, 205, 310
Telogen 84
Telomerisation 84
Tensile strength 71, 144, 162, 186, 336
Termination 15, 23, 52, 78, 215

by chain-transfer 24
by coupling 24, 81, 253
by disproportionation 81, 255
in anionic polymerisation 29
in cationic polymerisation 29,
in free-radical polymerisation 82
in Zeigler polymerization 31
Terylene (Terene) 91, 166
Tetrafluroethylene 11, 181
Thermal decomposition 19, 215
Thermoforming 352, 359
Thermomechanical analysis 330
Thermoplastic 10, 170, 183, 231, 338
Thermosetting 10, 338, 355
Theta solvent 277, 281
Theta temperature 276, 281
Three-dimensional network 40, 187
Trans-isomer 137
Transparency 4, 140
Triaziridine 247
Triepoxide 247
Triol 173, 249

Ubbelohde suspended level viscometer 317
Ultracentrifugation 68, 318
Ultracentrifuge 301, 318
Uniaxial orientation 368
Unperturbed dimension 77, 275, 282
Unsaturated polyesters 167
Unzipping 168, 202
Urea-formaldehyde (UF) resins 190
Urethanes 232, 353
Urotropin 251
Use temperature 129, 334, 341
U.V light 20, 210
U.V. rays 210, 212, 213

Vacuum forming 360
Vapour phase osmometry 303, 309
Vapour (phase) pressure osmometer 310
Vinyl acetate 16, 54, 104, 149, 176
Vinyl chloride 13, 53, 104, 149, 178
Vinyl compounds 13, 16
Vinyl configuration 111
Vinyl monomers 41, 285
Vinyl plastics 349
Vinyl polymerisation 15
Vinylidene chloride 16, 150, 179
Virgin polymers (resins) 72, 139
Viscoelasticity 129
Viscometer(s) 317
Viscometry 313
Viscose 197, 368
Viscose solution 197
Viscosity 7, 52, 71, 136, 183
increase in 7, 40, 71, 114, 187
inherent 162, 277, 303
intrinsic 68, 277, 314
reduced 57, 125, 268, 307, 345
relative 69, 157, 315, 336
specific 31, 83, 134, 218, 318, 362
of concentrated polymer solutions 283
of dilute polymer solutions 264, 317
Viscosity-molecular weight constants 279
Void volume 296
Vulcanisation 240, 336
Vulcanising agents 351

Wet spinning 352, 365, 367

X-ray 131, 213
X-ray crystallography 132
X-ray diffraction pattern 131, 133

Young's modulus 144

Ziegler-Natta catalysts 31, 163, 185
Zimm method 324
Zimm plot 324
Zinc oxide 186, 242